Drug Discovery and Drug Development

Madhu Dikshit
Editor

# Drug Discovery and Drug Development

## The Indian Narrative

Springer

*Editor*
Madhu Dikshit
NCR Biotech Science Cluster
Translational Health Science and
Technology Institute
Faridabad, India

ISBN 978-981-15-8001-7 ISBN 978-981-15-8002-4 (eBook)
https://doi.org/10.1007/978-981-15-8002-4



This Springer imprint is published by the registered company Springer Nature Singapore Pte Ltd.
The registered company address is: 152 Beach Road, #21-01/04 Gateway East, Singapore 189721, Singapore

# Foreword

Drug development following their discoveries is a multidisciplinary effort that requires the involvement of disciplines like chemistry, biology, and pharmacology. The whole process of identification of candidates, synthesis, characterization, screening, and assays for therapeutic efficacy is a large effort that eventually culminates into the drug going for clinical trials. As the process of drug discovery has to engage the technological and procedural changes, it is always relevant to assess the progresses made of the recent past.

It was of special interest to the Indian National Science Academy to get a narrative of Indian efforts in drug discovery and development. The collection is edited by Dr. Madhu Dikshit, former Director of the Central Drug Research Institute, Lucknow under the title of *Drug Discovery and Drug Development: The Indian Narrative*. This is very relevant to times and provides a resource for the field of research.

For India, healthcare industry marks its global presence. It is being recognized as a destination of world-class healthcare. However, the current over burden that the healthcare system has to bear will require substantial investment from the government as well as the industry. While this book provides a landscape of traditional medical systems available within the country usable for healthcare systems, it also talks about drug discovery for tuberculosis, malaria, and leishmaniasis as to where India stands on drug discovery for the infectious diseases. A huge part of Indian population suffers from diabetes, and there is a need for efficient management of the disease with cheaper drugs usable by the population below the poverty line as well. Several chapters are dedicated to drug discovery for the treatment of diabetes, which will provide a summary of marketed drugs and the involvement of the industry. The Indian healthcare industry is poised to grow significantly over the next decade and the next. This growth should be able to balance the rise in 'lifestyle-oriented diseases' like diabetes and cardiac ailments.

Lastly, clinical trials form the final part of drug discovery coming in several phases. A chapter is dedicated to capture the changing regulatory regimen of clinical trials in India.

This book therefore has gathered a variety of articles that address various areas of drug discovery and will come in useful for researchers and students alike.

INSA
New Delhi, India

C. Shaha

# Preface

Even though Ayurveda, a scientific and rationale system of medicine, was practiced in India for the last several thousand years, in the first quarter of the twentieth century, majority of Indians had no easy access to medicines especially allopathic medicines. All the drugs were imported as there was hardly any chemical/pharmaceutical industry in the country. The credit goes to a few pioneering individuals who laid the foundation of chemical industry in India—like Bengal Chemicals (1901), Alembic Chemical Works in Baroda (1907), and Bengal Immunity (1919). As demand for medicines escalated due to the Second World War, several industries like Unichem Laboratories, Chemo Pharmaceuticals, Zandu Pharmaceutical Works Limited, Calcutta Chemicals, Standard Chemicals, Chemical Industrial and Pharmaceutical Laboratories (Cipla), and East India Pharmaceutical Works took root.

To put this in real perspective, during the era when the modern (western) science was evolving, the rest of the world especially Asian continent which boasted of two great civilizations namely Indian and Chinese, were in deep turmoil, both political and social. However, by the end of the eighteenth and the first quarter of the nineteenth century, a few notable Indians like Sir JC Bose, Ramanujan, and SN Bose had provided sparks to science education in India. Establishment of Tata Institute of Sciences (IISc), Bangalore (1909) and Banaras Hindu University (BHU), Varanasi (1916) heralded a new era in education and propagation of scientific thoughts. To be noted is the reality that much of the feted research in India took place in mathematics and theoretical physics which require minimal laboratory infrastructure, and not in biological sciences. Hence, at the time of our independence, while global pharmaceutical industry was galloping with new drug introductions, India did not have much to talk about research in drug discovery and development. The then foresighted government established Council of Scientific & Industrial Research (CSIR), and few labs were created to carry out much needed work in chemical/biological sciences related to drug research like Central Drug Research Institute (CDRI), National Chemical Laboratory (NCL), and Regional Research Laboratories (RRL) at Jammu and Hyderabad. These laboratories not only provided technology and support to the then nascent pharma industry but also laid the foundations of later innovative chemistry-driven pharma growth and expansion. However, the necessity to keep up with the fast pace of new biology made the GoI establish the Department of Biotechnology and also to promote education of biotechnology in the country.

The credit for the remarkable expansion of our pharma industry beginning in the third quarter of the last century must be ascribed to superb organic chemistry strengths within the country and to the new patent laws. These enabled pharma companies to take up innovation in manufacturing processes of known drugs toward inexpensive API and their marketing. However, the present reluctance of Indian Pharma to pursue high-risk, high-cost NDDR could only be ascribed to their risk averse mindsets and limited financial strengths. Unlike the chemistry-based innovative growth, NDDR requires sustained inputs spread over many years and abundant expertise in a plethora of critical biological disciplines such as pharmacology, biochemistry, toxicology, and laboratory animal science. Over the years, Indian academic researchers have vigorously pursued researches related to various aspects of disease biology and have sustained and fostered an environment where substantial impactful contributions to NDDR can emerge. When I was given the responsibility by INSA to collate a current picture of NDDR in India, it was a difficult decision to retain focus only on researches directed to OR of immense important for NDDR. Though globally India ranks well in terms of scientific output and had made significant contributions in basic sciences, successful outcomes from NDDR have been rather limited due to the very nature of NDDR which requires sustained, goal-driven, focused, continual support/inputs of various disciplines over the years. Many a times, excellent observations/studies from academics fail to move forward due to lack of interdisciplinary support and thus become infructuous.

In this light to be noted and appreciated is the foresighted decision of the Government of India to create in 1951 the Central Drug Research Institute, which not only firmly established an early school of NDDR but also provided several new drugs. ormeloxifene (centchroman, marketed as Saheli) and arteether (marketed as EMal) are of notable market success.

An attempt has been made in this book to capture Indian momentum across some major disease areas. As both academic institutions and pharma industry are equal partners in the growth and sustenance of modern NDDR, any narrative of NDDR would have been one sided without incorporating researchers from pharma industry, and we have been successful to have contributions from many major Indian pharmaceutical establishments.

I would like to acknowledge first the towering presence and influence of Dr Nitya Anand, former director of CDRI, on the early NDDR scenario of the country. I also acknowledge the long consultative discussions with Dr Venkateshwarlu and Dr K Nagarajan which helped giving a shape to this book. I also thank the support received from all of my authors, Dr AK Singhvi, INSA Vice President, and not the least my husband Dr Dinesh Kumar Dikshit.

Faridabad, India Madhu Dikshit

# About the Book

Historically, Indians' contributions to science have been rather phenomenal. From the early astronomical concepts mentioned in Rig Veda (2000 BCE), Aryabhatta's (624 CE) contributions to the heliocentric theory, proposing the idea that the moon reflects the light of the sun, great astronomical concepts of Varahamihira (476 CE) and Brahmagupta (598 CE), surgery especially plastic surgery as propagated by Sushruta (sixth century BCE) to the discovery of wireless (1894) predating that of Marconi, by Sir J C Bose, mathematical concepts (1918) of S Ramanujan, FRS, discovery and use of urea stibamine (1920) for the treatment of leishmaniasis by Sir U Brahmachari, FRS, contributions of Indian science are well acknowledged.

Presently, India ranks as the ninth highest contributor of scientific publications (15,51,015, H index 570) as per Scimago and thirteenth by Nature index. India also has a thriving pharma industry valued at 33 billion USD. However, this growth story masks the lack of successful outcomes in indigenous new drug inductions. Given the very nature of new drug discovery research and astronomical cost of a new drug introduction (approximately 1–2 billion USD), the obvious lack of financial prowess has negatively impacted NDDR in the Indian Pharma sector.

Given a wide body of efforts being made in both Academia and Pharma sectors, the Indian National Science Academy of India has thought it appropriate to commission a compilation of recent Indian efforts in NDDR. The present book, edited by Dr. Madhu Dikshit, the former director of CSIR-Central Drug Research Institute, Lucknow, has prominent Indian researchers, having a ringside view of the topic, compiling and summarizing major Indian contributions in several prominent disease areas. This book gives a snapshot of Indian core competence and gaps in NDDR.

# Contents

# About the Editor

**Madhu Dikshit** A well laureled researcher and the former director of the CSIR-Central Drug Research Institute, Lucknow, India, has made significant contributions to the general area of molecular pharmacology with special interest in redox biology. Dr Dikshit's researches have impacted the area of neutrophil biology, by enhancing our understanding how nitric oxide modulates neutrophil free radical generation, NETosis, chemotaxis, phagocytosis, apoptosis, and also differentiation. Her pioneering work has helped delineate the molecular mechanisms involved in the translocation of NOS to the phagosomal compartment and the inflammatory potential of neutrophil extracellular traps.

At CSIR-CDRI, she had initiated and led the successful anti-thrombotic drug discovery program from which two NCEs advanced to preclinical stage, and one of these has recently received approval to conduct Phase I clinical trial.

Dr Dikshit has published more than 170 well-cited papers in peer-reviewed journals, has eight patents, and has guided more than 30 PhD students. Presently, she is holding the National Chair at Translational Health Sciences & Technology Institute (THSTI), Faridabad and heading its Non-Communicable Disease Programme.

# Traditional Medical System (TMS) for Sustainable Healthcare in India

1

Pulok K. Mukherjee, Subhadip Banerjee, Chandra Kant Katiyar, Shivani Sharma, and Naibedya Chattopadhyay

## 1.1 Introduction to Global Trends in Traditional Medical System (TMS)

Traditional medicine has a long history. The World Health Organization (WHO) recognizes that traditional medicine (TM) is a vital part of the global health care system. As per WHO recommendation, TM should be a part of health and wellness and promote safe and effective use. It can be done by proper regulation, research and integrate products, practitioners and practice into TM health systems. This can be assigned that the sum total of the knowledge, skills and practices based on the theories, beliefs, and experiences indigenous to different cultures, whether explicable or not, used in the maintenance of health, as well as in the prevention, diagnosis, improvement or treatment of physical and mental illnesses. The terms complementary/alternative/nonconventional medicine are used interchangeably with traditional medicine in some countries (Anon 2004).

Popularity of TM is rapidly spreading in industrialized and developing countries. Thirty percent to 50% of the total herbal medicinal consumption is done by Chinese traditional herbal medicine. Sixty percent of children with high fever resulting from malaria use herbal medicines in Ghana, Mali, Nigeria, and Zambia, as the first-line of treatment. Over 50% of the population have used complementary or alternative medicine in Europe, North America, and other industrialized regions. Seventy-five percent of people living with HIV/AIDS use TM in San Francisco, London, and

P. K. Mukherjee (✉) · S. Banerjee
School of Natural Product Studies, Jadavpur University, Kolkata, India

C. K. Katiyar
Emami Ltd., Kolkata, India
e-mail: ck.katiyar@emamigroup.com

S. Sharma · N. Chattopadhyay
Division of Endocrinology, CSIR-Central Drug Research Institute, Lucknow, India
e-mail: n_chattopadhyay@cdri.res.in



M. Dikshit (ed.), *Drug Discovery and Drug Development*,
https://doi.org/10.1007/978-981-15-8002-4_1

South Africa. Seventy percent of the population in Canada, and 90% of the population in Germany have used a natural remedy at some point in their life (WHO traditional medicine strategy: 2014–2023). Annual expenditure on alternative medicine is US$ 230 million in the UK. Annually over US$ 60 billion is the global market for herbal medicines currently and is growing steadily.

Randomized clinical trials have been satisfactorily carried out for many uses of acupuncture, some herbal medicines, and therapies. However, there is an urgent need to validate the safety and efficacy of herbal therapies at clinical level. The growing herbal market and its great commercial benefit can project a threat to biodiversity through the uncontrolled harvesting of the raw material for herbal medicines and natural products used in healthcare which may lead to the destruction of their natural habitats and resources and extinction of endangered herbal plant species. The international and by most national patent laws are also inadequate to protect traditional knowledge and biodiversity (Ubale 2011).

Twenty-five percent of modern medicines are made from plants. Postoperative pain, nausea during pregnancy, chemotherapy, dental pain along with anxiety, panic disorders, and insomnia can be relieved using marginal side effects with the use of acupuncture. While *Tai Ji* techniques can help the elderly reduce their fear of falls and Yoga can reduce asthma attacks (Burton et al. 2015).

*Artemisia annua* is a good example used in traditional medicine of China has been found to be effective against resistant malaria. This has made a breakthrough in prevent of millions of deaths annually, especially among children. The Medical Research Council in South Africa is conducting studies on the efficacy of the plant *Sutherlandia microphylla* in treating AIDS which is used as a tonic that increase energy, appetite and body mass in people with HIV. The World Health Organization (WHO) launched its first comprehensive traditional medicine strategy in 2002 to promote safe, effective, and affordable traditional medicine. The strategy is designed to assist countries to develop national policies on the evaluation and regulation of TM practices; ensure availability and affordability of TM including essential herbal medicines; and promote therapeutically sound use of TM by providers and consumers.

WHO supported clinical studies on antimalarial medicines in three African countries. The studies revealed efficacy of herbal antimalarial treatment. Collaborations with Ghana, Mali, Kenya, Uganda, Nigeria, Burkina Faso, the Democratic Republic of the Congo, and Zimbabwe in the research and evaluation of herbal treatments for HIV/AIDS, malaria, sickle cell anemia and diabetes mellitus, are underway. WHO is providing technical support to the government of Tanzania for the production of antimalarial medicines derived from the Chinese herb *Artemisia annua*. Local production of the medicine is expected to bring the price of a single dose down from US $6 or $7 to a more affordable $2. In the Philippines, WHO has facilitated the development and introduction of traditional and alternative health care curricula. In China, Mongolia, and Vietnam, training workshops have been organized on the use of traditional medicines for selected diseases and disorders. It is promoting the use of traditional medicines over one-third of the population in developing countries which lack access to essential medicines. The provision of safe and effective TM therapies could become a critical tool to increase access to health care (2013). Several countries have a national regulation on herbal medicines. The legislative control of medicinal plants different countries and diverse approaches have been adopted with regard to licensing, dispensing, manufacturing, and trading.

The limited scientific evidence on the safety and efficacy of TM as well as other considerations make it imperative for governments to, formulate national policies and regulations for a proper use of TM, integration into national health care systems. It should also establish regulatory mechanisms to control the safety and quality of products and practice of TM. Apart from these awareness about safe and effective TM therapies among the public and consumers should be promoted to, cultivate and conserve medicinal plants to ensure their sustainable use (Burton et al. 2015).

In India, Ayurveda developed by day-to-day life experiences is a part of its cultural heritage. Besides *Ayurveda*, the Indian Systems of medicine has several other complementary and alternative systems of medicine such as Homeopathy, *Siddha,* and *Unani* systems. These have been developed using plants and plant-based formulations for health care and disease treatments. About 2000 drugs of natural origin included in the Indian *Materia medica* where almost all of which are derived from different traditional systems and folklore practices (Mukherjee et al. 2010).

## 1.2 Impact of TMS on Sustainable Healthcare

The increasing global population is engulfing forests and other resources around the world. They are being rapidly and often irreversibly depleted for energy, food, shelter, material goods, and drugs to meet the immediate needs of population. Plants are being used in TMS all over the world, which is either in crude or extract form, and represent the basis of primary health care for the foreseeable future (Cordell 2011). The end of twentieth century marked a renewed interest in TM. The resurgence of natural therapies was mainly due to the limitations of modern drugs to cure complex disorders and also the observation of their increasing side effects. Contemporary harvesting methods for medicinal plants are severely depleting these critical indigenous resources.

Maintenance of the availability of quality herbal raw materials on a sustainable basis is an as yet unappreciated aspect of public health care. In order to achieve these goals for prospective health care, and refurbish the well-being of the Earth, a change in idea is necessary. Irrespective of their source, traditional medicine may be regarded as a sustainable commodity. Several approaches toward enhancing the accessibility of safe and efficacious plant-based medicinal agents include integrated strategies to get information on botany, chemistry, and biology for medicinal plant along with its quality control. Such integrated include information systems involving metabolomics, DNA barcoding, nanotechnology, biotechnology, the application of new detection techniques for in-field analysis of medicinal plants for better quality of safe and reproducible biological agents.

### 1.2.1 Sustainability in Traditional Medical Practice

Sustainability in TM is found in different sociological domains with different notions. Bioprospecting in general and a practiced in Ayurveda with respect to the usage of animal or plants could be the ultimate key, which can ensure a sustainable biodiversity for our future generations. Bioprospecting, which is defined as sector of pharmacological breakthroughs meant for human utilization conceives and include, (1) cultivation (2) the production of secondary metabolites in bioreactors, and (3) chemical synthesis of the compounds as the method to deal with the sustainability issue in regards to the substrates derived from living beings. Sustainability in traditional medical practice of ancient periods is relevant to current practice despite the absence of historical evidences. Theoretical assumption may be regarded as the ascendants of novel researches in science. With this in mind and with empirical evidence of its efficacy existing information base should be relied upon to take the science of TM further. Natural resources for TM are limited and are not rapidly renewable. Any unfamiliarity of this condition will be injurious to both TM and the global environment. Ideas regarding sustainability is not an assumption and is based upon perception and developments; therefore, new ideas must be respected for the betterment of global community.

As a consequence of expanding urbanization, rapid reduction in natural resources is leading to global warming and disrupted natural habitat caused a threat to the sustainability of traditional medicine. Traditional medicine being entirely dependent on natural resources like herbs, minerals, and animal products, would get extinct if heritage of an alternative way is not considered well in time, Ayurveda presents some unique sustainable models where animals and plants are used without posing any threat to their existence. In the current context, these ideas may give us a new insight to refine our outlook at natural resources used in traditional medicine to save natural resources which are facing a threat to their existence.

### 1.2.2 Sustainable Healthcare in Ayurveda

Sustainability in healthcare can be presented in Ayurveda at multiple levels such as preventive care, primary healthcare, resource utilization, and novel applications which is an ecologically sustainable medicinal system. For past several decades, *Ayurveda* and other traditional medical systems have faced the problem of nonavailability of genuine raw material. Manufacturers of Ayurveda-based medicines, have unsuccessfully tried to meet the demands by seeking alternatives for the conventional herbs or other ingredients, which are either not available or have lost their identity. Overexploitation, changes in global environment, and loss of delicate ecosystems due to urbanization collectively resulted in convergence of gene pool, growth of species with poor endurance and consequent loss of such species has been increasingly seen. As the resources are not easily available in future, TM will become more expensive. Producers of TM are going for mass production without considering intricacy of plant cultivation and their utilization, as described traditionally due to demand-oriented market pressures and poor quality checks in the TM. Ayurveda with regard to their utilization of natural products for human welfare in an eco- and bio-friendly way that takes care of the human health with an equal concern for the surrounding environment (Mukherjee et al., 2012).

## 1.3 Ayurveda-Traditional System of Medicine in India

Ayurveda is the 'Science of life' which promotes wellness at physical, psychological, spiritual, and social realms of human existence (Mukherjee et al. 2012). This ancient healthcare system originated around 5000 years back in Vedic text Atharvaveda which contains various references to Ayurvedic medicine and allied aspects of healthcare. The growth and development of Ayurveda and other traditional systems of Indian Medicine was promoted by the Department of Indian Systems of Medicine and Homoeopathy (ISM&H) under the Ministry of Health and Family Welfare in March 1995. In November 2003, it was renamed as Department of AYUSH where each letter of the acronym 'AYUSH' represents officially recognized systems of medicine Ayurveda, Yoga and Naturopathy, Unani, Siddha, and Homeopathy. In 2014, Ministry of AYUSH was constituted for systematic development in all spheres of Indian Medicine and to bridge the prevailing gaps in healthcare delivery and outreach of health services by policy formulation, execution of programs for augmenting the domain, quality and outcomes in inclusive terms. *Sowa-Rigpa* has also been given recognition and added in AYUSH. The AYUSH systems are being promoted in India as the preferred systems of living and practice for a healthy India. The mission of AYUSH is to undertake and support activities of education and communication, human resource development, drug administration, medicinal plants research and development, international collaboration and enhancement of public information regarding AYUSH-based health services. For the mainstreaming of AYUSH services through a program called National Health Mission (Srivastava et al. 2016), the public perception regarding AYUSH is being developed through the utilization of the potentiality, strength, and revival of Ayurveda (Patwardhan 2014). For the development of AYUSH, COE organizations have been upgraded to National Accreditation Board for Hospitals (NABH), National Accreditation Board for Testing and Calibration Laboratories (NABL), Good Laboratory Practice (GLP), and Good Manufacturing Practice (GMP) levels to achieve higher standards (Katoch et al. 2017).

Standardization of Ayurvedic drugs for the purpose of effective quality control 645 monographs of quality standards of single drugs and 252 monographs of quality standards of multi-ingredient formulations are published in two parts of Ayurvedic Pharmacopoeia in 13 volumes. For 265 standardized formulations from classical texts are published in four volumes of the National Ayurvedic Formulary. Pharmacopoeial standards of identity, purity, and strength of ASU drugs have been envisaged like identification, chemical constituents and limits of heavy metals, pesticide residue, aflatoxins, and microbial load. Various scientific laboratories and Pharmacopoeial Laboratory for Indian Medicine (PLIM) an appellate laboratory under the provisions of Drugs and Cosmetics rules work on standardization and creating Standard Operating Procedures for Ayurvedic drugs using analytical tools (Mukherjee et al. 2010). Standard Operating Procedures (SOP) of manufacturing processes of formulations, assays, atlas of chromatography, Pharmacognosy atlas are being added to the quality standards of drugs to facilitate the testing procedures, estimation of marker compounds and phytochemical standard materials. The National Medicinal Plants Board (NMPB) under Ministry of AYUSH support policies and programs for the growth of trade, export, conservation, and cultivation of medicinal plants. They provide support for survey, conservation strategies of medicinal plants, herbal gardens, linkage for joint forest management programs, research, and development. Restrictions are enforced for rampant deforestation for the collection of raw materials of medicinal value from the wild sources (Katoch et al. 2017).

For promotion of industry-institutional collaboration Drugs & Pharmaceuticals Research Programme (DPRP) in 1994–1995 and to help the industry in promoting exports Pharmaceutical Export Promotion Council (Pharmexcil) constituted a National committee for Ayurveda medicines. The Ministry of AYUSH has focused its attention on standardization and quality control of Ayurvedic drugs for globalizing the Ayurveda system and its products. Testing for heavy metals, viz. mercury, arsenic, lead, and cadmium in all herbal Ayurvedic drugs is mandatory for export (Katoch et al. 2017). These measures aim for greater impetus to consumer awareness, consumer and doctor benefit, global acceptance and safety. In order to promote evidence-based healthcare documentation of clinical safety and efficacy of Ayurvedic medicines (Patwardhan 2011); clinical trials following ‘Good Clinical Practice’ (GCP) for AYUSH has been developed to generate quality data. The Central Council for Research in Ayurvedic Sciences (CCRAS) is working on coordination, formulation, development, and promotional research on Ayurveda which includes research on clinical, drug, and literature on Ayurvedic Sciences.

Apart from regulation and research increased awareness and improved access to traditional medicines is needed (Srinivasan and Sugumar 2017). The Ministry of AYUSH, through National AYUSH Mission (NAM) is enabling access to services, strengthening educational institutions, enforcement of quality control of drugs, and continuous availability of raw materials in the States/UTs (Srinivasan and Sugumar 2017).

## 1.4 Role of TMS in Health Care

India has a wide array of tradition knowledge systems related to health. While modern medical science and technology have gained dominance, practices using both old and new systems, co-exist. Unwritten knowledge passed on through the oral tradition and practiced is ‘folk practice’. Systematized and duly codified texts receive greater legitimacy as ‘traditional’ or ‘alternative’ systems. Modern medical research has started verifying traditional medicines and this has led to their increasing adoptions.

The acronym AYUSH represents Ayurveda, Yoga and Naturopathy, Unani, Siddha, Homeopathy, and Sowa-Rigpa for which a network of government services have been set up by the center and the states. Amchi or Tibetan Medicine in the western Himalayan region and acupuncture also receive state support in some regions. The folk medicine has survived with little or no official support. While anthropologists have documented folk knowledge, and botanists have documented medicinal plants and their folk uses, the official approach has been to view them as knowledge base on medicinal plants and herbs that can be exploited as economic resources for commerce.

Organization, training and practices of each form of health knowledge change with regional contexts. Such interaction between diverse forms of knowledge on health has been at an informal level but need to be viewed holistically as a 'health service system'. Health care starts from the home and goes through the primary and secondary level services to the tertiary level i.e., hospital. That it includes use of various knowledge systems and use of one or all of these are desirable as a part of continuum of care for realistic, cost-effective context-specific plans for health services, based on rational criteria of effectiveness, safety, accessibility, and affordability.

Three possible directions for strengthening AYUSH are, (a) contribution within the to strengthen AYUSH services through improved quality of infrastructure, human resources, supplies, R&D, and management verticals, (b) enhance interactions between the systems and encourage cross-referrals between them, by placing all services under one roof, and (c) develop integrated medicine for primary and secondary care protocols.

### 1.4.1 Role of National Rural Health Mission

The National Rural Health Mission (NRHM) has created the opportunity for co-locating services of the AYUSH at the CHCs and PHCs (leaving out yoga and naturopathy). This has been termed as the 'mainstreaming of AYUSH'. 'Revitalization of the local health traditions' is integral to NRHM strategies, but its modalities are yet to be finalized. During the appraisal of State Programme Implementation Plans for NRHM, only some states allocated substantial funds for co-location of AYUSH at rural health facilities as well as other activities, while others were not planning for such a strategy at all. Surprisingly, several states went beyond the activities envisaged in the NRHM framework to strengthen co-location. The NRHM has created an environment in which the states could undertake 'innovations'.

Commercialization of these systems as a part of the 'medical and health industry' that includes the providers is a loss making proposition in many ways. While these add to health care costs, they take away natural resources from the reach of local communities and destroy a non-commercialized model of health care as a community activity available to all. In an age when 'user fees' and service 'packages', 'medical tourism' and 'health spas' are becoming a corporatized model of health care, the folk herbalists and home remedies provide a diametrically opposite version for a holistic and equitable health service system.

### 1.4.2 Changing Scenario of TMS in Health Care

Under the NRHM, 4981 AYUSH doctors and 934 Paramedics have been recruited on contract for co-location. As reported by states, about 44% of DHs, 24% of CHCs, and 17.6% of PHCs have co-location of AYUSH providers. Sixteen have allocated 1–3% of their NRHM budgets for this component, four have budgeted 3–10% and 12 states have budgeted over 10%. Drugs, equipment and buildings are funded by the department of AYUSH, while the NRHM flexi pool funds the health service providers hired on contract for co-location. The IPHS provide the guide lines for the level of services at each facility, from subcenters to PHC to CHC to DH.

These guidelines give the HR requirement, space and building, medicines and equipment and of a herbal garden in the SC and PHC premises. Co-locations seem to be the only activity followed promptly across all states, but at varied levels. States such as, Gujarat, Rajasthan, Himachal Pradesh, and J&K, with strong AYUSH base in the public health have rolled out the recruitment of AYUSH doctors for PHCs and CHCs under the NRHM to a greater degree than others. However, while continuing to strengthen its AYUSH services, Kerala is reluctant to co-locate them. There are reports of the State Directorates/Cell of AYUSH not being involved in the activities of mainstreaming under NRHM, leading to loss of synergy and lack of technical supervision for the co-located personnel.

The role of AYUSH doctors and paramedics in the co-located facilities needs much more attention for quality service delivery. While the role of AYUSH and LHT can be significant in RCH, this has not been adequately worked out. *Punarnavadi Mandoor*, the anti-anemia Ayurveda medicine, is the only one to have been widely included in the program. AYUSH doctors are being given training in SBA in only six states and IMNCI in three states. Since both require procedures specific to Allopathy such as injections and episiotomy and prescription of allopathic medicines, the legal issues need to be dealt with.

Training of AYUSH doctors in managerial and public health functions is not adequately planned. Membership of AYUSH doctors in the SHS, DHS, and RKS (planning, management and monitoring bodies created under NRHM) at various levels has been reported by most States, but their level of involvement is not known. Many additional and innovative activities are planned across states but their micro planning and implementation needs much more technical and managerial assistance.

District level planning has been done on mainstreaming of AYUSH under NRHM in a few states (as per the CRM). Planning for ensuring adequacy of appropriate AYUSH drugs is lacking. There are various additional inputs planned by some states under the following heads:

1. **IEC/BCC:** Sensitization activities for the general public about AYUSH and LHT.

2. **Specialty clinics/wards:** Half the states mention special AYUSH clinics or wards, especially a Ksharasutra therapy wing for anorectal diseases and Panchkarma clinics for intensive and specialized treatment at the CHC or DH.
3. **AYUSH health programs:** States like Orissa, Punjab, and Andhra Pradesh write in the PIPs about school Yoga programs and Yoga camps. The Tripura PIP also mentions sensitization of Primary school teachers regarding importance of yoga,'Suposhanam'. The special nutrition program for the tribal women is stated in the Rajasthan PIP. Ayurveda Mobile Units is also an activity mentioned in the Rajasthan state PIP.
4. **Outreach activities:** Utilization of AYUSH doctors for the Mobile Medical Units in some States, such as Jharkhand, Himachal Pradesh, J&K, and Orissa. Call centers for AYUSH in Madhya Pradesh and Tripura is a major innovation mentioned in their PIPs.
5. **Establishment of AYUSH epidemic cells:** Tamil Nadu and Kerala are using AYUSH in public health for preventive activities and epidemic control, e.g., Homeopathy for responding to the Chikungunya outbreaks. RAECH (Rapid action epidemic cell of homeopathy) in Kerala is a major AYUSH activity highlighted in the state PIP.
6. **Local health traditions:** The IPHS prescribes the setting up of an herbal garden within the space available in the subcenter and PHC premises. Most state PIPs have not mentioned this activity in particular, however, some states have for example—Chhattisgarh PIP has mentioned an innovative activity—the 'Ayurveda Gram' concept; 'Dadi ma ka batua' is an innovative scheme stated in the J&K PIPs, which plans to include traditional home remedies in the AYUSH drug kit; Madhya Pradesh has an innovation called Gyaan ki Potli which too plans to include prevalent and useful local health traditions/remedies which are accessible and affordable for various ailments as a step forward for LHT revitalization and Haryana has planned for courses on Local health traditions for the unemployed youth.
7. **Development of Medicinal Plants Resources:** The National Medicinal Plant Board (NMPB), the State Medicinal Plant Boards (SMPB) and NRHM are working in collaboration to preserve the large medicinal flora and their utilization in the primary health care. The 'Ayurveda gram yojna' must be explored and innovatively followed in the States.

### 1.4.3 Management and Technical Strengthening

Almost half the states have planned some strengthening of management and technical support to the AYUSH services. States like Rajasthan mention in the PIPs of year 2008–2009 about the formation of the State AYUSH Monitoring Cell (SAMC) for AYUSH services. Chhattisgarh too has a separate technical wing in the SHSRC for AYUSH. On a similar pattern, under the NRHM, Kerala, Jharkhand, Mizoram, Tripura, Delhi and Goa have planned for establishing a resource center or a separate cell for AYUSH.

India, with its 'Traditional Medical System' can be a world leader in the field. India's rich and mature indigenous medical heritages include Ayurveda, Yoga, Unani, Siddha, Homeopathy, Naturopathy, and Tibetan/Amchi Medicine. The size of the AYUSH sector in India is impressive. There is tremendous growth in the AYUSH education sector. Table 1.1 provides the status of infrastructure facilities in AYUSH.

### 1.4.4 Major Gaps and Technical Assistance Needs

The AYUSH services in the public sector are getting strengthened. However, the requirements for making full use of the opportunity are not yet adequately conceptualized or planned for. Some concerns that emerge are: inadequate inputs for optimizing the collocation strategy; relocating doctors from the well-established AYUSH facilities means weakening of AYUSH since they lose independent space, and there is loss of services to patients; no plans to orient the allopathic doctors to the strengths and role of AYUSH and LHT. Their non-appreciation is based on ignorance of research findings at the frontiers of modern medicine and the experiential knowledge of common people. Assessment of the role of AYUSH personnel under NRHM and in the public health system as a whole for mainstreaming of AYUSH and revitalization of local health traditions in an integrated and comprehensive manner was done (Samal, 2015).

Inputs by AYUSH doctors in co-located facilities toward fulfilling the service, training and capacity building for National Health Programs and Public Health need to be well defined. Orientation of the health personnel other than that of the AYUSH systems for sensitizing them toward AYUSH and the local health traditions is required. This should enhance the cross-referral systems, thereby optimizing the provision of benefits for all systems to patients. Drugs and equipment need to be reflected in the PIPs, based on a needs—assessment and monitoring of supplies.

The challenge to the public health system is how it visualizes the place of AYUSH and LHT within the health service system of the country. International experience shows how viewing them as the base to build upon for a continuum of care, from home to community, to health centers and dispensaries, to hospitals; letting each system grow according to its own epistemological orientation; and cross-referral based on mutual appreciation and respect serves people the best. It is to be hoped that the NRHM will be able to foster this spirit.

## 1.5 Conservation of Traditional Knowledge

The conservation of traditional knowledge is done by TKDL program deals with the documentation of existing knowledge on Ayurvedic systems of medicine in the form of an original proprietary database, which is fully protected by national and international laws of intellectual property rights. The Traditional Knowledge Resource Classification (TKRC) has converted 140,000 pages of information, containing

**Table 1.1** Summary of infrastructure facilities under AYUSH

| Facility | Ayurveda | Unani | Siddha | Yoga | Naturopathy | Homeopathy | Sowa-Rigpa | Total |
|---|---|---|---|---|---|---|---|---|
| Hospitals | 3186 | 259 | 291 | 8 | 17 | 225 | 0 | 3986 |
| Beds | 43,358 | 3352 | 2051 | 185 | 682 | 6958 | 0 | 56,586 |
| Dispensaries | 17,102 | 1621 | 848 | 235 | 101 | 7259 | 33 | 27,199 |
| Registered practitioners (IQ and NIQ) | 443,704 | 51,110 | 9125 | N/A | 2485 | 293,455 | 0 | 799,879 |
| Registered practitioners | 294,162 | 38,672 | 5685 | N/A | 2349 | 246,792 | 0 | 587,660 |
| UG colleges | 393 | 52 | 10 | N/A | 26 | 221 | 0 | 702 |
| Admission capacity (UG) | 25,407 | 2945 | 580 | 0 | 1730 | 16,173 | 0 | 46,835 |
| PG colleges | 137 | 11 | 2 | 0 | 3 | 50 | 0 | 203 |
| Admission capacity (PG) | 4188 | 127 | 94 | 0 | 47 | 1080 | 0 | 5536 |
| Exclusively PG colleges | 3 | 3 | 1 | 0 | 0 | 2 | 0 | 9 |
| Admission capacity (exclusive PG) | 156 | 75 | 46 | 0 | 0 | 72 | 0 | 349 |
| Total AYUSH colleges | 396 | 55 | 11 | 0 | 26 | 223 | 0 | 711 |
| Total admission capacity | 29,751 | 3147 | 720 | 0 | 1777 | 17,325 | 0 | 52,720 |
| Manufacturing units | 7718 | 625 | 214 | 0 | 0 | 397 | 0 | 8954 |

36,000 formulations from 14 texts of Ayurveda. These documents have patent-compatible format in various languages, viz. translation of Sanskrit shlokas not only in Hindi but also in English, French, German, Spanish, Japanese, etc. The information includes names of plants, Ayurvedic description of diseases under their modern names, therapeutic formulations, etc.

## 1.6 Validation of Traditional Medicine

Ensuring therapeutic efficacy, safety, and rationalization of QC of TMs are critical and essential aspects for scientific validation (Mukherjee et al. 2017). The demands of traditional medicines are rising day by day and about three-quarters of the world's population currently use herbs and other forms of traditional medicines to treat disease. Even though marvelous advancement occurred in Allopathic system of medicines, there are many areas in which allopathic medicines have failed to prove its efficiency. Nowadays, people have more faith toward traditional medicines then modern medicines due to occurrence of many side effects of modern medicines. The prime reasons for acceptance of traditional medicines are accessibility, associability, and affordability in developing countries. Major thrust areas of research for validation of TM includes: (1) phytochemical and pharmacological screening, (2) phyto-informatics, (3) DNA-bar coding, (4) metabolomic study (5) reverse pharmacology, (6) safety evaluation, (7) quality control and standardization, (8) clinical evaluation, etc. (Mukherjee et al. 2017). Systematic investigations and standardization of medicinal plants needs to be done for quality, safety, and efficacy. Apart from analytical chemical fingerprinting techniques e.g., high-performance liquid chromatography (HPLC), high-performance thin-layer chromatography (HPTLC); macroscopic and microscopic studies, genetic finger-printing methods should be done. LC-MS/MS may be also used for standardization (Mukherjee et al. 2017). Since traditional medicines often have numerous active ingredients, it could be used to measure the whole body's response to the mixture of compounds (Patwardhan and Mashelkar 2009).

### 1.6.1 Chemo-Profiling and Standardization of Traditional Medicine

In order to assess the quality of herbal traditional medicine chemical profiling is required for standardization of herbs, marker compound quantification, spurious drug determination, comparative fingerprint analysis for consistency of stability and quality of TMs. Botanicals from wild sources have the greatest challenge for ensuring consistent product quality (Harwansh et al. 2014). Environmental factors including geographical location, availability of light and water, temperature variations, soil conditions, nutrients, and that affect the phyto-constituents present in plants. Factors like cultivation and harvesting techniques and storage practices determines the quality parameters for plant extracts and final product. Substantial variation in composition, quality and therapeutic effects are observed in botanical extracts obtained from crude plant material show (Harwansh et al. 2015). Therefore, standardization of herbal drugs includes certification of quality raw material, assessment of intermediates, and finished product. DNA-based markers can also be used in identification of inter/intra-species variations because the genetic composition is unique for each species and is not affected by age, physiological conditions and environmental factors.

According to the monographs of the American Herbal Pharmacopoeia (AHP), the use of single or multiple chemical markers is crucial for QC along with proper cultivation, collection and quality, optimum extraction and standardization of raw materials. The evaluation of herbal medicine should be done in a better way to get fruitful results (Mukherjee et al. 2009).

Chemical fingerprints relate the chemical composition to biological activity for product standardization and validation which can be used to authenticate plant material, identification, and quantification of active compounds (Mukherjee et al. 2007b). Marker components may be classified as active principles, active markers and analytical makers, while biomarkers may be defined as pharmacological active. Chemical makers are frequently used for assuring quality consistency of natural products derived from botanical sources. Ideally, the chemical markers chosen should be bioactive (Mukherjee and Wahile 2006). However, marker compounds may not be pharmacologically active all the time but their presence is well established in products with characteristic chemical features (Mukherjee et al. 2017). Biologically active reference standard (BRS) or pharmacologically active reference standard (PRS) are also marker compounds. Determination of single or several marker compounds by a developed method are becoming popular for the identification/authentication of herbal drug components (Katoch et al. 2017).

### 1.6.2 Plant Metabolomics

Metabolomics is the quantitative and qualitative assessments of 'whole-set of metabolites' that occur in plants. It may be defined as the systemic study of the individual chemical fingerprints of all the metabolites in a biological system, which are the final products of its gene expression, is known as metabolome (Lee et al. 2017). The plant secondary metabolites are health care products or lead compounds for new drug development which have renewed interest in pharmaceutical and nutraceutical research (Newman and Cragg 2007). A targeted metabolomics study involves characterization of a set of defined metabolites and usually combines NMR-MS techniques. An array of metabolites can be intercepted by leading the way to major revelation in our understanding of cell biology, physiology and medicine (Cox et al. 2014). Metabolomics study has diverse fields of application and can be divided into four areas: (1) the metabolomics profiling—the quantitative and qualitative estimation of a set of compounds (2) target compound analysis—the quantification of specific metabolites, (3) metabolomic fingerprinting—sample classification by rapid global analysis, and (4) metabolite chemo-analysis—the qualitative and quantitative analysis of all metabolites (Ulrich-Merzenich et al. 2007).

These may assist evidence-based phyto-therapeutics research which may lead to a change of paradigm in the development and application of multicomponent botanical therapeutics (Wang et al. 2017a).

Metabolites profiling can identify the plant secondary metabolites by comparing the nature of compounds as the output of sensors (analytical detectors) are known as 'profiling' which are then classified and statistically analyzed by chemometrics (Noteborn et al. 2000). Metabolite identification is based on their spectral peaks and calibration curves which comprehensively examines entire range of metabolites in a sample by the mutual application of various analytical techniques (Glassbrook and Ryals 2001). Metabolomics reveals the phenotypic changes in the function of metabolic systems. However, among the 'omic' sciences metabolomics presents the terminal view of the biological system (Gahlaut et al. 2013) Metabolomics and phytomics provide mechanisms for assuring consistent quality and efficacy of herbal medicine. Traditional medicines contain multiple phyto-constituents and pharmacological properties are complex systems as single medicinal plant extracts which are influenced by the time of collection, area of plant origin, and environmental conditions. Therefore, above mentioned strategies regarding various issues are needed for validation of TM (Mukherjee et al. 2007a).

## 1.7 Quality Assurance of Herbal Medicine

Herbal drugs have turned out to be an important part of new generation health system in many countries. It has been observed in recent times that herbal products are substituting the conventional medicines in many countries. Increased demand of herbal product leads to growing demand in global market (Gouveia et al. 2015). Quality of TMs is determined by identity, purity, content, physical, and biological properties which is importance for efficacy and safety of herbal products. Ensuring quality of herbal products begins from the field to the bedside of the patients, followed by post-marketing surveillances. WHO has developed technical guidelines on the safety and Quality Assurance (QA) of medicinal plants and herbal materials are essential to follow, which includes Good Agriculture Practices (GAP), Good Harvesting Practices (GHP), Good Storage Practices (GSP), Good Clinical Practices (GCP), Good Manufacturing Practices (GMP), and Good Laboratory Practices (GLP).

With the increase in the use of herbal medicine there is also an increase in the reports of toxicity and adverse reactions. Such undesirable reactions can be due to:

1. Side effects (usually these are predictable in nature)
2. Reactions which are resultant of overdose, over-duration, tolerance, dependence-addiction
3. Hypersensitivity, allergic, and idiosyncratic reactions
4. Mid-term and long-term toxic effects including liver, renal, cardiac, and neurotoxicity also genotoxicity and teratogenicity.

Herbal products, which are being marketed, are generally not thoroughly tested for their pharmacological and toxicological effects. In addition there has been a problem related to the quality of herbal products due to unexpected toxicity, which occurs due to use of poor quality of raw materials, misidentified herbs, adulterations, and contaminations. With the help of GMP, these quality issues can be addressed properly and manufacturing of herbal medicine can be improved. However, there are some issues, which cannot be neglected like some herbs which come from different countries that have different standards and regulations. Such matters have remained a problem so far. Due to reports of serious effects like hepatotoxicity, renal failure and allergic reactions, the regulatory authorities are facing questions regarding the safety of marketed herbal medicines. The World Health Organization has developed guideline for the monitoring of herbal safety within the existing pharmacovigilance framework (Shaw et al. 2012). The primary steps in quality assurance are as follows:

### 1.7.1 Classical Systematics in Species Authentication and the Utility of Good Agricultural and Collection Practice (GACP)

In many developed countries like Australia, Canada, Europe, and the USA, the quality assurance and control of the herbal ingredient (raw materials) is the main job of the product license holder who should ensure the efficacy and compliance of the herbs with the respective national regulatory framework. The sponsor of the herbal medicine, in turn, has to place specific systems and requirements for approval of ingredients (raw materials) through the supply (value) chain back to the ingredient source and manufacture to assure quality. Many but not all international firms tend to adopt vendor audit programs to qualify herb source and ingredient manufacture by stringent GMP protocols. With the outsourcing of authentic herbs the quality assurance of the herbal ingredients throughout the supply chain begins. To establish the purity of the ingredient, authenticating the starting material is the major requirement. Authentication is often difficult when the herbs are purchased from local market with no traceability of their origin, although now days due to proper understanding and knowledge, medicinal herbs are sourced either from organized cultivation or by wild-crafting (Govindaraghavan 2008).

### 1.7.2 Genomic Profiling and DNA Barcoding in Species Authentication

Plant authentication can be done by genomic profiling and DNA barcoding, which are considered to be complementary techniques to classical systematic approaches helping in unequivocal identification of plant species. DNA barcoding in plants was promoted by The Plant Working Group of CBOL (Consortium for the Barcode of Life) in 2004. It was observed that single locus barcode was not sufficient for authentication of plant species, so two-locus combination was recommended for authentication. These two-locus combinations are chloroplast genes maturase K (matK) and ribulose-1,5-bisphosphate carboxylase/oxygenase large subunit (rbcL). The need for inclusion of third locus to increase the rate of identification is currently under debate. Sufficient variability for species identification can be exhibited by Deoxyribonucleic acid barcodes and this system can be amenable to ease in referencing (Govindaraghavan 2008).

### 1.7.3 Macroscopic and Microscopic Characterization of Sorted Plant Materials

Medicinal plant authentication can be done based on macroscopic and microscopic character sets of sorted plant parts along with their phytochemical (metabolite) profiles. The pharmacopoeial monographs, mentions the identification of the species based on the morphological and microscopic characters of dried and sorted plant parts based on their roots, leaves, berries/fruits, bark, flowers, seeds and husks, or derived products such as gum/resins (Govindaraghavan 2008).

### 1.7.4 Phytochemical Profiling of Plant Parts as a Tool for Identification and Characterization

The phytochemical profiling tools constitute high-performance thin-layer chromatography (HPTLC), high-performance liquid chromatography (HPLC) in addition to gas chromatography (GC) with detectors like ultraviolet (UV) and visible (Vis) light-based photodiode array (PDA) with mass spectrometry (MS) (Govindaraghavan 2008). Chemical profiling is essential in order to assess the quality of traditional herbal preparations (Sen and Chakraborty 2017).

### 1.7.5 Guidelines for Good Plant Authentication and Identification Practice (GPAIP)

An outstanding good practices guideline regarding plant identification for herbal industry has been provided by Agriculture and Agri-Food Canada. With the help of this example, it has been recommended to the industries manufacturing herbal medicines that GPAIP should be followed as the herbal ingredients undergo change in every step of the process across the supply chain.

### 1.7.6 Assurance of the 'Purity' of Botanical Raw Materials: Impurity Profiling

To obtain good quality of herbal ingredient, the main focus should be on its purity, i.e., it should be free from impurities, which can ensure quality end product. There are many components, which fall under the category of impurities like heavy metals, pesticide residues, aflatoxins/mycotoxins, etc. The contaminated soil is the major source of heavy metal contaminations in medicinal plant species. It has been reported that over 500 plant species are known to accumulate heavy metals and in some species, heavy metal concentrations in aerial parts exceed critical toxicity levels (Govindaraghavan 2008).

## 1.8 Pharmaceuticals, Nutraceuticals, Phyto-Pharmaceuticals Inspired from Ayurveda

Healthcare research has shifted toward quality is an urgent need for direct traditional medicine-inspired natural product research (Cordell and Colvard 2012). Approximately, 70% of cancer and 42% of rheumatic patients use complementary and alternative medicine (Ernst and Cassileth 1998). Scientifically validated botanicals Ispaghula, Garlic, Ginseng, Ginger, Gingko, St. John's Wort, and Saw palmetto are regular in practice by modern physicians (Wetzel et al. 2003). Therefore correct identification and authentication of traditional medicinal herbs is essential as quality of the source determine the final product (Franz et al. 2011). Integrated strategies should be considered for the validation of Ayurvedic medicines at each and every step (Mukherjee et al. 2012). Globalization and reinforcement of Ayurvedic medicine is necessary for the establishment of the evidence-based healthcare claims. There has been a development in the education and communication to disseminate the sources of Ayurvedic knowledge interpretable in Western terminology, and a variety of texts are now available online. The National Institute of Indian Medical Heritage (http://niimh.nic.in/) under the CCRAS, Ministry of AYUSH has published e-Samhitas of all the main classical works on Ayurveda, such as the Samhitas (*Charaka Samhita, Sushruta Samhita*) and other books like *Astanga Hridaya, Ashtanga Sangraha,* and *Bhavaprakash Nighantu* which are available online (Mukherjee et al. 2017).

These traditional resources provide the products including herbs, herbal materials, herbal preparations, and finished herbal products that contain parts of plants, other plant materials, or combinations thereof as active ingredients (WHO 2013). Finished herbal products consist of herbal preparations made from one or more herbs (Bhatt 2016). In India, the AYUSH drugs are regulated by 'Ministry of AYUSH'. However, the regulatory requirements for phyto-pharmaceuticals are governed by the Central Drugs Standards Control Organization (CDSCO). By legal definition regulatory provisions for phyto-pharmaceuticals and regulatory submission requirements for scientific data on quality, safety, and efficacy to evaluate and permit marketing of a herbal drug on similar lines to synthetic, chemical moieties (MHFW 2016). Phyto-pharmaceutical drug is defined as purified and standardized fraction with defined minimum four bioactive or phytochemical compounds (qualitatively and quantitatively assessed) of an extract of a medicinal plant or its part, for internal or external use of human beings or animals for diagnosis, treatment, mitigation, or prevention of any disease or disorder but does not include administration by parenteral route (MHFW 2016). According to Schedule Y, Appendix I B it is mandatory to submit evidence-based data along with the application in order to conduct clinical trials or import or manufacture of a phyto-pharmaceutical drug in the country (MHFW 2016). These phyto-pharmaceutical drugs are considered as NDA which includes standard requirements for a new drug safety and pharmacological information, human studies, and confirmatory clinical trials.

The new regulation for phyto-pharmaceutical is in line with regulations in developed countries involving data generation on scientific evaluation (Narayana and Katiyar 2013). Now the Indian scientists would discover and develop phyto-pharmaceutical drugs for as yet unmet medical needs from herbal drugs (Bhatt 2016). CSIR-Central Drug Research Institute has developed phyto-pharmaceutical drugs including standardized extracts of *Bacopa monnieri* (CDRI 08), *Picrorhiza kurroa* (Picroliv), *Commiphora mukul* (gugulipid,) and *Dalbergia sissoo* for memory enhancement, liver health, lipid disorders, and bone health, respectively. These phyto-extracts were defined by chemical markers and pharmacologically active ingredients, pharmacognosy, preclinical efficacy assessments in appropriate disease models, preclinical pharmacokinetics, and safety studies.

## 1.9 Global Harmonization and Regulatory Status of Traditional Medical Systems (TMS)

The popularity of Traditional and Complementary Medicine (TCM) worldwide is increasing and specifically in some developing countries where the native healers play a significant role as main health providers in rural areas. For instance in Africa, Lao, People's Democratic Republic most of the populations have traditional health practitioners. About one-fourth Europeans use TCM while China, North Korea, South Korea India, and Vietnam have introduced traditional healthcare as a mainstream health system. The safety, quality, and efficacy of TM services cannot be assured if there is no appropriate regulation of practices and practitioners. However the equality is missing in the regulation of TCM products, practices, and practitioners in the pace of progress. Whereas regulation of herbal medicines is progressing faster compared to TCM practices, the practitioners are lagging behind. This situation presents a serious challenge as there is a lack of knowledge and experience regarding the formulation of meaningful national policies. This is causing absence of regulation and a lack of proper integration of TM services into the health service delivery system. The WHO Traditional Medicine Strategy 2014–2023 has been formulated responding to these needs and challenges identified which devotes more attention than its predecessor to health services and systems, products, practices, and practitioners (Burton et al. 2015).

## 1.10 Clinical Trial and Pharmacovigilance of TM

Clinical trial and pharmacovigilance of TM will be discussed under the broad areas including geriatric, metabolic, infectious and inflammatory diseases and cancer because their combined disease burden contributes to lion's share of diseases afflicting the Indian population. Geriatric diseases will consider Alzheimer's disease and Parkinson's disease, metabolic diseases will cover diabetes, hypertension, and osteoporosis, infectious diseases will include tuberculosis, malaria, leishmaniasis and AIDS, inflammatory diseases will consider chronic obstructive pulmonary disease and rheumatoid arthritis, and cancer will cover breast, prostate, and lung cancers. These broad disease areas and their specific disease types are reviewed with respect to clinical assessment of various TM products. Given the overall paucity of clinical trials in the area of TM, we considered all trials done anywhere in the world regardless of their quality.

### 1.10.1 Geriatric Diseases

#### 1.10.1.1 Alzheimer's Disease and Cognition Deficit

Alzheimer's disease (AD) is a neurodegenerative irreversible disease characterized clinically by progressive memory deficits, impaired cognitive function, and altered, inappropriate behavior. Evidence obtained from ethnobotanical studies suggests that various traditional medicinal plants used in China, India, Iran, the USA, Southwest Nigeria, and Europe have a salutary effect in AD-type neurodegeneration (Eckert 2010; Tewari et al. 2018). Ethnobotanical survey of 37 species of medicinal plants used for memory enhancement and anti-aging effects were found to belong to 28 families with potential anticholinesterase and neuroprotective actions (Dos Santos-Neto et al. 2006). Out of these plants, clinical trials have been carried out with *Ginkgo biloba*, *Salvia officinalis*, *Melissa officinalis*, *Papaver somniferum*, *Crocus sativus*, and *Bacopa monnieri* having therapeutic effects for the treatment of cognitive impairment of AD (Akhondzadeh et al. 2003, 2010; Tian et al. 2010; Farooqui et al. 2018). In addition, several polyherbal formulations including 'Di-tan decoction', Yi-Gan San, Ba Wei Di Huang Wan, and YishenHuazhuo decoction were found to have significant memory improvement in patients with AD (Dos Santos-Neto et al. 2006; Zhang et al. 2015). However, methodological limitations such as poor study design, relatively small sample size, and invalid statistical analysis have been noted (Tian et al. 2010). Given that multiple clinical trials have been carried out with *Ginkgo biloba* extract EGb 761 and Huperzine A (Chinese herb) for AD, meta-analysis of the trials that included randomized placebo-controlled modes showed high risk of bias (Yang et al. 2013; Gauthier and Schlaefke 2014).

In a double-blind, placebo-controlled cross-over study showed that acute supplementation of CDRI 08 (enriched in bacosides A and B) produced adaptogenic and nootropic effects (Benson et al. 2014). In addition, a couple of clinical trials confirmed that CDRI 08 has salutary effects on memory, mood, and mental alertness as the healthy volunteers treated with CDRI 08 displayed improved speed of visual information processing, learning rate and memory consolidation, decreased anxiety and improved performance on 'working memory' (the ability to hold information and actually do something with that information) and 'visual information processing' (a measure of the ability to sustain attention) as compared to placebo-treated group (Stough et al. 2001, 2008).

#### 1.10.1.2 Parkinson's Disease

Parkinson's disease (PD) is one of the most common age-related neurodegenerative disorders of the central nervous system (CNS), affecting about 3% of the population over the age of 65 worldwide. Reports from ethnobotanical surveys done in Asian countries including China, India, Japan, and Korea provide an inventory of herbal medicine for the treatment of PD (Wang et al. 2008; Song et al. 2012). These surveys suggest that in recent years herbal medicines such as *Acanthopanax*, *Alpinia*, and *Astragalus* (Li et al. 2013b) have attracted considerable attention to treat PD, particularly in China. Based on traditional Chinese medicines, the medicinal herbs belong to 24 genera and 18 families. These herbal medicines could be an alternative and valuable source for anti-PD drug discovery. In addition, couple of polyherbal formulations—Dangguijakyak-san (Hwang et al. 2011) and a modified Yeoldahanso-tang (Bae et al. 2011) showed anti-PD effects. Clinical trials have also been carried out with *Ginseng*, *Ginkgo biloba*, *Mucuna pruriens,* and *Banisteria caapi* which suggested neuroprotective properties in PD patients (Bega and Zadikoff 2014; Guanti et al. 1976). The meta-analysis included 19 randomized controlled trials involving 1371 participants to assess the efficacy and safety of Chinese herbal medicine (CHM) paratherapy in patients with PD. In this meta-analysis, patients receiving CHM adjunct therapy plus WCM (western conventional medication) exhibit significant improvement in their PD symptoms as evidenced by improvements in their UPDRS (Unified Parkinson Disease Rating Scale) scores compared to WCM controls in spite of some methodological limitations. According to the safety assessment of this meta-analysis, the CHM add-on therapy for PD is generally safe and well tolerated (Wang et al. 2012).

### 1.10.2 Metabolic Diseases

#### 1.10.2.1 Diabetes

Reports from ethnobotanical surveys done in Africa, South Asia, and China provide an inventory of herbal remedies for diabetes mellitus (DM), which suggest hundreds of herbs and botanicals in use globally. From these surveys, it appears that in India alone 37 medicinal plants belonging to 25 families are being used for the treatment of DM (Mohammed et al. 2015). In addition, at least nine polyherbal products are in the market having a significant sale in India. Clinical trials have been carried out with *Coccinia indica*, *Momordica charantia*, *fenugreek seeds*, *Azadirachta indica,* and *Ficus racemosa* in type 2 DM patients. However, all these studies suffer from limitations such as small sample size and trial duration, inappropriate randomization, lack of blinding and hence non-application of intention-to-treat analysis. Outcome measures were often not quantitative (Ocvirk et al. 2013). Meta-analyses were attempted with 108 trials examining 36 herbs/botanicals (single or in combination) involving >4000 patients with type 2 DM or impaired glucose tolerance (Choudhury et al. 2018). However, heterogeneity in terms of outcome measures and a small number of trials for each agent precluded formal meta-analyses.

#### 1.10.2.2 Hypertension and Associated Cardiovascular Diseases

Hypertension is a chronic visceral medical complication in which blood pressure (BP) in the arteries is elevated (Tabassum and Ahmad 2011). The use of herbal medicines to treat hypertension is growing all over the world, particularly across the developing countries because of their wide biological and medicinal activities, higher safety margins and lesser cost. For example, several polyherbal Chinese medicines including Tongxinluo, Xuefu Zhuyu decoction (XZD), and Zhen Wu decoction (ZWD) have been reported to be effective in treating mild-to-moderate hypertension (Wang et al. 2014, 2015b; Xiong et al. 2015b). Reports for ethnobotanical investigations/surveys done in many developing countries such as African nations and South Asia (India and Pakistan) utilize sustainable herbal plants for drug discovery (Davids et al. 2016; Malik et al. 2018). From an inventory, 27 medicinal plants from 22 families are used for the treatment of hypertension. Most commonly used antihypertensive species were *Apiaceae*, *Rosaceae* and *Papaveraceae* (Baharvand-Ahmadi et al. 2016). Clinical trials have been carried out with *Rauwolfia serpentina*, *Allium sativum*, *Ginkgo biloba*, *Hibiscus sabdariffa*, *Salicin* (the source of aspirin) from *Salix alba* (willow bark), *Crataegus laevigata*, *Terminalia arjuna*, *Achillea wilhelmsii*, *Centella asiatica*, and *Combretum micranthum* which indicate blood pressure (BP) lowering effect as well as safety of these plants in hypertensive patients (Mashour et al. 1998; Haji Faraji and Haji Tarkhani 1999; Walker et al. 2006; Tabassum and Ahmad 2011; Chrysant and Chrysant 2017; Seck et al. 2017).

A recent surge in the popularity of CHM has led to several clinical trials which reported efficacy of this kind of treatment in hypertensive patients. These trials ranged from case reports and case series to controlled observational studies and randomized clinical trials (Wang and Xiong 2012). A meta-analysis evaluating the effectiveness of ZWD and XZD included seven trials involving 472 hypertensive patients and 15 studies involving 1364 hypertensive patients, respectively. These herbal medicines are more effective in lowering BP, improving lipid profile, lowering homocysteine, and improving hemorheology when compared with clinically approved antihypertensive drugs (nifedipine, captopril, hydrochlorothiazide, valsartan, and amlodipine) used as monotherapy (Wang et al. 2015b; Xiong et al. 2015b). However, some limitations associated with studies included poor methodological quality, selective bias, small sample size, and inadequate reporting on clinical data (Xiong et al. 2015a, b).

Guggulsterone containing gugulipid showed significant lipid-lowering effect in 80% hyperlipidemic patients who participated in the trial; and the drug showed no adverse effect which resulted in its marketing approval from the Drug Controller General of India in 1986. A clinical trial in Indian patients with hyperlipidemia, upon treatment with 50 mg gugulipid b.d. and with a diet rich in fruits and vegetables reported significant lipid-lowering effect as compared to placebo-treated group. After 36 weeks of receiving a combination of diet and gugulipid, patients showed a lipid-lowering efficacy comparable to allopathic medicine and that too without any severe side effects (Singh et al. 1994). Gugulipid also showed lipid-lowering effect in a multicentric (held in seven Indian cities) open trials where 500 mg t.d.s. was administered for 8 weeks and the effect of gugulipid was better than clofibrate (the standard drug) particularly in enhancing HDL-cholesterol (Nityanand et al. 1989). However, hyperlipidemic patients eating the western diet failed to respond to gugulipid given for 8 weeks at 1000- or 2000-mg (Szapary et al. 2003). Whether prolonging gugulipid treatment with non-western diet (western diet is typically rich in fat-derived calorie) could have shown hypolipidemic effect in western patients remains conjectural. Nonetheless, available pieces of evidence are promising and warrant additional multicentric randomized placebo-controlled trials to confirm the hypolipidemic impact of gugulipid.

The standardized water extract of stem bark of *Terminalia arjuna* (Roxb. ex DC.) Wight and Arn was given to patients with chronic heart failure (CHF) in combination with standard medical therapies (angiotensin converting enzyme inhibitor or an angiotensin II type 1 receptor blocker and a beta-blocker) for 12 weeks to assess add-on efficacy of the extract in comparison to patients receiving only standard medical therapies. The findings showed that the extract was well tolerated but produced no change in the primary outcome measure, i.e., left ventricular ejection fraction, when added to evidence-based pharmacotherapy in patients of CHF. Among the secondary outcome measures, the extract helped to (a) preserve RBC catalase activity and functional capacity (distance covered in 6 min walk); and (b) increase antioxidant reserve and quality of life in some patients as compared to placebo patients (Maulik et al. 2016).

#### 1.10.2.3 Osteoporosis

Osteoporosis is characterized by low bone mass and microarchitectural deterioration resulting in increased fracture risk that occurs primarily with aging. The skeletal efficacy of a variety of medicinal plants tested thus far can be divided by their mechanisms that are either calcium regulatory, hormonal regulatory or bone remodeling regulatory types. Survey of medicinal plants in human use in China, India, and Korea has enlisted *Herba epimedii*, *Fructus ligustri lucidi*, and *Fructus psoraleae* for their skeletal efficacy demonstrated in preclinical setting. Polyherbal preparation made from CHM, OST-6, and BHH-10 reported safety and bone conserving effect in postmenopausal osteoporosis (PMO) (Irshad Ahmed et al. 2002; Cho et al. 2018). However, this trial suffers from limitations including small population size, short study duration, and nonavailability of compliance data.

FuFang with phytoestrogen-rich (icariin) epimedium is safe and effective in the prevention of postmenopausal bone loss based on a 5-year multicentre placebo-controlled study on 194 postmenopausal women, which underscores its potential in reducing fragility fracture (Deng et al. 2012). Anti-osteoporosis effect of FuFang was mediated by attenuating bone resorption via the increased production of anti-osteoclastogenic cytokines, osteoprotegerin. A meta-analysis of 12 random control trials involving 1816 patients observed significant bone mass density (BMD) increase at femur and spine over the placebo group (Wang et al. 2013). Future studies must be conducted with large sample size and longer duration for substantiating the current primary findings.

A standardized extract of leaves of *Dalbergia sissoo* rich in caviunin 7-*O*-[β-D-apiofuranosyl-(1 → 6)-β-D-glucopyranoside] showed safe bone conserving effect in patients with postmenopausal osteoporosis. The underlying mechanism appears to be significant suppression of the elevated pro-inflammatory cytokine, tumor necrosis factor-α in the osteoporotic patients (Meeta et al. 2019).

### 1.10.3 Infectious Diseases

#### 1.10.3.1 Tuberculosis

Ethnobotanical study reports suggest a rich source of herbal anti-tuberculosis drugs in Nigeria, China, India, Russia, and the UK. A total of 36 plants belonging to 20 families have been proposed for the management of tuberculosis in Nigeria. The most frequently used herbs were *Cola acuminate*, *Garcinia kola*, *Vitellaria paradoxa*, *Costus afer*, *Pycnanthus angolensis,* and *Aframomum melegueta* for TB treatment (Liu et al. 2008; Ogbole and Ajaiyeoba 2009; Wang et al. 2015a). Studies were done with *Allium sativum*, *Erythrina abyssinica*, *Abelmoschus esculentus*, *Adhatoda vasica,* and *Ocimum basilium* in pulmonary tuberculosis patients for their anti-mycobacterial activity, and were found to be useful as adjuvant therapies to improve the efficacy of conventional antimycobacterial therapies; to decrease their adverse effects; and to reverse multidrug resistance due to the genetic plasticity and environmental adaptability of mycobacterium (Sharifi-Rad et al. 2017).

A meta-analysis of 30 RCTs involving 3374 participants of MDR-TB given CHM combined with clinically approved chemotherapy, which showed that CHM plus chemotherapy arm had improved sputum bacteria conversion, lung lesions resorption, cavity closure, abnormal liver and kidney functions, and gastrointestinal symptoms compared with chemotherapy arm alone (Wang et al. 2015a).

#### 1.10.3.2 Malaria

Artemisinin, isolated from *Artemisia annua* is the most potent antimalarial drug for the discovery of which Tu Youyou received Nobel Prize for Physiology or Medicine in 2015. Artemisinin acts against all malaria-causing organisms in the genus *Plasmodium.* An ethnobotanical survey of Indian antimalarial plants identified a total of 22 species of plants belonging to 17 botanical families mostly from Assam. Out of these Verbenaceae, Menispermaceae, and Acanthaceae are the most commonly used to treat malaria and its associated symptoms (Namsa et al. 2011). Clinical studies were done with Quinine, *Cryptolepis sanguinolenta*, *Artemisia annua*, *Cochlospermum planchonii*, *Argemone mexicana*, *Vernonia amygdalina*, *Azadirachta indica*, and Ayush-64 for safety and efficacy of herbal preparations for their antimalarial preparation activity (Willcox and Bodeker 2004; Challand and

Willcox 2009; Tabuti 2008; Mueller et al. 2004; Willcox 2011). Data suggested the efficacy of these phyto-preparations in patients infected with *Plasmodium falciparum* and *Plasmodium vivax*. A plant decoction, AM-1 formulation composed of *Jatropha curcas*, *Gossypium hirsutum*, *Physalis angulata,* and *Delonix regia* was found to eliminate malarial parasites (*Plasmodium falciparum* and *Plasmodium malariae*) from the peripheral blood of infected subjects (Ankrah et al. 2003).

#### 1.10.3.3 Leishmaniasis

Based on ethnobotanical reports, 98 types of plants have been identified and recorded for their use in the treatment of three genera of Leishmania spp. The most commonly plants used as anti-leishmanial activity were *Artemisia species*, *Allium sativum*, *Achillea millefolium*, *Peganum harmala,* and *Thymus vulgaris* (Soosaraei et al. 2017). A number of plants have been assessed for anti-leishmanial effects including *Nyctanthes arbor-tristis*, *Withania somnifera* Dunal (ashwagandha) and *Allium sativum*, *Bidens pilosa* (*Asteraceae*) and *Punica granatum* (*Punicaceae*). The extracts of ashwagandha *bidens*, *Allium sativum pilosa* and *Punica granatum* have been reported to have a potent leishmanicidal effect (Panda and Luyten 2018; Garcia et al. 2010; Sharma et al. 2009). Furthermore, a randomized clinical trial suggested that *Juniperus excelsa* can be used as an adjuvant treatment in addition to cryotherapy for cutaneous leishmaniasis (CL) as it can decrease the duration and success rate of CL treatment without any significant adverse effects (Parvizi et al. 2017).

#### 1.10.3.4 Acquired Immunodeficiency Syndrome

Ethnobotanical evidences indicate that traditional medicines are commonly used by HIV-positive patients in Zimbabwe, South Africa, and Uganda (Gail et al. 2015; Monera and Maponga 2012; Lamorde et al. 2010). In ethnobotanical surveys, 75 plant species belonging to 66 genera and 41 families were found to be in use in treating one or more HIV/AIDS-related infections in the aforementioned African countries (Kisangau et al. 2007). Survey data has shown that *Moringa oleifera* Lam is commonly used for medicinal and nutritional purposes among HIV-positive patients in Zimbabwe (Monera and Maponga 2012). Studies carried out in HIV-infected patients using various medicinal plants such as *Allium sativum*, *Dicoma anomala*, *Aloe* spp., *Hypoxis hemerocallidea*, *Sutherlandia*, St. John's wort (*Hypericum perforatum*) and andrographolide from *Andrographis paniculata*, revealed serious adverse effects on the pharmacotherapies that are clinically approved for AIDS and thus raised important safety concerns for their use (Mugomeri et al. 2016; Mills et al. 2005; Langlois-Klassen et al. 2007; Calabrese et al. 2000).

In India, clinical studies of Boxwood (*Buxus sempervirens*), Andrographolide (*Andrographis paniculata*), and neem (*Azadirachta indica*), as well as Siddha combination therapy RAN (*Rasagandhi mezhuga, Amukkara chooranum*, and *Nellikkai lehyam*) were done (Fritts et al. 2008).

In India, a nonrandomized, placebo-controlled, investigator-blinded trial with a polyherbal formulation (PHF) was compared with highly active antiretroviral therapy (HAART) for safety and efficacy in treating HIV/AIDS. PHF consisted of ingredients from 58 different plant species appears to provide protection against AIDS development by delaying the kinetics of CD4 cell reduction leading to significant improvements in the T-cell profile of the infected subjects culminating in decreased viral load (Asokan et al. 2013). Eight herbal products were tested in nine randomized placebo-controlled trials involving 499 patients with HIV infection and full-blown AIDS patients. The results showed that preparations called SPV30, IGM-1 (CHM), and SH delayed the progression of HIV-related symptoms (Liu et al. 2005).

### 1.10.4 Inflammatory Diseases

#### 1.10.4.1 Chronic Obstructive Pulmonary Disease

Two polyherbal formulations including bushen naqi huoxue and Yufeining (YFN) (in China) and hochu-ekki-to (in Japan) were studied and found to be effective with no adverse effects in COPD patients and act as an anti-inflammatory agent (Hong et al. 2018; Ram et al. 2011). Clinical trials have also been carried out with *Panax ginseng*, *Salvia miltiorrhiza*, *Swasakasathirku churnam,* and *Hedera helix*. However, limitations including small sample size and poor methodological quality preclude these from being included in a meta-analysis (Guo et al. 2006).

Meta-analyses of 11 trials comprising total of 925 patients for CHMs add-on to SFP (Salmeterol and Fluticasone Propionate), showed efficacy and safety of these herbal medicines in COPD when compared with SFP alone. Use of 11 different CHM (seven herbal decoctions, one pill, and remaining three were capsules) showed increase in forced expiratory volume in 1 s, St. George's Respiratory Questionnaire (SGRQ) scoring and frequency of exacerbations. Out of 11 CHMs, Runfeijianpibushen decoction and Renshenbufei pills with SFP showed better effect in improving St. George's Respiratory Questionnaire (SGRQ) scoring when compared with SFP alone and other CHMs (herbal decoctions and capsules) (Chung et al. 2016).

#### 1.10.4.2 Rheumatoid Arthritis and Osteoarthritis

Ethnobotanical surveys on traditional medicinal plants for inflammation carried out in India, China, Korea, and Nigeria covered 485 plant species belonging to 100 families, traditionally used in arthritis. Among 100 plant families the ones with anti-inflammatory effects are Malvaceae, Leguminasae, Fabaceae, Euphorbiaceae, Compositae, Araceae, Solanaceae, Liliaceae, Apocynaceae, Lauraceae, and Rubiaceae (Choudhary et al. 2015). Clinical trials with Borage seed oil (*Borago officinalis*), Tripterygium wilfordii Hook F, curcumin (*Curcuma longa*), feverfew (*Tanacetum parthenium*), *Zingiber officinale*, and *Angelica sinensis* resulted in reduction in swelling and tender joint counts, shortened the duration of morning stiffness, and decreased erythrocyte sedimentation rate; and C-reactive protein and rheumatoid factor (Soeken et al. 2003; Chang et al. 1997; Wang et al. 2017b; Daily et al. 2016; Lee et al. 2014).

A meta-analysis of a CHM, Zhengqing Fengtongning (ZQFTN) from 11 studies on RA patients (508 patients received a combination of CHM and methotrexate and 448 patients received methotrexate only) showed that pain relief and morning stiffness in the combination group were significantly better than the methotrexate group alone. However, joint swelling and tender joint count were not different between the combination and methotrexate groups. Taken together, using ZQFTN combined with MTX appears to have better overall effects and lower side effects as compared with methotrexate alone in RA patients (Chen et al. 2015) (Wang et al. 2017b).

A standardized extract of *Boswellia serrata* containing 3-acetyl-11-keto-β-boswellic acid with β-boswellic acid was assessed in 48 patients of osteoarthritis having a minimum pain visual analog scale of >4. This was a randomized double-blind trial followed up for 120 days. Patients were given extract of *Boswellia serrata* b.d. in tablet form (dose 170 mg × 2). The extract significantly reduced joint pain and improved mobility along with decrease in the serum levels of C-reactive protein, a potent inflammatory marker. In addition, radiographic improvements were apparent from reduced osteophytes. Furthermore, the extract was well tolerated and devoid of any severe adverse effects (Majeed et al. 2019).

### 1.10.5 Cancer

#### 1.10.5.1 Breast Cancer

Ethnobotanical studies/surveys were done for numerous herbal drugs of China, and other Asian countries, having purported anticancer properties. Collectively, 72 plants that belong to 44 families are utilized for cancer treatment out of which most commonly used plants belong to *Compositae* and *Lamiaceae* (Chung et al. 2015; Abu-Darwish and Efferth 2018). Clinical trials were done on breast cancer patients with *Herba Scutellaria barbata* (HSB), Flavopiridol derived from *Dysoxylum binectariferum Hook.f.* (Meliaceae), *Panax ginseng*, *Camellia sinensis*, *Curcuma longa*, Soy-derived phytoestrogens, *Radix astragalus*, *Rhizoma atractylodis macrocephalae*, and *Angelica sinensis*. Findings of these trials showed boosting of the immune system, pain relief, anti-inflammatory effect, alleviation of fatigue, protection against chemotherapy-induced cyto- and gastrointestinal toxicity, and protection against other side effects from chemo- and radiotherapies (Liao et al. 2013; Safarzadeh et al. 2014; Yin et al. 2013). Meta-analysis studies with 33 RCTs included 2098 patients, in which 1066 patients received CHM as an adjunct therapy and 1032 received standard chemotherapy. Among CHMs, *Radix astragalus*, *Rhizoma atractylodis macrocephalae*, and *Angelica sinensis* were most frequently used herbs. Combined therapy (CHM combined with chemotherapy) significantly decreased adverse effects caused by chemotherapy alone, including nausea and vomiting at toxicity grade of III–IV, WBC reduction at toxicity grade of III–IV, and platelet reduction at toxicity grade of I–IV or III–IV (Zhu et al. 2016).

#### 1.10.5.2 Lung Cancer

Lung cancer is the most common malignancy worldwide and a leading cause of cancer-related deaths. Non-Small cell lung cancer (NSCLC) is the most common form of lung cancer, which accounts for approximately 85% of all lung cancer cases (Li et al. 2013a). An ethnopharmacological survey on herbal drugs used for the treatment of lung cancer was conducted in the Middle East, Africa, India, Turkey the Far East, and Europe. In these studies, 72 medicinal plants belonging to 44 families were used for the treatment of lung cancer, however, the most frequently used were decoctions of *Ephedra alata*, *A. dioscoridis*, and *A. palaestinum* (Abu-Darwish and Efferth 2018; Jaradat et al. 2016). Traditionally, phytochemicals such as Vinorelbine, Abraxane and herbal plants like, *Platycodon grandiflorum* (*Campanulaceae*), *Morus alba* (*Moraceae*), *Prunus armeniaca* (*Rosaceae*), *Rhus verniciflua* (*Anacardiaceae*), *Perilla frutescens* (*Labiatae*), *Stemona japonica* (*Stemonaceae*), *Tussilago farfara* (*Compositae*), and *Draba nemorosa* (*Brassicaceae*) have been used to treat lung cancer. These herbs are mainly used in lung cancer to reduce therapy-associated toxicity and cancer-related symptoms and sometimes to directly increase anticancer effects. Numerous types of polyherbal formulations used for lung cancer treatment such as Sheng-mai injection, Gu-jin Granule, Feiji Recipe, Dixiong Decoction, Liangxue Jiedu Houxue Decoction, QingjinRunfei Decoction, and Shenqi-Fuzheng injection increased the survival rate among patients (Yin et al. 2013; Raj Parikh et al. 2014; Safarzadeh et al. 2014).

A meta-analysis of 22 studies was conducted on 1819 participants. The participants included patients with non-small cell carcinoma and radiotherapy pneumonitis (RP) who were (a) administered herbal medicines (*Ophiopogonis radix*, *Adenophorae radix*, *Astragali radix*, and *Angelicae sinensis radix*) during radiotherapy and (b) patients who underwent radiotherapy without herbal medicines. The study observed that the number of patients who developed RP decreased significantly in the herbal medicine plus radiotherapy group as compared with the radiotherapy alone group. In addition, quality of life also significantly increased in the HM (herbal medicine) plus radiotherapy group compared with the radiotherapy alone group. A few studies assessed safety of herbal medicines and latter were found to be safe (Kim et al. 2018).

#### 1.10.5.3 Prostate Cancer

Ethnopharmacological/ethnobotanical information revealed 57 plant species belonging to 30 families are used for the treatment of prostate cancer. For the treatment of benign prostate hyperplasia *Juglans regia*, *Quercus infectoria*, *Sambucus ebulus,* and *Zea mays* were used for the treatment of symptoms (Jaradat et al. 2017). Clinical studies have been done in India, China, Africa, Australia, the USA, and other Western countries with medicinal herbs such as saw palmetto, pomegranate, soy isoflavones, *H. hemerocallidea*, *Camellia sinensis*, *Wedelia chinensis*, *Panax ginseng*, *Allium sativum*, *Ginkgo biloba*, and *Prunus africana* (pygeum). These herbs showed an antiproliferative effect against prostate cancer cells, suppressed androgenic activity and showed potential for reducing the risk of prostate cancer

(Safarzadeh et al. 2014; Abrams 2018; Yin et al. 2013; Gratus et al. 2009; Steenkamp 2003).

Other polyherbal CHM formulations such as Chai-Hu-Jia-Long, Gu-Mu-Li-Tang, Suan-Zao-Ren-Tang, Ban-Xia-Xie-Xin-Tang, and Ba-Wei-Di-Huang-Wan are used for improving the survival rate of metastatic prostate cancer patients. In addition, Chai-Hu-Jia-Long-Gu-Mu-Li-Tang relieved the symptoms of hypogonadism, including insomnia, hot flushes, and erectile dysfunction, but did not change serum testosterone levels. In patients with benign prostatic hyperplasia, Ba-Wei-Di-Huang-Wan improved nocturia or incomplete bladder emptying and also prevented osteoporosis (Liu et al. 2016). A meta-analysis comprising 14 randomized clinical trials and three open-label trials, involving 4280 patients, were analysed for *Serenoa repens* extract (Permixon) for the treatment of symptomatic benign prostatic hyperplasia (BPH). These trials were of different size (22–1100 patients) and duration (21–720 days). In this, Permixontrials for treating men with BPH showed a significant improvement in peak flow rate and reduction in nocturia above placebo (Boyle et al. 2004).

Review of clinical trials with TM indicates a clear deficiency in high quality trials in every disease area dealt here. All trials assessed safety and tolerability of TMs and these were largely established. However, efficacy of TMs alone or as an adjunct to allopathic medicine in majority cases appear to fall short of drawing a satisfactory conclusion due to the lack of large randomized controlled trials.

Regulation of TMs is in its infancy. Adverse events arising from consumption of TMs are attributable to factors including use of wrong species of plants due to incorrect pharmacognosy, adulterated plant materials and contamination with heavy metals, pesticides, and other hazardous substance beyond allowable limits. World-wide, regulators have applied stringent quality control of TMs to minimize the chance of adverse events caused by the aforementioned factors. However, adverse events related to the use of TMs are far more complex than in the case of mainstream pharmaceuticals. For example drug–drug interaction between TMs and conventional pharmaceuticals is an important safety factor for which the availability of data is scarce. Also, because unlike pharmaceutical drugs TMs do not carry disease cure claims, pharmacovigilance similar to that of the former is unlikely to work for TMs. Indeed, not much literature, if any is available on the pharmacovigilance of TMs.

## References

Abrams DI (2018) An integrative approach to prostate cancer. J Altern Complement Med 24 (9-10):872–880

Abu-Darwish MS, Efferth T (2018) Medicinal plants from near east for cancer therapy. Front Pharmacol 9:56

Akhondzadeh S, Noroozian M et al (2003) Melissa officinalis extract in the treatment of patients with mild to moderate Alzheimer's disease: a double blind, randomised, placebo controlled trial. J Neurol Neurosurg Psychiatry 74(7):863–866

Akhondzadeh S, Shafiee Sabet M et al (2010) A 22-week, multicenter, randomized, double-blind controlled trial of Crocus sativus in the treatment of mild-to-moderate Alzheimer's disease. Psychopharmacology 207(4):637–643

Ankrah NA, Nyarko AK et al (2003) Evaluation of efficacy and safety of a herbal medicine used for the treatment of malaria. Phytother Res 17(6):697–701

Anon (2004) New alternative medicine guide launched amidst increasing reports of adverse reactions. Bull World Health Organ 82(8):635–636

Asokan M, Lone IN et al (2013) Evident stabilization of the clinical profile in HIV/AIDS as evaluated in an open label clinical trial using a polyherbal formulation. Indian J Med Res 137 (6):1128–1144

Bae N, Ahn T et al (2011) The neuroprotective effect of modified Yeoldahanso-tang via autophagy enhancement in models of Parkinson's disease. J Ethnopharmacol 134(2):313–322

Baharvand-Ahmadi B, Bahmani M et al (2016) An ethnobotanical study of medicinal plants administered for the treatment of hypertension. J Renal Inj Prev 5(3):123–128

Bega D, Zadikoff C (2014) Complementary & alternative management of Parkinson's disease: an evidence-based review of eastern influenced practices. J Mov Disord 7(2):57–66

Benson S, Downey LA et al (2014) An acute, double-blind, placebo-controlled cross-over study of 320 mg and 640 mg doses of Bacopa monnieri (CDRI 08) on multitasking stress reactivity and mood. Phytother Res 28(4):551–559

Bhatt A (2016) Phytopharmaceuticals: a new drug class regulated in India. Perspect Clin Res 7 (2):59–61

Boyle P, Robertson C et al (2004) Updated meta-analysis of clinical trials of Serenoa repens extract in the treatment of symptomatic benign prostatic hyperplasia. BJU Int 93(6):751–756

Burton A et al (2015) Building WHO's global strategy for traditional medicine*. Eur J Integr Med 7:13

Calabrese C, Berman SH et al (2000) A phase I trial of andrographolide in HIV positive patients and normal volunteers. Phytother Res 14(5):333–338

Challand S, Willcox M (2009) A clinical trial of the traditional medicine Vernonia amygdalina in the treatment of uncomplicated malaria. J Altern Complement Med 15(11):1231–1237

Chang DM, Chang WY et al (1997) The effects of traditional antirheumatic herbal medicines on immune response cells. J Rheumatol 24(3):436–441

Chen XM, Huang RY et al (2015) Systemic review and meta-analysis of the clinical efficacy and adverse effects of zhengqing fengtongning combined with methotrexate in rheumatoid arthritis. Evid Based Complement Alternat Med 2015:910376

Cho Y, Lee S et al (2018) The efficacy and safety of herbal medicine BHH10 in postmenopausal women with osteoporosis: study protocol for a phase II, multicenter, randomized, double-blinded, placebo-controlled clinical trial. Trials 19(1):482

Choudhary M, Kumar V et al (2015) Medicinal plants with potential anti-arthritic activity. J Intercult Ethnopharmacol 4(2):147–179

Choudhury H, Pandey M et al (2018) An update on natural compounds in the remedy of diabetes mellitus: a systematic review. J Tradit Complement Med 8(3):361–376

Chrysant SG, Chrysant GS (2017) Herbs used for the treatment of hypertension and their mechanism of action. Curr Hypertens Rep 19(9):77

Chung VC, Wu X et al (2015) Effectiveness of Chinese herbal medicine for cancer palliative care: overview of systematic reviews with meta-analyses. Sci Rep 5:18111

Chung VC, Wu X et al (2016) Chinese herbal medicine and salmeterol and fluticasone propionate for chronic obstructive pulmonary disease: systematic review and network meta-analysis. Medicine (Baltimore) 95(20):e3702

Cordell GA (2011) Sustainable medicines and global health care. Planta Med 77(11):1129–1138

Cordell GA, Colvard MD (2012) Natural products and traditional medicine: turning on a paradigm. J Nat Prod 75(3):514–525

Cox DG, Oh J et al (2014) The utility of metabolomics in natural product and biomarker characterization. Biochim Biophys Acta 1840(12):3460–3474

Daily JW, Yang M et al (2016) Efficacy of turmeric extracts and curcumin for alleviating the symptoms of joint arthritis: a systematic review and meta-analysis of randomized clinical trials. J Med Food 19(8):717–729

Davids D, Gibson D et al (2016) Ethnobotanical survey of medicinal plants used to manage high blood pressure and type 2 diabetes mellitus in Bitterfontein, Western Cape Province, South Africa. J Ethnopharmacol 194:755–766

Deng WM, Zhang P et al (2012) Five-year follow-up study of a kidney-tonifying herbal Fufang for prevention of postmenopausal osteoporosis and fragility fractures. J Bone Miner Metab 30 (5):517–524

Dos Santos-Neto LL, de Vilhena Toledo MA et al (2006) The use of herbal medicine in Alzheimer's disease-a systematic review. Evid Based Complement Alternat Med 3(4):441–445

Eckert GP (2010) Traditional used plants against cognitive decline and Alzheimer disease. Front Pharmacol 1:138

Ernst E, Cassileth BR (1998) The prevalence of complementary/alternative medicine in cancer: a systematic review. Cancer 83(4):777–782

Farooqui AA, Farooqui T et al (2018) Ayurvedic medicine for the treatment of dementia: mechanistic aspects. Evid Based Complement Alternat Med 2018:2481076

Franz CM, Huch M et al (2011) Enterococci as probiotics and their implications in food safety. Int J Food Microbiol 151(2):125–140

Fritts M, Crawford CC et al (2008) Traditional Indian medicine and homeopathy for HIV/AIDS: a review of the literature. AIDS Res Ther 5:25

Gahlaut A, Shirolkar A et al (2013) beta-sitosterol in different parts of Saraca asoca and herbal drug ashokarista: quali-quantitative analysis by liquid chromatography-mass spectrometry. J Adv Pharm Technol Res 4(3):146–150

Gail H, Tarryn B et al (2015) An ethnobotanical survey of medicinal plants used by traditional health practitioners to manage HIV and its related opportunistic infections in Mpoza, Eastern Cape Province, South Africa. J Ethnopharmacol 171:109–115

Garcia M, Monzote L et al (2010) Screening of medicinal plants against Leishmania amazonensis. Pharm Biol 48(9):1053–1058

Gauthier S, Schlaefke S (2014) Efficacy and tolerability of Ginkgo biloba extract EGb 761(R) in dementia: a systematic review and meta-analysis of randomized placebo-controlled trials. Clin Interv Aging 9:2065–2077

Glassbrook N, Ryals J (2001) A systematic approach to biochemical profiling. Curr Opin Plant Biol 4(3):186–190

Gouveia BG, Rijo P et al (2015) Good manufacturing practices for medicinal products for human use. J Pharm Bioall Sci 7(2):87–96

Govindaraghavan S (2008) Quality assurance of herbal raw materials in supply chain: challenges and opportunities. J Diet Suppl 5(2):176–212

Gratus C, Damery S et al (2009) The use of herbal medicines by people with cancer in the UK: a systematic review of the literature. QJM 102(12):831–842

Guanti G, Mollica G et al (1976) rDNA and acrocentric chromosomes in man. I. rDNA levels in a subject carrier of a 8p/13p balanced translocation and in his unbalanced son. Hum Genet 33 (2):103–107

Guo R, Pittler MH et al (2006) Herbal medicines for the treatment of COPD: a systematic review. Eur Respir J 28(2):330–338

Haji Faraji M, Haji Tarkhani A (1999) The effect of sour tea (Hibiscus sabdariffa) on essential hypertension. J Ethnopharmacol 65(3):231–236

Harwansh RK, Mukherjee K et al (2014) Cytochrome P450 inhibitory potential and RP-HPLC standardization of trikatu--a Rasayana from Indian Ayurveda. J Ethnopharmacol 153 (3):674–681

Harwansh RK, Mukherjee PK et al (2015) Enhanced permeability of ferulic acid loaded nanoemulsion based gel through skin against UVA mediated oxidative stress. Life Sci 141:202–211

Hong M, Hong C et al (2018) Effects of the Chinese herb formula Yufeining on stable chronic obstructive pulmonary disease: a randomized, double-blind, placebo-controlled trial. Medicine (Baltimore) 97(39):e12461

Hwang DS, Kim HG et al (2011) Dangguijakyak-san, a medicinal herbal formula, protects dopaminergic neurons from 6-hydroxydopamine-induced neurotoxicity. J Ethnopharmacol 133(2):934–939

Jaradat NA, Al-Ramahi R et al (2016) Ethnopharmacological survey of herbal remedies used for treatment of various types of cancer and their methods of preparations in the West Bank-Palestine. BMC Complement Altern Med 16:93

Jaradat NA, Zaid AN et al (2017) Ethnopharmacological survey of medicinal plants practiced by traditional healers and herbalists for treatment of some urological diseases in the West Bank/ Palestine. BMC Complement Altern Med 17(1):255

Katoch D, Sharma JS et al (2017) Government policies and initiatives for development of Ayurveda. J Ethnopharmacol 197:25–31

Kim KI, Jun JH et al (2018) Oral administration of herbal medicines for radiation pneumonitis in lung cancer patients: a systematic review and meta-analysis. PLoS One 13(5):e0198015

Kisangau DP, Lyaruu HV et al (2007) Use of traditional medicines in the management of HIV/AIDS opportunistic infections in Tanzania: a case in the Bukoba rural district. J Ethnobiol Ethnomed 3:29

Lamorde M, Tabuti JR et al (2010) Medicinal plants used by traditional medicine practitioners for the treatment of HIV/AIDS and related conditions in Uganda. J Ethnopharmacol 130(1):43–53

Langlois-Klassen D, Kipp W et al (2007) Use of traditional herbal medicine by AIDS patients in Kabarole District, western Uganda. Am J Trop Med Hyg 77(4):757–763

Lee WY, Chen HY et al (2014) Treatment of rheumatoid arthritis with traditional chinese medicine. Biomed Res Int 2014:528018

Lee KM, Jeon JY et al (2017) Application of metabolomics to quality control of natural product derived medicines. Biomol Ther (Seoul) 25(6):559–568

Li SG, Chen HY et al (2013a) The efficacy of Chinese herbal medicine as an adjunctive therapy for advanced non-small cell lung cancer: a systematic review and meta-analysis. PLoS One 8(2): e57604

Li XZ, Zhang SN et al (2013b) Recent advances in herbal medicines treating Parkinson's disease. Fitoterapia 84:273–285

Liao GS, Apaya MK et al (2013) Herbal medicine and acupuncture for breast cancer palliative care and adjuvant therapy. Evid Based Complement Alternat Med 2013:437948

Liu JP, Manheimer E et al (2005) Herbal medicines for treating HIV infection and AIDS. Cochrane Database Syst Rev 3:CD003937

Liu Q, Garner P et al (2008) Drugs and herbs given to prevent hepatotoxicity of tuberculosis therapy: systematic review of ingredients and evaluation studies. BMC Public Health 8:365

Liu JM, Lin PH et al (2016) Complementary traditional Chinese medicine therapy improves survival in patients with metastatic prostate cancer. Medicine (Baltimore) 95(31):e4475

Majeed M, Majeed S et al (2019) A pilot, randomized, double-blind, placebo-controlled trial to assess the safety and efficacy of a novel Boswellia serrata extract in the management of osteoarthritis of the knee. Phytother Res 33(5):1457–1468

Malik K, Ahmad M et al (2018) Ethnobotany of anti-hypertensive plants used in Northern Pakistan. Front Pharmacol 9:789

Mashour NH, Lin GI et al (1998) Herbal medicine for the treatment of cardiovascular disease: clinical considerations. Arch Intern Med 158(20):2225–2234

Maulik SK, Wilson V et al (2016) Clinical efficacy of water extract of stem bark of Terminalia arjuna (Roxb. ex DC.) Wight & Arn. in patients of chronic heart failure: a double-blind, randomized controlled trial. Phytomedicine 23(11):1211–1219

Meeta A, Raut A et al (2019) A clinical study of a standardized extract of leaves of Dalbergia sissoo (Roxb ex DC) in Postmenopausal Osteoporosis. J Midlife Health 10(1):37–42

Mills E, Cooper C et al (2005) African herbal medicines in the treatment of HIV: hypoxis and sutherlandia. An overview of evidence and pharmacology. Nutr J 4:19
Mohammed A, Kumar D et al (2015) Antidiabetic potential of some less commonly used plants in traditional medicinal systems of India and Nigeria. J Intercult Ethnopharmacol 4(1):78–85
Monera TG, Maponga CC (2012) Prevalence and patterns of Moringa oleifera use among HIV positive patients in Zimbabwe: a cross-sectional survey. J Public Health Afr 3(1):e6
Mueller MS, Runyambo N et al (2004) Randomized controlled trial of a traditional preparation of Artemisia annua L. (Annual Wormwood) in the treatment of malaria. Trans R Soc Trop Med Hyg 98(5):318–321
Mugomeri E, Chatanga P et al (2016) Medicinal herbs used by hiv-positive people in lesotho. Afr J Tradit Complement Altern Med 13(4):123–131
Mukherjee PK, Wahile A (2006) Integrated approaches towards drug development from Ayurveda and other Indian system of medicines. J Ethnopharmacol 103(1):25–35
Mukherjee PK, Kumar V et al (2007a) Screening of Indian medicinal plants for acetylcholinesterase inhibitory activity. Phytother Res 21(12):1142–1145
Mukherjee PK, Rai S et al (2007b) Plants of Indian origin in drug discovery. Expert Opin Drug Discovery 2(5):633–657
Mukherjee PK, Sahoo AK et al (2009) Lead finding from medicinal plants with hepatoprotective potentials. Expert Opin Drug Discovery 4(5):545–576
Mukherjee PK, Venkatesh P et al (2010) Ethnopharmacology and integrative medicine - let the history tell the future. J Ayurveda Integr Med 1(2):100–109
Mukherjee PK, Nema NK et al (2012) Changing scenario for promotion and development of Ayurveda--way forward. J Ethnopharmacol 143(2):424–434
Mukherjee PK, Harwansh RK et al (2017) Development of ayurveda - tradition to trend. J Ethnopharmacol 197:10–24
Namsa ND, Mandal M et al (2011) Anti-malarial herbal remedies of northeast India, Assam: an ethnobotanical survey. J Ethnopharmacol 133(2):565–572
Narayana DA, Katiyar C (2013) Draft amendment to drugs and cosmetics rules to license science based botanicals, phytopharmaceuticals as drugs in India. J Ayurveda Integr Med 4(4):245–246
Newman DJ, Cragg GM (2007) Natural products as sources of new drugs over the last 25 years. J Nat Prod 70(3):461–477
Nityanand S, Srivastava JS et al (1989) Clinical trials with gugulipid. A new hypolipidaemic agent. J Assoc Physicians India 37(5):323–328
Noteborn HP, Lommen A et al (2000) Chemical fingerprinting for the evaluation of unintended secondary metabolic changes in transgenic food crops. J Biotechnol 77(1):103–114
Ocvirk S, Kistler M et al (2013) Traditional medicinal plants used for the treatment of diabetes in rural and urban areas of Dhaka, Bangladesh--an ethnobotanical survey. J Ethnobiol Ethnomed 9:43
Ogbole OO, Ajaiyeoba EO (2009) Traditional management of tuberculosis in Ogun State of Nigeria: the practice and ethnobotanical survey. Afr J Tradit Complement Altern Med 7 (1):79–84
Panda SK, Luyten W (2018) Antiparasitic activity in Asteraceae with special attention to ethnobotanical use by the tribes of Odisha, India. Parasite 25:10
Parvizi MM, Handjani F et al (2017) Efficacy of cryotherapy plus topical Juniperus excelsa M. Bieb cream versus cryotherapy plus placebo in the treatment of Old World cutaneous leishmaniasis: a triple-blind randomized controlled clinical trial. PLoS Negl Trop Dis 11(10):e0005957
Patwardhan B (2011) Ayurveda GCP guidelines: need for freedom from RCT ascendancy in favor of whole system approach. J Ayurveda Integr Med 2(1):1–4
Patwardhan B (2014) Envisioning AYUSH: historic opportunity for innovation and revitalization. J Ayurveda Integr Med 5(2):67–70
Patwardhan B, Mashelkar RA (2009) Traditional medicine-inspired approaches to drug discovery: can Ayurveda show the way forward? Drug Discov Today 14(15-16):804–811

Ram A, Balachandar S et al (2011) Medicinal plants useful for treating chronic obstructive pulmonary disease (COPD): current status and future perspectives. Fitoterapia 82(2):141–151
Safarzadeh E, Sandoghchian Shotorbani S et al (2014) Herbal medicine as inducers of apoptosis in cancer treatment. Adv Pharm Bull 4(Suppl 1):421–427
Samal J (2015) Role of AYUSH workforce, therapeutics, and principles in health care delivery with special reference to National Rural Health Mission. Ayu 36(1):5–8
Seck SM, Doupa D et al (2017) Clinical efficacy of African traditional medicines in hypertension: a randomized controlled trial with Combretum micranthum and Hibiscus sabdariffa. J Hum Hypertens 32(1):75–81
Sen S, Chakraborty R (2017) Revival, modernization and integration of Indian traditional herbal medicine in clinical practice: importance, challenges and future. J Tradit Complement Med 7 (2):234–244
Sharifi-Rad J, Salehi B et al (2017) Medicinal plants used in the treatment of tuberculosis - ethnobotanical and ethnopharmacological approaches. Biotechnol Adv
Sharma U, Velpandian T et al (2009) Evaluation of anti-leishmanial activity of selected Indian plants known to have antimicrobial properties. Parasitol Res 105(5):1287–1293
Shaw D, Graeme L et al (2012) Pharmacovigilance of herbal medicine. J Ethnopharmacol 140 (3):513–518
Singh RB, Niaz MA et al (1994) Hypolipidemic and antioxidant effects of Commiphora mukul as an adjunct to dietary therapy in patients with hypercholesterolemia. Cardiovasc Drugs Ther 8 (4):659–664
Soeken KL, Miller SA et al (2003) Herbal medicines for the treatment of rheumatoid arthritis: a systematic review. Rheumatology (Oxford) 42(5):652–659
Song JX, Sze SC et al (2012) Anti-Parkinsonian drug discovery from herbal medicines: what have we got from neurotoxic models? J Ethnopharmacol 139(3):698–711
Soosaraei M, Fakhar M et al (2017) Medicinal plants with promising antileishmanial activity in Iran: a systematic review and meta-analysis. Ann Med Surg (Lond) 21:63–80
Srinivasan R, Sugumar VR (2017) Spread of Traditional Medicines in India: results of national sample survey organization's perception survey on use of AYUSH. J Evid Based Complement Alternat Med 22(2):194–204
Srivastava A, Gope R et al (2016) Are village health sanitation and nutrition committees fulfilling their roles for decentralised health planning and action? A mixed methods study from rural eastern India. BMC Public Health 16:59
Steenkamp V (2003) Phytomedicines for the prostate. Fitoterapia 74(6):545–552
Stough C, Lloyd J et al (2001) The chronic effects of an extract of Bacopa monniera (Brahmi) on cognitive function in healthy human subjects. Psychopharmacology 156(4):481–484
Stough C, Downey LA et al (2008) Examining the nootropic effects of a special extract of Bacopa monniera on human cognitive functioning: 90 day double-blind placebo-controlled randomized trial. Phytother Res 22(12):1629–1634
Szapary PO, Wolfe ML et al (2003) Guggulipid for the treatment of hypercholesterolemia: a randomized controlled trial. JAMA 290(6):765–772
Tabassum N, Ahmad F (2011) Role of natural herbs in the treatment of hypertension. Pharmacogn Rev 5(9):30–40
Tabuti JR (2008) Herbal medicines used in the treatment of malaria in Budiope county, Uganda. J Ethnopharmacol 116(1):33–42
Tewari D, Stankiewicz AM et al (2018) Ethnopharmacological approaches for dementia therapy and significance of natural products and herbal drugs. Front Aging Neurosci 10:3
Tian J, Shi J et al (2010) Herbal therapy: a new pathway for the treatment of Alzheimer's disease. Alzheimers Res Ther 2(5):30
Ubale S (2011) Twelfth international congress of ethnopharmacology, kolkata (India). J Ayurveda Integr Med 2(3):160
Ulrich-Merzenich G, Zeitler H et al (2007) Application of the "-Omic-" technologies in phytomedicine. Phytomedicine 14(1):70–82

Walker AF, Marakis G et al (2006) Hypotensive effects of hawthorn for patients with diabetes taking prescription drugs: a randomised controlled trial. Br J Gen Pract 56(527):437–443

Wang J, Xiong X (2012) Outcome measures of chinese herbal medicine for hypertension: an overview of systematic reviews. Evid Based Complement Alternat Med 2012:697237

Wang X, Liang XB et al (2008) Therapeutic strategies for Parkinson's disease: the ancient meets the future--traditional Chinese herbal medicine, electroacupuncture, gene therapy and stem cells. Neurochem Res 33(10):1956–1963

Wang Y, Xie CL et al (2012) Chinese herbal medicine paratherapy for Parkinson's disease: a meta-analysis of 19 randomized controlled trials. Evid Based Complement Alternat Med 2012:534861

Wang ZQ, Li JL et al (2013) Chinese herbal medicine for osteoporosis: a systematic review of randomized controlled trails. Evid Based Complement Alternat Med 2013:356260

Wang J, Xiong X et al (2014) Chinese patent medicine tongxinluo capsule for hypertension: a systematic review of randomised controlled trials. Evid Based Complement Alternat Med 2014:187979

Wang M, Guan X et al (2015a) Chinese herbal medicine as adjuvant treatment to chemotherapy for multidrug-resistant tuberculosis (MDR-TB): a systematic review of randomised clinical trials. Tuberculosis (Edinb) 95(4):364–372

Wang P, Xiong X et al (2015b) Efficacy and safety of a traditional Chinese herbal formula Xuefu Zhuyu decoction for hypertension: a systematic review and meta-analysis. Medicine (Baltimore) 94(42):e1850

Wang M, Chen L et al (2017a) Metabolomics highlights pharmacological bioactivity and biochemical mechanism of traditional Chinese medicine. Chem Biol Interact 273:133–141

Wang X, Zu Y et al (2017b) Treatment of rheumatoid arthritis with combination of methotrexate and Tripterygium wilfordii: a meta-analysis. Life Sci 171:45–50

Wetzel MS, Kaptchuk TJ et al (2003) Complementary and alternative medical therapies: implications for medical education. Ann Intern Med 138(3):191–196

Willcox M (2011) Improved traditional phytomedicines in current use for the clinical treatment of malaria. Planta Med 77(6):662–671

Willcox ML, Bodeker G (2004) Traditional herbal medicines for malaria. BMJ 329 (7475):1156–1159

Xiong X, Li X et al (2015a) Chinese herbal medicine for resistant hypertension: a systematic review. BMJ Open 5(1):e005355

Xiong X, Wang P et al (2015b) Meta-analysis of the effectiveness of traditional Chinese herbal formula Zhen Wu Decoction for the treatment of hypertension. BMJ Open 5(12):e007291

Yang G, Wang Y et al (2013) Huperzine A for Alzheimer's disease: a systematic review and meta-analysis of randomized clinical trials. PLoS One 8(9):e74916

Yin SY, Wei WC et al (2013) Therapeutic applications of herbal medicines for cancer patients. Evid Based Complement Alternat Med 2013:302426

Zhang Y, Lin C et al (2015) Cognitive improvement during treatment for mild Alzheimer's disease with a Chinese herbal formula: a randomized controlled trial. PLoS One 10(6):e0130353

Zhu L, Li L et al (2016) Chinese herbal medicine as an adjunctive therapy for breast cancer: a systematic review and meta-analysis. Evid Based Complement Alternat Med 2016:9469276

**Pulok K. Mukherjee** is working as the director of Institute of Bioresources and Sustainable Development-Department of Biotechnology (IBSD-DBT), Ministry of Science and Technology, Govt of India, Imphal, Manipur, India. He was formerly the director of School of Natural Product Studies, Jadavpur University, Kolkata, India. His research/academic works highlights on traditional medicine-inspired drug discovery from Indian medicinal plants to make them available from 'Farm to Pharma'. He has to his credit more than 200 publications in peer reviewed impact journals and several patents. Prof. Mukherjee has authored/edited five books and 18 book chapters.

**Subhadip Banerjee** is a senior research fellow at School of Natural Product Studies, Jadavpur University, pursuing his doctoral research under Prof Pulok K. Mukherjee Director, IBSD-DBT. He obtained his M.Pharm (Ayurveda) in pharmacology in the year 2015. His work focuses on establishing integrated approaches for drug development from Indian traditional medicine to understand the synergy based on network pharmacology and metabolomics.

**Chandra Kant Katiyar** has worked at Ranbaxy Laboratories Ltd, Gurgaon as Director, Herbal Drug Research and Dabur India Ltd as Head and Vice President of Dabur Research and Development Centre, NCR Delhi. Currently Dr Katiyar is CEO Health Care (Technical) of Emami Ltd., Kolkata, India. He has made important contributions in quality control of Ayurvedic/Herbal drugs being member of Ayurvedic Pharmacopoeia Committee, Herbal Committee for Indian Pharmacopoeia and South East Asia Expert Panel on Dietary Supplements for United States Pharmacopoeia.He developed several top selling Ayurvedic/herbal OTC and ethical products besides pharmaceutical products in the field of neurology and oncology. He has also made significant contributions in new drug discovery from medicinal plants. He is member of ASU Drugs Technical Advisory Board, Governing body, All India Institute of Ayurveda, Delhi, Scientific Advisory Board, Central Council for Research in Ayurveda Sciences and Scientific Advisory Group, Indian Council of Medical Research. He is currently Vice Chairman of National Phytopharmaceutical Mission of Department of Bio Technology, Ministry of Science and Technology, Government of India.

**Shivani Sharma** is M.Sc. in Biotechnology. After finishing her M.Sc., she has received industrial training at Wockhardt Ltd., Aurangabad under the Biotech Consortium India Limited. Presently, she is pursuing her PhD as a Senior Research Fellow (CSIR-UGC) in the laboratory of Dr Naibedya Chattopadhyay.

**Naibedya Chattopadhyay** is PhD in Endocrinology. He is presently a chief scientist at CSIR-Central Drug Research Institute, Lucknow. His area of research is metabolic bone disease and mineral homeostasis. In this area, he is working for the past 26 years. He was closely associated with the discovery and development of the calcimimetic drug (an agonist of calcium-sensing receptor) for the management of hypercalcemic conditions during his tenure at the Harvard Medical School, Boston, MA, USA. He is also instrumental in the discovery and development of a standardized herbal preparation from the leaves of *Dalbergia sissoo* for accelerated healing of fracture (marketed as Reunion®). Very recently, he has developed a herbal technology from *Cassia occidentalis* L for preventing corticosteroid-induced bone loss (licensed to an industry for commercialization).

# Modern Drug Discovery and Development for TB: The India Narrative

# 2

Tanjore S. Balganesh, Jagadeesh J. Bhat, and Bheemarao Ugarkar

## 2.1 Approaches to TB Drug Discovery: The Past and the Present

### 2.1.1 A Historical Perspective of the TB Drug Regimen

India has the task of eliminating tuberculosis (TB) by 2025. This translates to curing about two million TB cases present today as well as reducing the TB infection rates rapidly (WHO Global Tuberculosis Report 2017). The 'standard treatment regimen' being administered today is a combination of drugs discovered and developed in the 1950s and 1960s. This regimen falls short of an ideal therapy in many ways including the requirement of prolonged treatment period of 6 months with unpleasant and toxic drug side effects (Yee et al. 2003). However, the characteristic of the TB patient population has changed considerably in the last few decades—coexistence of diabetes and HIV (human immunodeficiency virus) being the main drivers (Balganesh et al. 2008). In addition, India has a significant number of drug-resistant TB patients (Indian TB report 2018) who need novel drugs and regimens for faster and permanent cure. Thus there is an urgent unmet medical need for which TB drug discovery and development efforts globally and in India need to rise to this occasion.

The unique biological aspects of the disease causing microbe, *Mycobacterium tuberculosis* (MTB) and its relevance to discovering new drugs were unravelled in the course of the development of the standard treatment of tuberculosis. The first drug used for the treatment of tuberculosis was streptomycin (Murray et al. 2015). In a limited clinical trial in 1946 it became obvious that treating TB patients with a

T. S. Balganesh (✉) · J. J. Bhat
GangaGen Biotechnologies Pvt Ltd, Bangalore, India

B. Ugarkar
Medicinal Chemistry, AstraZeneca India P Ltd, Bangalore, India

B308, Sobha Jade, Jakkur, Bangalore, India



M. Dikshit (ed.), *Drug Discovery and Drug Development*,
https://doi.org/10.1007/978-981-15-8002-4_2

**Table 2.1** Chronology of the discovery of drugs in the Short Course Chemotherapy (SCC)

| Drug | First clinical use | Reference |
|---|---|---|
| Para amino salicylic acid | 1946 | Lehmann (1946) |
| Streptomycin | 1952 | Schatz et al. (1944) |
| Isoniazid | 1952 | Crofton (1959) |
| Pyrazinamide | 1952 | Yeager et al. (1952) |
| Ethambutol | 1961 | Doster et al. (1973) |
| Rifampicin | 1966 | Yuan and Simpson (2018) |

single drug like streptomycin, though led to initial clinical improvement, rapidly resulted in the appearance of 'streptomycin resistant' microbe and the reappearance of the disease (Doll 1998; Yoshioka 1998). This gave an early indication of a requirement of combination therapy which led to the introduction of isoniazid (I), rifampicin (R), and pyrazinamide (Z) within a decade. A fourth drug, ethambutol (E), was added to the regimen in 1961 (D'Ambrosio et al. 2015; Mitchison and Davies 2012) which became 'standard of care' which is still in use as the first line of defence. The introduction of these drugs was a trial and error exercise, and led to the hypothesis that the drugs in the combination served different roles towards 'curing' the disease. The Standard Regimen, also referred to as short course chemotherapy (SCC), consisted of the above four drugs (Table 2.1), given over a period of 6 months divided in two phases; (a) all four drugs were given for first 2 months, called the intensive phase, followed by (b) two drugs, I and R for the following 4 months, called the continuation phase (Mitchison 1992).

TB cure is defined as the eradication of the microbe from all regions of the body, which leads to the prevention of 'relapse' of the disease and its symptoms even after 12 months of stopping the treatment—this is referred to as achieving 'sterilisation' (Mitchison 1992). As hypothesised by Mitchison, apart from preventing the appearance of drug-resistant mutants, drugs in the combination were active on different populations of the MTB bacilli in the patient. I and R acted on both intracellular and extracellular tubercule bacilli, while Z played a role in eliminating the 'dormant' bacilli, thus contributing to the prevention of 'relapse'. The cure achieved with this 'drug combination' has set the tone and the bar for aiming better cure through the introduction of newer drugs. Although all the clinical trials that led to the standardisation of the therapy were carried out on patients with pulmonary TB, the same regimen also became the 'standard of care' for the treatment of extra-pulmonary TB.

The main challenge with the short course therapy is patient 'compliance': successful therapy requires continuing the treatment for 6 months. The main challenge is the 'safety' profile of the drugs that have significant unpleasant side effects (Simon et al. 1991) necessitating patient monitoring and counselling throughout the treatment period.

Before the advent of the 'drug resistance' against the standard regimen, new drug discovery efforts were largely directed towards finding novel anti-tuberculosis (anti-TB) molecules that can shorten duration of therapy. Understanding the physiology of the MTB bacilli in the different tissue niches had revealed the presence of the extracellular and the intracellular (within the alveolar macrophages) microbes which had differential sensitivity to the drugs in the regimen. Further studies on the physiology of the microbe also revealed that the hypoxic environment induces specific sets of genes which programmes 'adapted mycobacteria' with altered physiological properties, as well as a changed drug sensitivity pattern (Simon et al. 1991; Bagehi et al. 2003). Therefore, the working hypothesis developed was that new drugs must be active on different subpopulations of the microbe in the TB lesion, and this was necessary to achieve a faster cure.

The overall success rate of >90% achieved with the 'standard regimen' against drug sensitive TB infections (Rockwood et al. 2016) shifted the emphasis for new drugs from just the 'potency' perspective to the need on 'shortening of therapy' as the urgent unmet medical need, because shortening of treatment duration would increase patient compliance. Therefore the TPP (Therapeutic Product Profile) of a new anti-TB compound included the following critical properties:

- Compatibility with the standard drug regimen
- Compatibility with DOTS (Directly Observed Therapy, short course)
- Affordable
- Reduction in the treatment duration

Furthermore, it was also envisaged that the new drug will have to be 'trialled' as an add-on to the current regimen to investigate if the new combination was capable of reducing the treatment duration.

WHO declared TB a 'global emergency' in 1993 in response to the increasing prevalence of drug-resistant TB encountered globally (Klaudt 1994). Several independent factors contributed to the rapid dissemination of drug resistance in MTB clinical isolates, one of them being co-infection of HIV and TB. Treatment of patients infected with I and R resistant MTB immediately became a challenge; the loss of these two front-line drugs necessitated the use of less potent and toxic second-line drugs leading to not only more complex treatments but also a steep decline in the cure rates. Unfortunately, the drug discovery pipeline in 2000 was thin and the development cycle for introducing new drugs was, and is, complex—together they precipitated an unprecedented crisis, spurring reinforced drug discovery efforts worldwide.

## 2.2 TB Drug Discovery: Past and Present—A Perspective

Drug Discovery efforts leading to the advent of chemotherapy for treating TB was mainly driven by medicinal chemistry efforts. These efforts involved exploring the anti-TB activities of different scaffolds and their derivatives resulting in the discovery Para Amino Salicylate (Table 2.1). This was also the period where natural products obtained from various Streptomyces species and related genera were being investigated (Waksman et al. 1946). These efforts led to the identification of

Streptomycin, which was the first chemotherapeutic agent that was tested on TB patients. The most potent and useful drug Rifampicin was discovered in 1966. With time, drug discovery paths have changed dramatically with the coming of the genome—sequencing era. The changes and the fresh challenges are discussed under the 'modern drug discovery' section.

### 2.2.1 The Modern Day Drug Discovery Process

Target-based screening is driven by the availability of the genome sequences of multiple MTB strains including that of *Mycobacterium bovis* (BCG). The process usually starts with a 'target identification and validation' step. The availability of large databases of genome sequences from various microbes as well as the human counterpart has aided in choosing targets by detailed bioinformatic analysis (Vashist et al. 2012; Raman et al. 2008; Vaishali et al. 2019). This has helped in either identifying targets unique to MTB, or targets sufficiently different (by genome analysis or by their crystal structures) to allow the identification of selective compounds. A variety of MTB targets have been validated (shown to be essential for the viability of the microbe) using gene deletion techniques or by using chemical inhibitors, both in vitro and in vivo (Sassetti et al. 2003). Compounds identified through target-based screening, are labelled as 'hits' which may or may not have antimicrobial properties. GSK070 currently in phase 2 trials was initially discovered using the 'target screening approach' (Li et al. 2017).

Phenotypic screening involves screening compounds directly on the microbe, MTB or surrogate non-pathogenic mycobacteria like *M. smegmatis* (Msm) or BCG. This approach had yielded leads like TMC 207, which was progressed through the drug discovery process as bedaquiline and has been registered as Sirturo (Matteelli et al. 2010). Several other compounds such as Q203, BTZ043 (ClinicalTrials.gov, NCT02858973 and NCT02530710, 2018a, b) are in the late development phase (Phase 2). Compounds that have advanced through the drug discovery path were also identified through phenotypic screening on the MTB microbe and were progressed through different phases of drug discovery based on their inhibitory properties (MIC). These include compounds like delamanid and pretomanid. Delamanid and pretomanid (PA824) originated from a library of nitroimidazofuran derivatives that was part of a CIBA-GIEGY effort to discover anti-TB as well as antiparasitic drugs against entamoeba (Laurenzi et al. 2007).

It is interesting to note that the majority of compounds in the current clinical development phase have been identified through phenotypic screening and progressed through conventional medicinal chemistry efforts with limited contributions to their progression from the target perspective (Laurenzi et al. 2007; Singh and Mizrahi 2017). The limited success of target based lead discovery is not a reflection of the inability to identify inhibitors of 'essential target' function but is more of a reflection on the inability to convert the target inhibition into antitubercular activity. This limitation of translating enzyme inhibition to microbial inhibition has

**Fig. 2.1** Schematic representation of the Drug Discovery Path

been instrumental to shift focus of anti-TB drug discovery efforts to starting points of compounds that have antimicrobial activity (MIC).

*Schematic representation of the drug discovery process is shown below:*

Drug discovery is a stepwise process: the different phases are shown in Fig. 2.1, wherein the first two steps focus on discovering molecules, while the last two steps involve fine-tuning the chemical structure to meet acceptable potency, safety and physicochemical properties resulting in a molecule that is ready for clinical evaluation (candidate drug). The criteria for progressing molecules in each of these steps are carefully designed keeping in mind the final characteristics of the drugs (TPP) to treat TB infections. While the essentials of the process in terms of the critical points to transition from one phase to the next remains almost unchanged, modern drug discovery has benefitted from advances in technology as well as the learning from past experiences. This has in turn increased the success rates in our ability to find novel molecules and have increased the chances of identifying a clinical candidate with a greater potential for success as effective therapeutics.

### 2.2.2 Hit Identification

In a nutshell, hit identification is finding starting chemical molecules which are usually weak to moderately potent against specific biochemical targets or against MTB cultured cells. In the early years of anti-infective drug discovery soil extracts were tested to identify active molecules (Clardy et al. 2009). The reason for this is that soils contain bacteria and other microbes that secrete their metabolites in order to protect themselves from the invading pathogen. Soil samples from all over the world including from the marine floor were collected and tested for antibacterial leads. This approach has yielded a large number of early antibiotics such as penicillin, vancomycin, streptomycin, rifampicin, etc. This effort continues to be used for discovering new antibiotics (von Bubnoff 2006; Durand et al. 2019).

Screening for ‘starting points’ in the modern drug discovery process now relies on ‘libraries of compounds’ that have been collected either through in-house research efforts or chemicals synthesized specifically against chosen targets, or compounds designed using starting points known to yield drugs, e.g. drug like molecules. Hit identification effort is further enhanced by the availability and deployment of the genome sequences of multiple MTB strains that has created opportunities to identify and use novel and specific biochemical pathways as targets. Availability of the genome information has also facilitated rapid identification of the molecular target/pathway for compounds that have inhibitory activity (MIC) on the cultured microbe (Mendes and Blundell 2017; Chim et al. 2011). Testing ‘libraries of compounds' is not only applicable to biochemical screening but also to phenotypic screening (Manjunatha and Smith 2015). The phenotypic screening approach in the modern discovery context is used to find inhibitors of not only the 'multiplying' microbe but also against the MTB microbe in in vitro models corresponding to the different physiological states predicted in vivo.

Yet another hit identification technology that is impacting drug discovery in multiple ways is the significant increase in computational prowess, which has enabled 'structure-based screening' capability using appropriate software. This approach builds on the availability of the atomic structures/reliable models of the targets of interest and the ability to dock, and evaluate the binding energies of compounds that putatively bind to the desired pockets on the target. Compound structures that show high binding affinity in this approach are synthesized and are directly tested on the pathogen. Compounds found to be active are channelled into the outlined drug discovery path (Musa et al. 2009).

### 2.2.3 Lead Identification

The main aims of this phase of the programme can be summarized under the following broad headings and are applicable to compounds identified through either of the screening approaches:

- Increase the robustness of the 'hits' through structural modifications: establish a structure–activity relationship with respect to the kill kinetics of the compound.
- Understand the drug metabolism and pharmacokinetic properties of the compound—Establish a Mode of Action (MoA) of the molecule.

Lead identification has also been impacted by learning from the decades of drug discovery in the pharmaceutical industry. In 1997 Christopher Lipinsky, after analyzing various physical and structural characteristics of all the FDA-approved drugs, devised rules that aid in designing new drugs irrespective of the therapeutic areas (Walters 2012; Lipinski et al. 2001). These rules today are referred to as Lipinsky rules. While the original four TB drugs (I, R, Z, E) and most of the second-line TB drugs, like clofazimine, amikacin, cycloserine, ethionamide, etc., comply with these rules, bedaquiline, pretomanid, and delamanid, the recently discovered anti-TB drugs violate almost or all of these rules.

Current day drug discovery also lays significant emphasis on the quality of the 'Lead' molecules, especially regarding possible safety issues that may be embedded in the molecule. Thus, chemical functionalities like the 'nitro' group and reactive groups like Michael acceptors, acid chlorides, sulphonyl chlorides, etc., are avoided. In addition to potency, properties like solubility and metabolic stability are key parameters for a good lead (Lipinski et al. 2001). However, many of the anti-TB drugs in the current pipeline carry 'nitro' groups, e.g. pretomanid, delamanid, BTZ, where the nitro group is essential for its anti-TB activity. The recently introduced drug into our anti-TB armamentum, bedaquiline is highly lipophilic, is metabolically unstable (readily metabolized by CYP3A4) and is reported to prolong the QT interval (Koide et al. 2008). Moxifloxacin, a broad spectrum quinolone antibiotic which is also being used to treat MDR TB is known to prolong the QT interval (Fox and Menzies 2013; Van Heeswijk et al. 2014; Koide et al. 2008). Thus, it appears that the ideal lead molecule for anti-TB drug discovery remains elusive; however the potency of these compounds on MDR MTB strains and their acceptable safety margins are the properties driving the choice of these molecules as drugs.

### 2.2.4 Lead Optimization (LO)

This is an important stage of drug discovery process. Reaching LO stage entails that a robust chemical class has been identified with cellular activity and robust SAR and SPR (structure property relationship) characterized (Rajarshi 2013). The availability of information on the molecular interaction between the inhibitor and its target is helpful in optimizing both the physical properties as well as in weeding out 'problem chemical groups' on the molecule. A TB structural genomics consortium of publicly available atomic structures of a large number of proteins has been useful in facilitating optimization of the binding potency of molecules to selected targets (Chim et al. 2011).

The major aim of the 'LO' efforts is to identify one or two molecules from the lead molecule cluster and demonstrate efficacy in an animal model of the disease with acceptable PK, PD and safety properties. TB drug discovery today not only requires testing of the LO compounds in various in vitro models representing different physiological states of the MTB microbe but also in in vivo models that represent these different physiological states. Compounds reaching this stage of discovery are also tested in different combinations with the standard regimen to ensure compatibility, overall efficacy as well as to rule out any antagonism between the compounds.

### 2.2.5 Challenges of Anti-Mycobacterial Chemistry and Progressing Leads into Clinical Candidates

The challenge to discovering new anti-TB drugs is to be able to design molecules that are active against the pathogen in its multiple metabolic states, or in niche environments (Dartois and Barry 2013).

Knowledge of the MTB microbe's pathophysiology has established the presence of different physiological niches in human host where the microbes reside in different physiological states with altered response to drugs. Current drug discovery approaches use this information to develop several in vitro/ex vivo models to reproduce the MTB microbe target-gene functions in order to enhance the success in identifying and prioritizing high quality lead molecules. For example, all the compounds that have been introduced into the anti-TB regimen recently like bedaquiline, pretomanid, delamanid or those in late clinical development such as

BTZ169 or sutezolid are not only active on the extracellular and the intracellular microbe, but are also active on the non-replicating microbe—the non-replicating property being an adaptive/induced response of the microbe to conditioned growth environment (Shehzad et al. 2013; Kaul et al. 2011).

The mouse model has been the work horse for 'TB drug discovery' and has been a robust reflection of the cidal activity of compounds (Chen et al. 2017; Devis et al. 2007). However, this model is not a true reflection of the immune responses that are seen in humans following an MTB infection, hence a disconnect between results obtained in the mouse model vis a vis in human trials. This is best exemplified by the lack of correlation seen in the 'shortening of therapy' studies in moxifloxacin containing regimens between the mouse and human (Lanoix et al. 2016). Several new modified models have currently been developed that are expected to be a closer approximation to that of the human host (Zhan et al. 2017; Gumbo et al. 2015).

Modern drug discovery uses PKPD (pharmacokinetic and pharmacodynamic) data modelling as the major tool to progress compounds rapidly through the preclinical stages. Based on the model, the dose and the optimal dosing intervals are calculated and confirmatory data is generated in appropriate models of infection (Vaddady et al. 2010; Naveen Kumar et al. 2014). This model is the basis for planning the dose and the frequency of dosing in humans both for safety studies as well as for efficacy demonstration. Unfortunately, success in predicting therapy outcomes in an MTB infection using this PKPD modelling has been elusive. This is probably due to our incomplete understanding of the processes that govern 'cure'. The hallmark of an antibacterial compound is the correlation between the observed PK parameters and the efficacy in the animal model—this has been and continues to be a challenge for anti-TB drugs when moving from animal models to treating infected humans. While the PK parameters in the mouse model do indeed correlate with a dose-dependent reduction of microbial counts in the mouse lungs, these parameters have little value in predicting the extent of 'sterilisation' that can been achieved as well as the overall effect on the duration of therapy when translated to the human (Gengenbacher et al. 2017; Dartois 2014). One of the key factors that influence the ability of a drug to 'sterilize' the tubercle is the cellular architecture of the 'lesion' that plays a major role in modulating the entry of the drug into the lesion, this in turn induces/selects the multiple physiological states of the tubercule bacilli residing in the lesion; for e.g.: the presence of a low oxygen environment will result in the induction of the 'anaerobic response' in the microbe which results in the 'non-replicating' state (Lebardo et al. 2018).

Determination of the actual drug permeability into the infected tissues requires elaborate studies using radiolabelled drugs; additionally there is a dearth of standard data in different animal models and thus our inability to predict the real-life permeability in the human host. The classical examples are the drugs in the standard regimen of the first-line therapy. These are dosed more as a logistical convenience rather than their PK properties, e.g. isoniazid has a half-life of 1–3 h but is dosed once daily or twice weekly (Handbook of Anti-Tuberculosis agents 2008). However, knowing that these drugs have been effective, the effectiveness of these drugs is attributed to the slow multiplication time of the microbe and the Post Antibiotic

Effect (PAE) of the compound (Chan et al. 2004). In contrast, the newer drugs, surtero, delamanid or pretomanid are dosed based on their PK parameters, where an MIC and/or exposure against the multiplying bacteria has been taken as the PK parameter to achieve cure (Esposito et al. 2015; Tiberia et al. 2018).

Last but not the least is the increased emphasis in understanding the 'side effects' as well as the 'drug–drug interactions' of newer TB drugs in standard of care regimen. The ability to predict possible 'off target' interactions based on the extensive databases available on the behaviour of different classes of compounds in human, has made putative drug–drug interactions predictable. This in turn has helped in designing drug combinations which would not only be efficacious but would also be safer.

The current compounds/drugs in various phases of Clinical trials are shown in Table 2.2. It is interesting to note that while protein synthesis inhibitors are drawn from the oxazolidinone class, the other major target of several compounds is DprE1, an enzyme involved in cell wall biosynthesis.

### 2.2.6 Newer Therapeutic Approaches

Modern anti-TB drug discovery includes two novel approaches which have yielded interesting results in appropriate animal models of TB infection. The first can be broadly labelled as 'Adjunct therapy', which as the name suggests would be drugs administered in combination with any Standard of Care anti-TB regimen and works synergistically with the former. The second approach is the identification of compounds that are registered as drugs for the treatment *of* diseases other than TB, but also have potent activity on MTB.

- **Adjunct therapy**
  This approach includes drugs that are not microbicidal on their own but modulate host pathways that combat MTB. The MTB bacilli survival strategy in the human host is known to include influencing/suppressing various inflammatory and cell-mediated immunogenic pathways that have the potential to eliminate the microbe. Compounds/drugs that can relieve this inhibition or augment activity of these anti-TB pathways can play a major role in combating the disease. Several compounds/drugs with such a potential have been identified using animal models (Rayasam and Balganesh 2015). In fact, it has been suggested that this may be a potent approach to reduce the duration of therapy.
- **Repurposed drugs as anti-TB therapy**
  Several drugs that have been approved for treating indications other than TB have shown potent anti-TB activity in both in vitro and in vivo models (Mishra et al. 2018). The main advantage with these leads is the ability to 'fast track' these compounds through the early development stages, provided their safety profile is compatible with the dose and dosing regimen for the anti-TB indication.

**Table 2.2** Global new TB drugs pipeline 2018

| S. No. | Compound and class | Target |
|---|---|---|
| Phase 1 Clinical trial | | |
| 1 | BTZ-043<br>Benzothiazinone | DprE1 |
| 2 | TBI-166<br>Riminophenazines | Membrane<br>Not clear |
| 3 | PBTZ 169 (Macozinone)<br>Benzothiazinone | DprE1 |
| 4 | GSK-656 (070)<br>Oxaborole | Leucyl t-RNA synthetase (LeuRS) |
| 5 | TBA-7371<br>Azaindole | DprE1 |
| 6 | Conezolid<br>(MRX-4/MRX-1) | Oxaziolidinone protein synthesis |
| Phase 2 Clinical trial | | |
| 1 | OPC-167832<br>3,4-dihydrocarbostyril derivative | DprE1 |
| 2 | Telecebec (Q-203)<br>Imidazopyridine amide | qcrB subunit<br>cytochrome bc1 complex |
| 3 | Delpazolid (LCB01-0371)<br>Oxazolidinone | Protein synthesis |
| 4 | Sutezolid<br>Oxazolidinone | Protein synthesis |
| 5 | SQ-109<br>Ethylenediamine | MmpL3 |
| 6 | Macozinone<br>Benzothiazinone | DprE1 |
| Phase 3 Clinical trial | | |
| 1 | Bedaquiline (TMC-207)<br>Diarylquinoline | ATP synthase |
| 2 | Delamanid (OPC_67683)<br>Nitrodihydro-imidazooxazole | Cell wall synthesis |
| 3 | Pretomanid (PA-824)<br>Nitroimidazole | Cell wall synthesis |

Source: Working Group for New TB Drugs

## 2.3 A Paradigm Shift in TB Drug Therapy: Finding New Combinations

The standard tuberculosis treatment regimen since its design, validation and acceptance as the 'Standard of Care' has been made up of I, R, Z and E. This combination continues to be the therapy for the treatment of infection caused by drug-sensitive MTB strains. The 'standard regimen' was arrived at systematically through testing multiple combinations, and either by adding or omitting new drugs in clinical trials (Aquinas 1982). This regimen has been effective and has been the first-line treatment of TB for nearly four decades since its introduction. Unfortunately, this was also the period [1963–2002] where no new drugs were discovered for the treatment of tuberculosis even though WHO had called 'TB a global emergency' in 1993 because of the emergence of multiple drug-resistant strains.

A fresh attempt at finding novel treatments that can reduce the duration of therapy was made by the introduction of the 'quinolone antibiotic—oflaxacin' in a trial conducted at TRC, Chennai in 2002 (Tuberculosis Research Centre Chennai 2002). This study showed that an oflaxacin containing regimen for 4 or 5 months was comparable in efficacy to the standard regimen of 6 months suggesting the possibility of a reduced treatment duration. In addition, the available limited data also showed that while the oflaxacin containing regimen administered in the 5 month therapy did have some efficacy against isoniazid resistant strains, the outcome with strains that were resistant to rifampicin and isoniazid simultaneously was poor.

With the appearance of resistance against the 'standard regimen', the approach for the development of a new drug through the clinical stages was to add the new drug into the existing regimen. This was dictated by the following pragmatic considerations:

- Need for empirical treatment vs. 'drug sensitivity' based therapy: The isolation of the infecting TB microbe from the patient followed by drug sensitivity testing takes nearly 2 months (Dheda et al. 2013) thus necessitating empirical treatment before the drug sensitivity results could be obtained.
- Need for a combination therapy vs. monotherapy: The need to treat TB with a combination of more than one drug is critical as monotherapy has been shown to rapidly induce drug resistance.
- The assumption that because of the novel mode of action of the new drug, the new drug would be effective against all clinical strains of TB.

MDR TB is defined as MTB strains resistant to both I and R, and this resistance pattern makes the SCC ineffective because, the remaining two drugs in the regimen, Z is less effective against rapidly multiplying TB bacilli, whereas E has limited antimycobacterial activity. The second-line therapy for the treatment of such MDR strains consists of clofazimine, cycloserine, amikacin, ethionamide and pyrazinamide (Ramachandran 2019). Unfortunately, these drugs are mostly bacteriostatic, have severe side effects and require prolonged treatment for 24 months with a cure rate of only ~30–50% (Ramachandran and Swaminathan 2015). New drugs, bedaquiline and delamanid successfully completed the Phase 1 and 2 clinical developments in the year 2016 (Yang et al. 2019; Ferlazzo et al. 2018). The urgent need for an effective therapy for the treatment of MDR TB has prompted the reservation of these drugs for of MDR TB only.

For the first time in the history of TB drug development, bedaquiline and delamanid were trialled directly on patients infected with MDR strains in a cocktail of the background regimen. By 2007, thanks to the advent of the Line Probe Assay (Ruvandhi et al. 2017) and GeneXpert (Evans 2011) drug sensitivity testing protocols for detecting I and R resistance within 7 days were available. This enabled the segregation of patients into those infected with a DS and those infected with MDR strains. The ability to test new drugs in a cocktail of the background regimen on MDR patients and compare the efficacy with the background regimen opened up a completely new opportunity for developing new drugs for TB. This was a significant development because the currently available 'second-line regimen' is poorly effective requiring a 24 month therapy with limited success rates. The new drugs were trialled on MDR TB patients, as an 'add-on' to the 5–6 standard background drug cocktails to investigate efficacy (Lienhardt et al. 2010). The new treatment (cocktails of background regimen + delamanid or background regimen + bedaquiline) containing various cocktails were shown to be effective, and significantly better in comparison to the background therapy alone (Karekar and Marathe 2018; Olaru et al. 2017). However, even though a new regimen for the treatment of MDR TB became available, the side effects of the drugs in the background regimen continue to make the treatment problematic with concomitant non-compliance.

It was obvious by early 2000 that a completely new combination therapy was needed because of the rising number of MDR patients and the limitations of even the new cocktails made up of background therapy with the new drugs. Furthermore, shortening duration of therapy was an important milestone which was required to improve compliance. The anti TB-drug discovery pipeline in the year 2008–2009 included several novel and a few repurposed chemical entities. The novel compounds were pretomanid (Pre), bedaquiline (B), delamanid (D), which were in different stages of clinical development. In addition, the oxazolidinone class had three members, linezolid (LZ), sutezolid and AZD 5847, which were also in trials for the treatment of TB (Balasubramanian et al. 2014). Quinolones such as moxifloxacin (M), levofloxacin (L) and gatifloxacin (G) were yet another class being investigated as cocktails with standard care or second-line TB drugs (Johnson et al. 2006). During this unique 'timeframe', several novel compounds/drugs were available for development as anti-TB molecules providing opportunities to clinically evaluate new combinations. The TB Alliance pioneered this concept and tested a variety of cocktails against drug sensitive infections in an 'Early Bactericidal Activity' clinical study—a 14 day treatment trial (Joseph et al. 2011). This investigation demonstrated that the combination of Pre, M and Z to be clearly superior to the other combinations tested. Following this, GATB also launched a trial with Pre, M and Z on drug-resistant patients, in South Africa and Tanzania (Diacon et al. 2012).

It was envisaged that the new combinations should be able to treat DS as well as MDR infections, thus opening up the possibility of having a single simplified regimen to treat patients with DS, MDR and XDR infections. XDR is referred to MDR strains that are also resistant to the Quinolones and the injectables. Several cocktails, such as a combination of B, Pre, M, Z is being trialled on DS and MDR TB patients (Dawson et al. 2015), whereas a combination of B, Pre and L (linezolid) is being tried on MDR or XDR patients (ClinicalTrials.gov—NCT02333799, 2015). Other studies with cocktails containing B, P and L in combination with M or clofazimine aim to investigate shortening of therapy in MDR patients (Lebardo et al. 2018). A number of these trials are scheduled to be completed by 2020 paving the way for introducing new treatments. A novel combination of drugs with both improved efficacy, a shortened duration of therapy and suitable for both DS and MDR TB would be a major breakthrough.

Finally, testing cocktails of the newer compounds, which are under patent protection requires the consent from the concerned patent owners. The use of compounds in novel combinations would generally be covered by the respective patents and hence a discussion mechanism to obtain permission to conduct such trials needs to be streamlined.

## 2.4 Global and India Efforts on TB Drug Discovery

India has a number of global firsts as far as TB drug discovery is concerned. It was at the Tuberculosis Research Center (TRC), Chennai and currently named National Institute for Research in Tuberculosis (NIRT), an Institute under the Indian Council of Medical Research (ICMR) that a number of the early clinical studies to find a treatment for TB were carried out (Radhakrishna 2012). These studies led to recommendation of the current standard regimen of I, R, Z, and E. The very same institute was also responsible for bringing quinolone antibiotics into the anti-TB treatment to start shaping a new and effective second-line therapy. This establishes the fact that the clinical trial community in India is quite capable of bringing new molecules through the development path to the patient (Joseph et al. 2011).

India has pioneered academic research in the field of mycobacteria, Prof T R Ramakrishnan and colleagues at the Indian Institute of Science, Bangalore extensively investigated the metabolism of MTB (Ramakrishnan et al. 1972). Their pioneering work on the mechanism of action of I on mycobacteria was also one of the first investigative biochemical studies of a drug action in MTB. Along these lines, over the years several important aspects of the MTB physiology, metabolism and their significance in the disease process have been discovered in several laboratories in India (Taneja et al. 2010; Anishetty et al. 2005). Detailed research on the biology of a variety of targets has resulted in the proposal of several targets for drug discovery effort (Nagaraja et al. 2017; Kurthkoti and Varshney 2012). A steady stream of Indian academic publications reveals design and discovery of several families of novel chemical moieties with inhibitory effects on *M. bovis* and *M. semgmatis* as well as studies on MTB (Singh et al. 2017; Puneet Chopra and Meena 2003). Some of this work has also been extended to identifying novel targets using repurposed compounds (Mishra et al. 2018). In addition, a significant progress has also been reported in the computational studies on metabolic pathways leading to identifying new targets (Chandra 2009; Balganesh and Furr 2007; Vashist et al. 2012). Indian scientists have been at the forefront in obtaining molecular structures of several putative target proteins of mycobacteria (Singh et al. 2018). The availability of the structural details of several proteins has also spurred computational studies involving the mapping of the binding of libraries of compounds on the protein structure to identify novel chemical starting points (Bhagavat et al. 2017; Syal et al. 2017). The progress made and the plethora of studies published by the Indian scientists suggests an ample background for TB drug discovery to capitalize on. However, very few of these studies have been progressed into the drug discovery path. This is perhaps a reflection of several India specific challenges including infrastructure and the limited drug development experience.

Tuberculosis drug discovery involves working with a highly infectious organism that requires advanced biosafety containment facilities and animal facilities which are also compliant with safety requirements (Singh 2013). The availability and access to such facilities in India is a limiting factor. This limitation restricts drug discovery in academic institutions to either carrying out studies with only the purified components of biochemical assays, or to use surrogate microbes like *M. smegmatis*. Unfortunately, this surrogate microbe cannot be used to progress anti-TB drug discovery through studies involving the 'physiologically relevant' models or in the animal models. Given this scenario it is only expected that the main contribution from the academic research is in the early parts of discovery.

Drug discovery requires an integrated team with expertise from different disciplines apart from those mentioned above and including pharmacology and clinical development. It is difficult to envisage such a setup of a multidisciplinary team in an academic environment. Advances made in the academic institutions will need to be capitalized upon by partnering with dedicated drug discovery teams either in industry or in consortia consisting of global players which include both academic and industry partners like the European Union Frame Work programmes (Lång et al. 2010). At the same time, it is imperative that the understanding of the basic biology of the microbe and the host continue in the academic institutions as these are the key enablers for drug discovery programmes.

Historically the first contribution of Pharma based in India to the anti-TB pipeline was by CIBA-GIEGY, India, where researchers identified CGI17341, a nitroimidazole, the development of which was halted because of 'mutagenicity concerns' (Mukherjee and Boshoff 2011). Examples of pharma in India involved in the discovery of anti-TB drugs are Lupin Limited and AstraZeneca R and D India (AZI); these companies have played important roles in the discovery of novel drugs for TB. Sudoterb, a pyrrole derivative discovered at the Lupin Labs in India was found to be effective in the mouse models of TB infection with potential for the reduction of therapy duration. Sudoterb was designed and synthesised at the Lupin Labs. Thus both the discovery and development was an in-house effort. Sudoterb was progressed through Phase 1 studies in 2004 (Ginsberg 2010) and Phase 2 in 2013 [LL3858, Sudoterb https://newdrugapprovals.org/2017/12/05/ll-3858-sudoterb/]. However, the outcomes of these trials have not been made public. AZI's in-house research has contributed several molecules to clinical development. For example, AZD5847, an oxazolidinone moiety, was progressed to Phase 2 (Balasubramanian et al. 2014) while AZ 7371 an azaindole moiety, which was also discovered at AZI is being progressed through Phase 1 by the Global Alliance (Chatterji et al. 2014). AZI was also involved in the discovery of benzothiazinones, a novel class of anti-TB molecules (Makarov et al. 2009) whose derivative PBTZ-169 is currently in clinical trials (ClinicalTrials.gov—NCT03334734, 2017). Given the fact that the overall profit potential of anti-TB drugs globally is limited, it would be of interest to examine the reasons for the three companies, Ciba-Giegy India, Lupin

Limited and AstraZeneca India, to be involved in research activities towards finding new drugs for TB treatment. During 1980s and 1990s the India based Ciba-Giegy unit was involved in the design and discovery of novel drugs for the treatment of anaerobic infections both bacterial and parasitic from which the anti-TB compound CGI17341 was identified. Lupin on the other hand continues to be a market leader in the sales of first-line anti-TB drugs and had an in-house commitment to find new drugs for the treatment of TB, which would also help enhance their sales portfolio. Sudoterb was one such compound, which was identified in this process. AstraZeneca India had a remit for finding new drugs for the treatment of neglected diseases and the compounds described were discovered in the projects for TB. Fast forward 2015, all these 'drivers' have changed and none of these three companies are currently focused on discovering new molecules for the treatment of TB.

The current list of clinical trials on TB being conducted in India since 2016 is shown in Table 2.3. For a country with a dire need to find new therapeutic options for treating TB, the number of trials are indeed limited. However, it is heartening to note that some of these are indeed exploring novel approaches—like the use of 'adjunct therapy', VPM 1002—a vaccine trial, as well trials exploring the use verapamil and metformin to augment existing therapy. While the paucity of TB compounds in clinical trials may be a concern, the fact remains that the 'duration' taken to trial a new drug through Phase 2 and 3 is prolonged, but the trialling procedure and duration have not changed over several decades. What we urgently need, in addition to new drugs, are biomarkers that are predictive of successful treatment—this could be a single most productive 'game changer' for goals of TB eradication in the near future.

## 2.5 Going Forward: TB Drug Discovery and India

With Lupin Labs and AZI no longer contributing to the discovery of novel anti-TB molecules it becomes even more imperative that alternative avenues are explored to continue populating the anti-TB drug discovery pipeline. Funding agencies in India, DBT, DST, CSIR and ICMR continue to invest in both the discovery and early development of potential anti-TB compounds. However given the understanding of the bottlenecks in making this happen, the major hurdle is in converting laboratory-based research into a directed product-based research. It is well established that the highest risk phase of the drug discovery path is in the translation of laboratory in vitro and in vivo animal data into effective and safe medicines in man which includes the regulatory toxicology studies, followed by the safety studies in man, and finally efficacy studies in patients. The Govt. of India under the auspices of its multiple funding agencies has been funding schemes that bring specific partners belonging to industry and academic to develop products. However, there have been limited takers from the Indian large pharma in this endeavour. The need to attract big pharma in India into this initiative is because of the 'capabilities' that reside 'only' in these units. Lupin developed Sudoterb demonstrating its in-house capacity to bring such molecules through the discovery and development chain. Knowing that the

**Table 2.3** Some of the registered clinical trials on tuberculosis in India 2016 onwards

| S. No. | Title | CTRI No. | Phase |
|---|---|---|---|
| 1 | An open-label, non-randomized, two-stage, dose-finding study of verapamil [ir] tablet formulation in adult tuberculosis patients in continuation phase of anti-tuberculosis treatment | CTRI/2016/05/006928 | Phase 2 |
| 2 | Optimising the involvement of private practitioners in Tuberculosis care and control in India | CTRI/2017/09/009672 | Phase2/3 |
| 3 | A Multicenter Phase II/III Double-Blind, Randomized, Placebo Controlled Study To Evaluate The Efficacy And Safety Of VPM1002 In The Prevention Of Tuberculosis (TB) Recurrence In Pulmonary TB Patients After Successful TB Treatment In India | CTRI/2017/03/008266 | Phase2/3 |
| 4 | A Phase I/II Randomized, Open-label Trial to Evaluate the Pharmacokinetics, Safety, and Treatment Outcomes of Multidrug Treatment Including High Dose Rifampicin with or without Levofloxacin versus Standard Treatment for Pediatric Tuberculous Meningitis | CTRI/2017/03/008004 | Phase 1/2 |
| 5 | A randomized trial of therapy shortening for minimal tuberculosis with new WHO-recommended doses/fixed-dose-combination drugs in African/Indian HIV+ and HIV−children: SHINE study | CTRI/2017/07/009119 | Phase 3 |
| 6 | The evaluation of a standard treatment regimen of anti-tuberculosis drugs for patients with MDR-TB Version 6.2 dated Feb 2015 | CTRI/2017/09/009693 | Phase 3/4 |
| 7 | Phase IIb open label, parallel, randomized controlled clinical trial to evaluate the safety, tolerability, pharmacokinetics and anti-bacterial activity of High dose rifampicin versus Conventional dose of Rifampicin along with standard anti-tubercular therapy (ATT) in drug sensitive adult patients of pulmonary tuberculosis | CTRI/2017/12/010951 | Phase 2 |
| 8 | A phase IIB Open Label Randomized Controlled Clinical trial to Evaluate the antibacterial activity, pharmacokinetics, safety and tolerability of Metformin when given along with rifampicin, isoniazid, pyrazinamide and ethambutol in adults with newly diagnosed sputum positive pulmonary tuberculosis: an 8-week study | CTRI/2018/01/011176 | Phase 2 |
| 9 | A Phase 2, Open-label, Multicenter, Single-arm Study to Evaluate the Pharmacokinetics, Safety, Tolerability and Anti-mycobacterial Activity of TMC207 in Combination With a Background Regimen (BR) of Multidrug Resistant Tuberculosis (MDR-TB) Medications for the Treatment of Children and Adolescents 0 months to less than 18 years of Age Who Have Confirmed or Probable Pulmonary MDR-TB | CTRI/2018/03/012637 | Phase 2 |

Source: CTRI, India

commercial market for a TB drug is limited, especially when half the sales are through government agencies (Arinaminpathy et al. 2016) and the drugs have to be given in pre-existing combinations, de-risking the progression of novel compounds through the ‘translation phase’ becomes imperative to attract pharma partners. This de-risking can be achieved through partnerships with academic institutions which have the necessary expertise or with industry partners who have brought drugs through the early clinical phases. Partnership with big pharma can be discussed once there is successful transition of compounds through the early development phase.

Indian research establishments need to find such a translation capacity through novel models. In this context it is worth examining global models that have addressed this topic:

### 2.5.1 European Union Framework 6 and 7 Projects

New Medicines for TB (NM4TB) and more medicines for TB (MM4TB)—were multi-partner consortia, which included both publicly and privately funded institutions which came together under this umbrella to take advantage of the discovery work being carried out in the academic laboratories and probe the translational aspects with help from the industry partners. NM4TB delivered a novel benzothiozinone into clinics in which AZI was a key player in this consortia. MM4TB too delivered a derivative of benzothiazinones, BTZ043 (ClinicalTrials.gov—NCT03590600, 2018c), which is currently in the late stage clinical development.

### 2.5.2 TB Accelerator Programme

Yet another initiative—Bill and Melinda Gates Foundation, which is at the forefront of tackling the diseases of the developing and underdeveloped countries, brings together both academic and industry players as partners to accelerate the identification and development of novel compounds for the treatment of TB. Several academic institutions and pharmaceuticals are members of this effort.

India has its own example of such a successful collaborative effort. The development of RotaVac, a low cost, rotavirus vaccine through a collaborative ‘public–private partnership’ effort between the Department of Biotechnology and Bharat Biotech is indeed a major milestone achievement for India’s drug/vaccine research community (Glass et al. 2005). One of the significant contributors to this successful partnership was the detailed preclinical and early clinical development that has been carried out before the final Phase 3 trial. This preclinical work had been supported by various funding agencies and expertise both in India and the USA (Press Information Bureau, Government of India 2013). The question is, is such a model workable for developing anti-TB drugs? The involvement of industry in a less ‘commercially attractive’ area like tuberculosis will be generally driven by a ‘risk analysis’ on the chances of success. The RotaVac team was able to mitigate most of the risk involved with a network of global partners. This model is similar to the EU Framework or the TB Drug Accelerator models.

Two models India can experiment with: The first is exemplified by the University of Dundee Drug Discovery Unit who have been successful in developing new drugs for treating malaria (Norcross et al. 2016). The Dundee unit is made up of joint faculty from the University of Dundee with extensive domain expertise are also part of the Drug Discovery Unit. They undertake Drug Discovery programmes in a project mode and progress molecules through the late preclinical stage to clinic readiness. India has institutes like CSIR-CDRI (Council for Scientific and Industrial Research-Central Drug Research Institute), which can be mandated to take leads identified across different academic institutions in India and progress the molecules to be clinic ready. This would require a commitment of the faculty of the institute to form a project team which would be responsible for planning and implementing every aspect projects from conception to clinical ready stage. The second model would be for funding agencies to form 'facilitation groups' which are made of experienced drug hunters who are mandated to evaluate and progress a portfolio of starting points. The starting points would be from various academic institutes, and the facilitation group would generate data on the putative leads through CRO's (Contract Research Organisations) as needed. A variation on this model has been the Open Source Drug Discovery (OSDD) approach for finding new therapies for Neglected Diseases (Rayasam and Balganesh 2015; Ummanni et al. 2014). OSDD built a network of projects starting from lead discovery to clinical trial that was facilitated with expert panels and funding. The fundamental tenet of OSDD was that all data generated should be in the public domain. As of today, OSDD is a part of the India TB Research and Development Consortium (ITRDC) initiative (Koshy 2016) which in turn is being run under the auspices of ICMR (ICMR press release 2017), and is mandated to bring new therapies faster into India, as well as fast track projects from the Indian research community that can yield additional candidate drugs. It will be of interest to follow up the workings of this consortium in the coming years.

The regulatory framework in India has been active and has responded to the unmet needs. However innovative paths need to be found/improvised to enable new drugs to be tested in a regulatory acceptable manner (Vaidyanathan 2019). Several new drug combinations are being trialled globally. It will be helpful if this testing is accelerated in India which would enable faster drug approvals resulting in benefits to our patients in a timely manner. Given that the big pharma interests in developing new drug for TB in India is limited because of the narrow commercial potential of anti-TB compounds, it will finally rest on the appropriate agencies to take the responsibility of finding ways to not only trial the new compounds in the Indian scenario but also of bringing new drugs to TB patients in India.

Finally it has to be understood that there are two faces to Drug Discovery and Development—one dealing with relieving human suffering and the other, the commercial value of the investment. Drug companies are conscious of the Return on Investment (RoI) that can be generated from successful drug development. Antibiotics generally have limited commercial attractiveness as compared to therapies for metabolic diseases because of several reasons, including:

- the treatment course is short

- clinical trials are challenging
- new antibiotics will always be used as last resort

These challenges have led to many pharmaceutical companies exiting the 'anti-infective' therapy area. Taking cognizance of this, Governments in the US and the EU have launched several initiatives to incentivize research in the anti-infective area (Simpkin et al. 2017).

Another important factor to note is that the TB market is also impacted by the socio-economic status of the patient population and that the treatments are a combination therapy. This has led to the US FDA to formulate ways to incentivize research into finding new anti-TB molecules, as well as a 'voucher' system, for example, to help compensate for low RoI to pharma companies (Beith et al. 2009). India too needs to find ways to encourage investment into anti-TB drug development to incentivise large pharma companies to participate in this effort.

There is an overall need for integrated thinking involving strengthening discovery research, building consortia which include India's big pharma, finding ways to trial newer anti-TB drugs in India and finally incentivizing investment in TB research with schemes that help the commercialization. Such a discussion would involve several government and regulatory authorities while concerted and sustained discussion on the various issues would help overcome this bottleneck—it is high time we make this happen in India.

## 2.6 Closing the Loop

India has the maximum number of TB patients in the world and also has a high percentage of MDR patients. TB therapy in India has to adapt to tackle several aspects that contribute to the prevalence—awareness, access to treatments and counselling, diagnostics, drug sensitivity testing, prolonged treatment duration, affordability and a host of several related factors like nutrition, stigma, and long-term patient support to help patient compliance, each of which has a bearing on the ability to treat a TB patient successfully. Newer TB drugs have to be robust enough to be effective under some of the conditions mentioned above, for e.g. active against DS and MDR TB, and at the same time be cost effective. Diagnostic tests including drug sensitivity tests will play a major role in our ability to not only treat the patient and contain the disease, but will also be critical to ensure 'long-term effectiveness' of the drugs. Patient centric counselling and subsidised medical support is also essential to increase compliance. It must also be accepted that there are logistical challenges when dealing with the large numbers of TB cases, newer TB drugs need to be free from aggravating these pre-existing encumbrances.

Clearly there are several strings that need to be knitted to achieve progress. The disease poses an immense threat to the nation and has direct bearing on the health economics of the nation. The regimens in use fall short of being user friendly in terms of the need for reduced treatment duration and their side effects, yet the aim is to achieve a TB free India by 2025. Therefore, it is imperative that Indian scientists from both the academic and industry along with the appropriate executive machinery as well as a political commitment need to get together to achieve this task.

## References


Anishetty S, Pulimi M, Pennathur G (2005) Potential drug targets in Mycobacterium tuberculosis through metabolic pathway analysis. Comput Biol Chem 29:368–378

Anon (2008) Handbook of anti-tuberculosis agents. Tuberculosis 88:85–170

Aquinas M (1982) Short course chemotherapy for tuberculosis. Drugs 2:118–132

Arinaminpathy N et al (2016) The number of privately treated tuberculosis cases in India-an estimation from drug sales data. Lancet Infect Dis 16(11):1255–1260

Bagehi G, Jaya M, Tyagi S (2003) Hypoxia- responsive expression of Mycobacterium tuberculosis Rv3134c and devR promoters. Microbiology 149:2303–2305

Balasubramanian V et al (2014) Bactericidal activity and mechanism of action of AZD5847, a novel oxazolidinone for treatment of tuberculosis. Antimicrob Agents Chemother 58(1):495–502

Balganesh TS, Furr BJA (2007) Molecular approaches to target discovery - evaluating targets for antituberculosis drug discovery programmes. Infect Disord Drug Targets 7:120–126

Balganesh TS, Alzari PM, Cole ST (2008) Rising standards for tuberculosis drug development. Trends Pharmacol Sci 29(11):576–581

Beith A, Eichler R, Weil D (2009) Case study. In: Worldwide - incentives for tuberculosis diagnosis and treatment. Centre for Global Development, Washington, DC

Bhagavat R et al (2017) A genome-wide structure-based survey of nucleotide binding proteins in M. tuberculosis. Nat Sci Rep 7:12489

von Bubnoff A (2006) Seeking new antibiotics in the nature's backyard. Cell 127:867–869

Central TB Division (2018) India TB report 2018. Annual status report. Central TB Division, New Delhi

Chan C-Y et al (2004) In vitro post-antibiotic effects of rifapentine, isoniazid, and moxifloxacin against mycobacterium tuberculosis. Antimicrob Agents Chemother 48:340–343

Chandra N (2009) Computational systems approach for drug target discovery. Expert Opin Drug Discovery 4(12):1221–1236

Chatterji M et al (2014) 1, 4-Azaindole, a potential drug candidate for treatment of tuberculosis. Antimicrob Agents Chemother 58(9):5325–5331

Chen X et al (2017) Delamanid kills dormant mycobacteria in vitro and in a guinea pig model of tuberculosis. Antimicrob Agents Chemother 61(6):e02402–e02416

Chim N et al (2011) The TB Structural Genomics Consortium: a decade of progress. Tuberculosis (Edinb) 91(2):155–172

Clardy J, Fischbach MA, Currie CR (2009) The natural history of antibiotics. Curr Biol 19(11): R437–R441

ClinicalTrials.gov (2015) A phase 3 study assessing the safety and efficacy of bedaquiline plus PA-824 plus linezolid in subjects with drug resistant pulmonary tuberculosis. NCT02333799

ClinicalTrials.gov (2017) Multicenter, open, randomized study with active control (isoniazid) to evaluate the early antibacterial activity, safety and pharmacokinetics of the drug PBTZ169 (capsules 80 mg) when used in patients with first-diagnosed tuberculosis of the respiratory system with bacterial excretion and saved bacterial susceptibility to isoniazid and rifampicin. https://clinicaltrials.gov/ct2/show/NCT03334734

ClinicalTrials.gov (2018a) A dose-escalation study to evaluate safety, tolerability and pharmacokinetics of multiple doses of Q203 in normal healthy male and female volunteers. NCT02858973

ClinicalTrials.gov (2018b) A single ascending dose study of BTZ043. NCT02530710

ClinicalTrials.gov (2018c) A single ascending dose study of BTZ043. https://www.clinicaltrials.gov/ct2/show/NCT03590600

Crofton J (1959) Chemotherapy of pulmonary tuberculosis. Br Med J 1:1610–1614
D'Ambrosio L et al (2015) New anti-tuberculosis drugs and regimens: 2015 update. ERJ Open Res 1:00010-2015
Dartois V (2014) The path of anti-tuberculosis drugs: from blood to lesions to mycobacterial cells. Nat Rev Microbiol 12(3):159–167
Dartois V, Barry CE (2013) A medicinal chemists' guide to lead optimization for tuberculosis. Bioorg Med Chem Lett 23(17):4741–4750
Dawson R et al (2015) Efficiency and safety of the combination of moxifloxacin, pretomanid (PA-824), and pyrazinamide during the first 8 weeks of anti-tuberculosis treatment: a phase 2b, open-label, partly randomised trial in patients with drug-susceptible or drug-resistant pulmonary tuberculosis. Lancet 385(9979):1738–1747
Devis GR et al (2007) Evaluation of new antituberculosis drugs in mouse models. Antimicrob Agents Chemother 51:403–404
Dheda K, Ruhwold M, Theron G, Peter J, Yam WC (2013) Point of care diagnosis of tuberculosis: past, present and future. Respirology 18(2):217–232
Diacon AH et al (2012) 14-day bactericidal activity of PA-824, bedaquiline, pyrazinamide, and moxifloxacin combinations: a randomised trial. Lancet 380(9846):986–993
Doll R (1998) Controlled trials - the 1948 watershed. BMJ 317:1217–1220
Doster B et al (1973) Ethambutol in the initial treatment of pulmonary tuberculosis. Am Rev Respir Dis 107:177–190
Durand GA, Raoult D, Dubourg G (2019) Antibiotic discover: history, methods and perspectives. Int J Antimicrob Agents 53:371–352
Esposito S, Bianchini S, Blasi F (2015) Bedaquiline and delamanid in tuberculosis. Expert Opin Pharmacother 16(15):2319
Evans CA (2011) GeneXpert—a game-changer for tuberculosis control. PLoS Med 8(7):e1001064
Ferlazzo G et al (2018) Early safety and efficacy of the combination of bedaquiline and delamanid for the treatment of patients with drug-resistant tuberculosis in Armenia, India, and South Africa: a retrospective cohort study. Lancet Infect Dis 18(5):536–544
Fox GJ, Menzies D (2013) A review of the evidence for using bedaquiline to treat multi drug resistant tuberculosis. Infect Dis Ther 2:123–144
Gengenbacher M et al (2017) NOS2-deficient mice with hypoxic necrotizing lung lesions predict outcome of tuberculosis chemotherapy in humans. Sci Rep 7:8853
Ginsberg A (2010) Drugs in development for tuberculosis. Drugs 70:2201–2214
Glass RI et al (2005) Development of candidate Rotavirus Vaccines derived from neonatal strain in India. J Infect Dis 192:530–535
Gumbo T et al (2015) Nonclinical models for anti-tuberculosis drug development: a landscape analysis. J Infect Dis 211(Suppl 3):S83–S95
ICMR (2017) Press release
Johnson JL et al (2006) Early and extended early bactericidal activity of levofloxacin, gatifloxacin and moxifloxacin in pulmonary tuberculosis. Int J Tuberc Lung Dis 10(6):605–612
Joseph P et al (2011) Outcome of standardized treatment for patients with MDR-TB from Tamil Nadu, India. Indian J Med Res 133(5):529–534
Karekar SR, Marathe PA (2018) Current status of delamanid in the management of MDR tuberculosis. J Assoc Physicians India 66:72
Kaul A et al (2011) The challenge of new drug discovery for tuberculosis. Nature 469:483–490
Klaudt K (1994) TB-a global emergency. WHO, Geneva
Koide T et al (2008) Severe QT interval prolongation associated with moxifloxacin- a case study. Cases J 1:409
Koshy J (2016) TB drug project gets a lease of life. The Hindu
Kumar N et al (2014) Pharmacokinetics and dose response of anti-TB drugs in rat infection model of tuberculosis. Tuberculosis 94:282
Kurthkoti K, Varshney U (2012) Distinct mechanisms of DNA repair in mycobacteria and their implications in attenuation of the pathogen growth. Mech Ageing Dev 133:138–146

Lång H, Quaglio GL, Olesen OF (2010) Tuberculosis research in the European Union: past achievements and future challenges. Tuberculosis 90:1–6
Lanoix JP, Chaisson RE, Nuermberger EL (2016) Shortening tuberculosis treatment with flouroquinolones- lost in translation. CID 62:484–490
Laurenzi M, Ginsberg A, Spigelman M (2007) Challenges associated with current and future TB treatment. Infect Disord Drug Targets 7:105–119
Lebardo MDJ, Helena BIM, Clifton BE (2018) The present state of the tuberculosis drug development pipe-line. Curr Opin Pharmacol 42:81–94
Lehmann J (1946) para amino salicylic acid in the treatment of tuberculosis. Lancet 1:15–16
Li X et al (2017) Discovery of a potent and specific M. tuberculosis leucyl-t-RNA synthetase inhibitor- (S) 3- (aminoomethyl)-4-chloro-7-(2-hydoxyethoxy) benzo(c) (1,2) oxoborol-1(3H)-ol (GSK656). J Med Chem 60:8011–8026
Lienhardt C, Vernon A, Raviglione MC (2010) New drugs and new regimens for the treatment of tuberculosis - review of the drug development pipeline and implications for national programmes. Curr Opin Pulm Med 16:186–193
Lipinski CA, Lombardo F, Dominy BW, Feeney PJ (2001) Experimental and computational approaches to estimate solubility and permeability in drug discovery and development settings. Adv Drug Deliv Rev 46:3–26
Makarov V et al (2009) Benzothiazinones Kill Mycobacterium tuberculosis by blocking arabinan synthesis. Science 324(5928):801–804
Manjunatha UH, Smith PW (2015) Challenges and opportunities in TB drug discovery from phenotypic screening. Bioorg Med Chem 23:5087–5097
Matteelli A, Carvalho ACC, Dooley KE, Kritsk A (2010) TMC207- the first compound of a new class of potent anti-tuberculosis drugs. Future Microbiol 5(6):849–858
Mendes V, Blundell TL (2017) Targeting tuberculosis using structure-guided fragment-based drug design. Drug Discov Today 22(3):546–554
Mishra A et al (2018) An allosteric inhibitor of Mycobacterium tuberculosis ArgJ: implication to a novel combination therapy. EMBO Mol Med 10:e8038
Mitchison DA (1992) The Garrod lecture - understanding the chemotherapy of tuberculosis—current problems. J Antimicrob Chemother 29:477–493
Mitchison D, Davies G (2012) The chemotherapy of tuberculosis: past, present and future. Int J Tuberc Lung Dis 16:724–732
Mukherjee T, Boshoff H (2011) Nitroimidazoles for the treatment of TB: past, present and future. Future Med Chem 3(11):1427–1454
Murray JF, Schraufnagel DE, Hopewell PC (2015) Treatment of tuberculosis - a historical perspective. Ann Am Thorac Soc 12(12):1749–1759
Musa TL, Ioerger TR, Sacchettini JC (2009) The Tuberculosis Structural Genomics Consortium: a structural genomics approachto drug discovery. Adv Prot Chem Struct Biol 77:43–76
Nagaraja V, Godbole AA, Henderson SR, Maxwell A (2017) DNA topoisomerase 1 and DNA gyrase as targets for Drug Discovery. Drug Discov Today 22:510–518
Norcross NR et al (2016) Tri-substituted pyrimidines as efficacious and fast-acting antimalarials. J Med Chem 59(13):6101–6120
Olaru ID, Heyckendorf J, Andres S, Kalsdorf B, Lange C (2017) Bedaquiline-based treatment regimen for multidrug-resistant tuberculosis. Eur Respir J 49:1700742
Press Information Bureau, Government of India (2013) DBT announces phase III clinical trial results of rotavirus vaccine developed in India say vaccine demonstrates strong efficacy
Puneet Chopra LS, Meena YS (2003) New drug targets for Mycobacterium tuberculosis. Indian J Med Res 117:1–9
Radhakrishna S (2012) Contributions of The Tuberculosis Research Centre, Chennai in the field of epidemiology of tuberculosis. Indian J Tuberc 59:68–77
Rajarshi G (2013) Exploring structure activity relationships. Methods Mol Biol 993:81–84
Ramachandran G (2019) Pharmacokinetics of second-line anti-tuberculosis drugs. Indian J Pediatr 86:714

Ramachandran G, Swaminathan S (2015) Safety and tolerability of second line anti-tuberculosis medications. Drug Saf 38:253–269

Ramakrishnan T, Suryanarayana Murthy P, Gopinathan KP (1972) Intermediary metabolism of mycobacteria. Bacteriol Rev 36:65–108

Raman K, Yeturu K, Chandra N (2008) targetTB: a target identification pipeline for Mycobacterium tuberculosis through an interactome, reactome and genome-scale structural analysis. BMC Syst Biol 2:109

Rayasam GV, Balganesh TS (2015) Exploring the potential of adjunct therapy in tuberculosis. TIPS 36:506–513

Rockwood N, du Bruyn E, Morris T, Wilkinson RJ (2016) Assessment of treatment response in tuberculosis. Expert Rev Respir Med 10(6):643–654

Ruvandhi R et al (2017) Accuracy of line probe assays for the diagnosis of pulmonary and multidrug- resistant tuberculosis: a systematic review and meta-analysis. Eur Respir J 49:1601075

Sassetti CM, Boyd DH, Rubin EJ (2003) Genes required for mycobacterial growth defined by high density mutagenesis. Mol Microbiol 48:77–84

Schatz A et al (1944) Streptomycin, a substance exhibiting antibiotic activity against gram-positive and gram-negative bacteria. Proc Soc Exp Biol Med 55(1):66–69

Shehzad A, Rehman G, Ul-Islam M, Khattak WA, Lee YS (2013) Challenges in the development of drugs for the treatment of tuberculosis. Braz J Infect Dis 17(1):74–81

Simon GK, Lye MS, Ahmad N (1991) Side effects of short course tuberculosis chemotherapy. Med J Malaysia 46(1):88

Simpkin VL, Renwick MJ, Kelly R, Mossialos E (2017) Incentivising innovation in antibiotic drug discovery and development - progress, challenges and next steps. J Antib 70:1087–1096

Singh S (2013) Biosafety precautions in tuberculosis laboratory. Indian J Tuberc 60:135–137

Singh V, Mizrahi V (2017) Identification and validation of novel drug targets in Mycobacterium tuberculosis. Drug Discov Today 22(3):503–509

Singh P et al (2017) S-enantiomer of the anti-tubrcular compound S006-830 complements activity of front line TB drugs and targets biogenesis of Mycobacterium tuberculosis cell wall. AECS Omega 2:8453–8465

Singh A, Vijayan M, Varsheny U (2018) Distinct properties of a hypoxia specific paralog of single stranded DNA (SSB) protein in mycobacteria. Tubreculosis 108:16–25

Syal K et al (2017) Synthetic analogue of (p) ppGpp analogue is an inhibitor of stringent response in mycobacteria. Antimicrob Agents Chemother 61(6):e00443–e00417

Taneja NK, Dhingra S, Mittal A, Naresh M, Tyagi JS (2010) Mycobacterial transcriptional adaptation, growth and dormancy phenotype is triggered by Vitamin C. PLoS One 5(5):e10860

Tiberia S et al (2018) New drugs and perspectives for new anti-tuberculosis regimens. Pulmonology 24(2):86–98

Tuberculosis Research Centre Chennai (2002) A randomised clinical trial for treatment of smear positive pulmonary tuberculosis with regimens using Ofloxacin in the intensive phase. Indian J Tuberc 49:27

Ummanni R, Ramchandran S, Kumar A, Rayasam GV (2014) Innovation at OSDD – unconventional vs conventional approaches. Sci Report 51:20

Vaddady PK, Lee RE, Meibohm B (2010) In vitro pharmacokinetic, pharmacodynamic models in anti-infective drug development - focus on TB. Future Med Chem 2(8):1355–1369

Vaidyanathan G (2019) India's clinical-trial rules to speed up drug approvals. Nature

Vaishali P et al (2019) Mycobacterial genomics and structural bioinformatics: opportunities and challenges in drug discovery. Emerg Microb Infect 8:109–118

Van Heeswijk RP, Dannemann B, Hoetelmans RP (2014) Bedaquilne - a review of human pharmacokinetics and drug-drug interactions. JAC 69:2310–2318

Vashist R et al (2012) Crowd sourcing a new paradigm for Interactome driven drug target identification in Mycobacterium tuberculosis. PLoS One 7(7):e39808

Waksman SA, Reilly HC, Johnstone DB (1946) Isolation of streptomycin-producing strains of Streptomyces griseus. J Bacteriol 52(3):393–397
Walters WP (2012) Going further than Lipinski's rule in drug design. Expert Opin Drug Discovery 7(2):99–107
WHO (2017) WHO global tuberculosis report. WHO, Geneva
Yang L, Sun F, Zhang W (2019) Bedaquiline and delamanid in the treatment of multidrug-resistant tuberculosis; Promising but challenging. Drug Dev Res 80:98–105
Yeager RL et al (1952) Pyrazinamide (Aldinamide) in the treatment of pulmonary tuberculosis. Am Rev Tuberc Pulmo Dis 65(5):523–546
Yee D et al (2003) Incidence of serious side effects from first-line anti-tuberculosis drugs among patients treated for active tuberculosis. Am J Crit Care Med 167:1492
Yoshioka A (1998) Use of randomisation in the Medical Research Council's clinical trial of streptomycin in pulmonary tuberculosis in the 1940s. BMJ 317:1220–1223
Yuan T, Simpson SN (2018) Hit generation in TB drug discovery: from genome to granuloma. Chem Rev 118:1887–1916
Zhan L, Tang J, Sun M, Qin C (2017) Animal models for tuberculosis in translational and precision medicine. Front Microbiol 8:717

**Tanjore S. Balganesh** (TSB), trained as a medical microbiologist, possesses more than three decades of experience in antibacterial drug discovery, with two decades of experience in leading anti-TB research in the industry. TSB served as head of research at the AstraZeneca's (AZI) unit in Bangalore, India. The unit specialised in anti-mycobacterial drug discovery. AZI became one of the global leaders in anti-TB drug discovery bringing all the tools and experience of 'industrial mode drug discovery' towards finding novel treatments for the disease. AZI has delivered several clinical candidates for clinical development. TSB currently serves as President of GangaGen Biotechnologies Pvt Ltd., Bangalore.

**Jagadeesh J. Bhat** (JB) has over 18 years of industry experience in various segments of Biotechnology. At GangaGen Biotechnologies Pvt Ltd. (GBPL), JB has been involved in finding new therapies for the treatment of serious bacterial infections where he is currently a senior scientist. He led the formulation team which was involved in making protein preparations suitable for human therapy. JB also has extensive experience in establishing techniques suitable for up scaling bacteriophages for their use in phage therapy.

**Bheemarao Ugarkar** (BGU), trained as a synthetic and medicinal chemist, has over three decades of drug discovery experience in a number of disease areas. He joined the drug discovery team of AstraZeneca India in 2004 as Director and Head of Chemistry and DMPK departments to lead the discovery of new medicines for treating TB and Malaria. BGU was involved in building a world-class medicinal chemistry and computational chemistry team which delivered several clinical candidates. Currently, BGU is consulting with Open Source Pharma (OSPF) and a few academic institutes on drug discovery sciences.

# Indian Discovery Effort in the Quest of Novel Antibiotics

# 3

Mahesh V. Patel, Sachin S. Bhagwat, and Prasad K. Deshpande

## 3.1 Background

India's rise as a major pharmaceutical hub is a result of focused intellectual and financial investments made by several domestic pharmaceutical companies over the past few decades. India continues to manufacture and supply a significant fraction of the global demand for pharmaceutical products. Beginning mid-1970s and until 2005, when India did not accept product patents, the domestic pharmaceutical industry acquired significant expertise in the bulk drug manufacturing (APIs) and formulation processes. Armed with this expertise, India expanded its position as a significant global generic pharmaceutical player. While this is a well-known story, what is not widely recognized is that India quietly prepared itself from 1995 onward for the era post TRIPS, which implemented product patent with effect from January 2005. For the pharmaceutical industry, to remain competitive, this development threw a challenge of taking the path towards innovation. As a result, several domestic companies ventured into the discovery and development of innovative drugs. The chapter provides a brief account of India's contribution to the discovery and development of innovative antibacterial agents and compositions.

## 3.2 Need for Antibacterial Drug Discovery and Development

Globally, since a long time, bacterial infections have remained a significant cause for morbidity and mortality. Despite several efforts, the human race has registered limited success in its fight against bacteria. Although, structurally, bacteria are a simple living organism, they are also smarter. Bacteria reproduce quickly and in doing so; they often develop sophisticated mechanisms to overcome the effect of

M. V. Patel (✉) · S. S. Bhagwat · P. K. Deshpande
Drug Discovery Research, Wockhardt Research Center, Aurangabad, India
e-mail: mpatel@wockhardt.com



M. Dikshit (ed.), *Drug Discovery and Drug Development*,
https://doi.org/10.1007/978-981-15-8002-4_3

antibiotics that were earlier effective in killing them. Thus, there is a continual need for developing newer antibiotics to treat infections caused by resistant bacteria (API Synthesis International 2016). An exhaustive study sponsored by the UK Government has estimated that there is a continual need to discover effective antibiotics in coming years. Otherwise, the anticipated burden of deaths from antimicrobial resistance (AMR) could reach ten million lives each year by 2050, implying a cumulative cost of USD 100 trillion to the global economic output (based on a report drafted by Jim O'Neil). The problem in tackling AMR also lies in the fact that many of the major pharmaceutical companies in the US and Europe have abandoned antibacterial research. It is for this reason that companies developing newer antibacterial agents assume importance. As we read this chapter, what brings us solace is that several Indian companies have made their sincere efforts to address the AMR issue by engaging in the discovery of novel antibiotics.

## 3.3 Antibacterial Discovery and Development in India

Antibacterial research has been actively pursued in India for about 50 years. This consistent long-term interest highlights the fact that India continues to experience a higher burden of infectious diseases. Even during the resource-constrained period of pre-1970s, government entities in India undertook initiatives in antibiotic research. Few of the developed products were meant for domestic consumption and thus did not traverse through high standards of the US and European regulators. Most of that research was limited to publications on newer observations.

The scientific foundation of antibacterial drug discovery in India was genuinely laid in the early 1970s by the German pharmaceutical giant Hoechst, which initiated a well-funded and sustained microbial secondary metabolite screening program at its primary research centre in Mumbai. Post-1990s, several Indian pharmaceutical companies joined the foray of discovering new antibiotics. As observed in other parts of the world, the onus of discovering new antibiotics in India was primarily hinged on pharmaceutical companies, and their programs met with varying level of success. Indian antibiotic research was much helped by the previous contribution of companies in US, Europe and Japan. However, post 2000, these multinational companies did not sustain their discovery programs amid increasing complexity of discovering newer antibiotics effective against multidrug-resistant (MDR) pathogens and low economic returns generated by most new antibacterial drugs. At the same time, widespread AMR kept pounding the healthcare system, compromising its ability to fight infections. In particular, India and China are at a higher risk of a substantial infectious disease burden. Against this background, the narratives given below present a glimpse of 50 years of India's contribution to antibacterial drug discovery and development.

### 3.3.1 Hindustan Antibiotics Limited

Hindustan Antibiotics Limited, a Government of India owned company, initiated one of the first indigenous antibiotic research in India. This Pune-based firm pursued a fermentation-based antibiotic discovery program, which led to the discovery of several antifungal compounds, including hamycin (Thirumalachar 1966), a polyene antibiotic (Fig. 3.1) structurally close to amphotericin B. Hamycin exhibited an inhibitory concentration of 0.01 mg/L against *Candida albicans*. The drug was

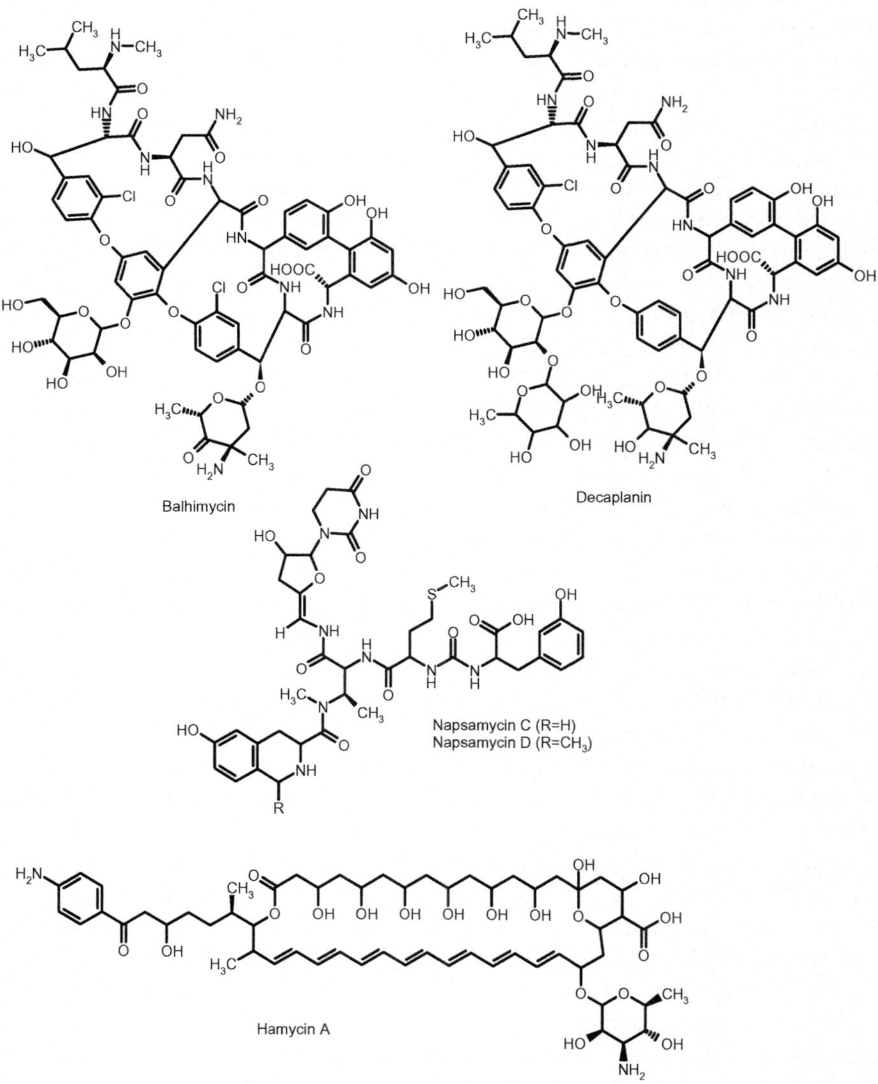


**Fig. 3.1** Structures of natural products discovered by Hoechst and Hindustan Antibiotics Ltd

used as a topical antifungal in the form of glycerine suspension for the treatment of oral thrush caused by *C. albicans* and otomycosis incited by *Aspergillus niger*; or in the form of lactose tablets for the treatment of vaginal moniliasis. Hamycin was launched in 1971, but is no longer marketed presently.

### 3.3.2 Hoechst India Limited

Hoechst, for the first time, brought a systematic natural product discovery culture in India by instituting a sustained antibacterial discovery program in the early 1970s. The German parent Hoechst had long demonstrated its capability of discovering novel antibacterial agents, with the earliest discovery of Salvarsan for treating syphilis in 1910. In India, Hoechst focused on the discovery of microbial fermentation-based novel antibiotics. For nearly 25 years (1972–1998), about 75 scientists at Hoechst's research centre in Mumbai isolated several soil microorganisms from samples collected from different parts of India. It was assumed that India's diverse range of climatic and soil conditions would give rise to diverse microorganisms capable of synthesizing novel antibiotics. Some of the clinically interesting antibiotics discovered were anti-MRSA (*methicillin-resistant Staphylococcus aureus*) agents (Fig. 3.1) such as balhimycin (Nadkarni et al. 1994), decaplanin (Sanchez et al. 1992), napsamycin (Chatterjee et al. 1994) and mersacidin (Chatterjee et al. 1992). Balhimycin and decaplannin were novel glycopeptides similar to the widely used vancomycin, while mersacidin and napsamycin belonged to a new structural class. The striking feature of napsamycin was its specific activity against *Pseudomonas aeruginosa*, a pathogen which is tough to tackle to date.

Even with active support of structural elucidation experts from Hoechst Germany, it took a couple of years for the revelations of complex structures of these novel antibiotics. There were several technical challenges in isolating pure antibiotics from the fermentation broth. One of the issues was that during the fermentation cycle, several closely similar structural analogues were co-produced and co-purified along with major antibiotic of interest. This created a considerable difficulty in establishing the structure of the key active component. Improvement of fermentation yield and the optimization of the downstream process to recover pure antibiotic from the fermentation broth took several months of effort. Structural complexity deterred Hoechst from undertaking further chemical optimization of these antibiotics, although they belonged to a novel class and offered the advantage of a new mechanism of action.

In subsequent years, several companies focused on natural products or fermentation-based antibacterial discovery. However, these efforts did not lead to success due to complexities associated with the development of natural products such as the lack of appropriate infrastructure, a dearth of scientific skills and heightened quality/analytical standards dictating the evaluation of highly purified preparation in clinical studies. Therefore, despite the initial excitement, natural product-based antibiotic discovery research met with a disappointing outcome. The challenge of transforming a natural product into a '*drugable antibiotic*' persists

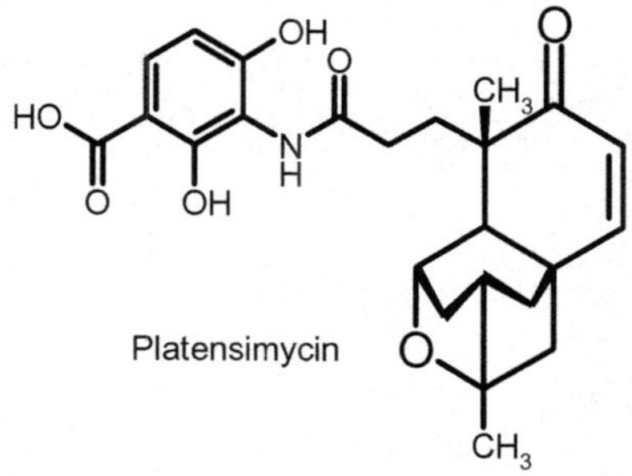


**Fig. 3.2** Platensimycin by Merck

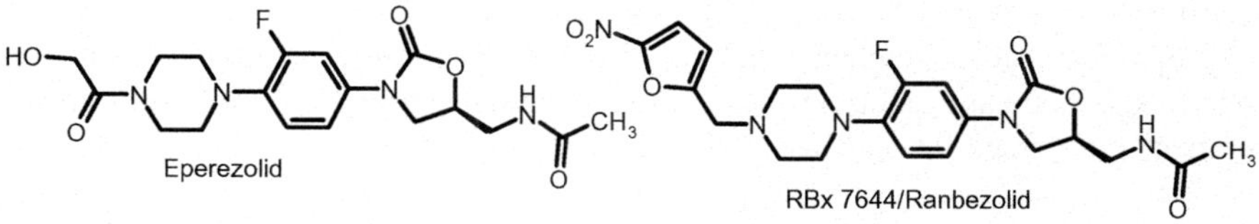


**Fig. 3.3** Modification of eperezolid by Ranbaxy

even in twenty-first century; a stark reality evident from the example of platensimycin discovered by Merck (Fig. 3.2), which ultimately remained merely a subject of scientifically interesting publication. Given the inherent pharmacological limitations linked with natural products, in recent years, not even a handful of clinically viable natural product-based antibiotics could be developed.

Taking a cue from this, in the 1990s, many global and Indian companies focused their discovery program on medicinal chemistry-based antibacterial drug discovery. These include Ranbaxy Laboratory, Dr Reddy's Laboratory, Wockhardt Limited, Orchid Pharmaceuticals, Zydus Cadila, Aurigene, Vyome Biosciences, Panacea Biotech, Bug Works and Vitas Pharma.

### 3.3.3 Ranbaxy Laboratories

For over 15 years, antibacterial research at Ranbaxy focused on fluoroquinolones, oxazolidinones and macrolide/ketolide class of agents. Ranbaxy's oxazolidinone research program optimized Pharmacia's lead compound eperezolid (PNU-100592, a backup clinical candidate for highly successful anti-MRSA drug linezolid). The company discovered an equipotent bio-isosteric replacement at piperazine ring of eperezolid in the form of substituted methyl-*N*-piperidine. Various five-member rings were attached with or without methylene bridge on to the piperazinyl-phenyl-oxazolidinone core, to provide potent oxazolidinone NCEs (Mehta 2001). The research produced two candidates for clinical development, MRSA-active ranbezolid (Fig. 3.3; RBX 7644) and MDR-TB-active RBX 8700.

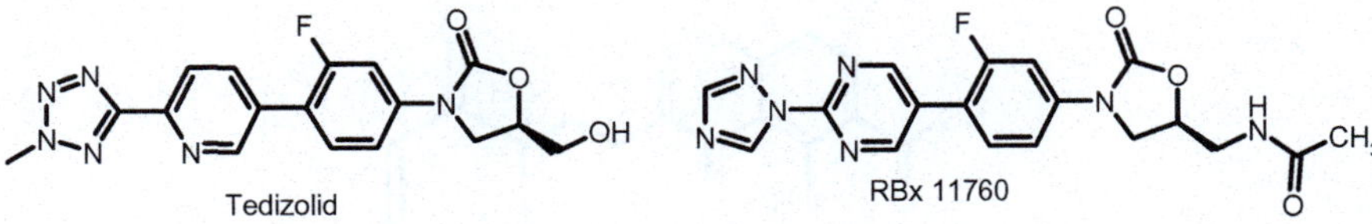


**Fig. 3.4** Modification of tedizolid by Ranbaxy

Ranbezolid showed in vitro MICs similar or slightly superior to linezolid (PNU-100766 Zyvox). Ranbezolid displayed activity against all anaerobes (Gram-positive and Gram-negative) (Das et al. 2005). Anti-anaerobes activity of ranbezolid was ascribed to not being a substrate of the efflux pump. Importantly, unlike other nitrofuran compounds, ranbezolid was reported to be free of DNA damaging activity in the Ames test, micronucleus test, chromosomal aberration test, and macromolecular synthesis in anaerobes test. Ranbezolid exhibited favourable pharmacokinetic and safety profile in the preclinical studies. The ranbezolid development progressed up to phase 1; however, further clinical development was not pursued.

Another biaryl oxazolidinone from Ranbaxy, RBX 11760 (Fig. 3.4) was inspired by a recently marketed compound tedizolid. RBX 11760 exhibited 2–4× lower MICs than linezolid which were comparable to that of tedizolid. It showed good oral bioavailability (60% and 72% in mouse and rat, respectively), with low plasma clearance and low to moderate volume of distribution in mouse and rat. It also displayed higher in vivo efficacy compared to tedizolid in the infection model (Barman et al. 2016).

RBX 11760 was also investigated for the possibility of treating *Clostridium difficile* infections, as it exhibited good MICs against *C. difficile* isolates, in the range of 0.5–1 mg/L. The drug showed concentration-dependent killing of *C. difficile* ATCC 43255 and *C. difficile* 6387 up to 2–4 × MIC (1–2 mg/L). However, further development of RBX 11760 was not pursued (Mathur et al. 2011).

Ranbaxy's macrolide research ended up identifying 2-fluoro-ketolide compounds. RBX 14255 (Fig. 3.5) was active against erythromycin- and clarithromycin-resistant *C. difficile* strains, including the epidemic BI/NAP1/027 strain. It exhibited better efficacy than metronidazole and vancomycin in the Golden Syrian hamster animal model, against *C. difficile* infection (Kumar et al. 2012). However, the development of RBX 14255 was halted in 2008, after the preclinical stage.

### 3.3.4 Dr Reddy's Laboratory

Dr Reddy's Laboratory (DRL) discovery activities focused on tetracycline and oxazolidinone class of antibacterial compounds. A methylthiocarbamate analogue of oxazolidinone, DRF 8417 (Fig. 3.6) progressed to preclinical studies. As an antibacterial agent, DRF 8417 was 2–4× more active than linezolid and its in vivo efficacy in Swiss albino mice was comparable to linezolid (Sreenivas et al. 2007).

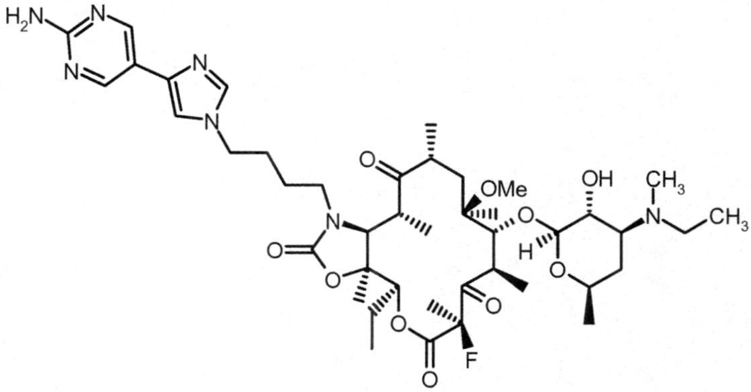


RBx 14255
AAC; 2012; 56(11); 5986

**Fig. 3.5** 2-Fluoroketolide by Ranbaxy

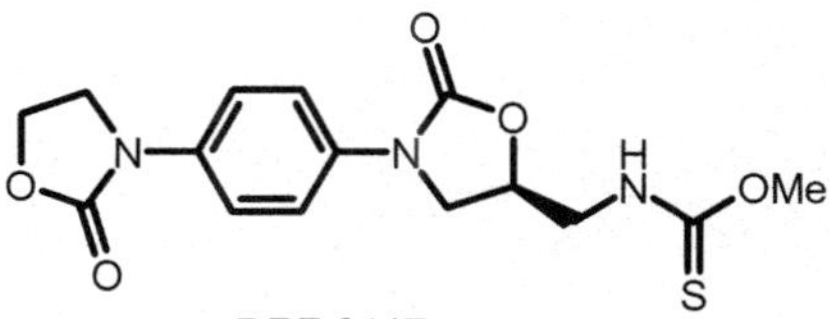


DRF 8417
(JAC; 2007; 60, 159)

**Fig. 3.6** Non-fluoro-phenyl thioacetamide by Dr Reddy's Laboratories

Additional reports on the progress of DRF 8417 are not available. As per company's website, from the oxazolidinone and 1,2,3-triazole program, four other compounds progressed up to preclinical stage: DRF 11057, DRF 13792, DRF 16048 and DRF 19440. Structures of these compounds are not available in the public domain as none of them progressed beyond preclinical studies.

### 3.3.5 Nicholas Piramal

In 1999, Nicholas Piramal acquired Hoechst's research centre in Mumbai and continued natural product-based antibacterial discovery program until 2015–2016. During this period, a novel antibiotic PM 181104 (Fig. 3.7) was developed jointly with the National Institute of Oceanography, Goa. PM 181104 has a complex structure and exhibited potent activity against MRSA and vancomycin-resistant enterococci (Mahajan 2009). It is a 23-member macrocyclic peptide isolated from bacterial species *Kocuria* (ZMA B1/MTCC 2569) fermented broth. The compound displayed extremely high potency against MRSA and *S. epidermidis* with MICs ranging from 0.00781 to 0.0625 mg/L.

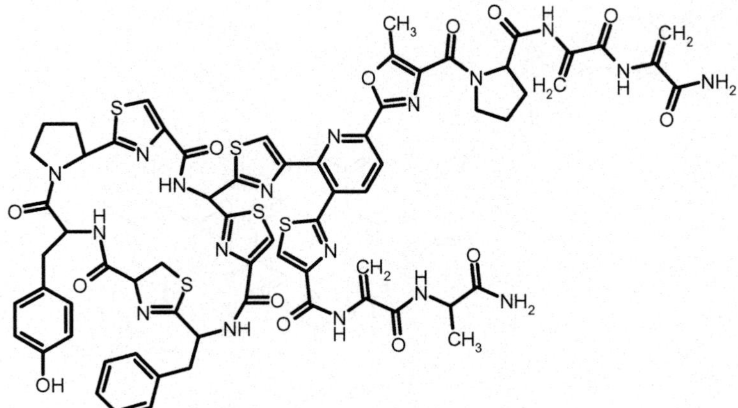

PM 181104
(EP 2010202)

**Fig. 3.7** Macrocyclic peptide by Nicholus Piramal

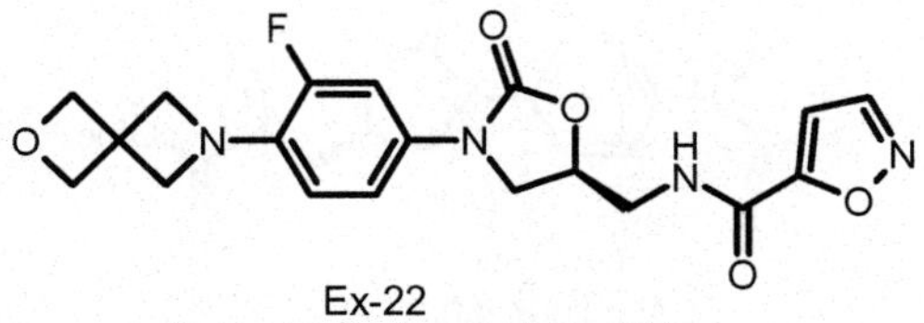

Ex-22
(Eur. J Med. Chem; 122, 475)

**Fig. 3.8** Spiro-oxazolidinone by Nicholus Piramal

In a systemic infection animal model of *S. aureus* E7 (MRSA), PM 181104 showed $PD_{100}$ at a dose of 5 mg/mL as compared to linezolid 25 mg/mL (EP 2010202). PM 181104 progressed up to the preclinical stage, after which no progress was reported.

As part of academia-industry collaborative research, Piramal and three other institutes, VIT University (Vellore), NMIMS (Mumbai), and National Chemical Laboratory (Pune), identified spiro analogues of oxazolidinone by replacing morpholine moiety of linezolid with a spiro ring: 2-oxa-6-aza spiro[3.3]heptane (Gadekar et al. 2016). This modification was aimed at minimizing the liability of oxidative metabolism associated with morpholine moiety in linezolid. Various further modifications at C5 oxazolidinone site displayed both antibacterial and antitubercular activity. The most potent compound was example-22 (Fig. 3.8), which was less active than linezolid.

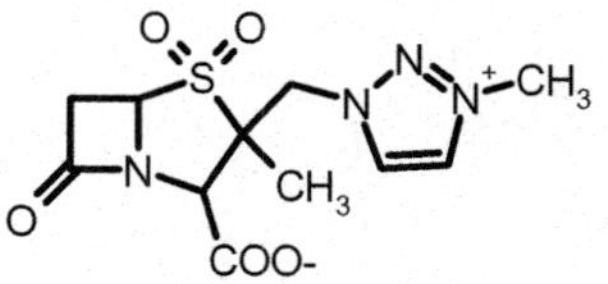


OCID101
Ex-1
(US 7687488)

**Fig. 3.9** Zwitterionic analogue of tazobactam by Orchid

### 3.3.6 Orchid Pharma

The discovery program at Orchid also started with oxazolidinone (OCID0050) class and then moved to the carbapenem-based discovery program, in collaboration with Merck Sharp Dohme. From publications and conference posters, it appears that Orchid's oxazolidinone NCE OCID0050 was once a promising candidate for development. Although the chemical structure of OCID0050 is not disclosed, the publication states that OCID0050 was a piperazinyl thioacetamide oxazolidinone (Paul-Satyaseela et al. 2009), two- to fourfold more active than linezolid, with activity even against resistant strains. There is no further information on the development of this compound.

Orchid's well-recognized strength in manufacturing the β-lactam class of drugs was leveraged in the discovery of novel β-lactamase inhibitor OCID 5090 (Fig. 3.9). A combination of β-lactam antibiotic cefepime and OCID101 [recently named as AAI 101 or enmetazobactam] entered global phase 3 clinical trial. The combination is being developed by Allecra, Germany which in-licensed the molecule from Orchid. OCID 5090 was described in US patents (Palanisamy 2010), along with various *N*-alkyl quaternary salts of tazobactam's 1,2,3-triazole. OCID 5090 is 2–8× more effective than tazobactam in lowering MIC of piperacillin against several class A β-lactamases producing *Enterobacteriaceae*. Its superior β-lactamase inhibitory activity reflected well in vivo, as 4:1 combination of piperacillin and OCID 5090 was 2–3× superior in terms of $ED_{50}$ than 4:1 piperacillin and tazobactam combination (Palanisamy 2014).

In a yet another in vitro study, MICs of piperacillin in combination with OCID 5090, at 1 mg/L concentration of OCID101, were found to be 1.0–4.0 mg/L against *E. coli* ($n = 6$); 2.0–8.0 mg/L against *K. pneumoniae* ($n = 6$); 2.0 mg/L against *E. cloacae* ($n = 2$); and 16.0–32.0 mg/L against *P. aeruginosa.*

In the in vivo mice infection model, a combination of piperacillin and OCID 5090 (4:1) was 2–2.5× superior to piperacillin and tazobactam (4:1) combination. For instance, $ED_{50}$ values were 22.62 (mg/kg b.w.) as compared with 37.22 (mg/kg b.w.) for *E. coli* MRO 10006; 24.88 (mg/kg b.w.) versus 58.81 (mg/kg b.w.) for *E. coli* MRO 10007, and 34.89 (mg/kg b.w.) versus 99.48 (mg/kg b.w.) for *K. pneumoniae* MRO 11008 (Paul-Satyaseela et al. 2014).

Rodent PK of OCID 5090 showed $AUC_{0\text{-}4}$ of 5.3 μg h/mL as compared with tazobactam AUC of 2.138 μg h/mL. However, the elimination half-life was almost the same for two β-lactamase inhibitors (0.272 h versus 0.269 h).

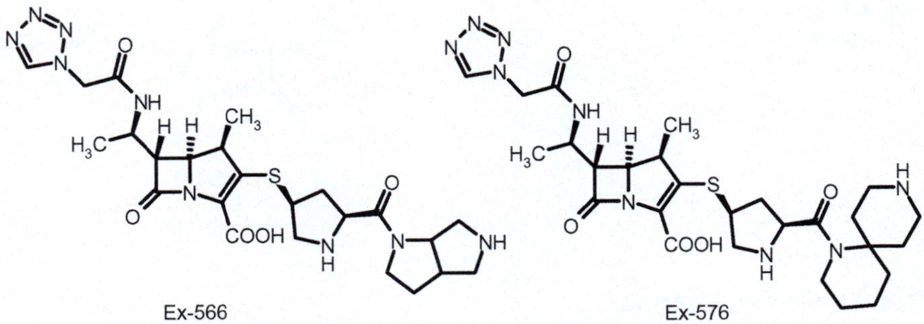


Ex-566
(WO 2017/158616)

Ex-576
(WO 2017/158616)

**Fig. 3.10** Carbapenem analogues by Orchid and Merck

OCID 5090 was also studied in combination with carbapenem such as imipenem. The US patent application (Palanisamy 2014) describes in vivo efficacy in mice systemic infection model involving KPC2 harbouring *K. pneumoniae*, where $ED_{50}$ of imipenem + OCID 5090 combination was 2.2 mg/kg (OCID 5090 at 64 mg/kg) while the same for the imipenem and tazobactam combination was 4 mg/kg (tazobactam at 64 mg/kg). Standalone imipenem provided $ED_{50}$ value of 8.9 mg/kg. The same patent also claims the use of OCID 5090 for the detection of β-lactamases, including KPC.

Merck & Co, in collaboration with Orchid, entered into the antibacterial discovery program with the objective of modifying the meropenem nucleus. Their collaborative work is published in patent application (Balasubramanian 2017), which lists the examples where the traditional hydroxyl ethyl substituent is replaced with a substituted aminoethyl group. Many compounds exhibited potent antibacterial activity against *P. aeruginosa* ATCC 27853 (meropenem susceptible strain), with MICs being in the range of 0.25–2.0 mg/L. The representative noteworthy examples are 566 and 576 (Fig. 3.10). These compounds showed MIC of 0.5 mg/L against *S. aureus* ATCC 29213, 0.5 mg/L against *K. pneumoniae* ATCC BAA 1705 expressing KPC2, 0.06 mg/L against *E. coli* ATCC 25922, and 0.25 mg/L against *P. aeruginosa*. No further development is reported for Merck-Orchid carbapenems.

A joint publication by Manipal College of Pharmaceutical Sciences and Orchid (Aaramadaka et al. 2007) described a series of urea-oxazolidinones, where piperazinyl nitrogen is a part of urea and thiourea.

### 3.3.7 Aurigene

Aurigene, a contract research laboratory established by Dr Reddy's Laboratories, explored the area of FAB1 [bacterial Fatty Acid Biosynthesis 1] inhibitors. Based on the literature disclosure, all Aurigene's FAB1 inhibitors appear to be modifications of aromatic α,β-unsaturated ketone nucleus of AFN-1252 [Korean company Affinium's lead compound, which was once the compound under detailed

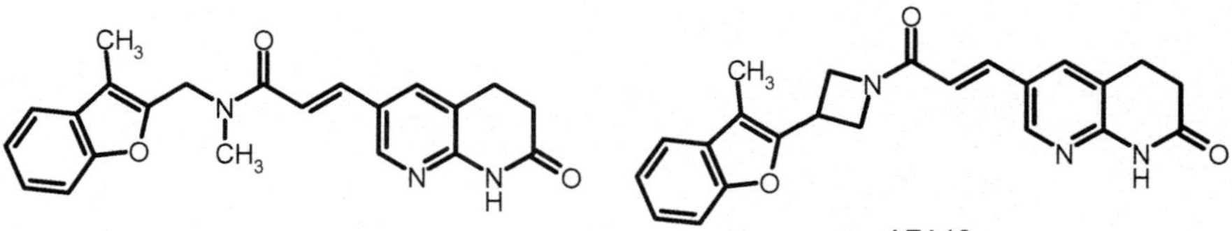


AFN-1252 AEA16

**Fig. 3.11** FAB1 inhibitor AEA16 by Aurigene

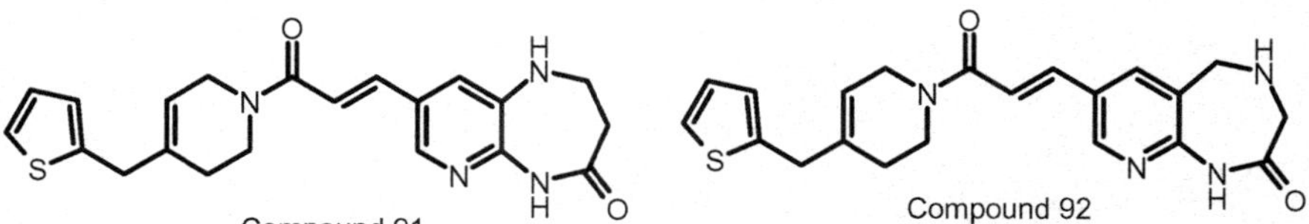


Compound 91 (US 9062002) Compound 92 (US 9062002)

**Fig. 3.12** FAB1 inhibitors by Aurigene and Pharmauji Sdn.BHD

investigation]. Aurigene's AEA16 (Fig. 3.11) progressed up to preclinical studies, and was structurally very close to AFN-1252. AEA16 showed improved mouse liver microsomal stability and PK properties as compared to AFN-1252. This property was ascribed to the closed ring structure in the form of azetidine present in AEA16, as compared to open *N*-methyl structure in AFN-1252 (Takhi et al. 2014). Extensive nonclinical studies have been published with AEA16. Cell-free FABI enzyme inhibition in *S. aureus* ($IC_{50}$) was reported to be 0.141 μM and MICs were in the range of 0.06–0.5 mg/L. In the murine systemic MRSA infection model, $ED_{50}$ of AEA16 was 0.90 mg/kg/day, while the corresponding value for the reference agent AFN-1252 was 2.3 mg/kg/day. However, the development of AEA16 was discontinued in 2015. In addition, a US patent (Takhi 2015) filed jointly by Aurigene and Pharmauji Sdn.BHD, Kuala Lumpur described piperdi-3-ene analogues. Many compounds from this series were fairly active-based on their $IC_{50}$s and MICs. For example, compound 91 (Fig. 3.12) showed MICs of 0.25–0.5 mg/L against *S. aureus*. No further progress on any of these compounds is reported.

### 3.3.8 Bugworks

Bugworks is a startup, established by the ex-employees of AstraZeneca India at Bengaluru. Its molecule BWC0977 is ready to enter full-fledged preclinical studies. Aimed at overcoming multidrug resistance and minimizing resistance development, BWC0977 targets bacterial topoisomerases—gyrase and topoisomerase IV with novel interactions. Two posters (Michael et al. 2019; Hameed et al. 2018a) describe the features of BWC0977; however, the exact structure is not available.

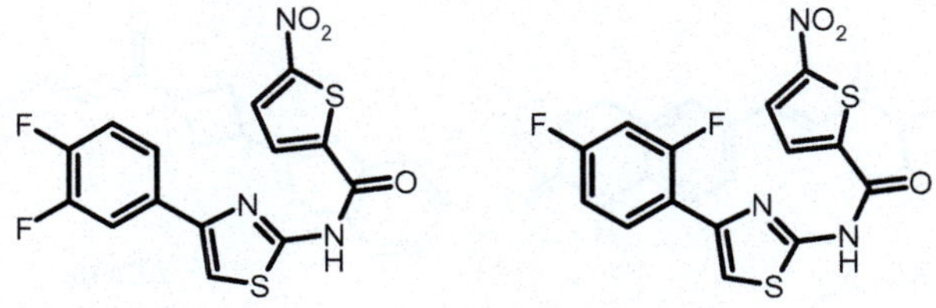

Compound 11 Compound 12

**Fig. 3.13** Nitrothiophene carboxamides by Bugworks

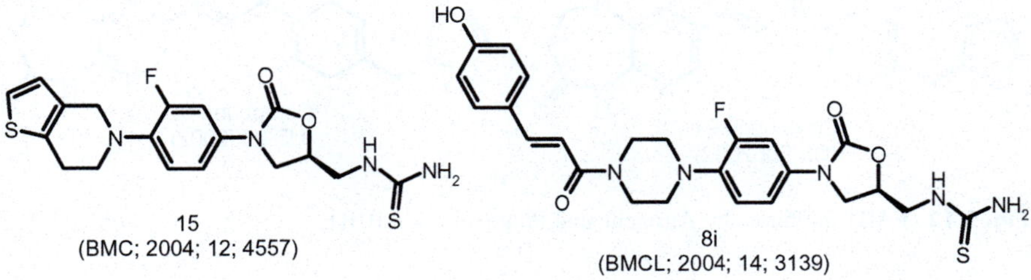

15
(BMC; 2004; 12; 4557)

8i
(BMCL; 2004; 14; 3139)

**Fig. 3.14** Thiacetamide oxazolidinones by Zydus Cadila

A separate publication (Hameed et al. 2018b) describes a novel narrow-spectrum nitrothiophene carboxamide (Fig. 3.13) series derived by using a structure-based design. These compounds were optimized to overcome efflux pump mediated resistance. Moreover, the compounds were claimed to be pro-drugs that require activation in *E. coli* by specific bacterial nitroreductases NfsA and NfsB. They were active against wild-type and multidrug-resistant clinical isolates of *E. coli*, *Shigella* spp., and *Salmonella* spp. The incorporation of nitro moieties in the structure of novel antibiotic, as exemplified by ranbezolid and more recently by nitrothiophene carboxamides, is indicative of the desperation of medicinal chemists for potency optimization at the cost of potential drugability concern.

### 3.3.9 Zydus Cadila

Zydus also worked in the oxazolidinone antibacterial class, for a brief period. According to publications, modifications were explored in the C5 position of oxazolidinone ring such as thioacetamide and thiourea and the morpholine/pierazinyl ring of linezolid/eperezolid analogues. Thienotetrahydropyridine oxazolidinone analogues (Lohray et al. 2004a) and substituted cinnamoyl piperazinyl oxazolidinone analogues (Lohray et al. 2004b) were synthesized and compared with linezolid and eperezolid for antibacterial activities. The most active compounds from both publications were 15 and 8i (Fig. 3.14). No further progress has been reported by Zydus Cadila.

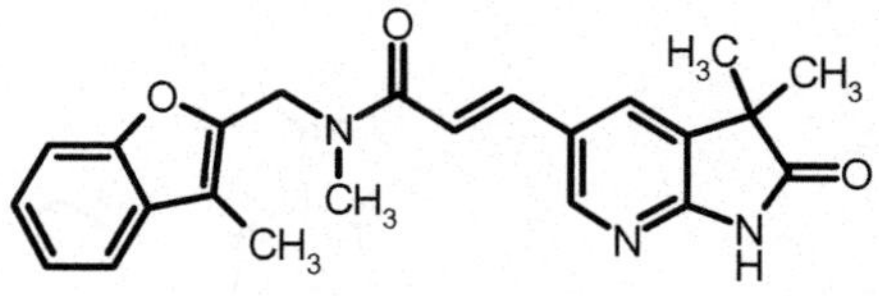


VT-02-00068
(US 9115149)

**Fig. 3.15** FAB1 inhibitor by Vitas Pharma

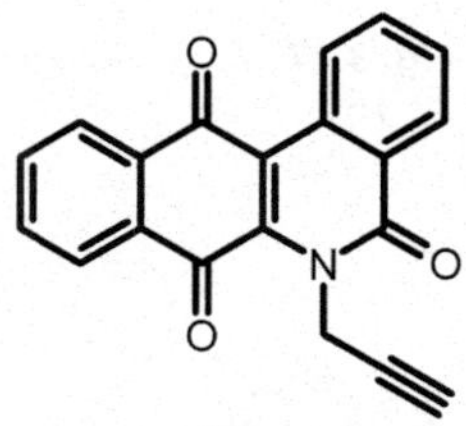


1f
(ACS Med. Chem Lett.)

**Fig. 3.16** MRSA active compound acting through novel mechanism by Vitas Pharma and IISER

### 3.3.10 Vitas Pharma

Vitas Pharma, a company based in Hyderabad, in collaboration with Cambridge University, reported another FABI inhibitor compound VT-02-00068 (Fig. 3.15), which is currently in the preclinical stage. An US patent (Rangarajan 2012) describes the features of this bacteriostatic compound, which is a modification of AFN-1252.

VT-02-00068 was studied extensively in preclinical studies. MICs of VT-02-00068 were reported to be in the range of 0.015–0.25 mg/L against several Gram-positive organisms. The compound was inactive against *E. faecalis*, *S. pneumoniae*, *H. influenzae*, *M. catarrhalis,* and *E. coli*. Metabolic stability in mouse liver microsome was established and a systemic infection model in mice with MRSA 33591 VT-02-00068 provided survival rates of 100%, 100% and 65% at intravenous doses of 30 mg/kg/body weight, 10 mg/kg/body weight and 3 mg/kg/body weight, respectively. Mutation prevention concentration was found to be 0.5 mg/L. Oral mouse PK properties were also characterized. As per Vitas Pharma's website, this compound is still in the preclinical toxicological evaluation.

Indian Institute for Scientific Education and Research (IISER), Pune and Vitas Pharma jointly reported natural product inspired redox active small molecule to overcome drug resistance in MRSA. From the publication (Khodade et al. 2014), compound 1f (Fig. 3.16) appears to be a potent MRSA inhibitor with a unique mode of action that involves enhancement of reactive oxygen species (ROS) levels in bacterial cell, thereby damaging DNA and causing cell death. The compound 1f has comparable or superior MICs than that of vancomycin. MICs against several MRSA strains are reported.

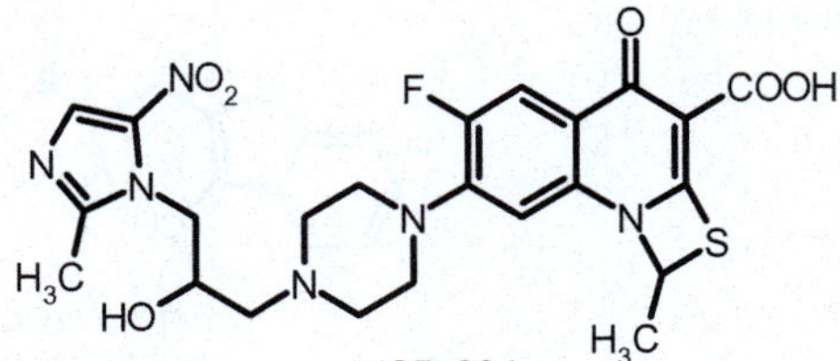

VCD-004
(US 10071103)

**Fig. 3.17** Fluoroquinolone analogue by Vyome Bioscience

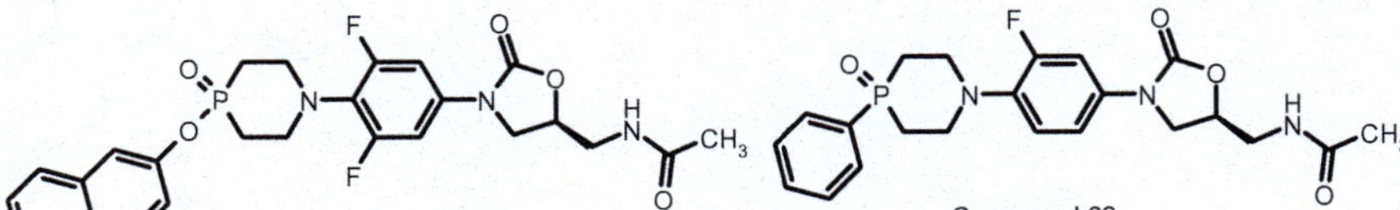

Compound 30
(WO 2008/090570)

Compound 62
(WO 2008/090570)

**Fig. 3.18** Phosphorus analogues of oxazolidinone by Panacea Biotech

### 3.3.11 Vyome Bioscience

As per Vyome Bioscience's website, four molecules as topical antibacterial agents are at the advanced stage of development, including: VB-1953 in the clinical phase 2 for resistant acne; VB-6395 at the preclinical stage for facial Gram-negative folliculitis; VB-9333 (Dual Action Rational Therapeutics NCE) at the preclinical stage for treating skin pathogen-mediated implant infection; and VB-9450 (Dual Action Rational Therapeutics NCE) at the preclinical stage for treating infections caused by antibiotic-resistant acne. The structures of these NCEs are not known. Vyome's research is mainly focused on catering to unmet needs in dermatology. The recent publication (Ghosh et al. 2018) describes VCD-004 (Fig. 3.17) and four other similar compounds. These are fluoroquinolones with 2-methyl-5-nitro-imidazole structure. Such chemical moieties are generally not considered suitable for systemic use of a compound to treat disseminated infections. VCD-004 shows high potency against resistant *P. acnes*, along with excellent in vivo efficacy. VCD-004 has improved mode of DNA gyrase binding and additional binding with QBP, which could translate in lowering the resistance development propensity (as indicated by low MPC/MIC ratio) as compared to clindamycin, the currently used drug. It has optimal skin penetration and a potent anti-inflammatory impact via the reduction of pro-inflammatory cytokine (IL-6), independent of its antibacterial action.

### 3.3.12 Panacea Biotech

Panacea worked in the area of oxazolidinones. PBL 2270 oxazolidinone was in the preclinical developmental stage, but the exact structure of this compound is not known. A patent application (Jain 2008) by Panacea describes modifications of linezolid at morpholine ring and by incorporating phosphorus atom in place of oxygen of morpholine. In vitro MICs of 15 compounds are reported to be comparable or superior to linezolid. The MIC of compounds 30 and 62 (Fig. 3.18) is one-fold lower than that of the linezolid in MSSA and MRSA strains and 4x lower in *E. faecalis* (susceptible) and *E. faecium* (vancomycin-resistant).

### 3.3.13 Central Drug Research Institute

Central Drug Research Institute (CDRI), a leading government research laboratory, is also active in the development of antibacterial agents. Publications (Sashidhara et al. 2015) from CDRI described a new class of hybrids synthesized using a pharmacophore hybridization approach. A series of novel hybrids possessing chalcone and thiazole moieties were synthesized and evaluated for their antibacterial activities. This class of agents exhibited potency against *S. aureus* and in particular compound 27 (Fig. 3.19) exhibited potent inhibitory activity relative to other synthesized hybrids.

Furthermore, the haemolytic and toxicity data demonstrated that compound 27 was non-haemolytic and nontoxic to mammalian cells. The in vivo studies utilizing the *S. aureus* septicaemia model, demonstrated that compound 27 was as potent as vancomycin (Sashidhara et al. 2015). Another publication (Yadav et al. 2015) from CDRI describes tricyclic dihydrobenzoxazepine and tetracyclic indole derivatives, which specifically target bacterial DNA ligases without cross pharmacological interaction with human DNA ligase. Unfortunately, no further systematic development of these compounds is reported to date.

### 3.3.14 Wockhardt Limited

For over 20 years, Wockhardt has focused on developing end-to-end antibacterial drug discovery capability to discover novel antibacterial agents; to address several

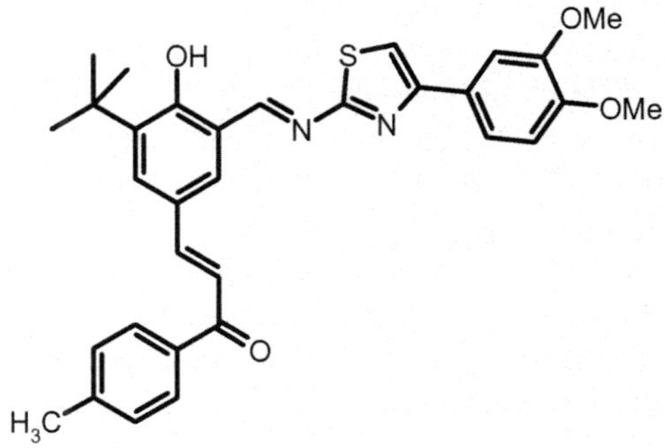


Compound 27

**Fig. 3.19** Chalcone and thiazole hybrid by CDRI

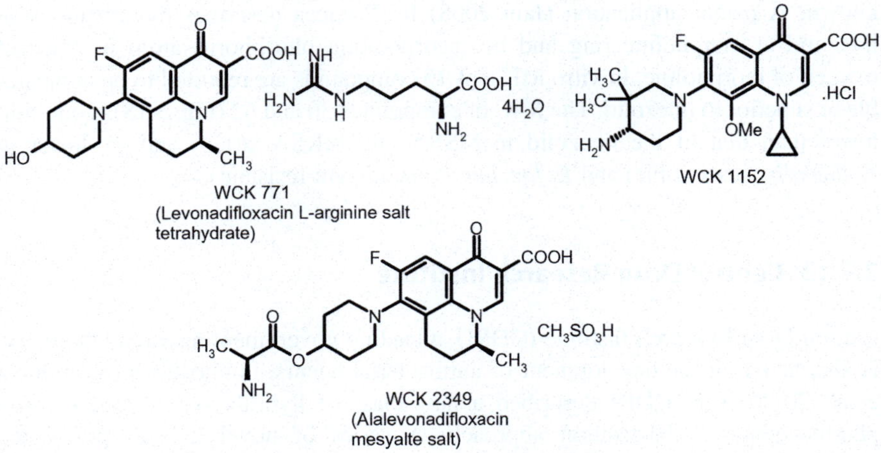


**Fig. 3.20** Fluoroquinolones by Wockhardt

unmet medical needs; and progress them to advance stages of clinical development. A discovery team of 125+ scientists at Wockhardt has invested its resources and efforts in designing new chemical entities (NCEs) aiming at clinically validated bacterial targets. This approach was guided by the fact that it ensures a reasonable level of *drugability* for a new drug candidate. Accordingly, discovery program at Wockhardt focused on targets such as bacterial DNA gyrase/topo IV inhibitors (fluoroquinolones), protein synthesis inhibitors (oxazolidinones and macrolide/ketolide), β-lactam β-lactamase inhibitor and non-β-lactam enhancers of β-lactam antibiotics in addition to β-lactamase inhibitors belonging to 1,6-diaza-bicycly-spiro [3.2.1] octanes class (DBO). Remarkably, from each class, the team was able to identify clinical candidates, with two of them gaining market approval and others progressing to the final Phase III stage of clinical development.

The salient chemical features of Wockhardt's NCEs are chiral molecules synthesized from noncommercial synthons. These molecules were put through discriminating biological tests at an early stage, instead of the sequential battery of traditional tests. The efforts have yielded six drug candidates; WCK 771/WCK 2349 (Fig. 3.20; for acute bacterial skin structure and soft tissue infections, Indian NDA approved); WCK 1152 (Fig. 3.20; for respiratory tract infections, development halted midway of phase 1); WCK 4873 (Fig. 3.22; for community-acquired pneumonia, global phase 2 completed successfully and slated to enter Phase III in India); WCK 4282 (for contemporary widely encountered MDR Gram-negative infections, global Phase III study to commence in second half of 2021) and WCK 5222 (for severe life-threatening XDR\MDR Gram-negative infections, Phase 3 scheduled in 2021). All the clinical phase NCEs of Wockhardt have been designated as QIDP (qualified infectious disease product) by the US FDA (Preston et al. 2019), based on their potential to treat a range of clinically important resistant pathogens.

Wockhardt's first clinical candidate WCK 771 (INN: levonadifloxacin; Fig. 3.20) represents a serendipitous find resulting from a quest for a safe broad-spectrum anti-MRSA drug with bactericidal action to address the gaps in the then prevailing well-established MRSA drugs—vancomycin/teicoplanin and daptomycin (de Souza et al. 2005). The major limitation perceived by the clinicians in the treatment of MRSA infections is the lack of oral therapy option and suboptimal performance (reported rates of clinical efficacy ~60%) of marketed anti-MRSA agents in the treatment of MRSA-caused pneumonia and blood stream infections. During 1999–2000, Wockhardt's investigations on a topical anti-acne cream, developed by the Japanese company Otsuka, based on a racemic mixture of nadifloxacin, revealed that its levorotatory isomer S-(-)-nadifloxacin shows excellent anti-MRSA activity and good tolerability in rodents when injected intravenously at relatively higher doses. Further research over the next 5 years led to the identification of injectable drug WCK 771, which is the first ever antibacterial agent in the form of amino acid arginine salt of S-(-)-nadifloxacin, designated as levonadifloxacin by WHO. WCK 771 was selected as a clinical candidate based on its superior injection site tolerability and safety at high intravenous doses. More than 3 years (2002–2006) of intensive prodrug-based research, involving the designing of >50 prodrugs, led to the discovery of the first ever antibacterial oral prodrug (alalevonadifloxacin, WCK 2349) (Patel 2007), which employs amino acid L-alanine as a prodrug moiety.

Another clinical candidate identified at Wockhardt was WCK 1152 (Fig. 3.20), 'S' enantiomer of racemic NCE 8-methoxy quinolone WCK 919. This compound showed potent activity against levofloxacin resistant *S. pneumoniae* strains (Al-Lahham et al. 2005) and progressed up to Phase-1 human clinical studies. On the other hand, 'R' enantiomer WCK 1153 showed unacceptable genotoxicity in preclinical studies. Even, the 'S' enantiomer WCK 1152 did not progress further because it caused unacceptable side effect of dose-dependent reversible visual disturbance, an uncommon adverse effect unlike that of the classical cardiac toxicity linked QT interval prolongation, usually associated with 8-methoxy fluoroquinolones.

To harness the advantage of anti-MRSA activity of levonadifloxacin and overcome vein irritation (phlebitis) at injection site observed with sodium salt, several amino acid salts such as L-lysine salt and L-arginine salt of levonadifloxacin (WCK 771) were prepared. The phlebitis free L-arginine salt of levonadifloxacin was selected as a clinical candidate. As an intravenous infusion, WCK 771 has undergone several Phase-1 and clinical pharmacology studies in the USA and Phase 2 and Phase 3 clinical trials in India. Based on successful completion of ABSSSI (acute bacterial skin structure and soft tissue infection) Phase 3 study in India, an NDA was filed with DCGI. In January 2020, Wockhardt received approval for manufacturing and marketing of levonadifloxacin and alalevonadifloxacin under the brand name EMROK.

Since levonadilfloxacin was not orally bioavailable due to its poor aqueous solubility and inefficient absorption, a prodrug approach was contemplated. Thus, amino acid ester-linked prodrugs of levonadifloxacin were explored for their improved oral administration potential, measured on the basis of oral bioavailability

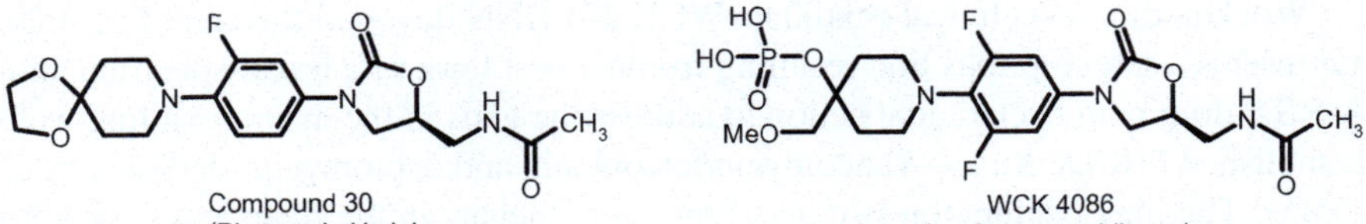

Compound 30
(Pharmacia UpJohn
US)

WCK 4086
(prodrug oxazoldinone)

**Fig. 3.21** Oxazolidinones by Wockhardt

(% of the orally-administered drug appearing in blood vs. injectable drug). Methane sulfonic acid salt of L-alanine ester of levonadifloxacin (WCK 2349) emerged as a prodrug of choice for oral delivery. It has >200 mg/mL aqueous solubility across the pH range encountered in the alimentary tract. L-alanine ester moiety helps for active absorption via intestinal amino acid receptors, and the ubiquitous presence of esterase enzymes efficiently releases the parent levonadifloxacin in the bloodstream. WCK 2349 has completed the Phase 1 clinical trial in the USA and Phase 2 and Phase 3 studies in India. Based on multiple nonclinical and clinical safety studies, both WCK 771 and WCK 2349 have emerged as safest fluoroquinolone known to date. (Bhagwat et al. 2019a, b; Appalaraju et al. 2020; Tellis et al. 2019).

Wockhardt's oxazolidinone program was aimed at inventing an NCE with a safety advantage over linezolid and PK commensurate to once-a-day dosing. A spiro ketal oxazolidinone series was explored to attain a better PK profile. This research was based on furthering a lead compound 30 reported by Pharmacia Upjohn (Yamada 1995). Compound 30 was less potent but exhibited good in vivo efficacy as an indication of good oral PK profile. The outcome of this oxazolidinone research was a phosphate ester prodrug of 4-hydroxy-4-methoxymethyl-piperidine oxazolidinone (PatilS 2007) (Fig. 3.21; WCK 4086) that exhibited superior PK, safety and tolerability over linezolid.

Wockhardt also worked in ketolide and 2-fluoroketolide antibacterial class, which was ridden with oral bioavailability, safety and tolerability issues. This was evident by the failure of several big pharma companies such as Johnson & Johnson, Aventis, GSK and Pfizer to progress their macrolide/ketolide lead candidates to Phase 3 or gain regulatory approval thereof. Wockhardt attempted to address these challenges by SAR driven improvement in potency, using hydrophilic moieties, without deploying 2-fluorine substituent. This approach was followed to minimize hepatic liability, which was ultimately achieved by introducing chiral methyl substituent in the side chain. This not only enabled adequate hepatic safety but also improved affinity to additional target in ribosomal domain II. Thus, a potent ketolide WCK 4873 (Fig. 3.22) active against high-level macrolide-resistant strains was identified (Deshpande et al. 2016a; Flamm et al. 2017; Rodvold et al. 2017; Takalkar 2016; Bader 2016). One of the salient features of Wockhardt ketolides is the replacement of traditional carbamate ring with chiral lactone ring, along with hydrophilic amidoxime arm in place of a conventional lipophilic *n*-butyl arm. The extra chiral centre arising from lactone arm adds specific direction and additional 'Z' stereochemistry of double bond of amidoxine function restricts free rotation. The overall

Fig. 3.22 Ketolide (Nafithromycin) by Wockhardt

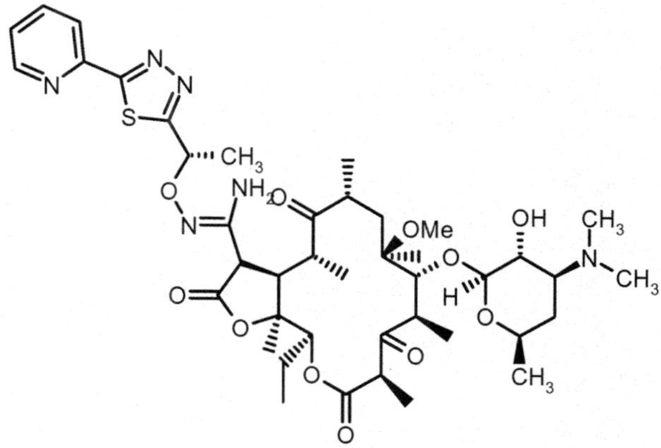


WCK 4873
(Nafithromycin)

hydrophilic character of the molecule provides a benefit of urinary elimination, which avoids excessive accumulation of the drug in the liver.

#### 3.3.14.1 1,6-Diaza-Bicycly-spiro[3.2.1] Octanes β-Lactam Enhancers

Aiming at overcoming MDR resistance in Gram-negative bacteria, the team at Wockhardt conceived two programs: (1) discovery of novel β-lactamase inhibitors targeting recently evolved β-lactamases, which are not inhibited by classical β-lactamase inhibitors (clavulanic acid, tazobactam, sulbactam and avibactam); and (2) finding compounds which could synergize with existing β-lactams without relying on β-lactamase inhibition, an approach based on novel β-lactam enhancer mechanism. Selection of novel β-lactamase inhibitors was contingent on the ability of NCEs to inhibit class D β-lactamases, specifically associated with newly emerged pathogen—*Acinetobacter baumannii*. On the other hand, to be designated as an enhancer, the β-lactamases stable non-β-lactam compounds were screened for PBP2 binding activity, which was later demonstrated to synergistically boosts the bactericidal effect of established β-lactam antibiotics primarily binding to PBP3. The non-β-lactam high-affinity PBP 2 binders (Fig. 3.23; based on DBO pharmacophore) represent the first ever completely man-made compounds selectively recognizing bacterial PBPs. Figure 3.24 depicts diverse naturally derived and semisynthetic PBP binding compounds and the year of their discovery, with non-β-lactam PBP binders shown in the centre.

Wockhardt's β-lactamase inhibitor program successfully identified two candidates: WCK 4234 and WCK 5061 (Bhagwat 2012). Both of these β-lactamase inhibitors are more potent than the marketed β-lactamase inhibitors, with regards to inhibition of serine-β-lactamases such as class C β-lactamases, which are widely prevalent in contemporary Gram-negative pathogens. It is noteworthy that WCK 4234 provided protection even against *Acinetobacter baumannii* associated serine-β-lactamases such as class D OXA carbapenemases and

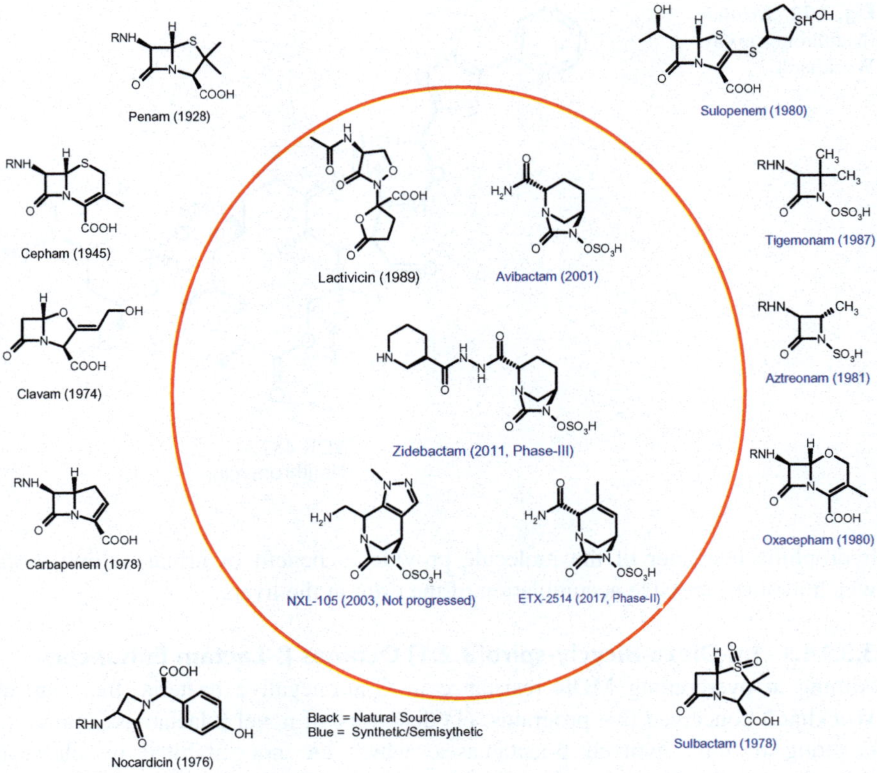


**Fig. 3.23** PBP binding β-lactam and non β-lactam pharmacophores

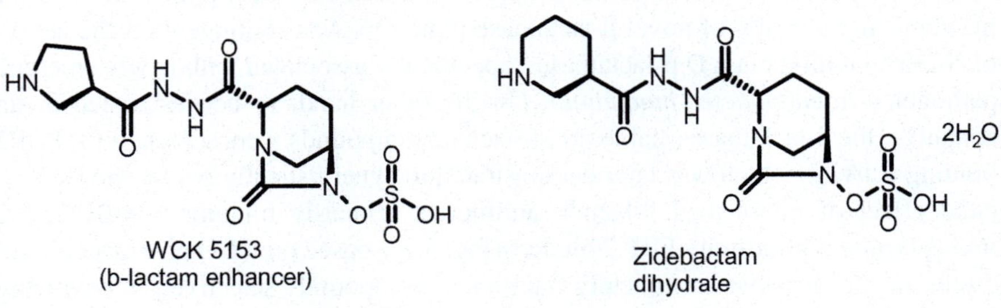


**Fig. 3.24** β-lactam enhancers by Wockhardt

Enterobacteriaceae associated class A and class D carbapenemases such as KPCs and OXA 48/181. However, these inhibitors were unable to inhibit metallo-β-lactamases (class B β-lactamase) due to the latter's significant structural and functional differences as compared to serine-β-lactamases. The combination of WCK 4234 with carbapenem antibiotics such as imipenem and meropenem provided in vivo efficacy against MDR *A. baumannii* infection (Patil 2014), not realizable with any of the marketed BL-BLI combination.

The second approach of identifying novel β-lactam enhancers emanated as an outcome of the team's realization that it would be a herculean task, if not impossible, to discover a β-lactamase inhibitor with an ability to inhibit all the four classes of β-lactamases. Unique in vitro and in vivo screening tactics employed in microbiology laboratories helped identify DBO-based NCEs showing high-affinity binding to Gram-negative PBP2. The medicinal chemistry team efficiently designed and synthesized several analogues to optimize their PBP 2 affinity, which reflected in terms of their standalone antibacterial activity. This led to the discovery of β-lactam enhancers WCK 5107 (zidebactam) and WCK 5153, both of which showed potent activity against a range of pathogens belonging to Enterobacteriaceae and *P. aeruginosa* (Deshpande et al. 2016b). The combination of PBP 3 binding diverse β-lactam antibiotics (such as penicillin, cephalosporin and monobactams) with PBP 2 binding β-lactam enhancers proved remarkably bactericidal to Gram-negative pathogens, irrespective of them expressing all four classes of β-lactamases and other non-enzymatic resistance mechanisms such as efflux and impermeability. The combination of zidebactam with fourth-generation cephalosporin cefepime was coded as WCK 5222. By now, several international publications describe the fascinating and ever-evolving antibacterial profile and safety features of WCK 5222 (Almarzoky Abuhussain et al. 2019; Avery et al. 2018, 2020; Kidd et al. 2020; Lepak et al. 2019; Livermore et al. 2017; Monogue et al. 2019; Mullane et al. 2019; Preston et al. 2019; Sader et al. 2017a, b).

The broadest spectrum of coverage of WCK 5222 is achieved through zidebactam-mediated pharmacodynamic enhancement (manifested as augmented bacterial killing in vitro and in vivo) of cefepime (Bhagwat et al. 2019c; Moya et al. 2019). In Gram-negative pathogens, WCK 5222 rapidly saturates all the essential PBPs such as PBP 1a/b, 2 and 3, leading to complete cessation of bacterial cell wall synthesis, causing pronounced bactericidal effect which is not possible with either cefepime alone or zidebactam alone or any other BL-BLI combinations. The rapidity of killing action of WCK 5222 parallels to that of carbapenems, the most powerful β-lactam antibiotics discovered to date. Studies demonstrating the rapid binding of cefepime and zidebactam to their respective PBP targets amid the presence of potent β-lactamases, explain the role of enhancer mechanism in surmounting multidrug resistance in diverse Gram-negative pathogens. Major drug regulatory authorities, US FDA, EMEA (Europe) and NMPA (China), have recognized the potential of WCK 5222 in addressing several toughest unmet needs, and have accordingly granted abridged development path involving the bypassing of conventional clinical Phase 2 study and directly undertaking clinical Phase 3 study. The enabling stand by regulatory agencies is a reflection of urgency caused by the worrying spread of AMR among Gram-negative pathogens and the inadequacy of current armamentarium.

Generally not highlighted, the development of a novel drug throws enormous challenges in API and formulation scale-up, which are required to be overcome efficiently to support time-bound clinical development. As an example, it took Wockhardt scientists more than a year to address multiple challenges involved in scaling up the manufacturing of sterile zidebactam with physical attributes such as a free-flowing nature and the absence of a tendency to form agglomerates. Such physical features were mandated by high-speed split-filling technology employed for manufacturing of WCK 5222 vials for Phase 3 study.

The highlights of antibacterial discovery programs undertaken in India over the past 50-odd years reveal that only a few molecules discovered could move to the clinical stage of development and even fewer novel antibacterial drugs progressed to an advanced stage of clinical development beyond India. The issue is no different at the international level, which is evident from the dwindling number of Phase 3 projects targeting MDR/XDR pathogens. Some of the discovery efforts were marred by a short-term and fragmented/patchy approach, which typically ended in reporting of sketchy in vitro data. Such a scenario highlights the issues of the clinical viability of the selected projects and the discovery team's ability to critically scrutinize the selected leads at every stage for go-no-go decision.

The successes of discovery programs critically rely on the scientific freedom bestowed to the team and the intra-team dynamics manifested in terms of good coordination in a spirit of harmony. Moreover, the attainment of discovery goals is dependent on access to resources provided by management. Wockhardt's remarkable success in antibacterial discovery is a result of several enabling elements that helped shape these discoveries. A few of these factors are discussed below:

## 3.4 Elements of Success in Wockhardt's Discovery Program

### 3.4.1 Scientific Freedom, Access to Resources and Organizational Capabilities

The discovery team was given complete scientific freedom on major issues and direction of the discovery program. This developed a trust between scientific leadership and management. Discovery core teams were assembled and an in-house animal housing facility was created to minimize the dependence on external sources. Existing organizational capabilities: scale-up team and a formulation research team (including respective GMP manufacturing teams) supported late-stage developmental activity such as large-scale synthesis of NCEs and NCE formulation development.

### 3.4.2 Team Building and Right Sizing

Discovery team was organized in two major sub-teams of chemistry and biology, to minimize hierarchy and create a free flow of information among scientist belonging to different disciplines. To sharpen in-house skills, relatively young and inexperienced scientists were encouraged to conduct experiments with drugs from known classes of antibiotics, to understand their features and compare findings with literature reports. Failed experiments were analyzed openly, which sometimes led to unexpected clues. A practice of sharing findings among teams was followed. This exercise led to a deeper understanding of the therapeutic area and generated a conducive milieu, which helped retain talent.

### 3.4.3 Understanding Gold Standard Antibiotics

The team undertook focused extensive experimentation on marketed gold standard drugs and newer agents in the global development pipeline. This imparted strength/weakness/opportunity analysis of standard drugs and developed experimental and interpretative skills, along with identifying an emerging unmet medical need. One of clinical candidate WCK 4282 (2 g cefepime combined with 2 g tazobactam in 1:1 proportion) is based on a finding that resistance mechanism of KPC can be neutralized by adding higher quantity of tazobactam to cefepime, but not with other injectable cephalosporins, while well-established marketed combination based on piperacillin employs tazobactam in a much lower 8:1 proportion.

### 3.4.4 Dynamics Between Classes of Antibiotics and Territorial Preference

Understanding about territorial preferences for certain classes of drugs was considered for the selection of antibacterial classes. For instance, as per market trend, fluoroquinolones are more prescribed in Japan, China and India for their wider indication profile and the dosing convenience (oral and injectable), compared to Europe or the USA. In Europe, carbapenems are preferred over cephalosporins while it is reverse in the USA. Some of the territorial preferences are due to differences in regulatory judgments of country-specific regulatory agencies. Considering the above dynamics, fluoroquinolone, oxazolidinone, macrolide and diazabicyclo octane classes were selected for the antibacterial research.

### 3.4.5 Differentiating Product Profile for Unmet Medical Needs

Unmet medical needs were identified for inventing newer agents with differentiating product profiles compared to existing products. For example, lack of bactericidal anti-MRSA agent, with IV to oral switchable option was identified as an unmet medical need. Fluoroquinolone class was selected for being bactericidal. Taking a clue from a Japanese publication, where unaffected MICs of (±)-nadifloxacin for Staphylococcal/MRSA isolates in light of increased MICs of other quinolones, WCK 771 and WCK 2349 projects were evolved deriving inspiration from nadifloxacin pharmacophore. The study revealed that affected quinolones such as ciprofloxacin and gatifloxacin were: (a) substrate of NorA+ efflux pump; and (b) preferentially bound to DNA topoisomerase IV, prone to mutation, whereas (±)-nadifloxacin and S-(-)-nadifloxacin are: (a) not a substrate of NorA+ efflux pump and; (b) prefer critical enzyme DNA gyrase as target over topoisomerase

IV. As a result, WCK 771 [L-arginine salt of (S)-nadifloxacin] was realized as the bactericidal intravenous anti-MRSA drug.

With a prodrug approach, taking a clue from marketed prodrug valacylovir, WCK 2349 [L-alanine ester of (S)-nadifloxacin mesylate] was identified as an oral anti-MRSA drug candidate. WCK 2349 did not generate any safety signal even at higher doses in animals and patients. Thus, the unmet medical need of IV plus oral bactericidal anti-MRSA agent was addressed.

Oxazolidinone class was also explored to evolve an anti-MRSA agent, as this class offered a novel mechanism of action leading to marketing success of linezolid. The unmet medical need identified was to overcome its class-specific limitations, such as toxicity arising from myelosuppression and MAO inhibition. To reduce toxicity, the research goals were identified as: (a) removing/reducing myelosuppression; and (b) optimizing PK profile to render once-a-day dose thereby reducing the daily drug load. For this, a specific toxicity screening model was developed and linezolid doses were identified to judge bone marrow safety of newer agents from this class. Simultaneously, in-house 'in vivo efficacious oxazolidinones' were screened through this model, ensuring exposures comparable to linezolid's 'bone marrow suppression inducing exposures'. The related impact on peripheral blood cell counts and bone marrow was examined and a structure-toxicity relationship was developed. The oxazolidinone WCK 4086 (phosphate prodrug of WCK 3023) was identified with features like linezolid comparable efficacy, good oral bioavailability, once-a-day dosing potential, metabolic stability in rat, dog and in human liver microsomes, and promising bone marrow safety.

For discovery program aimed at identifying respiratory antibiotics based on macrolide class, the mandatory differentiating attributes for a new product were identified as: (a) a new product must cover all types of resistances in respiratory pathogens; and (b) should minimize class-specific hepatotoxicity. Ketolide subclass of macrolide was chosen for research, as it is known to overcome the efflux mediated resistance mechanism. For other resistances, a screening test based on high potency against 'telithromycin-resistant pneumococci' was developed as an indicator of higher affinity towards the alternate binding site in domain II of ribosomal RNA. To ensure hepatic safety, a hepatotoxic dose of telithromycin was identified in rats based on a 7-day repeat dose toxicity study, while in parallel, liver and lung accumulations were measured. Potent ketolides with acceptable PK profile underwent this hepatic safety screening model. A comprehensive structure-toxicity relationship led to the identification of WCK 4873 (nafithromycin), which is active against the *ermB* gene harbouring *S. pneumoniae* that expresses a high level of macrolide resistance. Moreover, extensive preclinical PK studies helped establish a higher lung penetration of nafithromycin compared to the liver. The hepatic load was found to reduce further, as the drug has a renal elimination pathway ascribed to its hydrophilic character. Nafithromycin generates safe metabolites compared to telithromycin and solithromycin. In clinical studies, most of the attributes of nafithromycin were well realized.

While dealing with Gram-negative pathogens, since combinations based on novel β-lactamase inhibitors and β-lactam antibiotic have limitations of not protecting partner β-lactam antibiotic against β-lactamases of class B and D, treating pathogens harbouring all the four classes β-lactamases was identified as an unmet medical need. Designing novel universal β-lactamase inhibitor is fairly challenging and unrealizable in a reasonable period, because of the diversity of β-lactamase enzymes and varied mechanisms for hydrolysis of partner β-lactam antibiotics.

This problem was dealt with β-lactamase stable 'synergy driving NCE', which overcomes resistance through complementary PBP binding when combined with a suitable β-lactam antibiotic. A β-lactamase inhibitor, avibactam, belonging to diazabicyclo octane (DBO) class, was reported to have a weak affinity to Gram-negative PBP 2. Conceptually, it was thought that, if its PBP 2 affinity can be enhanced by 100× and then combined with a β-lactam antibiotic that binds to PBP 3, the combination could produce a potent synergy. Thus, novel DBO derivatives were tested for antibacterial activity against *E. coli* and Klebsiella strains. Standalone antibacterially active NCEs were further screened for concentration-dependent cell morphological changes for judging the extent of PBP 2 affinity. Among tested compounds, WCK 5107 (zidebactam) was identified as a potent selective PBP 2 binding agent. It displayed an antimicrobial spectrum that extended up to *P. aeruginosa*. The powerful affinity of the drug to PBP 2 was evident from its ability to promptly induce spheroplast formation (from characteristic rod form to oval spheroplasts) even at sub-MIC. Zidebactam synergized with a majority of penicillins, cephalosporins and monobactam aztreonam. Some of these combinations overcame all the known resistance mechanisms impacting β-lactam class of antibiotics. Its combination with cefepime (WCK 5222) became a novel mechanism-based clinical candidate. Another DBO analogue WCK 5153 showing features comparable to zidebactam with 2–4× improved potency against *P. aeruginosa* became a backup for WCK 5107.

### 3.4.6 Blending the Role of Experts and CROs with Internal Expertise

At various stages of drug development, specialized technical expertise is required. For instance, GLP toxicology studies need to be undertaken for IND/NDA filing as a regulatory requirement; independent microbiological studies are needed to confirm internal observations; radio-labelled drug synthesis, mass balance study, drug–drug interaction study are required at various stages of clinical development. Leading international experts are required for confirming the PK-PD adequacy of selected therapeutic dose for Phase 2 and Phase 3, and justifying breakpoints to regulatory bodies such as CLSI (USA) or EUCAST (Europe). At times, CROs and experts design their studies and interpret the results differently. Therefore, the role of the internal team leaders is to communicate constantly with external CROs/experts and harmonize the study findings to address critical data gaps. This collaborative aspect was handled successfully, which helped avoid regulatory delays.

### 3.4.7 The Role of Discovery Team in Supporting Clinical Development

Multidimensional understanding generated on a clinical candidate in preclinical development was leveraged in support of clinical development. For instance, safety, tolerability and metabolism-related observations helped in designing clinical study protocols. Similarly, blood-sampling time points and therapeutic dose to be employed for Phase 2/3 studies were identified on the basis of nonclinical PK-PD studies. The analytical team contributed to developing bio-analytical methods for plasma and body fluid samples collected to assemble human PK profile for the new drug.

### 3.4.8 Long-Term Engagement by Top Management

Some of the factors discouraging long-term commitments to antibacterial discovery are the increased cost of clinical development, complex regulatory requirements mandating indication-specific trials, and post-launch slower revenue growth for the first 4–5 years. Despite being aware of these factors, the long-term commitment showed by the top management contributed to the successful and meaningful outcome of the research.

### 3.4.9 De-risking Projects

Intending to minimize the probability of failures at the clinical developmental stage, certain de-risking approaches were an integral part of the Wockhardt discovery programs. The de-risking was ensured by focusing on time-tested, well-established antibacterial classes with known liabilities and developmental history providing clues for risk mitigation, rather than researching an entirely new class. The known classes have well-understood developmental path based on familiar PK/PD patterns, toxicity issues and difficulties encountered during development. The solutions to these issues were often known. Further de-risking comes from an approach of firmly relying on the outcome of laboratory-based experiments rather than theoretical predictions. Detailed in vitro and in vivo pharmacology helped in unravelling the finer aspects of development. For example, NCEs with higher MICs were evaluated in secondary and tertiary screens for unravelling other interesting features and generating multi-parametric SARs. Such an approach helped in understanding the dynamics between potency, PK and in vivo efficacy. Additional de-risking measures included avoiding over-reliance on cell-free screening assays ($IC_{50}$) and in-silico modelling rather than emphasis was put on efficacy, PK-PD investigations and safety assessment based on in vivo tests.

## 3.5 Urgency to Expand Antibiotic Discovery Initiatives in India

The combination of India centric and international factors contributes to the urgency for creating an efficient R&D based mechanism in India, to effectively deal with the threat posed by AMR. India ranks high on the charts of the infectious disease burden, and the judicious use of effective antibiotics is warranted. Bill and Melinda Gates Foundation surveillance has shown that India has one of the highest levels of resistance to all antibiotics (including carbapenems) and there is an alarming spread of multidrug-resistant organisms.

Taking note of the impact of AMR on health care, the WHO has identified AMR as a major healthcare threat. It has also identified several pathogens for which newer antibacterial agents are required urgently. As major pharmaceutical companies close their antibacterial discovery programs due to the unattractive business model as compared to other therapeutic areas, the pace of introduction of new antibiotics addressing unmet needs is quite slow. This situation calls for urgent measures to invigorate antibiotic discovery research capability within India, so that it would propel the country into the league of major antibiotic discovery hub, an area in which even China has yet to demonstrate the significant capability of progressing discoveries to an advanced stage of international development.

## 3.6 The Way Forward

The country must create self-sufficiency of safe and effective antibacterial agents. For this, high-calibre antibacterial discovery and development programs need to be supported by enabling national incentives/funding. It should be possible to identify projects worthy of national-level funding by employing objective criteria. The scrutiny of objective parameters such as clinical viability of the project, independent studies supporting the project concept, favourable review by international regulatory agencies and the safety outcomes from phase 1 study, serve as reliable tools to identify projects worthy of funding. These parameters would ensure that only high-impact projects are qualified for national funding. In this context, it is worth noting that the initiatives such as DST extending financial loans for undertaking clinical studies approved by DCGI provide the much-needed push incentive for antibiotic development.

As a longer-term strategy, two or three dedicated antibacterial research institutions competing with each other could be set up in India. This strategy would generate a competent national pool of scientists that would sustain and refine the art and science of antibiotic discovery and development.

Another area that needs careful attention pertains to building a dedicated competent regulatory mechanism for the evaluation and approval of novel antibiotics relevant to India. The regulatory team should have adequate representation of: (a) clinical microbiologists with a thorough background of diverse resistance mechanisms and pathogen-specific epidemiology; (b) clinicians with a deeper insight of management of infectious disease; and (c) PK/PD experts to judge the adequacy of selected clinical doses for a novel antimicrobial agent. A high-quality regulatory guidance would act as nudging force for the competing discovery teams, and would also help in implementing the course correction, as and when needed, during the transition of a new drug from nonclinical to clinical stage. Through these long-term measures, India would continue to serve its own need for antibiotics on an on-going basis while aspiring to become a global hub of antibiotic discovery and development.

**Acknowledgements** Authors sincerely thank Dr. Habil Khorakiwala, who graciously permitted inclusion of finer aspects of Wockhardt's drug discovery approach that might be of wider interest to the community of discovery scientists in India and elsewhere. We would also like to thank Dr. Vijay Patil, and Dr. Satish Bhavsar, who helped in preparing certain sections and chemical structures of novel antibacterials.

## References

Aaramadaka SKR, Guha MK, Prabhu G, Kini SG, Vijayan M (2007) Synthesis and evaluation of urea and thiourea derivatives of oxazolidinones as antibacterial agents. Chem Pharm Bull 55 (2):236–240. https://doi.org/10.1248/cpb.55.236

Al-Lahham A, De Souza NJ, Patel M, Rene Reinert R (2005) Activity of the new quinolones WCK 771, WCK 1152 and WCK 1153 against clinical isolates of Streptococcus pneumoniae and Streptococcus pyogenes. J Antimicrob Chemother 56(6):1130–1133. https://doi.org/10.1093/jac/dki361

Almarzoky Abuhussain SS, Avery LM, Abdelraouf K, Nicolau DP (2019) In vivo efficacy of humanized WCK 5222 (Cefepime-Zidebactam) exposures against carbapenem-resistant Acinetobacter baumannii in the neutropenic thigh model. Antimicrob Agents Chemother 63 (1):e01931. https://doi.org/10.1128/AAC.01931-18

API Synthesis International (2016) ACS Symposium. Recent Advances in Drug Development, 11–12 November 2016, Hyderabad, India. http://apisynthesisint.blogspot.com/

Appalaraju B, Baveja S, Baliga S, Shenoy S, Bhardwaj R, Kongre V, Dattatraya GS, Dhole T, Verma B, Mukherjee DN, Gupta S, Shanmugam P, Iravane J, Mishra SR, Barman P, Chopra S, Hariharan M, Surpam R, Pratap R, Joshi P, Khande H, Mane A, Jain R, Bhagwat S (2020) In vitro activity of a novel antibacterial agent, levonadifloxacin, against clinical isolates collected in a prospective, multicentre surveillance study in India during 2016-18. J Antimicrob Chemother 75(3):600–608. https://doi.org/10.1093/jac/dkz493

Avery LM, Abdelraouf K, Nicolau DP (2018) Assessment of the in vivo efficacy of WCK 5222 (Cefepime-Zidebactam) against carbapenem-resistant Acinetobacter baumannii in the neutropenic murine lung infection model. Antimicrob Agents Chemother 62(11):e00948. https://doi.org/10.1128/AAC.00948-18

Avery LM, Mullane EM, Nicolau DP (2020) Evaluation of the in vitro activity of WCK 5222 (cefepime/zidebactam) and currently available combination therapies against single- and double-carbapenemase producing Enterobacteriaceae: expanding the zone of hope. Int J Antimicrob Agents 55(2):105863. https://doi.org/10.1016/j.ijantimicag.2019.105863

Bader JC (2016) Pharmacokinetic-pharmacodynamic target attainment analyses to support nafithromycin (WCK 4873) dose selection for the treatment of community-acquired bacterial pneumonia. Paper presented at the ASM Microbe, Boston, USA

Balasubramanian G (2017) Carbapenem compounds. WIPO (PCT) Patent WO2017158616A1

Barman TK, Kumar M, Mathur T, Chaira T, Ramkumar G, Kalia V, Rao M, Pandya M, Yadav AS, Das B, Upadhyay DJ, Hamidullah KR, Raj VS, Singh H (2016) In vitro and in vivo activities of

a bi-aryl oxazolidinone, RBx 11760, against gram-positive bacteria. Antimicrob Agents Chemother 60(12):7134–7145. https://doi.org/10.1128/AAC.00453-16
Bhagwat S (2012) 1,6- diazabicyclo [3,2,1] octan- 7 - one derivatives and their use in the treatment of bacterial infections. WIPO (PCT) Patent WO2013030735A1
Bhagwat SS, Periasamy H, Takalkar SS, Chavan R, Tayde P, Kulkarni A, Satav J, Zope V, Patel M (2019a) In vivo pharmacokinetic/pharmacodynamic targets of levonadifloxacin against staphylococcus aureus in a neutropenic murine lung infection model. Antimicrob Agents Chemother 63(8):e00909. https://doi.org/10.1128/AAC.00909-19
Bhagwat SS, Nandanwar M, Kansagara A, Patel A, Takalkar S, Chavan R, Periasamy H, Yeole R, Deshpande PK, Bhavsar S, Bhatia A, Ahdal J, Jain R, Patel M (2019b) Levonadifloxacin, a novel broad-spectrum anti-MRSA benzoquinolizine quinolone agent: review of current evidence. Drug Des Devel Ther 13:4351–4365. https://doi.org/10.2147/DDDT.S229882
Bhagwat SS, Periasamy H, Takalkar SS, Palwe SR, Khande HN, Patel MV (2019c) The novel beta-lactam enhancer zidebactam augments the in vivo pharmacodynamic activity of cefepime in a neutropenic mouse lung acinetobacter baumannii infection model. Antimicrob Agents Chemother 63(4):e02146. https://doi.org/10.1128/AAC.02146-18
Chatterjee S, Chatterjee DK, Jani RH, Blumbach J, Ganguli BN, Klesel N, Limbert M, Seibert G (1992) Mersacidin, a new antibiotic from Bacillus. In vitro and in vivo antibacterial activity. J Antibiot (Tokyo) 45(6):839–845. https://doi.org/10.7164/antibiotics.45.839
Chatterjee S, Nadkarni SR, Vijayakumar EK, Patel MV, Ganguli BN, Fehlhaber HW, Vertesy L (1994) Napsamycins, new Pseudomonas active antibiotics of the mureidomycin family from Streptomyces sp. HIL Y-82,11372. J Antibiot (Tokyo) 47(5):595–598. https://doi.org/10.7164/antibiotics.47.595
Das B, Rudra S, Yadav A, Ray A, Rao AV, Srinivas AS, Soni A, Saini S, Shukla S, Pandya M, Bhateja P, Malhotra S, Mathur T, Arora SK, Rattan A, Mehta A (2005) Synthesis and SAR of novel oxazolidinones: discovery of ranbezolid. Bioorg Med Chem Lett 15(19):4261–4267. https://doi.org/10.1016/j.bmcl.2005.06.063
Deshpande PK, Tadiparthi R, Bhawsar SB, Patil VJ, Gupta SV, Pawar SS, Pavase LS, Kale RP, Deshmukh VV, Dhabade SK, Dekhane DV, Shaikh MU, Yeole RD, Patil KR, Rane VP, Trivedi BK (2016a) WCK 4873 (INN: Nafithromycin): structure activity relationship (SAR) identifying a lactone ketolide with activity against Telithromycin-resistant (Tel-R) Pneumococci (SPN) and S. pyogenes (SPY). Paper presented at the ASM MICROBE, Boston, USA
Deshpande P, Bhavsar S, Joshi S, Pawar S, Kale R, Mishra A, Jadhav S, Pavase L, Gupta S, Yeole R, Rane V, Ahirrao V, Bhagwat S, Patel MV (2016b) WCK 5107 (Zidebactam, ZID): structure activity relationship (SAR) of Novel Bicyclo acyl Hydrazide (BCH) pharmacophore active against gram-negatives including Pseudomonas (PA). Paper presented at the ASM MICROBE, Boston, USA
Flamm RK, Rhomberg PR, Sader HS (2017) In vitro activity of the novel lactone ketolide nafithromycin (WCK 4873) against contemporary clinical bacteria from a global surveillance program. Antimicrob Agents Chemother 61(12):e01230–e01217. https://doi.org/10.1128/AAC.01230-17
Gadekar PK, Roychowdhury A, Kharkar PS, Khedkar VM, Arkile M, Manek H, Sarkar D, Sharma R, Vijayakumar V, Sarveswari S (2016) Design, synthesis and biological evaluation of novel azaspiro analogs of linezolid as antibacterial and antitubercular agents. Eur J Med Chem 122:475–487. https://doi.org/10.1016/j.ejmech.2016.07.001
Ghosh S, Sinha M, Bhattacharyya A, Sadhasivam S, Megha J, Reddy S, Saini S, Singh H, Kumar D, Kaur SP, Mishra M, Usharani D, Ghosh S, Sengupta S (2018) A rationally designed multifunctional antibiotic for the treatment of drug-resistant acne. J Invest Dermatol 138(6):1400–1408. https://doi.org/10.1016/j.jid.2017.11.041
Hameed PS, Sharma S, Katagihallimath N, Jayaraman R, Balasubramanian V (2018a) A broad spectrum novel bacterial topoisomerase inhibitor (NBTI): killing kinetics, ADME and in vivo efficacy. Paper presented at the ASM Microbe Atlanta, 7–11 June 2018

Hameed PS, Bharatham N, Katagihallimath N, Sharma S, Nandishaiah R, Shanbhag AP, Thomas T, Narjari R, Sarma M, Bhowmik P, Amar P, Ravishankar R, Jayaraman R, Muthan K, Subbiah R, Ramachandran V, Balasubramanian V, Datta S (2018b) Nitrothiophene carboxamides, a novel narrow spectrum antibacterial series: mechanism of action and efficacy. Sci Rep 8(1):7263. https://doi.org/10.1038/s41598-018-25407-7

Jain R (2008) Novel antimicrobials. WIPO (PCT) Patent WO2008090570A1

Khodade VS, Sharath Chandra M, Banerjee A, Lahiri S, Pulipeta M, Rangarajan R, Chakrapani H (2014) Bioreductively activated reactive oxygen species (ROS) generators as MRSA inhibitors. ACS Med Chem Lett 5(7):777–781. https://doi.org/10.1021/ml5001118

Kidd JM, Abdelraouf K, Nicolau DP (2020) Efficacy of human-simulated bronchopulmonary exposures of cefepime, zidebactam and the combination (WCK 5222) against MDR Pseudomonas aeruginosa in a neutropenic murine pneumonia model. J Antimicrob Chemother 75 (1):149–155. https://doi.org/10.1093/jac/dkz414

Kumar M, Mathur T, Barman TK, Ramkumar G, Bhati A, Shukla G, Kalia V, Pandya M, Raj VS, Upadhyay DJ, Vaishnavi C, Chakrabarti A, Das B, Bhatnagar PK (2012) In vitro and in vivo activities of the novel Ketolide RBx 14255 against Clostridium difficile. Antimicrob Agents Chemother 56(11):5986–5989. https://doi.org/10.1128/AAC.00015-12

Lepak AJ, Zhao M, Andes DR (2019) WCK 5222 (Cefepime/Zidebactam) pharmacodynamic target analysis against metallo-beta-lactamase producing enterobacteriaceae in the neutropenic mouse pneumonia model. Antimicrob Agents Chemother 63:e01648. https://doi.org/10.1128/AAC.01648-19

Livermore DM, Mushtaq S, Warner M, Vickers A, Woodford N (2017) In vitro activity of cefepime/zidebactam (WCK 5222) against Gram-negative bacteria. J Antimicrob Chemother 72(5):1373–1385. https://doi.org/10.1093/jac/dkw593

Lohray BB, Lohray VB, Srivastava BK, Kapadnis PB, Pandya P (2004a) Novel tetrahydro-thieno pyridyl oxazolidinone: an antibacterial agent. Bioorg Med Chem 12(17):4557–4564. https://doi.org/10.1016/j.bmc.2004.07.019

Lohray BB, Lohray VB, Srivastava BK, Gupta S, Solanki M, Kapadnis P, Takale V, Pandya P (2004b) Oxazolidinone: search for highly potent antibacterial. Bioorg Med Chem Lett 14 (12):3139–3142. https://doi.org/10.1016/j.bmcl.2004.04.024

Mahajan GB (2009) Novel antibacterial compounds. Europe Patent EP2010202A2, 7 Jan 2009

Mathur T, Kumar M, Barman TK, Kumar GR, Kalia V, Singhal S, Raj VS, Upadhyay DJ, Das B, Bhatnagar PK (2011) Activity of RBx 11760, a novel biaryl oxazolidinone, against Clostridium difficile. J Antimicrob Chemother 66(5):1087–1095. https://doi.org/10.1093/jac/dkr033

Mehta A (2001) Oxazolidinone derivatives as antimicrobials disease. United States Patent US20020103186A1

Michael D, Huband JML, Amy A, Watters HK, Becker VR, Balasubramanian V, Flamm RK (2019) In vitro activity of BWC0977 (a novel bacterial topoisomerase inhibitor) and comparators against recent clinical and molecularly characterized enterobacteriaceae and non-fermenter isolates from the United States and Europe. Paper presented at the ECCMID, Amsterdam, 13–16 April 2019

Monogue ML, Tabor-Rennie J, Abdelraouf K, Nicolau DP (2019) In vivo efficacy of WCK 5222 (Cefepime-Zidebactam) against multidrug-resistant Pseudomonas aeruginosa in the neutropenic murine thigh infection model. Antimicrob Agents Chemother 63(7):e00233. https://doi.org/10.1128/AAC.00233-19

Moya B, Barcelo IM, Cabot G, Torrens G, Palwe S, Joshi P, Umarkar K, Takalkar S, Periasamy H, Bhagwat S, Patel M, Bou G, Oliver A (2019) In vitro and in vivo activities of beta-lactams in combination with the novel beta-lactam enhancers zidebactam and WCK 5153 against multidrug-resistant metallo-beta-lactamase-producing Klebsiella pneumoniae. Antimicrob Agents Chemother 63(5):e00128. https://doi.org/10.1128/AAC.00128-19

Mullane EM, Avery LM, Nicolau DP (2019) Comparative evaluation of the in vitro activities of WCK 5222 (cefepime/zidebactam) and combination antibiotic therapies against carbapenem-resistant Pseudomonas aeruginosa. Antimicrob Agents Chemother 64:e01669. https://doi.org/10.1128/AAC.01669-19

Nadkarni SR, Patel MV, Chatterjee S, Vijayakumar EK, Desikan KR, Blumbach J, Ganguli BN, Limbert M (1994) Balhimycin, a new glycopeptide antibiotic produced by Amycolatopsis sp. Y-86,21022. Taxonomy, production, isolation and biological activity. J Antibiot (Tokyo) 47(3):334–341. https://doi.org/10.7164/antibiotics.47.334

Palanisamy SU (2010) 2-substituted methyl penam derivatives. United States Patent US7687488B2

Palanisamy SU (2014) Compounds and their use. United States Patent US20140057888A1, 27 Feb 2014

Patel MV (2007) Prodrugs of benzoquinolizine-2-carboxylic acid. WIPO (PCT) Patent WO2007102061A2

Patil VJ (2014) Nitrogen containing compounds and their use. United States Patent US 2014/0296526

PatilS VJ (2007) Substituted piperidinophenyl oxazolidinones. United States Patent US8288416B2

Paul-Satyaseela M, Solanki SS, Sathishkumar D, Bharani T, Magesh V, Rajagopal S (2009) In vitro antibacterial activities of a novel oxazolidinone, OCID0050. J Antimicrob Chemother 64 (4):797–800. https://doi.org/10.1093/jac/dkp300

Paul-Satyaseela M, Koppolu K, Senthilkumar UP, Shapiro S (2014) Efficacy of the extended spectrum β-lactamase inhibitor, AAI101, combined with β-lactams in murine models of systemic Gram-negative infection. Paper presented at the 2014 ICAAC, Washington, USA

Preston RA, Mamikonyan G, DeGraff S, Chiou J, Kemper CJ, Xu A, Mastim M, Yeole R, Chavan R, Patel A, Friedland HD, Bhatia A (2019) Single-center evaluation of the pharmacokinetics of WCK 5222 (Cefepime-Zidebactam combination) in subjects with renal impairment. Antimicrob Agents Chemother 63(1):e01484. https://doi.org/10.1128/AAC.01484-18

Rangarajan R (2012) Heterocyclic compounds as inhibitors of fatty acid biosysnthesis for bacterial infections. United States Patent US 9115149 B2

Rodvold KA, Gotfried MH, Chugh R, Gupta M, Friedland HD, Bhatia A (2017) Comparison of plasma and intrapulmonary concentrations of nafithromycin (WCK 4873) in healthy adult subjects. Antimicrob Agents Chemother 61(9):e01096–e01017. https://doi.org/10.1128/aac.01096-17

Sader HS, Castanheira M, Huband M, Jones RN, Flamm RK (2017a) WCK 5222 (Cefepime-Zidebactam) antimicrobial activity against clinical isolates of gram-negative bacteria collected worldwide in 2015. Antimicrob Agents Chemother 61(5):e00072. https://doi.org/10.1128/AAC.00072-17

Sader HS, Rhomberg PR, Flamm RK, Jones RN, Castanheira M (2017b) WCK 5222 (cefepime/zidebactam) antimicrobial activity tested against Gram-negative organisms producing clinically relevant beta-lactamases. J Antimicrob Chemother 72(6):1696–1703. https://doi.org/10.1093/jac/dkx050

Sanchez ML, Wenzel RP, Jones RN (1992) In vitro activity of decaplanin (M86-1410), a new glycopeptide antibiotic. Antimicrob Agents Chemother 36(4):873–875. https://doi.org/10.1128/aac.36.4.873

Sashidhara KV, Rao KB, Kushwaha P, Modukuri RK, Singh P, Soni I, Shukla PK, Chopra S, Pasupuleti M (2015) Novel chalcone-thiazole hybrids as potent inhibitors of drug resistant Staphylococcus aureus. ACS Med Chem Lett 6(7):809–813. https://doi.org/10.1021/acsmedchemlett.5b00169

de Souza NJ, Gupte SV, Deshpande PK, Desai VN, Bhawsar SB, Yeole RD, Shukla MC, Strahilevitz J, Hooper DC, Bozdogan B, Appelbaum PC, Jacobs MR, Shetty N, Patel MV, Jha R, Khorakiwala HF (2005) A chiral benzoquinolizine-2-carboxylic acid arginine salt active against vancomycin-resistant Staphylococcus aureus. J Med Chem 48(16):5232–5242. https://doi.org/10.1021/jm050035f

Sreenivas K, Amarnath PV, Mallik A, Sarnaik H, Kumar NS, Takhi M, Trehan S, Kumar MS, Iqbal J, Rajagopalan R, Chakrabarti R (2007) In vitro and in vivo antibacterial evaluation of DRF 8417, a new oxazolidinone. J Antimicrob Chemother 60(1):159–161. https://doi.org/10.1093/jac/dkm116

Takalkar SS (2016) WCK 4873 (Nafithromycin): PK/PD analysis against S. pneumoniae (SPN) including telithromycin-non-susceptible (TEL-NS) strains through murine-lung infection model. Paper presented at the ASM Microbe, Boston, USA

Takhi M (2015) Substituted pyridine derivatives as FabI inhibitors. United States Patent US9062002B2

Takhi M, Sreenivas K, Reddy CK, Munikumar M, Praveena K, Sudheer P, Rao BNVM, Ramakanth G, Sivaranjani J, Mulik S, Reddy YR, Narasimha Rao K, Pallavi R, Lakshminarasimhan A, Panigrahi SK, Antony T, Abdullah I, Lee YK, Ramachandra M, Yusof R, Rahman NA, Subramanya H (2014) Discovery of azetidine based ene-amides as potent bacterial enoyl ACP reductase (FabI) inhibitors. Eur J Med Chem 84:382–394. https://doi.org/10.1016/j.ejmech.2014.07.036

Tellis M, Joseph J, Khande H, Bhagwat S, Patel M (2019) In vitro bactericidal activity of levonadifloxacin (WCK 771) against methicillin- and quinolone-resistant Staphylococcus aureus biofilms. J Med Microbiol 68(8):1129–1136. https://doi.org/10.1099/jmm.0.000999

Thirumalachar MJ (1966) Process of producing hamycin antibiotic and product produced. United States Patent US3261751A

Yadav N, Khanam T, Shukla A, Rai N, Hajela K, Ramachandran R (2015) Tricyclic dihydrobenzoxazepine and tetracyclic indole derivatives can specifically target bacterial DNA ligases and can distinguish them from human DNA ligase I. Org Biomol Chem 13 (19):5475–5487. https://doi.org/10.1039/c5ob00439j

Yamada H (1995) Oxazolidinone derivatives and pharmaceutical compositions containing them. WIPO (PCT) Patent WO1995025106A1

**Mahesh V. Patel** is involved in antibiotic discovery research since more than 45 years, including 21 years at Hoechst Research Centre, Mumbai, 4 years at Ranbaxy Laboratories, New Delhi and 22 years at Wockhardt Research Center, Aurangabad in Mumbai. His work at Hoechst led to the discovery of four novel antibiotics. Over the past 22 years (1998-till date), he has been leading the antibiotic discovery team at Wockhardt, Aurangabad culminating in to six clinical candidates for which US INDs have been accepted and QIDP (Qualified Infectious Disease Product) awarded for all of them. Discovery team recently gained first time ever market approval for two novel antibiotics in India. Dr. Patel is the inventor/co-inventor of more than 50 granted US Patents. He earned his Ph.D. degree from the University of Bombay in 1985 and postdoctoral research focused on HIV interaction with cell surface cofactors.

**Sachin S. Bhagwat** has 25 years of experience in the discovery and development of novel antibiotics. He began his discovery research career at Ranbaxy Laboratories, New Delhi (1995–1998) and later joined antibiotic discovery team at Wockhardt Research Center, Aurangabad (1998–till date) eventually leading the antibiotic pharmacology function. Dr. Bhagwat's research interest includes, historical discovery approaches, unmet need in bacterial infections, mechanism of action and resistance, Phase III dose identifying PK/PD studies, and international collaborative research. Dr. Bhagwat is the inventor/co-inventor in 40 + granted US/EU patents covering five major classes, fluoroquinolones, oxazolidinones, macrolide/ketolides, β-lactams and diazabicycloctane and has authored several research articles. He earned his Ph.D. degree in 2002 (Bhopal University) on pharmacodynamic efficacy determinants of quinolone antibacterials.

**Prasad K. Deshpande** obtained his Ph.D. in 1993 from Indian Institute of Chemical Technology, Hyderabad, India, in synthetic organic chemistry and later on postdoctoral research at North Dakota State University, USA (1993–1996). Having gained experience in optimizing manufacturing of active pharmaceutical ingredients, he led Wockhardt medicinal chemistry efforts in designing and synthesis of NCEs which culminated in various international patents, publications as well as clinical stage novel antibiotics.

# 4 New Drug Discovery and Development in India to Counter Malaria

Niti Kumar and Saman Habib

## 4.1 Introduction

Malaria is a vector-borne disease caused by the *Plasmodium* spp. in humans and animals. This protozoan infection has immense social and economic impact in endemic regions across globe. An estimated 219 million cases of malaria were reported in 2017 with highest prevalence in Africa (92%), followed by South-East Asia (5%) and Eastern Mediterranean (2%) region. According to the World Health Organisation's (WHO) World Malaria Report 2018, 15 countries in sub-Saharan Africa and India contribute to 80% of the global malaria burden. India contributes 90% of total malaria cases in South-East Asia and ~50% of *P. vivax* cases worldwide. Despite rigorous efforts in malaria elimination programmes, there are intermittent barriers which affect the long-term implementation of intervention strategies. Further, the emerging cases of drug resistance against current frontline drugs have raised a health alarm underscoring the need for alternative antimalarial intervention strategies. Other impediments in malaria elimination programmes are socioeconomic disparity, suboptimal diagnosis and treatment which often lead to transplacental transmission of parasite in pregnant mothers, noncompliance to drug regimen

This is CDRI communication no. 10130.

N. Kumar
Division of Molecular Parasitology and Immunology, CSIR-Central Drug Research Institute, Lucknow, India
e-mail: niti.kumar@cdri.res.in

S. Habib (✉)
Division of Molecular and Structural Biology, CSIR-Central Drug Research Institute, Lucknow, India
e-mail: saman_habib@cdri.res.in



M. Dikshit (ed.), *Drug Discovery and Drug Development*,
https://doi.org/10.1007/978-981-15-8002-4_4

which contributes to parasite recrudescence/relapse, and nonavailability of a potent vaccine against malaria. In the era of globalization, increased movement of people brings asymptomatic reservoir from endemic regions into non-endemic areas resulting in transmittance of resistance across geographical boundaries. Further, climate change has affected vector population size and density, vector survival, and parasite transmission rates.

For the treatment of malaria, WHO recommends Artemisinin combination therapy (ACT) regimens for 3-day treatment with following the drug combinations (1) artemether + lumefantrine; (2) artesunate + amodiaquine; (3) artesunate + mefloquine; (4) dihydroartemisin (DHA) + piperaquine (PPQ) (for children <25 kg weight) and artesunate + sulfadoxine-pyrimethamine. Delayed clearance upon treatment or "partial resistance" to Artemisinin has been detected in five countries of the Greater Mekong Subregion [Cambodia, Lao People's Democratic Republic (PDR), Myanmar, Thailand and Vietnam] (Imwong et al. 2017a, b; Ménard et al. 2018). Some areas of Greater Mekong Subregion have also reported resistance against ACT partner drugs, regardless of the presence of artemisinin partial resistance, thus leading to treatment failure. Delayed clearance time has been reported for dihydroartemisinin and piperaquine in Cambodia and South of Vietnam (Spring et al. 2015; Thanh et al. 2017); this is accompanied by significantly higher $IC_{50}$s of isolates with delayed clearance ($\geq$72 h) than those with normal clearance times for chloroquine, DHA, and PPQ (Thriemer et al. 2014). Resistance toward sulfadoxine-pyrimethamine has been reported in Republic of Congo (Alker et al. 2008), Sudan (A-Elbasit et al. 2006) and India (Kumar et al. 2015). There is also a geographical overlap of malaria co-infections with HIV and tuberculosis, with limited information available on how HIV or tuberculosis modifies therapeutic response to ACTs.

Even though India has made progress in controlling malaria, detection of malaria cases with co-infection of Dengue and Japanese Encephalitis Virus is of concern (Dev 2019; Sahu et al. 2016). Until 2016, the country remained under the category of malaria "controlling," not malaria "eliminating," nations (Newby et al. 2016). The National Strategic Plan for Malaria Elimination in India (2017–2022) has formulated effective disease management (diagnosis, preventive/curative interventions), surveillance (epidemiological, entomological) and awareness programmes to make India malaria-free by 2027. High-risk areas for malaria in India include Odisha, Chattisgarh, Jharkhand, Madhya Pradesh, Maharashtra, and north-eastern states (especially Assam, Tripura, Mizoram, and Meghalaya). Most cases of human malaria infections are reported for *P. falciparum* (prevalent in forest and peripheral areas) and *P. vivax* (prevalent in plains). The disease is transmitted by *Anopheles* spp. (*A. stephensi*, *A. culicifacies*, *A. fluviatilis,* etc.) which have heterogeneous distribution throughout the country. The various challenges for malaria elimination in India are: (a) shortage of skilled human resource to participate in and coordinate malaria elimination programmes, (b) limited access of health services at point-of-care (remote, conflict-affected or endemic areas), (c) ineffective vector control programmes and widespread resistance to insecticides, (d) movement of asymptomatic reservoir across states and international boundaries including from countries with reported artemisinin resistance, (e) lack of effective public–private partnership for monitoring antimalarial drug quality, efficacy which are critical for sustenance of antimalarial initiatives.

The National Malaria Drug Policy (2013) of India recommends that *P. vivax* be treated with chloroquine for 3 days and primaquine for 14 days. However, primaquine is contraindicated in pregnant women, breastfeeding mothers, infants (<6 months age) and patients with G6PD deficiency. Uncomplicated *P. falciparum* cases are to be treated with ACT (artesunate for 3 days + sulphadoxine-pyrimethamine for 1 day). This is to be accompanied by single-dose primaquine preferably on day 2. Due to reports of resistance to sulphadoxine-pyrimethamine in North-Eastern states of India, artemether-lumefantrine is recommended in these regions (not recommended during the first trimester of pregnancy and for children weighing <5 kg). For severe/complicated malaria, parenteral treatment regimens with artesunate/artemether/arteether/quinine injections followed by area-specific oral ACT or quinine + doxycycline or clindamycin are recommended.

Globally, efforts are being made by government sponsored antiparasitic drug screening programmes in academic institutions including the "open source" drug discovery platforms. These are strengthened with support and coordination offered by the not-for-profit organization—Medicines for Malaria Venture, Geneva. The Wellcome Trust (UK), Consortium for Parasitic Drug Development (USA) and the Bill and Melinda Gates Foundation also fund initiatives toward development of safe and efficacious next-generation medicines against drug-resistant parasitic infections. Ongoing research efforts involve generation of new antimalarials with the following features: (1) rapid action, delayed development of resistance, (2) knowledge of mode of action, (3) oral delivery with preferably single-dose cure for improved compliance, (4) action on multiple parasite stages for transmission blocking, (5) action on the liver stage, especially hypnozoite in the case of *P. vivax*. Drug formulations for improved bioavailability and ease of use in comatose patients are also being developed.

## 4.2 Antimalarials: The India Story

The urgent need to develop indigenous new fast-acting schizonticides was recognized with the resurgence of malaria and rise in drug-resistant cases in the country in the 1970s. Control and management of complicated *P. falciparum* malaria cases was also a concern. Efforts were made to control malaria by employing vector control strategies as well as investing in new drug discovery efforts. Some significant developments (Fig. 4.1) are discussed here.

### 4.2.1 α/β Arteether: A Successful Intervention for Complicated Malaria

Focused effort toward new chemical entities (NCE) resulted in the development of α/β arteether, a semisynthetic derivative of artemisinin, as a joint programme of two

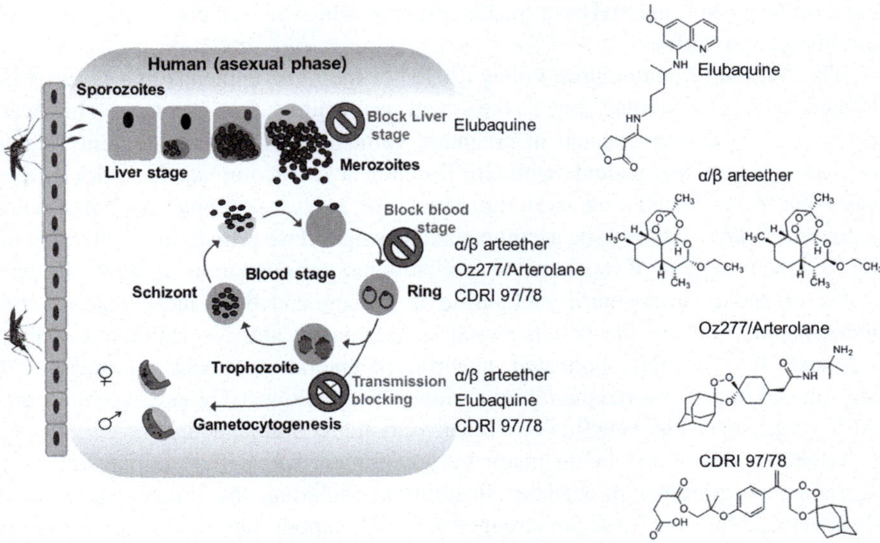


**Fig. 4.1** Sites of antimalarial interventions of compounds discovered and developed in India. Out of these, α/β arteether and Oz277/arterolone are in market. Known sites of intervention in blood stage are; cytosol (antifolates: pyrimethamine, proguanil; tRNA synthetases: cladologs, eEF2: DDD107498), plasma membrane (ATP4 inhibitors: KAE609), mitochondrion (DHODH inhibitor: DSM265; Cytb inhibitor: atovaquone), apicoplast (translational inhibitors: doxycycline, clindamycin) and food vacuole (inhibitors for hemozoin formation: chloroquine, mefloquine; protease inhibitors: WEHI-842, MG132), vesicle trafficking (PI4K inhibitor: MMV390048, UCT943)

CSIR laboratories-CIMAP and CDRI (R.A. Vishwakarma et al. Indian Patent no. 173947). Arteether is a mixture of β and α anomers (70:30 ratio). It was first reported as a fast-acting schizonticide in the rodent parasite *P. yoelii nigeriensis* MDR screen, exhibiting curative efficacy at 5 mg/kg × 4 days, i.m. (Dutta et al. 1989a). Its curative efficacy was also established in primate malaria models: *P. knowlesi* (12 mg/kg × 5 days) (Bajpai et al. 1989), *P. cynomolgi* (5 mg/kg × 3 days) (Dutta et al. 1989b), *P. fragile*—a cerebral malaria model (5 mg/kg × 3 days) (Bajpai et al. 1989); α/β arteether is also gametocytocidal (Tripathi et al. 1996). Following successful preclinical safety evaluation of the injectable formulation and clinical safety in a Phase I study at CDRI, trials were conducted for uncomplicated and complicated *P. falciparum* malaria. Phase II clinical trial with 51 patients at the Ispat General Hospital, Rourkela established proof of concept of clinical antimalarial efficacy of α/β arteether at the dose of 150 mg once daily for 3 days (i.m.) that killed parasites in the blood between 1 and 3 days and cleared fever between 1 and 6 days with no side effects (Mishra et al. 1995). Phase III trials were carried out in more than 500 uncomplicated and complicated *P. falciparum* patients in seven centers across the country [Jawaharlal Nehru Hospital, Bhilai; Malaria Research Centre/Government Medical College, Jabalpur; Malaria Research Centre,

New Delhi; Lady Hardinge Medical College, New Delhi; Central Reserve Police Force Base Hospital, Guwahati; Regional Malaria Research Centre (ICMR), Dibrugarh, Assam; Tata Main Hospital, Jamshedpur; Ispat General Hospital, Rourkela] (Mohanty et al. 1997; Mukim et al. 2011). The cure rate was found to be 93.3% in uncomplicated cases. In the 211 complicated disease cases treated with α/β arteether, only 14 expired of which ten died within 2 days before completing the 3-day arteether therapy (Asthana et al. 2001).

α/β arteether was approved for marketing by the Government of India, licensed to Themis Chemicals, and marketed as E-Mal in 1997. Its safety and 100% efficacy was confirmed in post-marketing surveillance in 2003. E-Mal was included in the National Drug Policy on Malaria, Ministry of Health and Family Welfare, Govt. of India in 2010 and continues as preferred treatment for complicated/cerebral malaria cases in India. The drug is also being exported to seven nations in Africa. From the single brand E-Mal in 1997, α/β arteether grew to 175 products in June 2016. This expansion resulted in a desirable fall in its market price (from ~Rs. 400 in 1997 to ~Rs. 80 in 2016), making α/β arteether an affordable antimalarial for public health.

### 4.2.2 Primaquine Derivative Elubaquine/Aablaquine for *P. vivax*

Primaquine (PQ), an 8-aminoquinoline that was first synthesized in the USA in 1946 remained the only transmission blocking antimalarial till the recent launch of Tafenoquine (Lacerda et al. 2019). Although PQ is effective against all exoerythrocytic forms of the parasite, its low efficacy against endo-erythrocytic parasites necessitates co-administration with a blood schizonticide in a 14-day treatment for anti-relapse therapy and radical cure of *P. vivax* (or *P. ovale*)-infected patients (Vale et al. 2009). However, PQ is contraindicated in infants, pregnant women and causes hemolytic anemia in individuals with G6PD deficiency. The latter is a special problem in India as populations residing in malaria-endemic regions of eastern India also have a higher incidence of G6PD deficiency. Another adverse effect of PQ is methemoglobinemia, a pathological condition arising from abnormal accumulation of methemoglobin, the product of auto-oxidation of the hemoglobin iron core.

A primaquine derivative, elubaquine/bulaquine/aablaquine (CDRI 80/53) was developed at CDRI, Lucknow. The drug had proven gametocytocidal efficacy in monkeys infected with *P. cynomolgi*; a single administration of elubaquine at 1.25 mg/kg blocked parasite oocyst development after 24 h and at 3.75 mg/kg within 5 h (Puri and Dutta 2005), thus preventing transmission through infected *Anopheles stephensi* mosquitoes. Gametocytocidal activity of elubaquine is more potent and faster than PQ. The improved safety profile of the drug over PQ is indicated by low methemoglobin accumulation in human volunteers (CDRI document on 80/53 1997). After 7 days of administration, elubaquine-induced methemoglobin levels ranged between 2.29% and 3.02%, whereas identically administered PQ increased methemoglobin from 3.97% to 16.32%.

Phase II clinical trials with elubaquine, carried out on 697 patients infected with *P. vivax*, showed that the drug given orally for 5 days at 25 mg/kg had a similar pattern of relapse as PQ at 15 mg/kg (Valecha et al. 2001). A safety and tolerability comparison of PQ and elubaquine conducted on 141 *P. vivax*-infected patients in Thailand (Krudsood et al. 2006) reported that four G6PD-deficient patients treated with PQ experienced a significant fall in hematocrit beyond the 7 day treatment period, whereas the three elubaquine-treated G6PD deficient patients did not exhibit a significant change in hematocrit leading to the conclusion that elubaquine does not cause clinically significant hemolysis. Pharmacokinetic analysis of elubaquine in different animal species has suggested species-specific differences (Mehrotra et al. 2007). The improved G6PD-related safety profile of elubaquine needs to be validated in a larger patient pool together with generation of its complete in vivo pharmacokinetic and metabolite profile for understanding the observed differences with PQ. Elubaquine was licensed to Piramal Enterprises Ltd. in 1999. It is not being marketed at present.

The most recent addition to the drug arsenal against *P. vivax* relapse is tafenoquine, an 8-aminoquinoline marketed by GlaxoSmithKline. Although its long half-life (2–3 weeks) allows a single oral dose to clear hypnozoites, it still has the problem of causing hemolysis in G6PD deficient patients. The United States Food and Drug Administration (FDA) approved single-dose tafenoquine for radical cure (prevention of relapse) of *P. vivax* malaria in 2018. Initial development of tafenoquine (WR 238605) was done by the Walter Reed Army Research Institute, Washington DC during which time collaborative primate malaria studies were conducted at CDRI, Lucknow and showed that tafenoquine was more effective at $ED_{50}$ level compared to PQ (Puri and Dutta 2003).

Exploration of quinolines for new antimalarials continues. A series that acts on both chloroquine-sensitive and -resistant parasites was identified (Indian patent 2291/DEL/2013) and a molecule from the series with demonstrated efficacy in the *P. yoelii* N67-mouse model and *P. cynomolgi*-Rhesus macaque model is currently under preclinical safety evaluation at CDRI (unpublished data). Several 4-aminoquinolines with in vitro activity against chloroquine-resistant *P. falciparum* and curative efficacy in *P. berghei*-infected BALB/c mice have also been recently reported (Singh et al. 2016).

### 4.2.3 Synthetic Endoperoxides

As ACTs started gaining ground as first-line treatment for malaria in many parts of the world, malaria chemotherapy moved to counter the vagaries of *Artemisia annua* production, supply chain and extraction costs. Progress was made using the endoperoxide sesquiterpene artemisinin and its derivatives (artemether, arteether, artesunate etc.). Chemists at CDRI synthesized substituted 1,2,4-trioxanes that were extensively screened against MDR *P. yoelii nigeriensis* rodent parasite followed by efficacy evaluation in *P. cynomolgi* and *P. knowlesi* in rhesus monkey model (Singh et al. 2007a, b) (US Patent, 6316493). The gametocytocidal action of two molecules in this series was also confirmed. CDRI 97/78 and CDRI 99/411 were identified as potent oral antimalarials, found safe in safety pharmacology and toxicology studies and subsequently licensed to IPCA Pharmaceuticals for clinical development. CDRI 97/78 is an equipotent hemisuccinate derivative of CDRI 97/63 and is rapidly metabolized to the latter. The compound was approved for Phase I clinical trials in India; Phase Ia (single ascending dose safety and PK) conducted in PGIMER, Chandigarh found the compound safe in human volunteers (Shafiq et al. 2014).

Another fully-synthetic peroxide, OZ277 (RBx11160, Arterolane, of the Dispiro 1,2,4-Trioxolane series) was synthesized by Vennerstrom and coworkers (Vennerstrom et al. 2004) (U.S. Patent number 7371778) as a fast-acting blood schizonticide. Arterolane exhibited single-dose curative efficacy against *P. berghei* ANKA infection in mice at 30 mg/kg dose. OZ277 is also effective against artemisinin-resistant *P. falciparum* Cam3.$I^{R539T}$ parasites (isolate from Cambodia with K13-propeller mutation R539T) at pharmacologically relevant concentrations (Baumgartner et al. 2017). A multicenter, randomized Phase II trial of Arterolane with 230 patients from four centers in Thailand, India, and Tanzania (mainland and Zanzibar) who received 50, 100, or 200 mg of arterolane (once daily for 7 days) showed that it was rapidly acting, effective, and safe (Valecha et al. 2010). A subsequent Phase III trial (Toure et al. 2016) with fixed-dose combination of arterolane maleate with piperaquine phosphate showed that the combination had comparable efficacy with the artemether-lumefantrine combination for treatment of uncomplicated *P. falciparum* malaria. Arterolane-piperaquine (Synriam) was launched by Ranbaxy in 2012 and is currently marketed by SunPharma. Artefenomel (OZ439), a novel trioxolane, was designed to bypass the initial low exposure liability related with OZ277 (Charman et al. 2011). OZ439 is currently a front-runner candidate in combination with ferroquine to allow for once-daily dosing in fewer doses than the current 3-day ACT regimen (Sanofi and MMV partnership).

### 4.2.4 Exploration of Leads from Traditional Knowledge

The fact that two major antimalarial compounds—quinine and artemisinin were isolated from plants as a result of scientific exploration of traditional knowledge has provided impetus to further reconnaissance of plant-derived molecules with clues from Ayurveda/Unani systems and folk medicine. Although several reports using plant extracts for assaying antimalarial activity have been published (Bagavan et al. 2011; Kantamreddi et al. 2009; Panda and Luyten 2018; Shankar et al. 2012; Simonsen et al. 2001), there has been little progress in identifying bioactive molecules.

The Central Council of Ayurveda and Siddha patented Ayush-64, a combination of four plants namely *Alstonia scholaris* (aqueous extract of bark), *Picrorhiza kurroa* Royle (aqueous extract of rhizome), *Swertia chirata* (aqueous extract of whole plant), and *Caesalpinia crista* Linn (powder of seed pulp) for its antimalarial activity. However, when the combination was tested in *P. vivax* patients in a non-crossover, randomized clinical trial at the Malaria Research Centre in collaboration with National Anti-Malaria Programme (Valecha et al. 2000), the results showed a much lower cure rate with Ayush-64 (1 g oral dose, three times a day for 5–7 days) as compared to chloroquine (1500 mg oral, over 3 days). Leaf paste of *Nyctanthes arbor-tristis* Linn ("Parijat" or "Harsingar") was tested in 120 *P. vivax* and *P. falciparum* patients at the MA Podar Hospital, Mumbai. The paste given orally three times a day for 7–10 days was reported to cure 92 (76.7%) patients within 7 days, another 20 patients by 10 days, and the remaining had to be cured by standard antimalarial therapy (Karnik et al. 2008). Activity-guided RPHPLC fractionation of the ethanol extract of *Nyctanthes* leaves suggested that more than one active compound defines potency and that iridoid glycosides are the most probable phytoconstituents responsible for anti-plasmodial activity (Kumari et al. 2012).

The dried rind of *Punica granatum*, promoted as OMARIA, is being used in Odisha for prophylaxis and cure of malaria (Lekana-Douki et al. 2012). In a study conducted in Italy, the preparation was subjected to activity-guided fractionation of the methanolic extract of fruit rind and in vitro activity was found to be associated with fractions enriched in tannins-punicalagins, ellagic acid and its glycoside. However, both the methanolic extract and the fraction did not show in vivo efficacy in the murine malaria model (Dell'Agli et al. 2009). This was attributed to lower bioavailability and possible conversion of ellagitannins into inactive metabolites in mice. The lack of efficacy in the mouse malaria model makes it difficult to take forward a preclinical analysis of OMARIA, although it continues to be used to treat patients in Odisha.

Curcumin was reported as an antimalarial by Reddy et al. (2005) as it inhibited chloroquine-resistant *P. falciparum* growth in culture ($IC_{50}$ of 5 μM) and reduced parasitemia by 80–90% when administered orally (100 mg/kg for 5 days) to *P. berghei*-infected mice. Subsequently, a combination of α,β-arteether (E-Mal) and curcumin was shown to result in better survival rates of *P. berghei*-infected mice, and a 3-day oral regimen of curcumin with a single injection of α,β-arteether at 750 μg or 1.5 mg per mouse led to 100% survival and complete protection of animals against recrudescence (Nandakumar et al. 2006), possibly through enhancement of TLR-2 mediated innate immune response (Vathsala et al. 2012). Further, a PLGA-based nanoformulation of curcumin was developed to improve therapeutic index (Dende et al. 2017); PLGA-curcumin improved bioavailability of curcumin and had comparable efficacy with native curcumin at a 15-fold lower concentration; PLGA-curcumin prevented the breakdown of the blood–brain barrier and inhibited the sequestration of parasitized-RBCs and $CD8^+$ T cells in the brain. It also inhibited mRNAs for inflammatory cytokines and chemokine receptor CXCR3 and activated the anti-inflammatory cytokine IL-10 and was proposed for use as an adjunct in antimalarial therapy.

### 4.2.5 Formulations

Novel drug delivery approaches have been attempted for increasing the bioavailability of drugs, reducing dose or providing an alternative route of administration. A polymeric lyotropic liquid crystalline formulation of arteether-lumefantrine gave a prolonged release of both drugs and conferred complete protection with no mortality at 1/40th of the therapeutic dose in *P. berghei* ANKA infected mice, suggesting the possibility of single-shot therapy (Dawre et al. 2018). Arteether nanoemulsions and self-nanoemulsifying drug delivery systems have also been tested (Dwivedi et al. 2015, 2014). Atovaquone that targets parasite mitochondrial cytochrome B is marketed in combination with proguanil as Malarone (GSK). Atovaquone nanoparticles formulation given with proguanil showed approximately two-fold improved bioavailability in rats compared to Malarone (R) with significant dose reduction in Peter's 4-day suppressive tests in mice (Darade et al. 2018). The bioavailability of lumefantrine in a solid dispersion formulation is enhanced many folds when assessed in a randomized, open-label study in healthy volunteers (Jain et al. 2017). Nanostructured lipid carriers (NLC) for artemether-lumefantrine oral therapy with significant reduction in dose have been formulated (Prabhu et al. 2016b) and an NLC formulation of the drug combination has also been made for intravenous therapy in cerebral malaria patients (Prabhu et al. 2016a). A solid self-microemulsifying drug delivery system (SMEDDS) formulation of artemether–lumefantrine capable of maintaining plasma concentration of lumefantrine above the minimum effective concentration for approximately 4 days has been reported (Bhandari et al. 2017). A long circulatory PEGylated liposomal formulation of the gametocytocidal ionophore maduramicin has shown enhanced antiparasitic activity compared to the free drug (Raza et al. 2018).

Specific targeting to infected RBCs has been attempted by using chitosan particles derivatized with dehydroascorbic acid (DHA) (Shafi et al. 2017). DHA competes with glucose for binding to the mammalian glucose transporter GLUT-1 which is over-expressed in *Plasmodium*-infected human RBCs. The chitosan-DHA particles were superior in terms of uptake and extent of preferential targeting to infected RBCs in vitro. Specific targeting to infected RBCs is not usually required as antimalarial drugs in use reach plasma concentrations high enough to sustain therapeutic threshold concentrations in erythrocytes; however, targeting to infected cells can result in dose reduction and aid in overcoming efflux-mediated drug resistance.

## 4.3 Exploring New Biological Targets for Drug Discovery and Design

Parasite resistance to antimalarial drugs is generated by accumulation of mutations which result in (1) reduced uptake or increased efflux of drug, (2) altered protein–drug interaction, (3) metabolic bypass of targeted pathway, and (4) upregulation of stress–response pathways. Concerted efforts are required to develop a comprehensive understanding (biochemical, structural, and molecular) of critical pathways to evaluate druggability of validated biological targets and design specific inhibitors. In the subsequent sections, we discuss some of the potential drug targets/pathways that have been investigated by researchers in India.

### 4.3.1 Fatty Acid Synthesis (FAS-II Pathway)

The mammalian host has type I fatty acid synthases (FAS-I) which are multifunctional proteins whose distinct domains catalyze different steps involved in fatty acid synthesis. In contrast, the malaria parasite possesses type II fatty acid synthesis system (FAS-II) whereby individual reactions are catalyzed by separate FAS enzymes. The FAS-II pathway involves preparation, initiation, and elongation phases for fatty acid synthesis with the parasite relict plastid, the apicoplast, being the site of FAS-II mediated synthesis. Surolia and coworker (Surolia and Surolia 2001) reported the antimalarial activity of Triclosan [5-chloro-2-(2,4-dichlorophenoxy)phenol]. Triclosan is an antimicrobial biocide that is known to inhibit fatty acid synthesis in bacteria (McMurry et al. 1998). Triclosan exhibited $IC_{50}$ value of 0.7 μM in in vitro *P. falciparum* cultures and curative dose of 38 mg/kg in the *P. berghei*-mouse model. The authors also reported that Triclosan specifically inhibits enoyl-ACP reductase which is required for fatty acid synthesis in apicoplast (Surolia and Surolia 2001). Mutational analysis of enoyl-ACP revealed differences in the binding of Tricolsan to parasite and bacterial enzyme (Kapoor et al. 2004). Through structure-based approach, 2′-subsituted analogs of Triclosan were designed and tested for their blood stage antimalarial activities; some of the scaffolds showed nanomolar activities in enzyme based inhibition of *Pf*ENR and were similar to Triclosan-mediated enzyme inhibition (Kapoor et al. 2009). Another class of enoyl-ACP reductase inhibitors, bromo-benzothiophene carboxamides were designed, but their antimalarial activities were not better than Triclosan (Banerjee et al. 2011). Gene knock-out studies have subsequently revealed that FASII is dispensable in blood stages; however, it is critical for liver stages and sporozoite development in the mosquito midgut. This suggested that the observed blood stage antimalarial activity of FAS-II inhibitors was likely to be due to alternative biological targets (Shears et al. 2015), although inhibition of growth of late liver stage parasites by Triclosan was reported (Singh et al. 2009). Nevertheless, a comprehensive SAR for FAS-II inhibitors might aid in identification of inhibitors of parasite multiplication in the liver and survival of hypnozites in *P. vivax* (Schrader et al. 2013).

### 4.3.2 Aminoacyl tRNA Synthetases

*Plasmodium* has active protein translation in the cytoplasm and also in its organelles—the mitochondrion and apicoplast. A critical component of the translation machinery is aminoacyl tRNA synthetases (aaRS) that function to charge tRNA with the cognate amino acid for its incorporation in the growing polypeptide chain. All parasite aaRS are nuclear-encoded and localize to the parasite cytoplasm (16 aaRS), apicoplast (15 aaRS), mitochondrion (1aaRS); four aaRS exhibit dual localization in the cytoplasm and apicoplast (Habib et al. 2016). Structure elucidation of aaRS led to structure-based design and evaluation of specific inhibitors (Khan et al. 2011, 2013a, b). Initial screening of small-molecule library led to identification of hit molecules with promising methionyl-tRNA synthetase inhibition and antimalarial activities. This included REP3123 and REP8839 which are known to inhibit bacterial methionyl-tRNA synthetases and also exhibit antimalarial activity ($IC_{50}$ values ~150 nM) (Hussain et al. 2015). Over the years, in silico and structural biology based approaches have helped to map the druggable site for aaRS (Manickam et al. 2018). For instance, ATP binding site inhibitors (cladosporin), molecules binding to 3′-end of tRNA pocket (halofuginone and borrelidin) and molecules binding to tRNA editing site (benzoxaborole) have been investigated (Goodman et al. 2016; Jain et al. 2015; Khan et al. 2014). Amongst these, cladosporin (an antifungal metabolite) has shown potent antimalarial activity against blood and liver stages of the malaria parasite ($IC_{50} < 100$ nM). The scaffold includes 6,8-dihydroxylisocoumarin ring joined to tetrahydropyran group with a methyl moiety and is known to specifically target lysyl-tRNA synthetase (KRS) over the mammalian homolog (Khan et al. 2014). Recently, a library of stereoisomers of cladosporin (cladologs) were synthesized and assessed for their antimalarial activity (Das et al. 2018). With systematic changes at chiral centers, the authors could map the critical attributes which increased the antimalarial potency of cladologs with specificity over human KRS. In a recent study, another cladosporin-based scaffold showed potent in vivo efficacy with $ED_{90}$ of 1.5 mg/kg (oral administration) in rodent malaria model along with selective inhibition of *Pf*KRS over human KRS (Baragana et al. 2019). Cytosolic and organellar tRNA synthetases, thus offer an exciting opportunity for identification of novel antimalarials.

### 4.3.3 DNA Gyrases

DNA gyrases (type II Topoisomerases) play a critical role in maintenance of DNA topology during bi-directional *ori* replication of the circular apicoplast genome in the parasite (Nagano et al. 2014; Weissig et al. 1997). The *P. falciparum* nuclear genome encodes GyrA and GyrB proteins that harbor an N-terminus signal and transit sequence for apicoplast targeting. GyrA has DNA cleavage and DNA wrapping domains, while GyrB has an N-terminus ATPase domain with DNA binding domain and GyrA-interacting domain at the C-terminus. Bacterial DNA gyrase inhibitors (ciprofloxacin, novobiocin) also inhibit parasite growth. Novobiocin was shown to inhibit ATPase activity of GyrB, albeit with binding affinity different from bacterial gyrase probably due to the presence of unstructured domains in *P. falciparum* GyrB. Novobiocin treatment specifically altered apicoplast DNA integrity in comparison to nuclear and mitochondrial DNA (Raghu Ram et al. 2007). Similarly, coumermycin was shown to inhibit the ATPase activity of *P. falciparum* GyrB in a dose-dependent manner (Dar et al. 2007). This confirmed that apicoplast DNA replication can be a potential site of intervention. Recently, acriflavine with known antibacterial and anticancer activity has been shown to have anti-plasmodial activity in both in vitro *P. falciparum* culture and the *P. berghei*-mouse model. Acriflavine impairs DNA replication by specifically inhibiting apicoplast GyrB

(Dana et al. 2014). Other DNA gyrase inhibitors include coumarin-triazole based compounds (Yadav et al. 2018); however, their comprehensive SAR and evaluation of in vivo efficacy is required.

### 4.3.4 Noncanonical Structures at Telomeric Ends

Besides protein targets, noncanonical nucleic acid secondary structures especially at telomeric ends can be sites for intervention. Telomeric and sub-telomeric regions have high enrichment of G-rich sequences which have the propensity to form G-quadruplex structures in *Plasmodium* (Bhartiya et al. 2016). Telomere architecture and dynamics of the parasite and host have remarkable differences in telomere length (*Plasmodium* 1–2 kb, human 10–15 kb), telomerase size (*P. falciparum* telomerase is 2.5 times larger than the human homolog), and diversity of proteins binding to telomeric and sub-telomeric regions (Figueiredo and Scherf 2005; Figueiredo et al. 2005). Therefore, disturbing telomere structure and function through G-quadruplex interacting ligands can be explored as an alternative strategy for antimalarial intervention. The life of drugs that target critical parasite biosynthetic pathways is limited by the progressive accumulation of mutations in protein targets leading to drug resistance. Hence, a probable advantage of pharmacological targeting of noncanonical structures in telomeric ends can be a delay in emergence of resistance, as acquiring mutations that disturb secondary structures in telomeres is likely to have a dramatic fitness cost. G-quadruplex interacting ligands like bisquinolinium derivatives of 1,8-naphthyridine and pyridine, bisbenzimidazole carboxamide derivatives, benzothiazole hydrazones of furylbenzamides show in vitro antimalarial activity with reasonable selectivity index (Anas et al. 2017). These ligands cause shortening of parasite telomeres, alter the transcriptional dynamics of sub-telomeric genes and induce DNA damage. However, a comprehensive SAR is essential to identify "best in class" so that they can be taken forward for evaluation of in vivo efficacy.

### 4.3.5 ATP Transporters

Amongst various transporters, parasite plasma membrane localized *Pf*ATP4 has emerged as a promising drug target (Dennis et al. 2018). This is a P-type ATPase that converts energy from ATP hydrolysis into cation transport (sodium export is coupled with proton import). Multiple chemotypes like KAE609 (spiroindolone), MMV007275 and MMV0011567 (carboxamides from MMV malaria box), NF-Pf4492 (aminopyrazole), 21A092 (pyrazoleamide), and (+)-SJ733 (dihydroisoquinolone) identified by international groups showed promising inhibition of *Pf*ATP4 activity and parasite growth. Of these, KAE609 (also known as cipargamin) kills *P. falciparum* and *P. vivax* parasites in vitro at $EC_{50}$ values $<$10 nM (Rottmann et al. 2010; White et al. 2014). KAE609 also inhibits gametocyte and oocyst development in mosquitoes. KAE609 has rapid parasite clearing ability and good pharmacokinetic–pharmacodynamic properties and has progressed through Phase I and IIa clinical trials (Leong et al. 2014). In India, molecular docking studies revealed that 1,3-benzoxazine derivatives of phytophenol eugenol and isoeugenol bind to *Pf*ATP4. These compounds also showed promising antimalarial activity due to disruption of sodium homeostasis (Sharma et al. 2018). A recent report has indicated that disturbing AMP homeostasis in *Plasmodium* by overexpression of adenosine 5′ monophosphate deaminase (AMPD) is inimical to parasite survival suggesting that allosteric activators of *Pf*AMPD could be designed and evaluated as antiparasitic agents (Nagappa et al. 2019).

### 4.3.6 Heme Biosynthesis and Detoxification Pathways

*Plasmodium* possesses a heme biosynthesis pathway despite having access to host-derived heme (Nagaraj and Padmanaban 2017; Sato et al. 2004; Surolia and Padmanaban 1992). Parasite-synthesized heme serves as a cofactor for mitochondrial cytochromes that support electron transport. The heme biosynthesis pathway is not essential in blood stages, but is critical for liver and mosquito stages of the parasite life cycle (Goldberg and Sigala 2017; Ke et al. 2014; Nagaraj et al. 2013). Heme biosynthesis inhibitors will be ineffective in blood stages but might serve as transmission blockers and/or prophylactic agents. Researchers have argued that heme biosynthesis may not be a good drug target (Koreny et al. 2013). In fact, proteins involved in heme detoxification are good drug targets to explore. Two major antimalarials chloroquine and artemisinin act on heme polymerization and chloroquine also inhibits hemoglobin degradation (Chugh et al. 2013). *Pf*HDP (heme detoxification protein), a major protein involved in hemozoin formation (nontoxic crystalline form of heme) has been suggested as a critical drug target. Through screening of Maybridge library, the identified hit ML-2 (1-(3,4-dihydronaphthalen-2-yl)-4-[3-(trifluoromethyl)phenyl]piperazine) showed dose-dependent parasite inhibition (Gupta et al. 2017). Quinolines and peroxide-based compounds have been shown to inhibit hemozoin formation (Fong and Wright 2013; Pandey et al. 1999; Verma et al. 2016). Some synthetic peroxides that show promising antimalarial activity against chloroquine-sensitive and resistant strains include 1,2-dioxane; 1,2,4-trioxanes; 1,2,4-trioxalanes and their hybrid molecules like trioxaquines (Chauhan et al. 2010; Yadav et al. 2014).

### 4.3.7 Proteases/Disaggregases

*Plasmodium* spp. have evolved efficient protein degradation machinery which helps in maintenance of protein homeostasis through efficient removal of misfolded or aggregated species. For instance, Clp proteases (ATP-dependent disaggregase machinery) are localized in different parasite organelles (mitochondria and apicoplast) (El Bakkouri et al. 2010; Jain et al. 2013). In silico screening of inhibitors of Clp proteases and comprehensive SAR will help in identification of potential hits

(Mundra et al. 2017). Other proteases include subtilisin-like protease, cysteine protease (falcipain), aspartyl protease (plasmepsin). These proteases are involved in hemoglobin digestion, protein trafficking, parasite invasion of RBC and egress from infected RBCs and hepatocytes (Moura et al. 2009; Prasad et al. 2012). In light of the existing knowledge about their enzymatic reactions and 3D-structure, these proteases are being explored as drug targets (Deu 2017; Mishra et al. 2019; Nasamu et al. 2017; Pino et al. 2017). Collaborations between ICGEB-New Delhi and researchers in Canada and Italy have identified initial hits targeting falcipain-2 and apicoplast ClpP protease (Chakka et al. 2015; Mundra et al. 2017; Rizzi et al. 2011). Isoquinolines, hydroxyethylamines-based active pharmacophores have been designed against plasmodial proteases (Batra et al. 2003; Gupta et al. 2017; Singh et al. 2019). Molecular dynamics based approach is also employed for design of PEXEL-based mimetics against parasite plasmepsin (Bedi et al. 2016). Through structure-guided drug discovery, KNI compounds initially discovered as inhibitors of HIV protease were modified to inhibit vacuolar plasmepsins (Mishra et al. 2018). KNI scaffold consists of allophenylnorstatine [[(2S,3S)-3-amino-2-hydroxy-4-phenylbutyricacid] (Apns)]. Alkylamino analog and phenylacetyl tripeptides exhibited promising antimalarial activity with minimal toxicity in human cells (Mishra et al. 2018). Apart from proteases, the ubiquitin-proteasome machinery of the parasite is being explored as a target for antimalarial intervention. A peptidyl inhibitor MG132 with a P2 leucine inhibited both cysteine protease and ubiquitin-proteasome system activities in parasite extracts and also strongly inhibited recombinant falcipains (Prasad et al. 2013). MG132 was highly selective for inhibition of *P. falciparum* indicating the possibility of generating dual-target inhibitors of malaria parasites.

### 4.3.8 Other Unique Pathways/Targets

Additional biological pathways explored for future antimalarial intervention include proteins involved in genome maintenance. For instance, *Pf*Alba (Acetylation lowers binding affinity) superfamily proteins whose acetylation state affects their DNA binding ability and consequently affects transcription have been studied. *Pf*Alba3 interacts with the epigenetic modifier Sir2a and occupies telomeric/sub-telomeric regions including the *var* gene promoters for transcriptional regulation of antigenic variation/virulence genes (Goyal et al. 2012, 2016). Similarly, another DNA binding protein, Origin Recognition Complex (ORC) binds to the sub-telomeric regions and regulates the expression of antigenic variation genes (Deshmukh et al. 2012; Gupta et al. 2008). Telomere/sub-telomere structure and function is also influenced by the epigenetic machinery (histone acetylases, deacetylases, methylases, and demethylases). Phylogenetic analyses of *Plasmodium* histone acetylases and deacetylases suggests that many of them are close to prokaryotic acetylases and deacetylases (Kanyal et al. 2018), and thus may be further evaluated as targets.

Research groups are also investigating protein folding, assembly and processing machineries. This includes the diverged protein folding machinery which has acquired potential noncanonical roles to provide survival advantage to parasite (Bhartiya et al. 2015). Amongst the diverse class of chaperones or heat shock proteins (HSPs), HSP40 family has structurally diverse domains, exhibit differential organellar localization (few even exported into RBC), and execute noncanonical functions (Behl et al. 2019; Kulzer et al. 2012; Kumar et al. 2010). Establishing assays for their biochemical (protein folding and oligomer remodeling) activities will help to validate compounds which specifically inhibit parasite chaperones.

The biogenesis of [Fe-S] clusters by the SUF pathway has also been proposed as a unique target, as this pathway is absent in humans but is conserved in bacteria, protozoa, and plants. In the malaria parasite, components of the SUF and ISC systems are largely encoded by the nucleus and localize to the apicoplast and mitochondrion, respectively (Kumar et al. 2011). The SUF pathway is essential for parasite survival in the blood and the mosquito stages (Charan et al. 2017; Gisselberg et al. 2013). No inhibitors of the SUF pathway are currently known but the possibility of chemically inhibiting the first enzyme (cysteine desulfurase SufS) of the pathway has been suggested (Charan et al. 2014). Besides biochemical characterization, genetic tools for generation of knock-out (KO) parasite strains are being employed to assess the essentiality and phenotypes of KO parasite. These attempts give insights into biological roles of uncharacterized *Plasmodium* proteins which may help identify new targets for future drug discovery/design (Al-Nihmi et al. 2017; Jaijyan et al. 2016; Mastan et al. 2017).

## 4.4 Repositioning of Molecules for Antimalarial Activity

The possibility of repurposing FDA-approved drugs for other uses such as antimalarials has been addressed by several groups. Screening of drug libraries in blood stage *P. falciparum* culture and liver stage in *P. berghei* model identified potential drug candidates for malaria (Chong et al. 2006; Derbyshire et al. 2012). These potential candidates belonged to diverse categories such as proton pump inhibitor, immunosuppressant, antihypertensive, ovulatory stimulant, antimicrobial, bone resporption inhibitor with varied $IC_{50}$ values ranging from nanomolar (cyclosporine: 17 nM; telmisartan: 25 nM), submicromolar (clomiphene citrate: 0.219 μM; raloxifene hydrochloride and pentamidine isethionate: ~0.5 μM) and micromolar (lopinavir, rotanavir, azithromycin:~2.0 μM) (Pazhayam et al. 2019). Though the primary drug targets of these candidates are known in humans, in some cases their corresponding orthologs are absent in the malaria parasite. This suggests novel mechanisms of drug action in parasites. These drug candidates can be further evaluated for multistage activity (liver stage and gametocytocidal activity). Ivermectin, currently used against lymphatic filariasis, inhibits *Plasmodium* nuclear import/export (Panchal et al. 2014) and kills the parasite in the mosquito and human liver stages (Mendes et al. 2017). Drug repurposing efforts can help fast-track the optimized leads into next-generation antimalarials. Apart from this, open source drug discovery efforts supported by MMV and GSK have provided access to chemically diverse libraries (Malaria Box and Pathogen Box) to determine mode of action and generate SAR around specific hit molecules. A recent study has used the Malaria Box compounds and identified novel modes of action such as defects in apicoplast segregation, inhibition of host cell egress/invasion with future work to be focused on target identification (Subramanian et al. 2018).

## 4.5 Gaps and Future Measures

Drug discovery for infectious diseases such as malaria requires an intense and sustained effort for generating novel active chemical scaffolds as leads for new drugs to counter emerging drug resistance. It is therefore imperative that a consolidated effort involving medicinal chemists with computational and parasite biologists is made to explore new validated protein targets for designing molecules and generating compound libraries around specific active sites or protein-protein interfaces. Screening models for evaluation in murine malaria exist in several laboratories in the country and the primate malaria model (*P. cynomolgi*—a model for *P. vivax* like relapse malaria, and *P. knowlesi*) is maintained at CDRI, Lucknow and used to evaluate new leads. The absence of a continuous in vitro *P. vivax* culture system presents constraints; also, the diversity of *P. vivax* strains in schizont/hypnozoite ratios and relapse times presents a challenge to standardization. Newer models such as humanized-(athymic) nude mouse that support the replication of asexual blood stages of *P. falciparum* in human erythrocytes would aid in evaluation of efficacy against falciparum malaria and pharmacokinetics/pharmacodynamics of leads (McCarthy et al. 2016a). It will also be of use in setting up a model for assaying drug-induced hemolysis in G6PD-deficient human blood which is a major issue in the widespread use of 8-aminoquinolines such as primaquine and tafenoquine.

In recent years, evaluation of clinical efficacy of drug candidates after completion of preclinical studies has been conducted in malaria-naïve volunteers infected with blood stage parasites or sporozoites. This combined Phase I/early Phase II study helps to rapidly obtain human pharmacokinetic and pharmacodynamics data and also identifies the correct dosing regimen for causal prophylaxis and/or cure (McCarthy et al. 2016b, 2011; Nyunt et al. 2009; Sulyok et al. 2017). Another advantage of the controlled human malaria infection (CHMI) approach, with infections initiated by mosquito bite or blood stage parasites, is that if volunteers are screened for antimalarial immunity in advance, higher parasite clearance rates due to preexisting immunity will not lead to an overestimation of drug efficacy (Stanisic et al. 2018). CHMI induced by *P. falciparum* blood stages for assessment of parasite clearance by schizonticidal drugs also allows precise quantitation of the number of parasites initiating infection in each study volunteer. However, adopting this clinical trial mode in resource-poor nations poses ethical challenges. A study on volunteers in a controlled infection trial in Kenya (Njue et al. 2018) revealed that financial compensation was among the strongest motivations for participation in the trial raising the possibility of exploitation of communities. If such CHMI trials are to happen in India, new ethical guidelines defining the threshold of risks must combine with a strong rationale for conducting the trial.

Since interest in the search for new antimalarials, or for that matter drugs for most infectious diseases including tuberculosis, is primarily restricted to public-funded institutions in India, government-aided discovery and development efforts must be strengthened. Existing pre-clinical regulatory test facilities need upgradation and should be made more accessible so that costs of developing anti-infectives are minimized. Identifying an industry partner for taking a "drug product" through clinical trials remains challenging. Since the fruits of this enterprise are not necessarily high on profit, privately owned pharmaceutical industry keeps away. Serious consideration must also be given to public sector manufacturing for drugs for infectious diseases. There are lessons to be learnt from the remarkable success of E-Mal for which scientists took only process patents for isolation and synthesis. Initially licensed to one company, it is now produced by more than 180 companies. The accompanying dramatic fall in prices of E-Mal (Misra 2017) ensures that patients who need it are also able to afford it. It is for this reason that antimalarials and drugs for other infectious diseases such as tuberculosis and leishmaniasis have to be considered "public goods" and not "commodities," and renewed effort must be made to bring together research groups with diverse expertise to work toward identification and development of new drugs and formulations.

## References

A-Elbasit IE, Elbashir MI, Khalil IF, Alifrangis M, Giha HA (2006) The efficacy of sulfadoxine-pyrimethamine alone and in combination with chloroquine for malaria treatment in rural Eastern Sudan: the interrelation between resistance, age and gametocytogenesis. Tropical Med Int Health 11:604–612

Alker AP, Kazadi WM, Kutelemeni AK, Bloland PB, Tshefu AK, Meshnick SR (2008) dhfr and dhps genotype and sulfadoxine-pyrimethamine treatment failure in children with falciparum malaria in the Democratic Republic of Congo. Tropical Med Int Health 13:1384–1391

Al-Nihmi FM, Kolli SK, Reddy SR, Mastan BS, Togiri J, Maruthi M, Gupta R, Sijwali PS, Mishra S, Kumar KA (2017) A novel and conserved *Plasmodium* sporozoite membrane protein SPELD is required for maturation of exo-erythrocytic forms. Sci Rep 7:40407

Anas M, Sharma R, Dhamodharan V, Pradeepkumar PI, Manhas A, Srivastava K, Ahmed S, Kumar N (2017) Investigating pharmacological targeting of G-quadruplexes in the human malaria parasite. Biochemistry 56:6691–6699

Asthana OP, Srivastava JS, Kamboj VP, Valecha N, Sharma VP, Gupta S, Pande TK, Vishwanathan KA, Mahapatra KM, Nayak NC et al (2001) A multicentric study with arteether in patients of uncomplicated falciparum malaria. J Assoc Physicians India 49:692–696

Bagavan A, Rahuman AA, Kaushik NK, Sahal D (2011) In vitro antimalarial activity of medicinal plant extracts against *Plasmodium falciparum*. Parasitol Res 108:15–22

Bajpai R, Dutta GP, Vishwakarma RA (1989) Blood schizontocidal activity of a new antimalarial drug, arteether (alpha/beta), against *Plasmodium knowlesi* in rhesus monkeys. Trans R Soc Trop Med Hyg 83:484

Banerjee T, Kapoor N, Surolia N, Surolia A (2011) Benzothiophene carboxamide derivatives as novel antimalarials. IUBMB Life 63:1111–1115

Baragana B, Forte B, Choi R, Nakazawa HS, Bueren-Calabuig JA, Pisco JP, Peet C, Dranow DM, Robinson DA, Jansen C et al (2019) Lysyl-tRNA synthetase as a drug target in malaria and cryptosporidiosis. Proc Natl Acad Sci U S A 116:7015–7020

Batra S, Sabnis YA, Rosenthal PJ, Avery MA (2003) Structure-based approach to falcipain-2 inhibitors: synthesis and biological evaluation of 1,6,7-trisubstituted dihydroisoquinolines and isoquinolines. Bioorg Med Chem 11:2293–2299

Baumgartner F, Jourdan J, Scheurer C, Blasco B, Campo B, Maser P, Wittlin S (2017) In vitro activity of anti-malarial ozonides against an artemisinin-resistant isolate. Malar J 16:45

Bedi RK, Patel C, Mishra V, Xiao H, Yada RY, Bhaumik P (2016) Understanding the structural basis of substrate recognition by *Plasmodium falciparum* plasmepsin V to aid in the design of potent inhibitors. Sci Rep 6:31420

Behl A, Kumar V, Bisht A, Panda JJ, Hora R, Mishra PC (2019) Cholesterol bound Plasmodium falciparum co-chaperone 'PFA0660w' complexes with major virulence factor 'PfEMP1' via chaperone 'PfHsp70-x'. Sci Rep 9:2664

Bhandari S, Bhandari V, Sood J, Jaswal S, Rana V, Bedi N, Sehgal R, Tiwary AK (2017) Improved pharmacokinetic and pharmacodynamic attributes of artemether-lumefantrine-loaded solid SMEDDS for oral administration. J Pharm Pharmacol 69:1437–1446

Bhartiya D, Chandramouli B, Kumar N (2015) Co-evolutionary analysis implies auxiliary functions of HSP110 in *Plasmodium falciparum*. Proteins 83:1513–1525

Bhartiya D, Chawla V, Ghosh S, Shankar R, Kumar N (2016) Genome-wide regulatory dynamics of G-quadruplexes in human malaria parasite *Plasmodium falciparum*. Genomics 108:224–231

Chakka SK, Kalamuddin M, Sundararaman S, Wei L, Mundra S, Mahesh R, Malhotra P, Mohmmed A, Kotra LP (2015) Identification of novel class of falcipain-2 inhibitors as potential antimalarial agents. Bioorg Med Chem 23:2221–2240

Charan M, Singh N, Kumar B, Srivastava K, Siddiqi MI, Habib S (2014) Sulfur mobilization for Fe-S cluster assembly by the essential SUF pathway in the *Plasmodium falciparum* apicoplast and its inhibition. Antimicrob Agents Chemother 58:3389–3398

Charan M, Choudhary HH, Singh N, Sadik M, Siddiqi MI, Mishra S, Habib S (2017) [Fe-S] cluster assembly in the apicoplast and its indispensability in mosquito stages of the malaria parasite. FEBS J 284:2629–2648

Charman SA, Arbe-Barnes S, Bathurst IC, Brun R, Campbell M, Charman WN, Chiu FC, Chollet J, Craft JC, Creek DJ et al (2011) Synthetic ozonide drug candidate OZ439 offers new hope for a single-dose cure of uncomplicated malaria. Proc Natl Acad Sci U S A 108:4400–4405

Chauhan SS, Sharma M, Chauhan PM (2010) Trioxaquines: hybrid molecules for the treatment of malaria. Drug News Perspect 23:632–646

Chong CR, Chen X, Shi L, Liu JO, Sullivan DJ Jr (2006) A clinical drug library screen identifies astemizole as an antimalarial agent. Nat Chem Biol 2:415–416

Chugh M, Sundararaman V, Kumar S, Reddy VS, Siddiqui WA, Stuart KD, Malhotra P (2013) Protein complex directs hemoglobin-to-hemozoin formation in *Plasmodium falciparum*. Proc Natl Acad Sci U S A 110:5392–5397

Dana S, Prusty D, Dhayal D, Gupta MK, Dar A, Sen S, Mukhopadhyay P, Adak T, Dhar SK (2014) Potent antimalarial activity of acriflavine in vitro and in vivo. ACS Chem Biol 9:2366–2373

Dar MA, Sharma A, Mondal N, Dhar SK (2007) Molecular cloning of apicoplast-targeted *Plasmodium falciparum* DNA gyrase genes: unique intrinsic ATPase activity and ATP-independent dimerization of PfGyrB subunit. Eukaryot Cell 6:398–412

Darade A, Pathak S, Sharma S, Patravale V (2018) Atovaquone oral bioavailability enhancement using electrospraying technology. Eur J Pharm Sci 111:195–204

Das P, Babbar P, Malhotra N, Sharma M, Jachak GR, Gonnade RG, Shanmugam D, Harlos K, Yogavel M, Sharma A et al (2018) Specific stereoisomeric conformations determine the drug potency of cladosporin scaffold against malarial parasite. J Med Chem 61:5664–5678

Dawre S, Pathak S, Sharma S, Devarajan PV (2018) Enhanced antimalalarial activity of a prolonged release in situ gel of arteether-lumefantrine in a murine model. Eur J Pharm Biopharm 123:95–107

Dell'Agli M, Galli GV, Corbett Y, Taramelli D, Lucantoni L, Habluetzel A, Maschi O, Caruso D, Giavarini F, Romeo S et al (2009) Antiplasmodial activity of Punica granatum L. fruit rind. J Ethnopharmacol 125:279–285

Dende C, Meena J, Nagarajan P, Nagaraj VA, Panda AK, Padmanaban G (2017) Nanocurcumin is superior to native curcumin in preventing degenerative changes in Experimental Cerebral Malaria. Sci Rep 7:10062

Dennis ASM, Lehane AM, Ridgway MC, Holleran JP, Kirk K (2018) Cell swelling induced by the antimalarial KAE609 (Cipargamin) and other PfATP4-associated antimalarials. Antimicrob Agents Chemother 62:e00087

Derbyshire ER, Prudencio M, Mota MM, Clardy J (2012) Liver-stage malaria parasites vulnerable to diverse chemical scaffolds. Proc Natl Acad Sci U S A 109:8511–8516

Deshmukh AS, Srivastava S, Herrmann S, Gupta A, Mitra P, Gilberger TW, Dhar SK (2012) The role of N-terminus of *Plasmodium falciparum* ORC1 in telomeric localization and var gene silencing. Nucleic Acids Res 40:5313–5331

Deu E (2017) Proteases as antimalarial targets: strategies for genetic, chemical, and therapeutic validation. FEBS J 284:2604–2628

Dev N (2019) An infection cocktail: malaria, dengue, chikungunya and Japanese encephalitis. Trop Dr 49:42–43

Dutta G, Bajpai R, Vishwakarma R (1989a) Antimalarial efficacy of arteether against multiple drug resistant strain of *Plasmodium yoelii nigeriensis*. Pharmacol Res 21:415–419

Dutta G, Bajpai R, Vishwakarma R (1989b) Comparison of antimalarial efficacy of artemisinin (qinghaosu) and arteether against *Plasmodium cynomolgi* B infection in monkeys. Trans R Soc Trop Med Hyg 83:56–57

Dwivedi P, Khatik R, Khandelwal K, Srivastava R, Taneja I, Raju KSR, Dwivedi H, Shukla P, Gupta P, Singh S (2014) Self-nanoemulsifying drug delivery systems (SNEDDS) for oral delivery of arteether: pharmacokinetics, toxicity and antimalarial activity in mice. RSC Adv 4:64905–64918

Dwivedi P, Khatik R, Chaturvedi P, Khandelwal K, Taneja I, Raju KS, Dwivedi H, Singh SK, Gupta PK, Shukla P et al (2015) Arteether nanoemulsion for enhanced efficacy against *Plasmodium yoelii nigeriensis* malaria: an approach by enhanced bioavailability. Coll Surf B Biointerf 126:467–475

El Bakkouri M, Pow A, Mulichak A, Cheung KL, Artz JD, Amani M, Fell S, de Koning-Ward TF, Goodman CD, McFadden GI (2010) The Clp chaperones and proteases of the human malaria parasite *Plasmodium falciparum*. J Mol Biol 404:456–477

Figueiredo L, Scherf A (2005) *Plasmodium* telomeres and telomerase: the usual actors in an unusual scenario. Chromosom Res 13:517–524

Figueiredo LM, Rocha EP, Mancio-Silva L, Prevost C, Hernandez-Verdun D, Scherf A (2005) The unusually large *Plasmodium* telomerase reverse-transcriptase localizes in a discrete compartment associated with the nucleolus. Nucleic Acids Res 33:1111–1122

Fong KY, Wright DW (2013) Hemozoin and antimalarial drug discovery. Future Med Chem 5:1437–1450

Gisselberg JE, Dellibovi-Ragheb TA, Matthews KA, Bosch G, Prigge ST (2013) The SUF iron-sulfur cluster synthesis pathway is required for apicoplast maintenance in malaria parasites. PLoS Pathog 9:e1003655

Goldberg DE, Sigala PA (2017) *Plasmodium* heme biosynthesis: to be or not to be essential? PLoS Pathog 13:e1006511

Goodman CD, Pasaje CFA, Kennedy K, McFadden GI, Ralph SA (2016) Targeting protein translation in organelles of the Apicomplexa. Trends Parasitol 32:953–965

Goyal M, Alam A, Iqbal MS, Dey S, Bindu S, Pal C, Banerjee A, Chakrabarti S, Bandyopadhyay U (2012) Identification and molecular characterization of an Alba-family protein from human malaria parasite *Plasmodium falciparum*. Nucleic Acids Res 40:1174–1190

Goyal M, Banerjee C, Nag S, Bandyopadhyay U (2016) The Alba protein family: structure and function. Biochim Biophys Acta 1864:570–583

Gupta A, Mehra P, Dhar SK (2008) *Plasmodium falciparum* origin recognition complex subunit 5: functional characterization and role in DNA replication foci formation. Mol Microbiol 69:646–665

Gupta P, Mehrotra S, Sharma A, Chugh M, Pandey R, Kaushik A, Khurana S, Srivastava N, Srivastava T, Deshmukh A et al (2017) Exploring heme and hemoglobin binding regions of *Plasmodium* Heme Detoxification Protein for new antimalarial discovery. J Med Chem 60:8298–8308

Habib S, Vaishya S, Gupta K (2016) Translation in organelles of Apicomplexan parasites. Trends Parasitol 32:939–952

Hussain T, Yogavel M, Sharma A (2015) Inhibition of protein synthesis and malaria parasite development by drug targeting of methionyl-tRNA synthetases. Antimicrob Agents Chemother 59:1856–1867

Imwong M, Hien TT, Thuy-Nhien NT, Dondorp AM, White NJ (2017a) Spread of a single multidrug resistant malaria parasite lineage (PfPailin) to Vietnam. Lancet Infect Dis 17:1022–1023

Imwong M, Suwannasin K, Kunasol C, Sutawong K, Mayxay M, Rekol H, Smithuis FM, Hlaing TM, Tun KM, van der Pluijm RW (2017b) The spread of artemisinin-resistant *Plasmodium falciparum* in the Greater Mekong subregion: a molecular epidemiology observational study. Lancet Infect Dis 17:491–497

Jaijyan DK, Verma PK, Singh AP (2016) A novel FIKK kinase regulates the development of mosquito and liver stages of the malaria. Sci Rep 6:39285

Jain S, Rathore S, Asad M, Hossain ME, Sinha D, Datta G, Mohmmed A (2013) The prokaryotic ClpQ protease plays a key role in growth and development of mitochondria in *Plasmodium falciparum*. Cell Microbiol 15:1660–1673

Jain V, Yogavel M, Oshima Y, Kikuchi H, Touquet B, Hakimi MA, Sharma A (2015) Structure of Prolyl-tRNA Synthetase-Halofuginone complex provides basis for development of drugs against malaria and toxoplasmosis. Structure 23:819–829

Jain JP, Leong FJ, Chen L, Kalluri S, Koradia V, Stein DS, Wolf M-C, Sunkara G, Kota J (2017) Bioavailability of lumefantrine is significantly enhanced with a novel formulation approach, an outcome from a randomized, open-label pharmacokinetic study in healthy volunteers. Antimicrob Agents Chemother 61:e00868–e00817

Kantamreddi VS, Parida S, Kommula SM, Wright CW (2009) Phytotherapy used in Orissa state, India for treating malaria. Phytother Res 23:1638–1641

Kanyal A, Rawat M, Gurung P, Choubey D, Anamika K, Karmodiya K (2018) Genome-wide survey and phylogenetic analysis of histone acetyltransferases and histone deacetylases of *Plasmodium falciparum*. FEBS J 285:1767–1782

Kapoor M, Gopalakrishnapai J, Surolia N, Surolia A (2004) Mutational analysis of the triclosan-binding region of enoyl-ACP (acyl-carrier protein) reductase from *Plasmodium falciparum*. Biochem J 381:735–741

Kapoor N, Banerjee T, Babu P, Maity K, Surolia N, Surolia A (2009) Design, development, synthesis, and docking analysis of 2′-substituted triclosan analogs as inhibitors for *Plasmodium falciparum* Enoyl-ACP reductase. IUBMB Life 61:1083–1091

Karnik S, Tathed P, Antarkar D, Gidse C, Vaidya R, Vaidya A (2008) Antimalarial activity and clinical safety of traditionally used *Nyctanthes* arbor-tristis Linn. Indian J Tradit Knowl 7:330–334

Ke H, Sigala PA, Miura K, Morrisey JM, Mather MW, Crowley JR, Henderson JP, Goldberg DE, Long CA, Vaidya AB (2014) The heme biosynthesis pathway is essential for *Plasmodium falciparum* development in mosquito stage but not in blood stages. J Biol Chem 289:34827–34837

Khan S, Sharma A, Jamwal A, Sharma V, Pole AK, Thakur KK, Sharma A (2011) Uneven spread of cis- and trans-editing aminoacyl-tRNA synthetase domains within translational compartments of *P. falciparum*. Sci Rep 1:188

Khan S, Garg A, Camacho N, Van Rooyen J, Kumar PA, Belrhali H, Ribas de Pouplana L, Sharma V, Sharma A (2013a) Structural analysis of malaria-parasite lysyl-tRNA synthetase provides a platform for drug development. Acta Crystallogr D Biol Crystallogr 69:785–795

Khan S, Garg A, Sharma A, Camacho N, Picchioni D, Saint-Leger A, Ribas de Pouplana L, Yogavel M, Sharma A (2013b) An appended domain results in an unusual architecture for malaria parasite tryptophanyl-tRNA synthetase. PLoS One 8:e66224

Khan S, Sharma A, Belrhali H, Yogavel M, Sharma A (2014) Structural basis of malaria parasite lysyl-tRNA synthetase inhibition by cladosporin. J Struct Funct Genom 15:63–71

Koreny L, Obornik M, Lukes J (2013) Make it, take it, or leave it: heme metabolism of parasites. PLoS Pathog 9:e1003088

Krudsood S, Wilairatana P, Tangpukdee N, Chalermrut K, Srivilairit S, Thanachartwet V, Muangnoicharoen S, Luplertlop N, Brittenham GM, Looareesuwan S (2006) Safety and tolerability of elubaquine (bulaquine, CDRI 80/53) for treatment of *Plasmodium vivax* malaria in Thailand. Korean J Parasitol 44:221–228

Kulzer S, Charnaud S, Dagan T, Riedel J, Mandal P, Pesce ER, Blatch GL, Crabb BS, Gilson PR, Przyborski JM (2012) *Plasmodium falciparum*-encoded exported hsp70/hsp40 chaperone/co-chaperone complexes within the host erythrocyte. Cell Microbiol 14:1784–1795

Kumar A, Tanveer A, Biswas S, Ram EV, Gupta A, Kumar B, Habib S (2010) Nuclear-encoded DnaJ homologue of *Plasmodium falciparum* interacts with replication ori of the apicoplast genome. Mol Microbiol 75:942–956

Kumar B, Chaubey S, Shah P, Tanveer A, Charan M, Siddiqi MI, Habib S (2011) Interaction between sulphur mobilisation proteins SufB and SufC: evidence for an iron-sulphur cluster biogenesis pathway in the apicoplast of *Plasmodium falciparum*. Int J Parasitol 41:991–999

Kumar A, Moirangthem R, Gahlawat SK, Chandra J, Gupta P, Valecha N, Anvikar A, Singh V (2015) Emergence of sulfadoxine-pyrimethamine resistance in Indian isolates of *Plasmodium falciparum* in the last two decades. Infect Genet Evol 36:190–198

Kumari P, Sahal D, Jain SK, Chauhan VS (2012) Bioactivity guided fractionation of leaves extract of *Nyctanthes arbor tristis* (Harshringar) against *P. falciparum*. PLoS One 7:e51714

Lacerda MV, Llanos-Cuentas A, Krudsood S, Lon C, Saunders DL, Mohammed R, Yilma D, Batista PD, Espino FE, Mia RZ (2019) Single-dose tafenoquine to prevent relapse of *Plasmodium vivax* malaria. N Engl J Med 380:215–228

Lekana-Douki JB, Bhattacharya D, Zatra R, Toure-Ndouo FS (2012) Indian anti-malaria OMARIA is effective against African drug resistant *P. falciparum* field isolates and laboratory strains; without toxicity. Int J Clin Med 3:1

Leong FJ, Li R, Jain JP, Lefevre G, Magnusson B, Diagana TT, Pertel P (2014) A first-in-human randomized, double-blind, placebo-controlled, single- and multiple-ascending oral dose study of novel antimalarial Spiroindolone KAE609 (Cipargamin) to assess its safety, tolerability, and pharmacokinetics in healthy adult volunteers. Antimicrob Agents Chemother 58:6209–6214

Manickam Y, Chaturvedi R, Babbar P, Malhotra N, Jain V, Sharma A (2018) Drug targeting of one or more aminoacyl-tRNA synthetase in the malaria parasite *Plasmodium falciparum*. Drug Discov Today 23:1233–1240

Mastan BS, Narwal SK, Dey S, Kumar KA, Mishra S (2017) *Plasmodium berghei* plasmepsin VIII is essential for sporozoite gliding motility. Int J Parasitol 47:239–245

McCarthy JS, Sekuloski S, Griffin PM, Elliott S, Douglas N, Peatey C, Rockett R, O'Rourke P, Marquart L, Hermsen C et al (2011) A pilot randomised trial of induced blood-stage *Plasmodium falciparum* infections in healthy volunteers for testing efficacy of new antimalarial drugs. PLoS One 6:e21914

McCarthy JS, Marquart L, Sekuloski S, Trenholme K, Elliott S, Griffin P, Rockett R, O'Rourke P, Sloots T, Angulo-Barturen I et al (2016a) Linking murine and human *Plasmodium falciparum* challenge models in a translational path for antimalarial drug development. Antimicrob Agents Chemother 60:3669–3675

McCarthy JS, Ruckle T, Djeriou E, Cantalloube C, Ter-Minassian D, Baker M, O'Rourke P, Griffin P, Marquart L, Hooft van Huijsduijnen R et al (2016b) A Phase II pilot trial to evaluate safety and efficacy of ferroquine against early *Plasmodium falciparum* in an induced blood-stage malaria infection study. Malar J 15:469

McMurry LM, Oethinger M, Levy SB (1998) Triclosan targets lipid synthesis. Nature 394:531–532

Mehrotra N, Lal J, Puri SK, Madhusudanan KP, Gupta RC (2007) In vitro and in vivo pharmacokinetic studies of bulaquine (analogue of primaquine), a novel antirelapse antimalarial, in rat, rabbit and monkey--highlighting species similarities and differences. Biopharm Drug Dispos 28:209–227

Ménard D, Clain J, Ariey F (2018) Multidrug-resistant *Plasmodium falciparum* malaria in the greater Mekong subregion. Lancet Infect Dis 18:238–239

Mendes AM, Albuquerque IS, Machado M, Pissarra J, Meireles P, Prudencio M (2017) Inhibition of *Plasmodium* liver infection by Ivermectin. Antimicrob Agents Chemother 61:e02005–e02016

Mishra S, Asthana O, Mohanty S, Patnaik J, Das B, Srivastava J, Satpathy S, Dash S, Rath P, Varghese K (1995) Effectiveness of α, β-arteether in acute falciparum malaria. Trans R Soc Trop Med Hyg 89:299–301

Mishra V, Rathore I, Arekar A, Sthanam LK, Xiao H, Kiso Y, Sen S, Patankar S, Gustchina A, Hidaka K et al (2018) Deciphering the mechanism of potent peptidomimetic inhibitors targeting plasmepsins - biochemical and structural insights. FEBS J 285:3077–3096

Mishra M, Singh V, Singh S (2019) Structural insights into key *Plasmodium* proteases as therapeutic drug targets. Front Microbiol 10:394

Misra A (2017). https://thewire.in/science/arteether-malaria-cdri-csir-research-funding

Mohanty S, Mishra S, Satpathy S, Dash S, Patnaik J (1997) α, β-Arteether for the treatment of complicated falciparum malaria. Trans R Soc Trop Med Hyg 91:328–330

Moura PA, Dame JB, Fidock DA (2009) Role of *Plasmodium falciparum* digestive vacuole plasmepsins in the specificity and antimalarial mode of action of cysteine and aspartic protease inhibitors. Antimicrob Agents Chemother 53:4968–4978

Mukim A, Shukla M, Patel J, Patel H, Naik S, Pandyas N, Sanghani K (2011) A phase III clinical trial of alpha, beta-arteether injection 150 mg/ml in patients of *P. falciparum* malaria. J Indian Med Assoc 109:597–599

Mundra S, Thakur V, Bello AM, Rathore S, Asad M, Wei L, Yang J, Chakka SK, Mahesh R, Malhotra P et al (2017) A novel class of Plasmodial ClpP protease inhibitors as potential antimalarial agents. Bioorg Med Chem 25:5662–5677

Nagano S, Lin TY, Edula JR, Heddle JG (2014) Unique features of apicoplast DNA gyrases from *Toxoplasma gondii* and *Plasmodium falciparum*. BMC Bioinformatics 15:416

Nagappa LK, Singh D, Dey S, Kumar KA, Balaram H (2019) Biochemical and physiological investigations on adenosine 5′ monophosphate deaminase from *Plasmodium* spp. Mol Microbiol 112:699. https://doi.org/10.1111/mmi.14313

Nagaraj VA, Padmanaban G (2017) Insights on heme synthesis in the malaria parasite. Trends Parasitol 33:583–586

Nagaraj VA, Sundaram B, Varadarajan NM, Subramani PA, Kalappa DM, Ghosh SK, Padmanaban G (2013) Malaria parasite-synthesized heme is essential in the mosquito and liver stages and complements host heme in the blood stages of infection. PLoS Pathog 9:e1003522

Nandakumar DN, Nagaraj VA, Vathsala PG, Rangarajan P, Padmanaban G (2006) Curcumin-artemisinin combination therapy for malaria. Antimicrob Agents Chemother 50:1859–1860

Nasamu AS, Glushakova S, Russo I, Vaupel B, Oksman A, Kim AS, Fremont DH, Tolia N, Beck JR, Meyers MJ et al (2017) Plasmepsins IX and X are essential and druggable mediators of malaria parasite egress and invasion. Science 358:518–522

Newby G, Bennett A, Larson E, Cotter C, Shretta R, Phillips AA, Feachem RG (2016) The path to eradication: a progress report on the malaria-eliminating countries. Lancet 387:1775–1784

Njue M, Njuguna P, Kapulu MC, Sanga G, Bejon P, Marsh V, Molyneux S, Kamuya D (2018) Ethical considerations in Controlled Human Malaria Infection studies in low resource settings: experiences and perceptions of study participants in a malaria Challenge study in Kenya. Well Open Res 3:39

Nyunt MM, Hendrix CW, Bakshi RP, Kumar N, Shapiro TA (2009) Phase I/II evaluation of the prophylactic antimalarial activity of pafuramidine in healthy volunteers challenged with *Plasmodium falciparum* sporozoites. Am J Trop Med Hyg 80:528–535

Panchal M, Rawat K, Kumar G, Kibria KM, Singh S, Kalamuddin M, Mohmmed A, Malhotra P, Tuteja R (2014) *Plasmodium falciparum* signal recognition particle components and anti-parasitic effect of ivermectin in blocking nucleo-cytoplasmic shuttling of SRP. Cell Death Dis 5:e994

Panda SK, Luyten W (2018) Antiparasitic activity in Asteraceae with special attention to ethnobotanical use by the tribes of Odisha, India. Parasite (Paris, France) 25:10

Pandey AV, Tekwani BL, Singh RL, Chauhan VS (1999) Artemisinin, an endoperoxide antimalarial, disrupts the hemoglobin catabolism and heme detoxification systems in malarial parasite. J Biol Chem 274:19383–19388

Pazhayam NM, Chhibber-Goel J, Sharma A (2019) New leads for drug repurposing against malaria. Drug Discov Today 24:263–271

Pino P, Caldelari R, Mukherjee B, Vahokoski J, Klages N, Maco B, Collins CR, Blackman MJ, Kursula I, Heussler V et al (2017) A multistage antimalarial targets the plasmepsins IX and X essential for invasion and egress. Science 358:522–528

Prabhu P, Suryavanshi S, Pathak S, Patra A, Sharma S, Patravale V (2016a) Nanostructured lipid carriers of artemether–lumefantrine combination for intravenous therapy of cerebral malaria. Int J Pharm 513:504–517

Prabhu P, Suryavanshi S, Pathak S, Sharma S, Patravale V (2016b) Artemether lumefantrine nanostructured lipid carriers for oral malaria therapy: enhanced efficacy at reduced dose and dosing frequency. Int J Pharm 511:473–487

Prasad R, Atul, Soni A, Puri SK, Sijwali PS (2012) Expression, characterization, and cellular localization of knowpains, papain-like cysteine proteases of the *Plasmodium knowlesi* malaria parasite. PLoS One 7:e51619

Prasad R, Atul, Kolla VK, Legac J, Singhal N, Navale R, Rosenthal PJ, Sijwali PS (2013) Blocking *Plasmodium falciparum* development via dual inhibition of hemoglobin degradation and the ubiquitin proteasome system by MG132. PLoS One 8:e73530

Puri SK, Dutta GP (2003) Blood schizontocidal activity of WR 238605 (Tafenoquine) against *Plasmodium cynomolgi* and *Plasmodium fragile* infections in rhesus monkeys. Acta Trop 86:35–40

Puri SK, Dutta GP (2005) Plasmodium cynomolgi: gametocytocidal activity of the anti-malarial compound CDRI 80/53 (elubaquine) in rhesus monkeys. Exp Parasitol 111:8–13

Raghu Ram EV, Kumar A, Biswas S, Kumar A, Chaubey S, Siddiqi MI, Habib S (2007) Nuclear gyrB encodes a functional subunit of the Plasmodium falciparum gyrase that is involved in apicoplast DNA replication. Mol Biochem Parasitol 154:30–39

Raza M, Bharti H, Singal A, Nag A, Ghosh PC (2018) Long circulatory liposomal maduramicin inhibits the growth of *Plasmodium falciparum* blood stages in culture and cures murine models of experimental malaria. Nanoscale 10:13773–13791

Reddy RC, Vatsala PG, Keshamouni VG, Padmanaban G, Rangarajan PN (2005) Curcumin for malaria therapy. Biochem Biophys Res Commun 326:472–474

Rizzi L, Sundararaman S, Cendic K, Vaiana N, Korde R, Sinha D, Mohmmed A, Malhotra P, Romeo S (2011) Design and synthesis of protein-protein interaction mimics as *Plasmodium falciparum* cysteine protease, falcipain-2 inhibitors. Eur J Med Chem 46:2083–2090

Rottmann M, McNamara C, Yeung BK, Lee MC, Zou B, Russell B, Seitz P, Plouffe DM, Dharia NV, Tan J et al (2010) Spiroindolones, a potent compound class for the treatment of malaria. Science 329:1175–1180

Sahu PS, Sahu M, Ambu S (2016) A review of concurrent infections of malaria and dengue in Asia. Asian Pac J Trop Biomed 6:633–638

Sato S, Clough B, Coates L, Wilson RJ (2004) Enzymes for heme biosynthesis are found in both the mitochondrion and plastid of the malaria parasite *Plasmodium falciparum*. Protist 155:117–125

Schrader FC, Glinca S, Sattler JM, Dahse HM, Afanador GA, Prigge ST, Lanzer M, Mueller AK, Klebe G, Schlitzer M (2013) Novel type II fatty acid biosynthesis (FAS II) inhibitors as multistage antimalarial agents. ChemMedChem 8:442–461

Shafi H, Reddy DVS, Khan T, Ranjan R, Srivastava A, Vaishya S, Sharma T, Siddiqui MI, Habib S, Misra A (2017) Dehydroascorbate-derivatized chitosan particles for targeting antimalarial agents to infected erythrocytes. Int J Pharm 524:205–214

Shafiq N, Rajagopalan S, Kushwaha HN, Mittal N, Chandurkar N, Bhalla A, Kaur S, Pandhi P, Puri GD, Achuthan S et al (2014) Single ascending dose safety and pharmacokinetics of CDRI-97/78: first-in-human study of a novel antimalarial drug. Mal Res Treat 2014:372521

Shankar R, Deb S, Sharma BK (2012) Antimalarial plants of northeast India: an overview. J Ayurveda Integr Med 3:10–16

Sharma V, Amarnath N, Shukla S, Ayana R, Kumar N, Yadav N, Kannan D, Sehrawat S, Pati S, Lochab B et al (2018) Benzoxazine derivatives of phytophenols show anti-plasmodial activity via sodium homeostasis disruption. Bioorg Med Chem Lett 28:1629–1637

Shears MJ, Botte CY, McFadden GI (2015) Fatty acid metabolism in the Plasmodium apicoplast: drugs, doubts and knockouts. Mol Biochem Parasitol 199:34–50

Simonsen HT, Nordskjold JB, Smitt UW, Nyman U, Palpu P, Joshi P, Varughese G (2001) In vitro screening of Indian medicinal plants for antiplasmodial activity. J Ethnopharmacol 74:195–204

Singh C, Chaudhary S, Kanchan R, Puri SK (2007a) Conversion of antimalarial drug artemisinin to a new series of tricyclic 1,2,4-trioxanes1. Org Lett 9:4327–4329

Singh C, Kanchan R, Sharma U, Puri SK (2007b) New adamantane-based spiro 1,2,4-trioxanes orally effective against rodent and simian malaria. J Med Chem 50:521–527

Singh AP, Surolia N, Surolia A (2009) Triclosan inhibit the growth of the late liver-stage of *Plasmodium*. IUBMB Life 61:923–928

Singh S, Agarwal D, Sharma K, Sharma M, Nielsen MA, Alifrangis M, Singh AK, Gupta RD, Awasthi SK (2016) 4-Aminoquinoline derivatives: synthesis, in vitro and in vivo antiplasmodial activity against chloroquine-resistant parasites. Eur J Med Chem 122:394–407

Singh S, Rajendran V, He J, Singh AK, Achieng AO, Vandana, Pant A, Nasamu AS, Pandit M, Singh J et al (2019) Fast-acting small molecules targeting malarial Aspartyl Proteases, Plasmepsins, inhibit malaria infection at multiple life stages. ACS Infect Dis 5:184–198

Spring MD, Lin JT, Manning JE, Vanachayangkul P, Somethy S, Bun R, Se Y, Chann S, Ittiverakul M, Sia-ngam P et al (2015) Dihydroartemisinin-piperaquine failure associated with a triple mutant including kelch13 C580Y in Cambodia: an observational cohort study. Lancet Infect Dis 15:683–691

Stanisic DI, McCarthy JS, Good MF (2018) Controlled human malaria infection: applications, advances, and challenges. Infect Immun 86:e00479

Subramanian G, Belekar MA, Shukla A, Tong JX, Sinha A, Chu TTT, Kulkarni AS, Preiser PR, Reddy DS, Tan KSW et al (2018) Targeted phenotypic screening in *Plasmodium falciparum* and *Toxoplasma gondii* reveals novel modes of action of Medicines for Malaria Venture Malaria Box molecules. mSphere 3:e00534

Sulyok M, Ruckle T, Roth A, Murbeth RE, Chalon S, Kerr N, Samec SS, Gobeau N, Calle CL, Ibanez J et al (2017) DSM265 for *Plasmodium falciparum* chemoprophylaxis: a randomised, double blinded, phase 1 trial with controlled human malaria infection. Lancet Infect Dis 17:636–644

Surolia N, Padmanaban G (1992) de novo biosynthesis of heme offers a new chemotherapeutic target in the human malarial parasite. Biochem Biophys Res Commun 187:744–750

Surolia N, Surolia A (2001) Triclosan offers protection against blood stages of malaria by inhibiting enoyl-ACP reductase of *Plasmodium falciparum*. Nat Med 7:167–173

Thanh NV, Thuy-Nhien N, Tuyen NT, Tong NT, Nha-Ca NT, Dong LT, Quang HH, Farrar J, Thwaites G, White NJ et al (2017) Rapid decline in the susceptibility of *Plasmodium falciparum* to dihydroartemisinin-piperaquine in the south of Vietnam. Malar J 16:27

Thriemer K, Hong NV, Rosanas-Urgell A, Phuc BQ, Ha DM, Pockele E, Guetens P, Van NV, Duong TT, Amambua-Ngwa A, D'Alessandro U, Erharta A (2014) Delayed parasite clearance after treatment with Dihydroartemisinin-Piperaquine in *Plasmodium falciparum* malaria patients in central Vietnam. Antimicrob Agents Chemother 58:7049

Toure OA, Valecha N, Tshefu AK, Thompson R, Krudsood S, Gaye O, Rao BHK, Sagara I, Bose TK, Mohanty S et al (2016) A phase 3, double-blind, randomized study of arterolane maleate-piperaquine phosphate vs artemether-lumefantrine for falciparum malaria in adolescent and adult patients in Asia and Africa. Clin Infect Dis 62:964–971

Tripathi R, Dutta GP, Vishwakarma RA (1996) Gametocytocidal activity of alpha/beta arteether by the oral route of administration. Am J Trop Med Hyg 54:652–654

Vale N, Moreira R, Gomes P (2009) Primaquine revisited six decades after its discovery. Eur J Med Chem 44:937–953

Valecha N, Devi CU, Joshi H, Shahi V, Sharma V, Lal S (2000) Comparative efficacy of Ayush-64 vs chloroquine in vivax malaria. Curr Sci 78:1120–1122

Valecha N, Adak T, Bagga A, Asthana O, Srivastava J, Joshi H, Sharma V (2001) Comparative antirelapse efficacy of CDRI compound 80/53 (Bulaquine) vs primaquine in double blind clinical trial. Curr Sci 80:561–563

Valecha N, Looareesuwan S, Martensson A, Abdulla SM, Krudsood S, Tangpukdee N, Mohanty S, Mishra SK, Tyagi PK, Sharma SK et al (2010) Arterolane, a new synthetic trioxolane for treatment of uncomplicated *Plasmodium falciparum* malaria: a phase II, multicenter, randomized, dose-finding clinical trial. Clin Infect Dis 51:684–691

Vathsala PG, Dende C, Nagaraj VA, Bhattacharya D, Das G, Rangarajan PN, Padmanaban G (2012) Curcumin-arteether combination therapy of *Plasmodium berghei*-infected mice prevents recrudescence through immunomodulation. PLoS One 7:e29442

Vennerstrom JL, Arbe-Barnes S, Brun R, Charman SA, Chiu FC, Chollet J, Dong Y, Dorn A, Hunziker D, Matile H et al (2004) Identification of an antimalarial synthetic trioxolane drug development candidate. Nature 430:900–904

Verma S, Pandey S, Agarwal P, Verma P, Deshpande S, Saxena JK, Srivastava K, Chauhan PM, Prabhakar YS (2016) N-(7-Chloroquinolinyl-4-aminoalkyl) arylsulfonamides as antimalarial agents: rationale for the activity with reference to inhibition of hemozoin formation. RSC Adv 6:25584–25593

Weissig V, Vetro-Widenhouse TS, Rowe TC (1997) Topoisomerase II inhibitors induce cleavage of nuclear and 35-kb plastid DNAs in the malarial parasite *Plasmodium falciparum*. DNA Cell Biol 16:1483–1492

White NJ, Pukrittayakamee S, Phyo AP, Rueangweerayut R, Nosten F, Jittamala P, Jeeyapant A, Jain JP, Lefevre G, Li R et al (2014) Spiroindolone KAE609 for falciparum and vivax malaria. N Engl J Med 371:403–410

Yadav N, Sharma C, Awasthi SK (2014) Diversification in the synthesis of antimalarial trioxane and tetraoxane analogs. RSC Adv 4:5469–5498

Yadav N, Agarwal D, Kumar S, Dixit AK, Gupta RD, Awasthi SK (2018) In vitro antiplasmodial efficacy of synthetic coumarin-triazole analogs. Eur J Med Chem 145:735–745

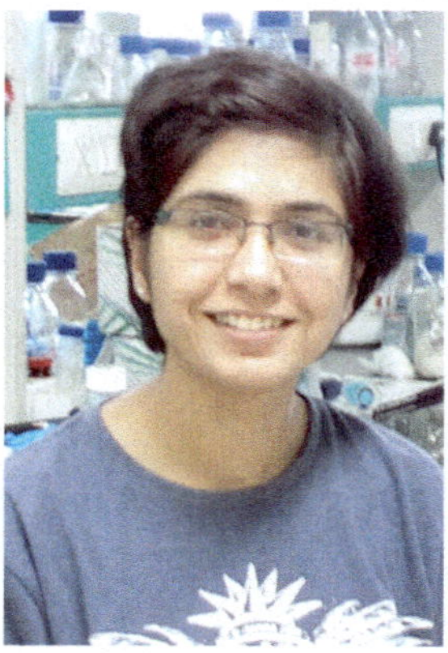

**Niti Kumar** conducted her postdoctoral research work on ribosome-nascent chain complexes at Max Planck Institute of Biochemistry, Munich, Germany. She is now an independent group leader at CSIR-Central Drug Research Institute, Lucknow, India. Her lab is interested in understanding how human malaria parasite is able to keep its metastable and aggregation-prone proteome in soluble and functional state. Her lab conducts research on the protein quality control machinery (protein folding and degradation machinery) in human malaria parasite. She is also involved in antimalarial drug screening and understanding their mode of action.

**Saman Habib** is coordinator of the Parasitic Diseases Area and also manages preclinical development of candidate antimalarial leads at CSIR-Central Drug Research Institute, Lucknow, India. Her group works on *Plasmodium* organelle biology with focus on protein translation machinery, genome organization, replication and repair, and [Fe-S] cluster assembly pathways in the apicoplast and mitochondrion of the malaria parasite. She has also investigated host genetic factors, particularly polymorphisms in human host molecules involved in adhesion, rosetting, and invasion of the malaria parasite in susceptibility/resistance to falciparum malaria in areas of differential disease endemicity in India.

# Modern Drug Discovery and Development in the Area of Leishmaniasis

5

Neena Goyal, Manveer Patel, and Sanjay Batra

## 5.1 A Brief Introduction About the Disease and the Causative Agent

Parasites of the genus Leishmania cause leishmaniasis in humans and in a variety of vertebrate hosts that include canids, rodents, marsupials and hyraxes (McGwire and Satoskar 2013). In humans, clinical manifestations produced by the parasite comprise of the visceral (VL) and tegumentary forms like cutaneous (CL), diffuse cutaneous (DCL) and mucocutaneous (MCL) leishmaniasis (WHO 2019). However, in many cases, these infections remain asymptomatic (Ostyn et al. 2011; Andrade-Narvaez et al. 2016). VL (kala-azar) is the most severe form of leishmaniasis characterized by prolonged fever, pancytopenia, splenomegaly, weight loss and hyper-gammaglobulinemia; and leads to death if left untreated. Approximately, 20 *Leishmania* species are pathogenic for humans and they spread by the bite of an infected sandfly (*Phlebotomus* in the Old World and Lutzomyia in the New World) (de Freitas et al. 2016). The disease has a wide geographical occurrence, covering 98 countries and territories worldwide (WHO 2017) with 350 million

N. Goyal (✉)
Biochemistry Division, CSIR - Central Drug Research Institute, Lucknow, India

Academy of Scientific and Innovative Research, CSIR - Human Resource Development Centre, Ghaziabad, Uttar Pradesh, India

M. Patel
Medicinal and Process Chemistry Division, CSIR - Central Drug Research Institute, Lucknow, India

S. Batra
Academy of Scientific and Innovative Research, CSIR - Human Resource Development Centre, Ghaziabad, Uttar Pradesh, India

Medicinal and Process Chemistry Division, CSIR - Central Drug Research Institute, Lucknow, India



M. Dikshit (ed.), *Drug Discovery and Drug Development*,
https://doi.org/10.1007/978-981-15-8002-4_5

people at risk. The annual case incidence ranges from 70,000 to one million and reported deaths range from 26,000 to 65,000 (de Vries et al. 2015). The disease is prevalent both in South and the Central America, Southern Europe, Africa, Middle East, Central Asia and Indian subcontinent. Leishmaniasis, constitutes a major global public health problem showing an increasing burden over the last decade (Alvar et al. 2012). Leishmaniasis has been identified as a category I disease (emerging and uncontrolled) by the World Health Organization (WHO), and the resolution 43.18 of World Health Assembly (WHA) recognizes it to be a major public health concern (WHO 2004).

*Leishmania* is a dimorphic parasite (Protozoa, Kinetoplastida, Trypanosomatidae) that resides in two forms in the sand fly vector and the mammalian host (Molyneux and Killick-Kendrick 1987). Whereas in the sandfly, the parasite lives as flagellated promastigote in the gut lumen, in mammalian host, it multiplies as aflagellated amastigotes within the macrophages (Handman 1999). The conversion from the promastigote (procyclic, metacyclic) to the amastigote stage and vice versa is essential for parasite survival and pathogenesis, and is allied to the parasite's response to the environmental changes during their transmission from poikilothermic, hematophagous insects to homeothermic mammals and vice versa. The amastigote survival within macrophage is facilitated by mechanisms such as tolerance to stress, morphogenesis, resistance to complement lysis, changes in surface molecules and its ability to modulate many effecter and accessory functions of host macrophages. Similarly, promastigotes with in mid gut of sand fly undergo changes to develop into metacyclic forms, which is able to cause infection in mammalian host. Thus, this differentiation process involves major morphological changes in parasites and the retooling of several metabolic processes (Kima 2014).

It is worth mentioning that recently several excellent reviews pertaining to different aspects including drug discovery against leishmaniasis have been published (Freitas-Junior et al. 2012; Nagle et al. 2014; Ferreira and Andricopulo 2018; Alcântara et al. 2018; Vijayakumar and Das 2018; Burza et al. 2018; Lindoso et al. 2018; Sundar et al. 2019; Reguera et al. 2019). Therefore, present review discusses the disease more in Indian context and includes various medicinal chemistry efforts made worldwide (reports published after 2015) for identifying new antileishmanial agents.

## 5.2 Global Distribution and Status of the Disease in India

There is an estimated 0.7–1.2 million cases of CL and 0.2–0.4 million cases of VL in the endemic regions on five continents (Alvar et al. 2012). Often leishmaniasis are zoonoses, which affect the poors in rural and natural areas, where many domestic and wild reservoir hosts and sand fly vectors maintain the infection (Ready 2013). It is recorded that the worldwide VL pandemic has affected at least 55 countries (Pigott et al. 2014), located primarily in South America, East Africa and in South-East Asia. It is also reported that out of an estimated 50,000–90,000 new cases of VL which occur worldwide each year, only an approximately 25–45% are disclosed to WHO. In 2017, more than 95% of new cases reported to WHO were from ten countries including Bangladesh, Brazil, China, Ethiopia, India, Kenya, Nepal, Somalia, South Sudan and Sudan (World Health Organization Global Health Observatory 2017). Historically the VL outbreaks are known to occur in developing or agricultural countries (WHO 2019). A comprehensive update on VL describing the endemicity, treatments and pitfalls was recently presented (Alves et al. 2018).

In India, VL is endemic majorly in the state of Bihar and parts of West Bengal, Jharkhand and East Uttar Pradesh. It was earlier disclosed that Bihar, an east Indian state with a population of 110 million carries an estimated 40% of the world's VL burden (Bora 1999; WHO 2011). The detection of active cases carefully, revealed that in Bihar, the annual VL incidence per 100,000 population ranges from 298 to 380 cases (Mondal et al. 2009). Further, it was reported that after recovery from VL, about 50% of patients in Sudan and approximately 10% in Indian subcontinent develop Post Kala-azar Dermal Leishmaniasis (PKDL) (Zijlstra et al. 2003). PKDL is a dermal manifestation characterised by macular, maculopapular or nodular rash and this form of disease is very difficult to treat, especially in some patients in East Africa (Desjeux et al. 2013). Moreover, PKDL patients are suggested to be a reservoir for transmission of VL (Sen Gupta 1968).

It is reported that in more than 35 countries, HIV-VL coinfection prevails (Alvar et al. 2008). Subjects of HIV not only carries increased risk of developing VL but also have high parasite load, poor therapeutic response, and increased chances of relapse. As per the WHO report, one third of all HIV patients worldwide are located in the regions where leishmaniasis is endemic (WHO 1995). Initially, it was a disease of Mediterranean region; however, it is now increasing in numbers in India, Brazil and Ethiopia (Burza et al. 2014; Cota et al. 2014; Mengesha et al. 2014). Although it is estimated that India has prevalence of HIV-VL coinfection in the range from 2% to 5.6%, there is lack of records for it (Redhu et al. 2006; Mathur et al. 2006). It is assumed that Bihar state though has low prevalence of HIV (between 0.22% and 0.33%), its high population translate into an estimate of 300,000 people with HIV/AIDS living there (Pandey et al. 2012).

## 5.3 Elimination Program in India and Its Challenges

In 2005, the Ministers of Health of Bangladesh, India and Nepal affirmed intercountry cooperation and cross border collaboration to eliminate VL by 2015 i.e. reduce the incidence to less than 10 per 10,000 people (WHO 2005) and PKDL incidence to 0 by 2018 (Govt. of Nepal-report 2009). Conversely, for rest of the world, where VL is mainly zoonotic and caused by other parasite species, the WHO did not set any elimination deadline but aims for 100% detection and treatment of human cases. As per the resolution 60.13 of WHA in 2007, WHO is supposed to update the epidemiological evidence and provide technical assistance in initiation, maintenance, and expansion of leishmaniasis control programs. For achieving the elimination of VL, as strategized and updated in 2012 (WHO 2012), it is essential to perform early diagnosis and complete case management, effective disease and vector surveillance, social mobilization and building partnerships, and clinical and operational research.

In addition, the governments of India, Bangladesh, Nepal, Bhutan and Thailand signed a memorandum of understanding for setting an ambitious goal of eliminating VL as a public health problem (at subdistrict level in India and Bangladesh, and district-level in Nepal and Bhutan) by or before the end of 2017 (WHO South-East Asia 2014). Presently, the cases of VL in Bhutan and Thailand are currently very low whereas after a decade after the launch of the Regional Strategic Framework for VL Elimination in 2005, there is a substantial (more than 75%) reduction of incident of VL cases in the Indian subcontinent (WHO 2015). In contrast, Nepal reached the targeted low incidence level in 2014 and has maintained it for 2 years (WHO 2015). Notably, in Bangladesh and India too, the target-level incidence of 90% and 70% reduction in endemic subdistricts, respectively by 2015 was achieved (WHO 2015).

## 5.4 Present Chemotherapy and Its Limitations

In the absence of vaccine, the chemotherapy remains the most practical and effective way to treat all three major forms of leishmaniasis. Unfortunately however, not many drugs are available to treat Leishmaniasis and each drug has its own benefits and limitations. The mainstay of Leishmania chemotherapy included sodium stibogluconate (SSG) and meglumineantimoniate, the pentavalent antimony at dose 10 mg/kg for 10 days but since the reports for treatment failure against VL emerged in the endemic area of Bihar, India (Tiuman et al. 2011; Mukhopadhyay et al. 2011) its use became restricted. The major reason for antimony resistance was attributed to its widespread misuse, which was endorsed by a study which disclosed that only 26% of the patients were treated according to the WHO guidelines (Sundar et al. 1994, 2019; Sundar and Chatterjee 2006). The other major drawbacks are their parentral administration, constant monitoring due to hepatic and cardiotoxicity. It is also not recommended for HIV-VL coinfected patients since they are less efficacious and more toxic due to presence of antimony (Diro et al. 2014). Resistance together with restricted benefits (Ponte-Sucre et al. 2017) has discontinued the prescriptions for antimonials in the Indian subcontinent for the treatment of VL. Nonetheless in Africa for VL, it is recommended as combination therapy with paromomycin whereas as a monotherapy for PKDL. However, antimonials as monotherapy or in combination with other drugs are still the drugs of choice for CL and MCL (WHO 2010).

Pentamidine (**1**) and amphotericin B (AmB) (**2**) (Fig. 5.1) are listed as the second line treatments for leishmaniasis. In India, pentamidine was used as alternate for treating cases of refractory VL. However, side effects such as insulin-dependent diabetes mellitus and resistance to it were deterrant for its further usage (Jha et al. 1991). Other side effects include pain at injection site, abscess, nausea, vomiting, myalgia, headache, dizziness, hypotension, etc. (Sundar and Chakravarty 2015). In spite of such side effects, the drug is prescribed for secondary prophylaxis in HIV-VL coinfection cases (Diro et al. 2015). In India, by 1990 it was replaced with amphotericin B deoxycholate as the recommended treatment (Sundar 2001).

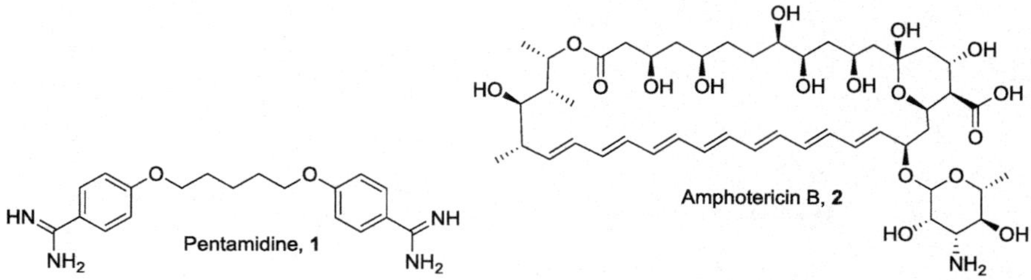


**Fig. 5.1** Structures of Pentamidine and Amphotericin B

AmB, a known antifungal agent, cures the leishmania (100%) at dosage between 0.75 and 1.00 mg/kg via 15–20 infusions either daily or on alternate days (Thakur et al. 1999). However, due to very low oral bioavailability of the drug, the parenteral administrations are done in systemic infections, which require hospitalization. Due to its insolubility in aqueous media, it is mixed with deoxycholate micelles (fungizone) to form the micelles which are unstable in the bloodstream and release AmB that is absorbed particularly in kidneys thereby causing nephrotoxicity (Guo and Working 1993; Espuelas et al. 2003). To avoid these toxic effects, lipid formulations of AmB were developed and they provide targeted drug delivery to most affected organs like liver and spleen. Three commercially available formulations are used worldwide against VL viz. Liposomized AmB (L-AmB), AmB lipid complex, and AmB cholesterol dispersion. Among them, the L-AmB has been used extensively in the Indian subcontinent as its single dose of 10 mg/kg has an efficacy of 95% thereby making it the treatment of choice for VL (Sundar et al. 2010). In contrast, in the Mediterranean region, South America, Sudan and Ethiopia, a higher dose (18–50 mg/kg) is recommended to treat complicated Leishmaniasis (Tamiru et al. 2016; Ritmeijer et al. 2011). This drug is also being recommended for the treatment of HIV-VL coinfection at a dose of 4 mg/kg for 10 doses (days 1–5, 10, 17, 24, 31, and 39) up to a total dose of 40 mg/kg (WHO 2010). Though the reports of treatment failure with AmB are rare, it is disclosed that in both immune-competent and immune-suppressed patients the resistant strain can be isolated (Srivastava et al. 2011; Morizot et al. 2016). Very recently, it was suggested that owing to the widespread use of single dose of L-AmB in the Indian subcontinent, monitoring of the efficacy of the regimen with the drug and early detection of treatment failures for preventing development and spread of L-AmB resistance (Sundar et al. 2019). The first case of AmB-resistance has already appeared in India (Purkait et al. 2012; Srivastava et al. 2011), where isolated parasites showed clear resistance to drug (Purkait et al. 2012).

Miltefosine (**3**), originally discovered as an anticancer agent, was repurposed to be the first oral drug for treating VL and CL in India in 2002 together with Germany, Colombia and Bangladesh (Dorlo et al. 2012; Rahman et al. 2011) (Fig. 5.2). It is considered to be the third-line anti-leishmanial drug (Croft et al. 2003). Although it is orally efficacious, it too has major limitations which include high cost (Sundar and Chakravarty 2013), gastrointestinal side effects and most important teratogenicity

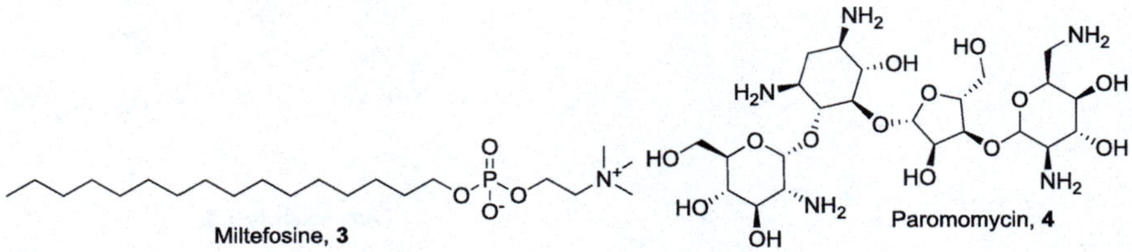


**Fig. 5.2** Structures of Miltefosine and Paromomycin

and long lifespan, which results in growing concerns for increased clinical relapse (Prajapati et al. 2012). This drug cannot be advised to pregnant women (Wadhone et al. 2009). The appearance of miltefosine resistance in India was reported in 2012, as the relapse rate was doubled (Sundar et al. 2012). In Nepal and Bangladesh, the relapse rate has reached upto 20% and 15%, respectively (Rijal et al. 2013; Rahman et al. 2011). Indeed, since 2014, miltefosine was replaced by single dose (10 mg/kg) L-AmB treatment as the treatment of choice in the elimination program in the Indian subcontinent.

Paromomycin (PM) (**4**), is an aminoglycoside and is used an alternative drug for treating Leishmaniasis in the form of both parenteral and topical formulations (Krause and Kroeger 1994) (Fig. 5.2). PM showed 94.6% cure rate against VL in a Phase III clinical trial study conducted in India (Sundar et al. 2007) and was approved for VL treatment in the Indian subcontinent in 2006. However, PM was ineffective in curing PKDL and had an efficacy of 37.5% only (Sundar et al. 2014). In East Africa multicentre trial, PM was found significantly less effective than SSG (Musa et al. 2012) though for CL, systemic PM had excellent cure rates in Brazil (Correia et al. 1996) and low efficacy for MCL (Romero et al. 1996). Due to its limited use, resistance with PM in VL is not known though two drug-resistant *L. aethiopica* isolates were obtained from patients of relapsed Leishmaniasis (Chakravarty and Sundar 2010).

### 5.4.1 Combination Therapy

Successful use of combination therapy for infectious diseases such as malaria, HIV and tuberculosis, provides awareness for the use of combination therapies in Leishmaniasis too as it is believed that it may offer substantial benefits, including overcoming resistance. The rationales for using combination therapy in Leishmaniasis include (a) increase in the efficacy at low dose due to synergistic or additive activity of two drugs with different mechanism of action, (b) reduce the treatment duration hence the cost, and (c) reduce the chances of emergence of resistance to antileishmanial drugs (Olliaro et al. 2005). It is reported that in East Africa, a combination of PM and SSG for 17 days' resulted in higher cure and survival rates compared to $Sb^V$ alone (Melaku et al. 2007). In another in vivo study it was reported that the combination showed the potentiation of miltefosine activity with AmB as well as PM (Sundar et al. 2008). According to WHO recommendations

(WHO 2010) for combination therapy in Indian VL patients, single dose of L-AmB together with miltefosine (Sundar et al. 2011a) or a single dose of L-AmB plus PM (Sundar et al. 2011b) is suggested to control the disease though possibility of Leishmania developing resistance to this combination cannot be ruled out (Hendrickx et al. 2016). Although it is hypothesized that it is more difficult for Leishmania to acquire resistance to combination therapy over monotherapy, resistance to combinations is distinct possibility (especially when PM is one of the partner drugs) (Ponte-Sucre et al. 2017). Combination therapy has also been cited to be an attractive option for treating the PKDL (Ramesh et al. 2014).

## 5.5 Discovery of New Chemical Entities for Antileishmanial Chemotherapy and Present Scenario of the Drug Pipeline

### 5.5.1 Pipeline (Leads Under Development)

As evident from the preceding text, the present chemotherapy mostly involves prolonged parenteral administration, which leads to complications like high cost, poor patient compliance, blood-borne disease from unsanitary conditions and the need for hospitalization. Therefore, new drug discovery efforts are directed towards developing new oral therapies. In this context, repositioning or repurposing of existing drugs with new applications for VL or new formulations of known antileishmanial drugs, have become increasingly common (Ashburn and Thor 2004; Aubé 2012; Andrews et al. 2014; Klug et al. 2016; Wu et al. 2019). It is known that use of drug repositioning strategy essentially circumvents several development phases to develop NCE drugs thereby shortening the time period required for development. Thus repurposing offers a faster path to new drug at much reduced cost.

#### 5.5.1.1 Nitroimidazoles

Several compounds have been explored to discover new chemical agent for treating leishmaniasis. The most pursued chemical class is of nitroimidazoles and its fused heterocyclic analogues. One of the nitroimidazoles, the fexinidazole (**5**), which has displayed significant activity against trypanosomiasis (Jameson et al. 2010) has also shown promising antileishmanial effect (Fig. 5.3). In clinical trials, the efficacy of fexinidazole was evaluated for the treatment of VL patients in Sudan but only 21% patients exhibited cure in 6 months' follow-up (DNDi 2013). The combination of fexinidazole and miltefosine for the treatment of VL patients in Eastern Africa was also assessed by DNDi (DNDi 2016). Unfortunately, the use of fexinidazole for

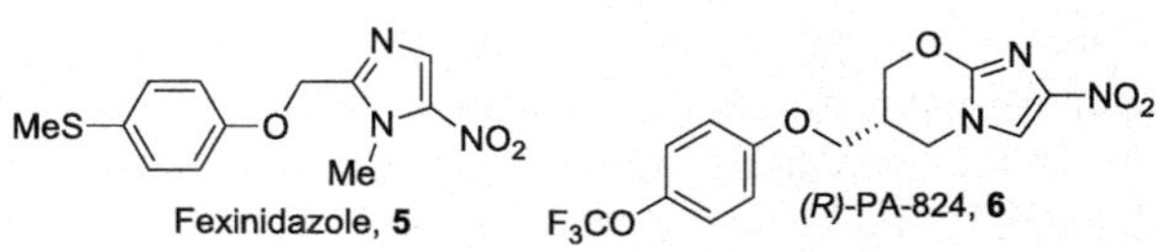


**Fig. 5.3** Structures of nitroimidazole derivatives, Fexinidazole and (R)-PA-824

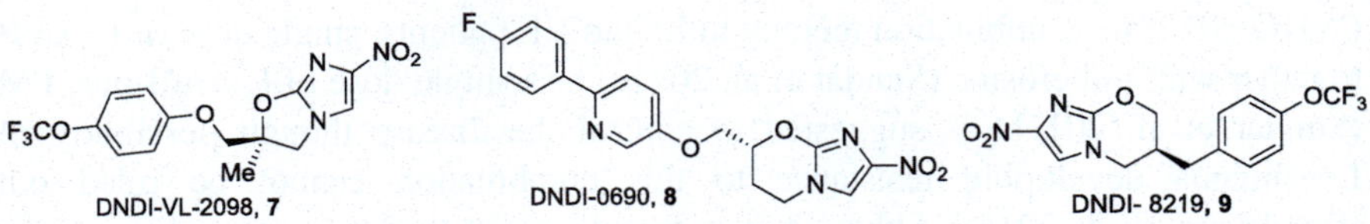


**Fig. 5.4** Structures of nitroimidazole-based antileishmanial agents DNDI-2098, DNDI-0690, DNDI-8219

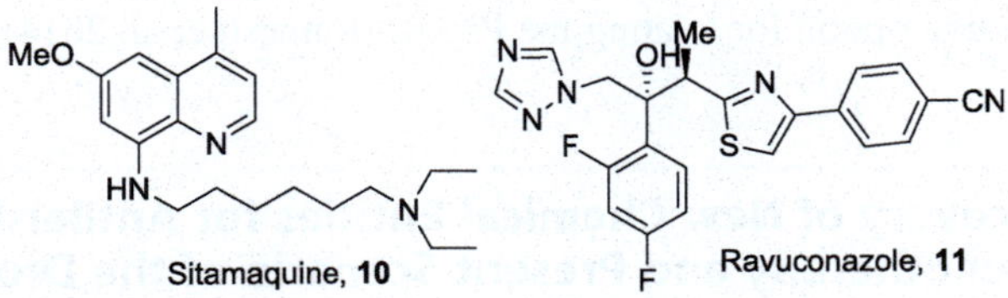


**Fig. 5.5** Structures of Sitamaquine and Ravuconazole

treating leishmaniasis as monotherapy was abandoned due to its ineffectiveness in Phase II clinical trial (https://clinicaltrials.gov/ct2/show/NCT01980199).

Another derivative of this class (*R*)-PA-824 (**6**) (also called pretomanid), which is a 4-nitroimidazooxazine with potent antibacterial activity, has also displayed significant in vitro and in vivo leishmanicidal activity (Patterson et al. 2013). It was shown to have an additive effect when administered in combination with fexinidazole. Thus, it was proposed that (*R*)-PA-824 is a promising candidate for late lead optimization for VL and may have potential for future use in combination therapy with fexinidazole.

Another nitroimidazo-oxazoles derivative DNDI-2098 (**7**) identified to be active against *L. donovani* was selected for development (Gupta et al. 2015; Thompson et al. 2016) (Fig. 5.4). However, due to the toxicity issue, this compound was abandoned and replaced with DNDI-0690 (**8**) as the preclinical candidate (Petri e Silva et al. 2016; Thompson et al. 2017). This compound was selected for Phase I single ascending dose study in healthy volunteers in 2018. A clinical trial application for the Phase I was submitted to UK authorities in February 2019 (DNDi portfolio).

In further development, to have a back-up molecule of this class, a library of 900 compounds based on pretomanid was synthesized taking into consideration both the enantiomer forms and screened for VL. The structure–activity relationship (SAR) studies and detailed in vitro/in vivo profiling resulted in identification of (6*R*)-2-Nitro-6-[4-(trifluoromethoxy)phenoxy]-6,7-dihydro-5*H*-imidazo[2,1-*b*][1,3] oxazine (DNDI-8219) (**9**) as a new lead for VL (Thompson et al. 2018) (Fig. 5.4).

#### 5.5.1.2 Sitamaquine

Sitamaquine (**10**) is an 8-aminoquinoline derivative that showed antileishmanial activity against *L. donovani* in hamster model (Fig. 5.5). The compound went upto Phase II clinical trial in India and Kenya. In India, the cure rate at the dose of 1.5 mg/ kg and 2.0 mg/kg was observed to be 87% and 100%, respectively. However, due to nephrotoxicity at a dose >2.0 mg/kg and reduced efficacy of 85% at the dose of 2.0 mg/kg for 21 days, further development of this drug for VL was abandoned (Jha et al. 2005).

#### 5.5.1.3 Ravuconazole

Ravuconazole (**11**), a second generation oral triazole antifungal drug, was able to inhibit the proliferation of *L. amazonensis* promastigotes and intracellular amastigotes as demonstrated in a recent study (Fig. 5.5). In vitro experiments indicated $IC_{90}$ of 8–10 μmol/L for promastigotes and intracellular amastigotes. Furthermore, ravuconazole exposure induced significant alterations in the shape of *Leishmania* with a "crumbled" appearance, an altered mitochondrial morphology and function with loss of ATP, and an accumulation of un-metabolized lipids. It was suggested that the required ravuconazole concentrations can be achieved in humans without harm. However further work is required to demonstrate its efficacy in the in vivo system (Teixeira de Macedo et al. 2018).

#### 5.5.1.4 Aminopyrazoles

The discovery of a novel series of pyrazole-based antileishmanial compounds with excellent in vitro potency against *L. infantum* was reported (Mowbray et al. 2015). Several compounds from this series show equivalent in vitro potency against *L. donovani* and high levels of in vivo efficacy (>90%) against *L. infantum* in a hamster model of VL. The initial hit was identified by random screening of collection of 9500 compounds belonging to diverse subset, carefully chosen to represent the broad chemical space of Pfizer's compound collection. Initially, the pyrazole amides (**12**) with cyclopropyl group at 4-position were prepared which displayed good bioactivity but poor metabolic stability (Fig. 5.6). Further optimization led to identification of a series of pyrazole ureas (**13**) of which the benzyl urea derivatives were found to display submicromolar activity with no cytotoxicity but high levels of instability against hamster liver microsomes (HamLM). Replacing the benzyl group with piperidine moiety not only gave compounds (**14**) having increased antileishmanial efficacy but also had high HamLM stability. Altering piperidine with piperazine offered compounds (**15**), that possessed excellent levels of antileishmanial activity and SI combined with good metabolic stability, and an improvement in HamLM stability. The better activity and stability of piperazine derivative over piperidine was attributed to the presence of the extra nitrogen atom, which could block a site of oxidative metabolism that is specific to hamster (HamLM).

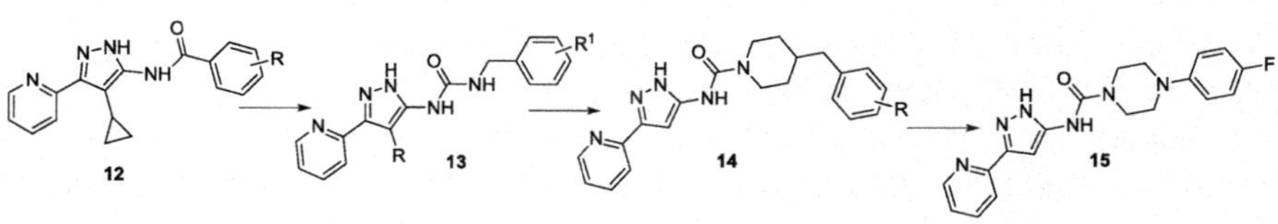


**Fig. 5.6** Optimization of 5-aminopyrazoles for antileishmanial effect

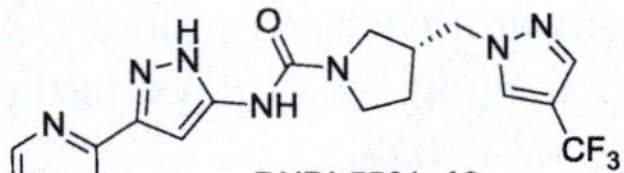


**Fig. 5.7** DNDI-5581, 5-aminopyrazole selected for clinical trials

The hamsters infected with *L. infantum* were treated with representative example at 50 mg/kg BID orally for 5 days. Treatment with pyrazole urea analogue **15** resulted in 92.7% and 95% reduction in parasite burden in liver and spleen respectively, without any signs of toxicity. The compound was compared to miltefosine which when dosed orally at 40 mg/kg once for 5 days, showed 97.8% (liver) and 99.6% (spleen) reduction in parasite burden. Pharmacokinetic studies in hamsters with a single oral dose of 50 mg/kg demonstrated that compound **15** rapidly achieved good levels of exposure.

Several additional advanced and differentiated second generation leads have been identified from the aminopyrazole series of which the most advanced compound is DNDI-5561 (**16**) (Fig. 5.7). The preclinical trial application package studies to enable a Phase I clinical trial was expected to start in early 2019 and completed by the end of 2019 (https://www.dndi.org/diseases-projects/portfolio/dndi-5561/).

### 5.5.2 Medicinal Chemistry Efforts to Identify New Chemical Compounds as Antileishmanials

There are multiple chemical classes, which are core units of several compounds displaying the antileishmanial efficacy (Walkinshaw 2014). In 2015, Sangshetti et al. have summarized the various chemical motifs that have displayed antileishmanial efficacy. Therefore, the citations published after 2015 are only being discussed within the scope of this article.

The synthesis of a series of 3-nitroimidazo[1,2-*a*]-pyridine derivatives, bearing a phenylthio (or benzylthio) moiety at position 8 of the scaffold was disclosed recently (Fersing et al. 2019). Based on the in vitro assay, compound **17** was identified to be the hit ($IC_{50} = 1$–$2.1$ μM) against *L. donovani*, *L. infantum* and *L. major* with low cytotoxicity for the human HepG2 cell line ($CC_{50} > 100$ μM) in comparison to several reference drugs such as miltefosine, fexinidazole, eflornithine and benznidazole ($IC_{50} = 0.6$–$13.3$ μM). This compound had a low reduction potential ($E^{\circ} = -0.63$ V) and was shown to be selectively bioactivated by the *L. donovani* type 1 nitroreductase (NTR1). The compound was neither mutagenic as assessed by Ames test nor genotoxic as examined by comet assay and showed poor microsomal stability though the main metabolite (sulfoxide) was also active and nonmutagenic.

Beside fused-imidazo class of compounds, fused pyrazole class of compounds are also reported as potent antileishmanial agents. A series of pyrazolo[1,5-*c*] pyrimidines was synthesized and investigated for their antileishmanial property against *L. major*. Several compounds displayed better activity than the standard miltefosine in the in vitro evaluation against promastigotes. Having the 7-methoxy substituent on 2-phenylpyrazolo[1,5-*c*]pyrimidine as in compound **18** was the most

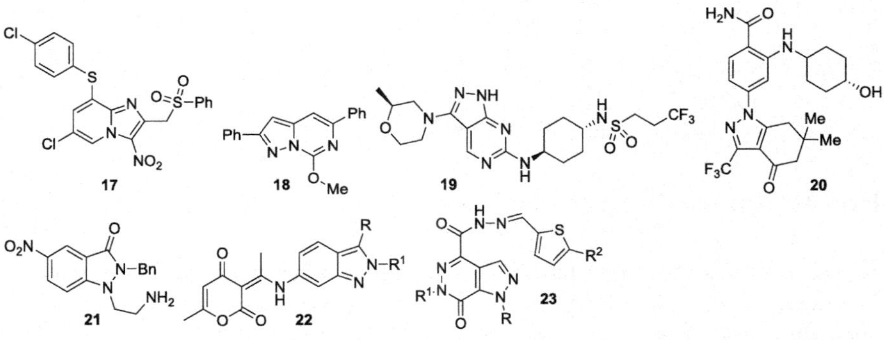

**Fig. 5.8** Antileishmanial agents from fused-imidazo and fused pyrazole class of compounds

suited for bioactivity with $IC_{50}$ values of 1.1 μM and 1.96 μM against promastigotes and amastigotes, respectively, with no apparent toxicity either via oral or ip route (Fig. 5.8). From the reverse docking approach, the putative target of the active compound was suggested to be the *L. major* pteridine reductase (Atta et al. 2017).

The discovery of a lead compound DDD853651/GSK3186899 (**19**) from a series based on pyrazolopyrimidine scaffold was reported recently (Wyllie et al. 2018) (Fig. 5.8). It was active against *L. donovani* in an intra-macrophage assay with an $EC_{50}$ value of 1.4 μM, and showed good selectivity against mammalian THP-1 host cells ($EC_{50}$ value >50 μM) which was comparable with the clinically used drugs miltefosine and paromomycin ($EC_{50}$ values of 0.9 μM and 6.6 μM, respectively). This compound was active in the cidal axenic amastigote assay ($EC_{50}$ value of 0.1 μM) and at a concentration of 0.2 μM, was cytocidal at 96 h. It demonstrated less than tenfold variation in potency against a panel of *Leishmania*-derived lines and was also more active in a panel of *Leishmania* lines using human peripheral blood mononuclear cells as host cells. Further studies found it to be efficacious in a mouse model of visceral leishmaniasis; has suitable physicochemical, pharmacokinetic and toxicological properties required for further development. Detailed studies on the mode of action revealed that the compound acts principally by inhibiting the parasite cdc-2-related kinase 12 (CRK12) which was defined to be a druggable target for VL. This compound is now progressing to Phase I clinical trials.

The antileishmanial activity of a series of novel tetrahydroindazoles preparations, which was based on the known Hsp90 agent SNX-2112 was disclosed by Kanwar et al. 2017. Several compounds displayed comparable or better leishmanicidal activity as compared to miltefosine. The most active analogue **20** displayed in vitro activity against *L. donovani* amastigotes ($IC_{50} = 0.42$ μM).

Several 2-benzyl-5-nitroindazoles were also prepared and evaluated for trypanonocidal activity (Fonseca-Berzal et al. 2018). The most active compounds were additionally screened for the antileishmanial efficacy against the promastigotes of *L. amazonensis*. Although the activity of these analogues was not very encouraging, but compound **21** exhibited significant in vitro activity with an $IC_{50}$ value of 1.04 μM and SI value of 228 (Fig. 5.8).

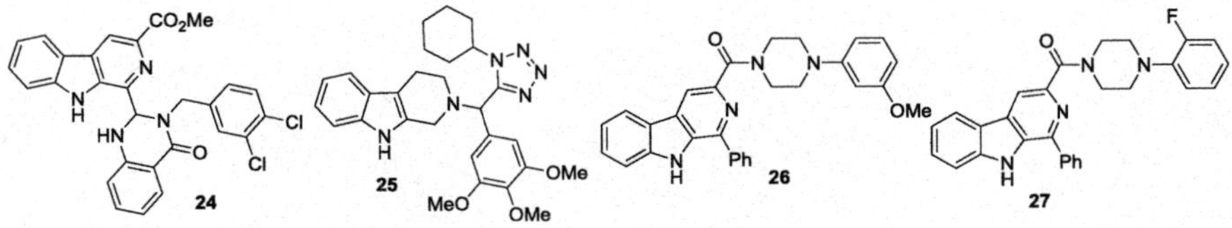

**Fig. 5.9** Antileishmanial agents bearing β-carboline core

The synthesis and antileishmanial evaluation of indazole–pyrone hybrids against *L. donovani* axenic and intra-macrophage amastigotes revealed that out of all the analogues analysed, only one compound **22** displayed moderate activity with $IC_{50}$ of values of 2.48 ± 1.02 μM and 2.25 ± 1.89 μM against the axenic and intra-macrophage amastigotes, respectively (Ghozlani et al. 2019) (Fig. 5.8).

The synthesis and in vitro antileishmanial efficacy of pyrazolo[3,4-*d*]pyridazin-7-one derivatives **23** against *L. amazonensis* promastigote and axenic amastigote forms showed that none of the analogues displayed better activity than the standard drug (Jacomini et al. 2016) (Fig. 5.8).

Compounds bearing β-carboline core are ascribed with a variety of bioactivities. Several substituted β-carbolines are reported to display significant antileishmanial activity too. The β-carboline-quinazolinone hybrid was reported to inhibit leishmanial trypanothione reductase (LdTR) (Chauhan et al. 2015). When assayed against promastigotes of *L. donovani* the $IC_{50}$ values of these compounds were in the range of 3.3–9.9 μM. The most active compound **24** was evaluated against amastigotes of *L. donovani* and was found to display the $IC_{50}$ values of 4.3 μM with good SI (Fig. 5.9).

In continuation, they reported a series of 2,3,4,9-tetrahydro-β-carboline tetrazole derivatives via Ugi reaction which were discovered to be good antileishmanial chemotype (Purohit et al. 2017). Several derivatives exhibited significant in vitro activity against the promastigote ($IC_{50}$ from 0.59 ± 0.35 to 31 ± 1.27 μM) and intracellular amastigote forms ($IC_{50}$ from 1.57 ± 0.12 to 17.6 ± 0.2 μM) of *L. donovani*, which was comparable with standard drugs miltefosine and SSG. The most active compound **25** with $IC_{50}$ values of 2.81 μM and 1.75 μM against the promastigotes and intracellular amastigotes, respectively was studied for its in vivo antileishmanial activity in the *L. donovani*/golden hamster model at a dose of 50 mg/kg × 5 days via ip route and was found to display 75.04 ± 7.28% inhibition of splenic parasite burden (Fig. 5.9). The preliminary PK profile of the compound was also evaluated.

A molecular hybridization approach was adopted to synthesize a series of (1-phenyl-9*H*-pyrido [3,4-*b*]indol-3-yl)(4-phenylpiperazin-1-yl)methanone derivatives which were assessed for the cytotoxicity and antileishmanial efficacy against *L. infanum* and *L. donovani* (Penta et al. 2018). Though several compounds showed activity, compounds **26** and **27** were the most potent against the *L. infantum* and *L. donovani*, respectively and this activity was comparable to that of miltefosine and pentamidine. The $EC_{50}$ of **26** against *L. infantum* was 2.89 ± 0.34 μM for promastigotes (SI = 71) and 2.8 ± 0.13 μM for axenic amastigotes (SI = 73),

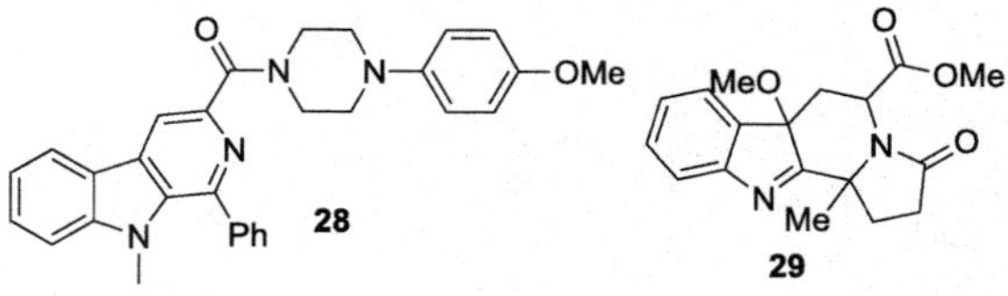


28

29

**Fig. 5.10** β-carboline-containing antileishmanial agents

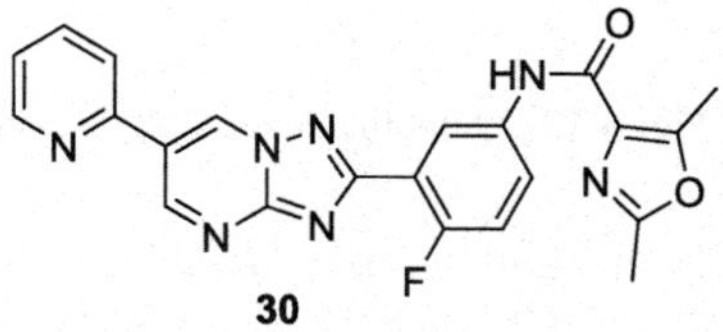


30

**Fig. 5.11** Fused-triazole derivative with antileishmanial activity

whereas the $EC_{50}$ of **27** against *L. infantum* was 3.47 ± 0.17 μM for promastigotes (SI = 144.1), 2.8 ± 0.1 μM for axenic amastigotes (SI = 178.6) and 4.0 ± 0.6 for intracellular amastigotes (SI ≥ 125) (Fig. 5.9).

These workers extended the study by preparing the *N*-methyl analogues of the compounds and discovered that there was no correlation between the antileishmanial efficacies of the corresponding unsubstituted analogues (Penta et al. 2019). Unlike earlier report, herein the compound **28** bearing 4-methoxy group on the phenyl ring was the most potent with $EC_{50}$ of 0.92 μM against axenic amastigote of *L. donovani* as compared to pentamidine and miltefosine ($EC_{50}$ 1.6 and 2.8 μM) (Fig. 5.10). Indeed this compound was also most potent analogue against *L. infantum* with $EC_{50}$ value of 1.4 μM against axenic amastigotes.

A diversity oriented synthesis of a library of β-carboline-quinazolinone hybrids (**29**) and its antileshmanial assessment by targeting the *L. donovani* trypanothione reductase (LdTR) leading to apoptosis was also reported (Ramu et al. 2017) (Fig. 5.10). The most active analogues were demonstrated to have relatively better inhibition of the promastigotes as compared to the standard drug AmB. The inhibitory potential of the series was further supported by in silico analysis of protein–ligand interactions, which indicated high binding efficiency towards the catalytic pocket of LdTR.

Triazole is another class of nitrogen heterocycle, which has been probed for the antileishmanial property. The compound GNF 6702 (**30**), a fused triazole derivative, was reported as the selective inhibitor of the kinetoplastid proteasome (Khare et al. 2016) (Fig. 5.11). Its action is via noncompetitive mechanism and is well tolerated in mice. As compared to miltefosine, it causes superior reduction in liver parasite burden in the infected mice. The $EC_{50}$ of compound **30** in *L. donovani* was 18 ± 1.8 nM and oral administration of 10 mg/kg BID for 8 days in mice showed that the free concentration of the compound (fraction unbound in plasma = 0.063) stayed above the $EC_{99}$ value of *L. donovani* (the concentration inhibiting 99% of intra-macrophage parasite growth in vitro) for the duration of the dosing period. It was discovered that 90% reduction in parasite burden in the mouse model was realized when the mean free compound plasma concentration during treatment equalled 0.94-fold multiple of the *L. donovani* $EC_{90}$ value. Treating the event of

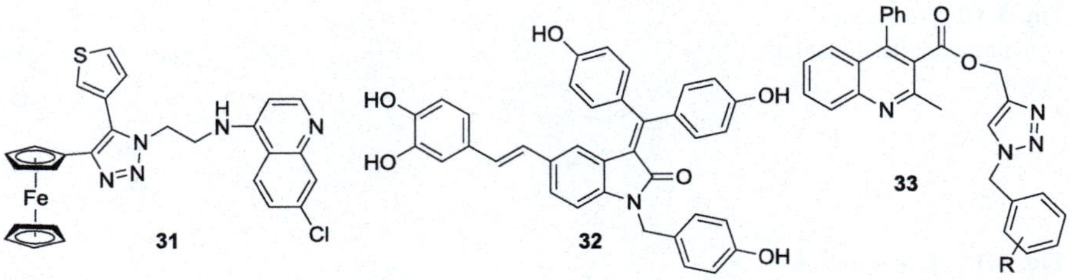


**Fig. 5.12** Quinoline and polyhydroxylatedoxindole-based antileishmanial agents

footpad infection of BALB/c mice with the dermatotropic *L. major* strain, with **30** at 10 mg/kg twice daily resulted in five-fold decrease in footpad parasite burden and a reduction in footpad swelling. Further it was reported that both 3 and 10 mg/kg twice-daily regimens of compound **30** gave better result as compared to 30 mg/kg once-daily miltefosine ($P < 0.01$), which translates into an approximately two-fold higher miltefosine plasma concentration in mice than that observed in clinical dosing.

A series of ferrocenyl-quinoline bearing triazole ring was prepared and assessed for antileishmanial activity against *L. donovani* (Yousuf et al. 2016). All compounds were investigated for efficacy against *L. donovani* and the best compound **31** from the series has $IC_{50}$ value of 21 μM (equipotent to miltefosine) against promastigotes and 8 μM against amastigotes (Fig. 5.12). Mechanistically, it was suggested that the heterocyclic rings present in the molecule damage parasite membranes via altering the sterol composition of the membrane, or block sterol biosynthesis by inhibiting the enzymatic activity of sterol 14α-demethylase (CYP51), leading to parasite death.

Subsequently, this group reported the synthesis and antileishmanial efficacy of polyhydroxylated oxindole derivative **32**, which showed promising in vitro activity against amastigotes and promastigotes of *L. donovani* with $IC_{50}$ values of 1 μM and 15 μM, respectively (Yousuf et al. 2018) (Fig. 5.12). It was suggested that this compound induces the parasite death by promoting oxidative stress, accumulating cytosolic lipids, altering the cellular integrity finally leading to apoptosis.

The synthesis and bioevaluation of certain quinoline-based triazoles (**33**) as antileishmanial agents was reported though none of the compound was found to display better activity profile as compared to miltefosine that was used as standard in the bioassay (Upadhyay et al. 2018) (Fig. 5.12).

A series of 1,2,3-triazole and thiosemicarbazone hybrids having comparable or superior activity to miltefosine was disclosed (Temraz et al. 2018). Two of the most potent compounds **34** and **35** showed nanomolar $IC_{50}$s against promastigotes of *L. major* (227.4 nM and 140.3 nM, respectively, vs. 7.8 μM for miltefosine) (Fig. 5.13). Their antiamastigote $IC_{50}$s were 1.4 mM and 1 mM, respectively, which were six- and eight-folds the activity of miltefosine ($IC_{50}$ 8.09 mM). The compounds were found to be safe when administered orally. It was proposed that these compounds act via an anti-folate mechanism specifically by interacting with pteridine reductase.

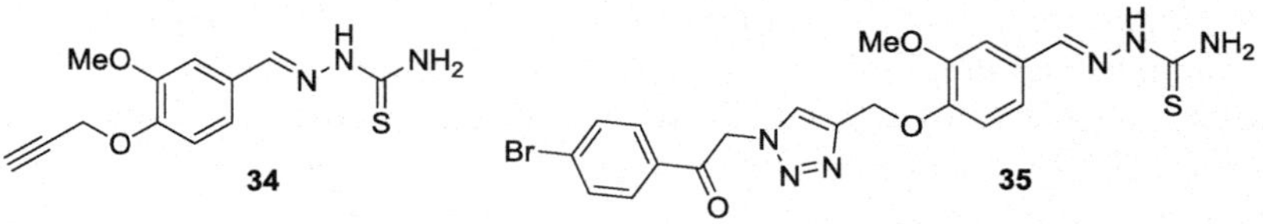


**Fig. 5.13** Thiosemicarbazone-based antileishmanial agents

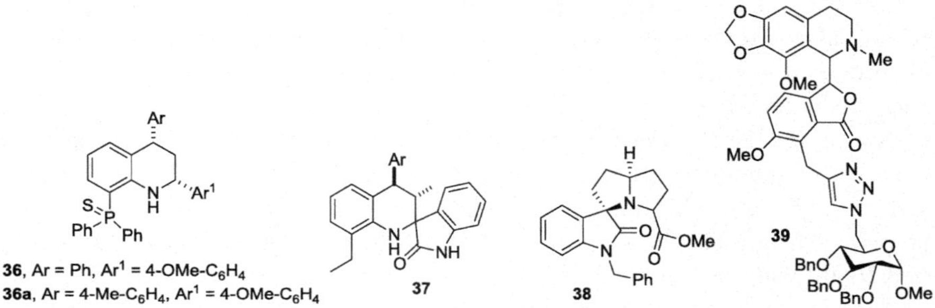


**Fig. 5.14** Antileishmanial agents bearing tetrahydroquinoline, pyrrolizidine or tetrahydroisoquinoline cores

Several compounds belonging to tetrahydroquinoline and isoquinoline were recently reported to show antileishmanial property. The synthesis of hybrid tetrahydroquinoline and quinoline derivatives with phosphorated groups via multi-component reaction between phosphorus substituted anilines aldehyde and styrene was recently reported (Tejería et al. 2019). The compounds were assessed for their antileishmanial efficacy against promastigotes and amastigotes of *L. infantum*, and three functionalized tetrahydroquinolines and one quinoline exhibited equipotent activity to the standard drug AmB with close selective index (SI between 43 and 57) against *L. infantum* amastigotes. The most active tetrahydroquinolyl phosphinesulfide **36** with an $EC_{50}$ value ($0.61 \pm 0.18$ mM) and selective index (SI $= 56.87$) was close to the standard drug AmB ($0.32 \pm 0.05$ mM). One of the compounds **36a** from the series displayed remarkable inhibition of *Leishmania* topoisomerase IB (Fig. 5.14).

Recently, the synthesis and antileishmanial activity of spirodihydroquinoline-oxindoles resulted in evaluation of ten compounds at different stages of the life cycle of *L. braziliensis* which causes cutaneous leishmaniasis in South America (Leañez et al. 2019). Among them, the 8′-ethyl-4′-(4-hydroxy-3-methoxyphenyl)-3′-methyl-3′,4′-dihydro-1′H-spiro[indoline-3,2′-quinolin]-2-one (**37**) termed as JS87 inhibited the growth of promastigotes without affecting the host cells viability, and decreased the number of intracellular amastigotes (Fig. 5.14). It was suggested that this spiro compound acts via alterating the parasite internal regulation by disrupting the regulatory volume decrease (RVD), and affecting the sterol biosynthetic pathway by modulating squalene epoxidase (SE) enzyme.

Earlier too, a novel spirooxindole derivative, *N*-benzyl-2,2′α-3,3′,5′,6′,7′,7α,α-′-octahydro-2-methoxycarbonyl-spiro[indole-3,3′-pyrrolizidine]-2-one (**38**), was demonstrated to inhibit *L. donovani* topoisomerase IB (*Ld*TopIB) and killed both, the wild-type and drug-resistant parasite strains (Saha et al. 2016) (Fig. 5.14). It inhibited catalytic activity of *Ld*TopIB competitively and interfered with formation of the drug-DNA-enzyme covalent complex. From the fluorescence studies it was shown that compound **38** stoichiometrically binds to *Ld*TopIB. The compound was reported to be highly cytotoxic to promastigotes of *L. donovani* and it was suggested to induce apoptosis-like cell death. It was also shown to effectively clear the amastigote forms of wild-type and drug-resistant parasites from infected mouse peritoneal macrophages but was less effective on host macrophages. The compound elicited good in vivo efficacy at doses of 5 and 10 mg/kg and nearly eliminated the parasite burden in the liver and spleen.

A molecular hybridization approach to combine noscapine, a tetrahydroisoquinoline derivative and carbohydrates resulted in preparation of a series of 7-O-noscapine analogues, which were assessed for their antileshmanial property (Mishra et al. 2019). It was observed that one of the compounds **39** has $IC_{50}$ of 8.36 μM that was marginally better than that of miltefosine (10.4 μM) (Fig. 5.14).

The natural extracts and compounds based on them have served as source of a myriad of bioactive agents. Several natural compounds and synthetic analogues prepared using them as the basis have been reported for antileishmanial activity in various publications. In his review, Oryan (2015) described various plant extracts, natural compounds and oils exhibiting antileishmanial efficacy. A comprehensive review assimilating the different prepared extracts and compounds from marine macroalgae along with their antileishmanial activity was presented (Yamthe et al. 2017). Chandrasekar et al. (2018) reviewed the biological role of flavonoids and terpenoids in the treatment of visceral leishmaniasis. They included the source (plant, fruit, etc.) of these chemicals and their bioactivity against the type of *Leishmania* spp. Panda and Luyten reviewed the chemical compounds derived from the Astraceae plants for their antiparasitic activity including Leishmaniasis. Several reports of terpenes and flavonoids displaying the antileishmanial efficacy especially from plants present in the Odisha state of India were included in the compilation (Panda and Luyten 2018). The benzene extract of *C. deodara* that contained 1.29% of Linalool was found to display strong antileishmanial activities within a dose 25–200 μg/mL culture with non-significant haemolytic activities and significant immunomodulant activities against the host cells (Narayan et al. 2017). The antileishmanial efficacy of *C. cassia* bark dichloromethane fraction (CBD) was investigated and it was observed that the fraction had significant direct parasiticidal activity with partial modulation of Th1 immune response (Afrin et al. 2019). Very recently, the iterative phenotypic screening and fractionation approach to identify potent and selective antileishmanial hits from Hypha Discovery's fungal extract library, resulted in identification of a novel oxidised bisabolane sesquiterpene that showed activity in an infected cell model and was reported to disrupt multiple processes using a metabolomic approach (Mbekeani et al. 2019). The leishmanicidal activity of (+)-dehydroabietylamine derivatives functionalized at C7 and/or C12 against *L. infantum*, *L. donovani*, *L. amazonensis* and *L. guyanensis* was studied. It was observed that most of the benzamides displayed activities at low concentration

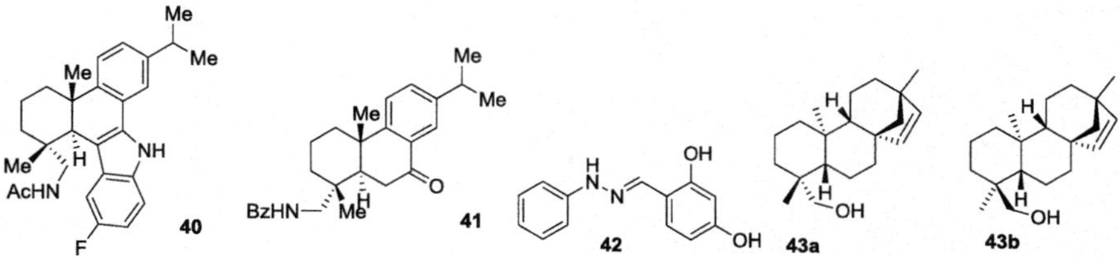


**Fig. 5.15** Examples of natural compounds as antileishmanial agents

against cultured promastigotes of *Leishmania* spp. ($IC_{50}$ = 2.2–46.8 μM), without cytotoxicity on J774 macrophage cells. Compound **40** was disclosed to be the most active against promastigotes of *L. infantum*, *L. donovani*, and *L. amazonensis* but showed some cytotoxicity on J774 cells (Fig. 5.15). Another compound **41** displayed potent in vitro activity against amastigotes of *L. infantum* ($IC_{50}$ = 2.5 μM, SI > 80) with almost 23-fold better potency than the standard miltefosine (Dea-Ayuela et al. 2016).

The molecular modelling, computational docking and in vitro analysis to explore the antileishmanial effect of resveratrol analogues were investigated (da Silva et al. 2018). In order to generate new analogues against trypanothione reductase enzyme, the C=C bond was changed to C=N in the stilbenoid nucleus and the bioactivity was assessed. The most active compound **42** displayed the $IC_{50}$ of 41.47 ± 3.10 and 7.33 ± 0.92 μM (promastigotes *L. amazonensis* and *L. braziliensis*) and 21.62 μM (16.29–28.70) and 21.74 μM (16.95–27.88) (amastigotes of *L. amazonensis* and *L. braziliensis*). It was observed that the OH group potentiated the antileishmanial activity.

Recently, it was reported that beyerenols are potential candidates for cutaneous leishmaniasis chemotherapy via topical application (Murillo et al. 2019). Based on the in vitro assay of *ent*-beyerene diterpenes and their derivatives, it was disclosed that beyerenols **43a** and **43b** showed $EC_{50}$ of 4.6 ± 9.4 μg/mL and 5.3 ± 9.4 μg/mL against amastigotes of *L. brazilensis*, with SI of 5.1 and 7.7, respectively (Fig. 5.15). In vivo experiments with topically applied bereyenols showed cure in 50% of hamsters infected with *L. brazilensis*. The respective creams contained beyerenol **43a**, 0.81%, w/w or beyerenol **43b**, 1.96%, w/w, suggesting that these agents may be helpful in treating cutaneous leishmaniasis.

The antileishmanial assessment of a small library of di- to tetra-peptides with *s*-triazine moiety at the N-terminal and the C-terminal in the form of either ethyl ester or amide was reported (Khattab et al. 2018). The other two positions of *s*-triazine were substituted with dimethoxy, dimorpholino, or dipiperidino groups. The evaluation of compounds resulted in identification of four dipeptide amide derivatives displaying better activity against promastigote or amastigote of *L. aethiopica* than the reference miltefosine. Indeed, the most active analogue **44** with $IC_{50}$ = 1.4 μM (against promastigotes) was about five-fold more active than miltefosine ($IC_{50}$ = 7.8 μM) (Fig. 5.16). However, all compounds displayed inferior activity as compared to the standard AmB.

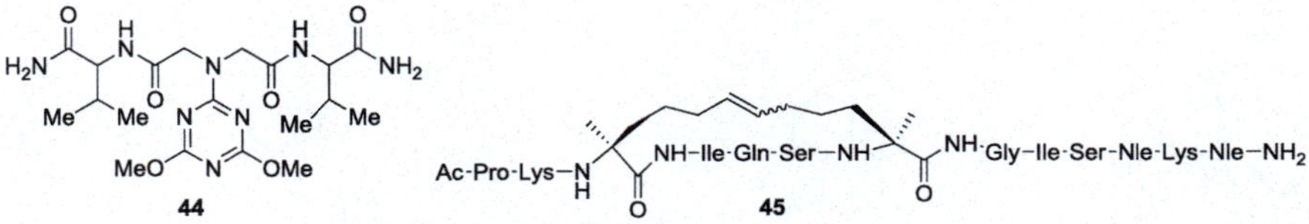

**Fig. 5.16** Peptide-based antileishmanial agents

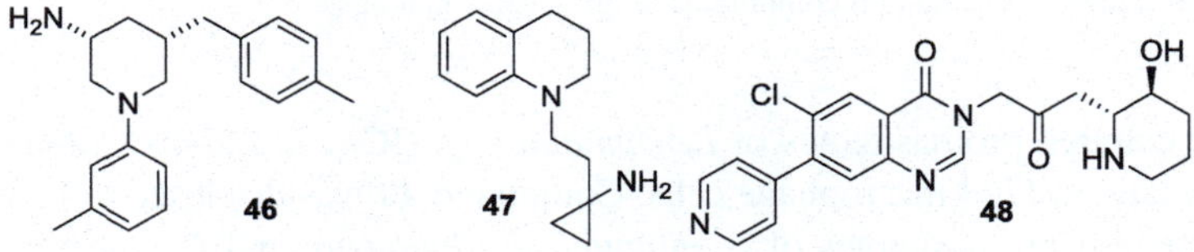

**Fig. 5.17** Antileishmanial hits from computational studies

It was previously reported that linear and lactam-bridged 13-residue peptides derived from ana-helical region making up part of the dimeric interface of *L. infantum* TR (*Li*-TR) prevents reduction in trypanothione by disrupting enzyme dimerization (Ruiz-Santaquiteria et al. 2018). They then showed that i,i þ 4 side-chain cross-linking with an all-hydrocarbon staple stabilizes the helical structure of these peptides and significantly improves their resistance to protease cleavage as compared to previous linear and cyclic lactam analogues. They found that replacing the amide bridge by the hydrocarbon staple at the same cyclization positions generated new derivatives which similarly inhibited oxidoreductase activity of the enzyme but unexpectedly stabilized the TryR homodimer. The most proteolytically stable peptide **45** covalently linked to oligo-arginines elicited potent in vitro leishmanicidal activity against *L. infantum* parasites (Fig. 5.16).

Besides, there are miscellaneous chemical compounds belonging to different structures, which have shown significant to moderate antileishmanial property. These are being included in the following section. A high-throughput screening and quantum mechanics approach to identify leishmanial TR activity of imipramine analogues was reported (Pandey et al. 2016a). During the study 93,328 imipramine analogues were docked in three sequential modes out of which, 98 displayed better docking score than the standard ligand. On subjecting these hits to ADME, toxicity, binding energy calculation and docking validation, two compounds (**46** and **47**) were identified as the best ligands. However no in vitro evaluation was performed (Fig. 5.17).

They also assessed the LdTR inhibitory activity of 8813 febrifugine analogues via docking studies and found that 108 analogues displayed better docking scores than the two standard ligands (Pandey et al. 2016b). To assess the toxicity and binding free energy, these 108 febrifugine analogues and the standard inhibitor clomipramine were subjected to ADMET, QikProp and MM-GBSA study. The two best ligands (FFG7 and FFG2) were then subjected to molecular dynamics simulation, on the basis of which it was suggested that 6-chloro-3-[3-(3-hydroxy-2-piperidyl)-2-

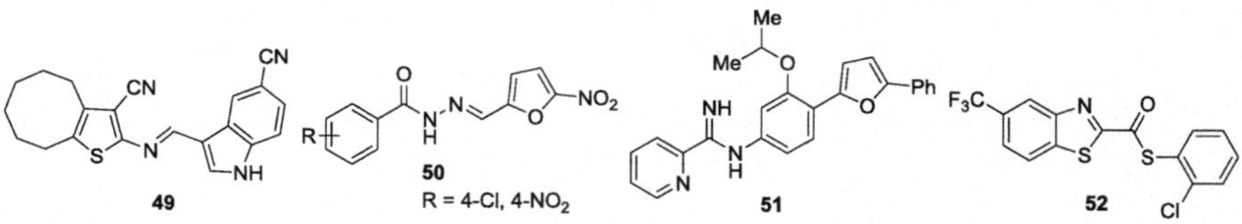

**Fig. 5.18** Miscellaneous heterocyclic compounds with antileishmanial effect

oxo-propyl]-7-(4-pyridyl) quinazolin-4-one (**48**) could be a bioactive compound to treat *Leishmania* (Fig. 5.17).

A series of hybrid compounds bearing cycloalka[*b*]thiophene and indole units were prepared and assessed for their cytotoxicity and in vitro antileishmanial efficacy against promastigotes of *L. amazonensis* (Félix et al. 2016). Several compounds displayed antileishmanial activity ($IC_{50}$ lower than 10.0 μg/L) better than the reference drugs (tri- and pentavalent antimonials). The $IC_{50}$ value for the best compound **49** was 2.1 μg/mL, with no toxicity against the human erythrocytes even at 400 μg/mL (Fig. 5.18). The antileishmanial activity was suggested to be associated with DNA fragmentation of *L. amazonensis* promastigotes.

The antileishmanial activity of the derivatives of nitro heterocycle against promastigotes and amastigotes *L. infantum* was reported by inducing death of the parasite via apoptosis and increased production of NO by infected cells (Petri e Silva et al. 2016). The most active compounds (**50**) from the series were the ones bearing 4-chloro or the 4-nitro group on the phenyl ring (Fig. 5.18). In comparison to the $EC_{50}$ of the AmB of 0.22 μM, the $EC_{50}$ of the chloro-substituted analogue was 0.72 μM and that of nitro-substituted derivative was 0.58 μM. However, the $CC_{50}$ of the nitro-derivative was 10.9 μM as compared to 0.06 μM of the chloro-derivative.

The synthesis and evaluation of mono-arylimidamide derivatives resulted in discovery of six compounds that exhibited sub-micromolar potency against both intracellular *L. donovani* and *L. amazonensis* amastigotes, and three of these compounds also displayed SIs of 25 or greater for the parasites compared to J774 macrophages (Zhu et al. 2016). The most active compound **51** of the series showed superior activity against amastigotes of *L. amazonensis* and *L. donovani* with $IC_{50}$ values of 130 nM and 310 nM, respectively (Fig. 5.18). In vivo efficacy results indicated that compound **51** reduced liver parasitemia by 46% in *L. donovani*-infected mice at an oral dose of 100 mg/kg/day for 5 days.

The synthesis and bioevaluation of *S*-aryl benzothiazole-2-carbothioate as antileishmanial agents was reported (Dar et al. 2016). It was discovered that a few analogues such as **52** have moderate activity against *L. donovani* promastigotes with an $IC_{50}$ value of 190.3–193.0 μM (Fig. 5.18).

To identify the chemical starting point, a high-throughput screen (HTS) of approximately 600,000 small molecules for discovering growth inhibitors of the promastigote form of *Leishmania* was carried out (Ortiz et al. 2017). The investigation identified close to 2700 compounds which inhibited growth by over 65% at 10 μM concentration. To identify a potent analogue against the intra-macrophage amastigote form with limited toxicity towards the host macrophages, these actives

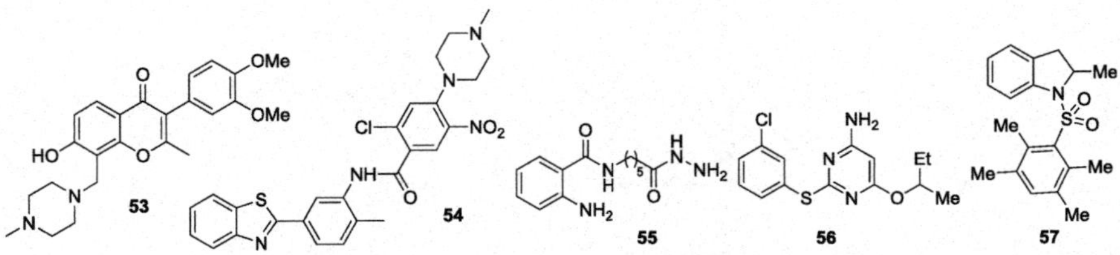


**Fig. 5.19** Miscellaneous compounds with significant antileishmanial activity

were investigated further. This two-step screening strategy resulted in identification of nine unique chemical scaffolds within the collection, including two previously described antileishmanials. They further investigated compounds **53** and **54** for in vitro absorption, distribution, metabolism, excretion, and in vivo pharmacokinetics (Fig. 5.19). Both compounds were found to be orally bioavailable, affording plasma exposures above the $EC_{50}$ concentration for at least 12 h. The result of the in vivo assay in a murine model of cutaneous leishmaniasis showed both compounds to be efficacious.

A library of 2-aminobenzoyl amino acid hydrazide derivatives and quinazolinones derivatives obtained from them was prepared and assessed for its activity against promastigotes *L. aethiopica* (Khatta et al. 2017). In general, it was discovered that the 2-benzoyl amino acid hydrazide showed higher inhibitory effect than the corresponding quinazoline derivative. Among the prepared compounds; 2-amino-*N*-(6-hydrazinyl-6-oxohexyl)benzamide **55** ($IC_{50} = 0.051$ μM) displayed 154-fold more activity as compared to the reference drug miltefosine ($IC_{50} = 7.832$ μM), but was less active than amphotericin B ($IC_{50} = 0.035$ μM) (Fig. 5.19). Further, this compound was reported to be safe and well tolerated orally up to 250 mg/kg and parenterally up to 100 mg/kg with no histopathological changes in liver, spleen and kidney in the experimental animals.

In an effort to identify new antileishmanial agents, Costi and coworkers evaluated the thioether derivatives in whole-cell screening assays against *Leishmania* (Saccoliti et al. 2017). One of the compounds RDS 777 (6-(sec-butoxy)-2-((3-chlorophenyl)thio)pyrimidin-4-amine) **56** ($IC_{50} = 29.43$ μM), was able to impair the defense mechanism of the parasite against the reactive oxygen species by inhibiting the trypanothione reductase (TR) with high efficiency (Ki $0.25 \pm 0.18$ μM) (Fig. 5.19). The X-ray structure of *L. infantum* TR in complex with RDS 777 suggested that the compound binds to the catalytic site and is involved in hydrogen bonding with the residues more involved in the catalysis, namely Glu466′, Cys57 and Cys52, thereby inhibiting the trypanothione binding and avoiding its reduction.

The assessment of in vitro antiparasitic activity of a library of synthetic benzenesulfonyl derivatives of heterocycles with drug-like properties was reported (Pagliero et al. 2017). Four compounds showed $IC_{50}$ between 0.25 and 3.0 μM against *L. donovani* with low cytotoxicity and out of them (1-(2,3,5,6-tetramethylphenylsulfonyl)-2-methylindoline) **57** was equipotent to the reference drug miltefosine (Fig. 5.19).

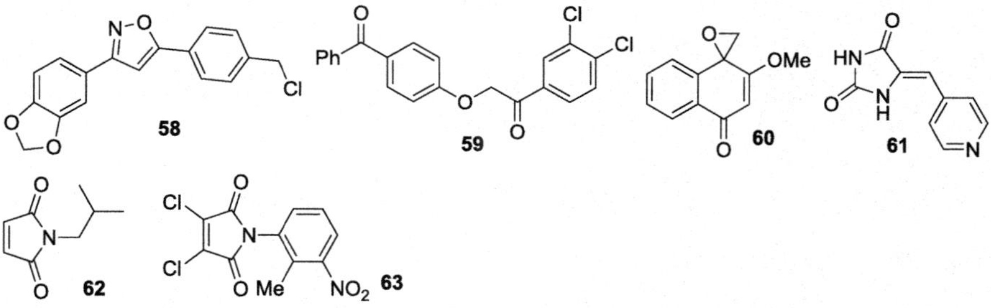

**Fig. 5.20** Diverse chemotypes as antileishmanial agents

The 3,5-disubstituted isoxazoles which were prepared to explore their trypanocidal activity were also screened for their leishmanicidal activity (da Rosa et al. 2017). Most of the compounds displayed $IC_{50}$ of >100 μM except compound **58**, which showed $IC_{50}$ value of 5.08 μM and SI value of 35.8 against *L. amazonensis* amastigotes (Fig. 5.20).

A series of 4-substituted ethers of benzophenone was synthesized and screened in vitro against the promastigotes of *L. major* using pentamidine as standard drug (Arshia et al. 2018). Among the series, 15 compounds displayed antileishmanial activity with $IC_{50}$ values within the range of 1.94–82.30 μg/mL and compound **59** with an $IC_{50}$ value of 1.94 μg/mL was the most active analogue.

The effectiveness of a combination of epoxymethoxylawsone **60** and meglumineantimoniate against *L. amazonensis* intracellular amastigotes was assessed (Oliveira et al. 2018) (Fig. 5.20). The $IC_{50}$ value of epoxymethoxylawsone (7.41 ± 0.2 μM) was 1.7-fold more than that of meglumineantimoniate (4.43 ± 0.25 μM), which became pronounced after 48 h of exposure ($IC_{50}$ value of epoxymethoxylawsone 0.40 ± 0.001 μM). Cytotoxicity evaluation revealed that epoxymethoxylawsone ($CC_{50}$ = 40.05 μM) was better than the meglumineantimoniate (24.14 ± 2.6 μM). Treatment of the paw lesion in infected BALB/c mice with epoxymethoxylawsone resulted in remarkable 27% reduction ($p < 0.05$) of the lesion size, for all administrated doses, compared to the control group. It was suggested that this compound furthers merits for the preclinical screening.

The bioisosteres of imidazolidinone were synthesized and evaluated their antileishmanial property (Ramu et al. 2019). All compounds (**61**) possessed activity against extracellular as well as intracellular *L. donovani* parasites in nanomolar concentrations with SI of more than 1000 (Fig. 5.20). The antileishmanial efficacy of these compounds was attributed to the inhibition of kinases by increase in intracellular sodium levels in the parasites.

The in vitro antileishmanial efficacy and cytotoxicity of a series of maleimides resulted in identification of several compounds displaying bioactivity better than the standard AmB and pentamidine (Fan et al. 2018). Two compounds **62** and **63** with the $IC_{50}$ < 0.0128 μg/mL and $IC_{50}$ < 0.0128 μg/mL, respectively were the most potent analogues from the series (Fig. 5.20).

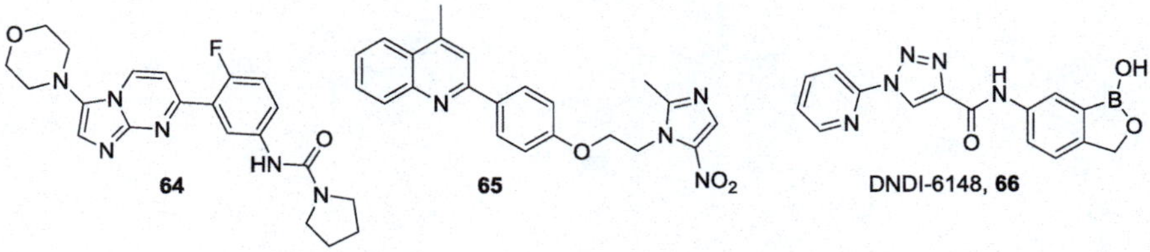


**Fig. 5.21** Potent antileishmanial agents

Very recently, discovery of a preclinical candidate GSK3494245/DDD01305143 compound **64** was disclosed (Wyllie et al. 2019) (Fig. 5.21). This compound was the outcome of the repurposing and further modifications of compounds identified to treat trypanosomiasis. The compound **64** has $EC_{50}$ of 1.6 μM with good solubility and in vitro metabolic stability ($CL_{int}$ = 0.8 mL/min/g) and was negative for the Ames test. At 25 mg/kg the compound caused 95% reduction in parasitemia and the dose-ranging studies resulted in curative dose of 25 mg/kg BID × 10 days. Compound **64** was reported to act via inhibiting the chymotrypsin-like activity catalysed by the β5 subunit of the *L. donovani* proteasome.

In another report, synthesis of quinolone-metronidazole hybrid and their assessment as antileishmanials was reported (Upadhyay et al. 2019). The best compound **65** from the series had $IC_{50}$ value of 5.42 μM (promastigotes) and 3.75 μM (intracellular amastigotes), respectively (Fig. 5.21). This compound inhibited the parasite burden in the liver and spleen (>80%) of infected mice. It was suggested that compound **65** triggers oxidative stress, which induces bioenergetic collapse and apoptosis of the parasite by decreasing ATP production and mitochondrial disintegration.

Benzoxaborole is a versatile boron-heterocyclic scaffold which has found remarkable utility in different medicinal chemistry program. Compounds bearing this core subunit have been reported as antibacterial, antifungal, anti-protozoal, antiviral as well as anti-inflammatory agents (Nocentini et al. 2018). Benzoxaboroles have also been listed to be potent antiparasitic agents against the diseases including malaria, leishmania and trypanosomiasis. There are several benzoxaboroles which have shown remarkable activity against leishmania (Jacob et al. 2018). The DNDI-6148 (**66**), a benzoxaborole derivative is waiting for approval for the Phase I clinical trial (Fig. 5.21). The mechanism of action of this compound is still not known, but it elicited antileishmanial activity even against the strains, which are resistant to known drugs. Indeed, there are two more analogues DNDI-5421 and DNDI-5610, which have been identified to be the back-up compounds from this class.

In the hit to lead portfolio of DNDi, there is another project which investigates the compounds from the L 2015 series. The lead compound DNDI-6588 displayed in vivo efficacy both in rat and hamster model (Available with: https://www.dndi.org/diseases-projects/portfolio/leish-l205-series/). Unfortunately, the structure of this series could not be traced in the literature.

In addition, there are reports of metal-based compounds, which have shown significant antileishmanial efficacy. It was reported that a binuclear cyclopalladated

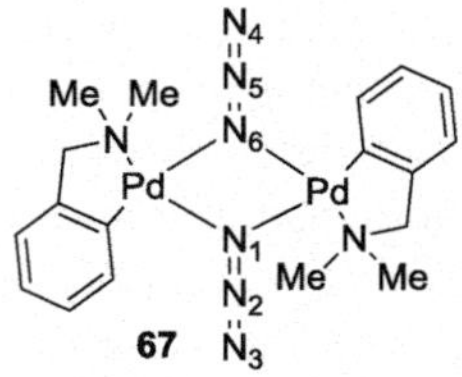

67

**Fig. 5.22** Metal-based antileishmanial agent

complex containing Pd(II) and *N,N*-dimethylbenzylamine (Hdmba), [Pd(dmba)(μ-$N_3$)]$_2$ **67** displayed potent effect against *L. amazonensis* (Velásquez et al. 2017). The compound was found to inhibit promastigote growth ($IC_{50}$ = 13.2 ± 0.7 μM) and inhibits the growth of intracellular amastigotes in in vitro incubated macrophages ($IC_{50}$ = 10.2 ± 2.2 μM) without a cytotoxic effect when tested against peritoneal macrophages (50% cytotoxic concentration = 506.0 ± 10.7 μM). The compound showed 80% reduction in parasite load compared to infected and nontreated animals in the in vivo assays against *L. amazonensis* in BALB/c. In addition, compared to amphotericin B treatment, compound **67** did not show any side effects, which was validated by the analysis of plasma levels of different hepatic and renal biomarkers (Fig. 5.22). Furthermore, compound **67** was able to inhibit *L. donovani* topoisomerase 1B (*Ld*topo1B) suggesting its activity via inhibiting this important target.

## 5.6 Drugs in Use in New Delivery Systems

Another approach that has been explored for the treatment of leishmaniasis is the use of controlled delivery systems, such as liposomes and nanoparticles as this mode provides a greater efficacy and safety. Adsorption or encapsulation of drug in such carriers often reduces the dose and adverse reactions of conventional formulations (Petitti et al. 2008). Additionally they may address the issues of low aqueous solubility of drugs and degradation of the drug in the biological fluids. Interestingly, alternative routes to parenteral drug delivery (e.g. nasal, pulmonary, and topical) may also improve patient compliance. In this context, the focus of treating *Leishmania* relies on delivering the drug directly to macrophages, using appropriate nanosized delivery systems (Jain and Jain 2013). It is presumed that nanostructured delivery systems improve the leishmanicidal activity of the agents (Bruni et al. 2017). A targeted approach requiring modification of the nanocarrier surface is predicted to increase parasite selectivity though of such sequencing and scale-up approaches is considered to be very challenging.

A liposomal formulation of AmB, Ambisome, is already in clinical use to treat VL, showing better efficacy and safety than AmB deoxycholate. The stearylamine liposome formulation containing SSG developed by Roychoudhury et al. (2011) showed efficacy against strains of *L. donovani* resistant to SSG. Shio et al. (2014) revealed that another drug, oleylphosphocholine (OlPC), formulated as liposomes, was able to kill the intracellular amastigotes of *L. major* and *L. mexicana* in macrophages. A poloxamer P407-based paromomycin-containing micelle nanogel system is disclosed to possess negligible toxicity and has effective antileishmanial activity against *L. major* and *L. infantum* promastigotes. It was suggested that the formulation provides controlled, effective, and safe delivery of PM in mice. Miltefosine loaded in PLGA–PEG nanoparticles also showed a fourfold increase in in vitro activity and in vivo antileishmanial efficacy and bioavailability as compared to free drug (Kumar et al. 2016).

The most challenging task was to achieve oral administration of AmB. Since AmB is absorbed via passive diffusion through the intestinal membrane, this was considered to be limitation. Moreover, AmB is relatively unstable at the acidic pH of the gastrointestinal tract, and under oral administration could be degraded prior to absorption. It is known to undergo extensive P-gp efflux, which further lowers its oral bioavailability. Several nanosystems have been developed to improve the oral bioavailability of AmB (Torrado et al. 2013). Soya lecithin (Lec)-based biodegradable nanocarriers loaded with AmB were prepared and the surface of these Lec–AmB nanoparticles was modified with PEG and Tween 20 to improve their stability and biological activity (Javed et al. 2015).

Lipid AmB formulations potentially suited for oral administration were also developed. The formulation, named iCo-010, had been developed based on mono- and diglycerides with or without a lipophilic derivative of vitamin E, D-α-tocopheryl polyethylene glycol 1000 succinate (TPGS) (Wasan et al. 2010). It stabilizes AmB in simulated gastric and intestinal fluids and exhibited a significant antileishmanial activity in a VL-infected murine model (Hnik 2016).

Another interesting approach to prepare lipid systems for oral delivery is that of nano-cochleates, cigar-shaped nanostructures composed of negatively charged lipid bilayers (usually PS), bridged by a divalent cation, generally calcium (Zarif 2005). Nano-cochleates containing AmB or AmB deoxycholate were tested in vitro against *L. chagasi*, and results showed similar activity as that of AmB (Sesana et al. 2011).

Lipid–polymer hybrid nanoparticles (LPNPs) loading AmB with the anionic core made up of PLGA polymer and TPGS surfactant and the shell made of cationic stearylamine lipid are also reported. In vivo experiments demonstrated significantly stronger parasite growth inhibition in animals treated with AmB–LPNPs than in those treated with carrier without stearylamine or AmBisome formulation (Asthana et al. 2015).

Another approach for targeted drug delivery is the use of functionalized carbon nanotubes (f-CNTs), which have been tested as AmB carriers in experimental VL (Prajapati et al. 2011). It is reported that the AmB attached to functionalized carbon nanotubes has a significantly greater leishmanicidal activity relative to the conventional amphotericin B in *L. donovani*-infected hamsters. Further, the chemical synthesis of f-CNT-AmB is reported to involve covalent coupling rather than biological molecules, thereby suggesting that its production could be cheaper than the existing L-AmB. Likewise attaching betulin, a pentacyclic triterpenoid, to f-CNT improved its leishmanicidal effect and showed lower toxicity (Saudagar and Dubey 2014).

Unfortunately, the above citations are of academic interest only, as none of the compounds has reached the stage for the clinical trial.

## 5.7 Role of Pharma Companies and Central Funding Agencies in Drug Discovery and Development Against VL

*Leishmania* being a neglected poor man's disease has not attracted assistance from pharma companies in India or abroad. The population affected with VL, mostly survive on an income, which is $2.0 or less per day (WHO 2002) which makes the development of antileishmanial commercially unviable operation for any pharmaceutical company until it is a public enterprise.

The central funding agencies in India though funds the basic research in disease biology, are non-committal on any drug discovery and development project. The Melinda Gates foundation through DNDi has been supporting different discovery projects across the globe. In order to make some headway in the drug discovery, DNDi has proposed to initiate a consortium wherein a few centres would be involved in the synthesis of compounds and others would perform the screening. If required, the international stakeholders under the project would pitch in to assist the advanced drug discovery process.

Some of the key players operating in the leishmaniasis treatment market include Sanofi S.A., Sequus Pharmaceuticals Inc., Paladin Labs Inc., Enzon Pharmaceuticals Inc., Gland Pharma Limited, and Lifecare Innovations Private Limited.

In an extension to their previous agreement with WHO in 2016, Gilead Sciences, Inc. is donating 380,400 vials of AmBisome between 2016 and 2021. The 5-year collaboration and funding of US$ 20 million by Gilead Sciences, Inc. are proposed to provide access to diagnosis and treatment to the populations affected by VL in key endemic countries such as Ethiopia, India, Nepal, Bangladesh, South Sudan and Sudan. Financial contribution made by Gilead Sciences, Inc. has also helped WHO to expand, control, and reinforce surveillance in several endemic areas.

Very recently, Wellcome Trust has pledged over £10 million to the DND*i* to develop new treatments for leishmaniasis. This 3-year partnership is directed towards enabling DND*i* and Wellcome to develop new combinations of entirely new orally acting chemical entities. The programme brings together several R&D partners, including the University of Dundee, Celgene, GSK, Pfizer, TB Alliance and Takeda Pharmaceutical Company Limited. A portfolio of lead series, preclinical and clinical drug candidates, originating from different chemical classes with different mechanisms of action against *Leishmania* parasites have been built by these partners. The focus of this 3-year programme is to evaluate 10 candidates for advanced preclinical evaluation and finally selecting two all-oral new chemical entities for testing as a combination in Phase II trial.

## 5.8 Challenges and Solutions

The drug discovery in modern times mostly relies on biochemical targets, which are specific, suitably validated and have proven their significance in the disease process. Currently however, there are only a few targets including LdTR, Pteridine reductase, *Ld*-dipeptidyl carboxypeptidase, *Ld*nitroreductase, reported for *Leishmania*. Essentially, it is required that the chemical compound, which displays some degree of influence on the target during the in vitro assay, displays some order of activity in the in vivo assay. However, it has been often found that chemical compounds, which elicit potent response in the in vitro system do not show corresponding efficacy in the in vivo assays. *Leishmania* is further characterized via 3-stage assessment system wherein the first screening is performed on promastigotes followed by ex-vivo system which include the intracellular amastigote within macrophages followed by in vivo evaluation in the hamster model (mostly done via ip route though oral route is more useful). In India, most of the screening is pursued in *L. donovani*, therefore it is prudent to evaluate the active analogues in the east African strains too due to their higher degree of virulence. Thus, it remains a challenge for the compound to be universally effective against all species of *Leishmania*.

The other challenge is that the *Leishmania* parasite resides in spleen and liver and therefore the drug accumulation should be in these two organs. In case the drug when administered orally is metabolized too fast or has less oral bioavailability, results in decreased effectiveness. Further, till date there is no option available for treating PKDL. The situation is complex as no bioassay exists to test any drug against this stage of the disease. Therefore, there is a need to develop at least a bioassay which may allow discovering compounds to treat PKDL.

It has been widely accepted that for the drug discovery for the tropical diseases, the partnership between the academia, industry and non-profit organizations is essential (Ferrins and Pollastri 2018). Since the development of drug in this area is highly restricted; there is limitation even to invite public-private partnership (PPP). Interestingly the compounds which are being pushed to the antileishmanial pipeline are mainly identified either through random approaches or are repurposed and optimized from other disease areas. Therefore, the importance of repurposing of bioactives and random screening approach could be the most suited option for discovering new chemical architecture as bioactive for this neglected disease. Indeed, more Open Source Drug Discovery programs with crowd sourcing need to be encouraged. The patenting may be a factor for developing any drug, but for *Leishmania*, encouragement should be there to take up any active compound, even if it is known, for the development purpose.

## 5.9 Conclusions

Despite remarkable advances made during the last decade, the antileishmanial drug discovery still remains a challenge. The newer chemicals bearing diverse structures have reached the Phase I clinical trial stage, but it is to be closely watched how they respond in the patients after Phase I clinical trial. It is experienced that though several chemotypes display potent in vitro antileishmanial effect, did not retain corresponding efficacy in the in vivo system. The *Leishmania* parasite is known to reside majorly in liver and spleen. Therefore to show in vivo effect, the compound should essentially reach these sites. It has been found that if the compound has low bioavailability or does not clear the first pass metabolism it fails to show potent in vivo efficacy. A few biochemical targets exist but seldom a compound that has displayed activity against these enzymes elicited profound in vivo efficacy. Interestingly, most of the discoveries of new chemical scaffold in this area had been via random screening or repurposing, hence such activities to identify new hits needs to be encouraged. AmBisome has been successful to restrict the VL in India to some extent. However, the new cases of leishmaniasis in eastern Uttar Pradesh reported last year has raised several concerns towards elimination. Therefore, the research for developing new antileishmanial agents should be encouraged and this can only be taken forward with financial assistance from the Governmental agencies.

*CDRI Communication No. 9865*

## References


Afrin F, Chouhan G, Islamuddin M, Want MY, Ozbak HA, Hemeg HA (2019) *Cinnamomum cassia* exhibits antileishmanial activity against *Leishmania donovani* infection *in vitro* and *in vivo*. PLoS Negl Trop Dis 13:e0007227

Alcântara LM, Ferreira TCS, Gadelha FR, Miguel DC (2018) Challenges in drug discovery targeting TriTryp diseases with an emphasis on leishmaniasis. Int J Parasitol Drugs Drug Resist 8:430–439

Alvar J, Aparicio P, Aseffa A et al (2008) The relationship between leishmaniasis and AIDS: the second 10 years. Clin Microbiol Rev 21:334–359

Alvar J, Vélez ID, Bern C et al (2012) Leishmaniasis worldwide and global estimates of its incidence. PLoS One 7:e35671

Alves F, Bilbe G, Blesson S et al (2018) Recent development of visceral leishmaniasis treatments: successes, pitfalls, and perspectives. Clin Microbiol Rev 31:e00048–e00018

Andrade-Narvaez FJ, Loría-Cervera EN, Sosa-Bibiano EI, Van Wynsberghe NR (2016) Asymptomatic infection with American cutaneous leishmaniasis: epidemiological and immunological studies. Mem Inst Oswaldo Cruz 111:599–604

Andrews KT, Fisher G, Skinner-Adams TS (2014) Drug repurposing and human parasitic protozoan diseases. Int J Parasitol Drugs Drug Resist 4:95–111

Arshia AF, Ghouri N, Kanwal KKM, Perveen S, Choudhary MI (2018) Synthesis of 4-substituted ethers of benzophenone and their antileishmanial activities. R Soc OpenSci 5:171771

Ashburn TT, Thor KB (2004) Drug repositioning: identifying and developing new uses for existing drugs. Nat Rev Drug Discov 3:673–683

Asthana S, Jaiswal AK, Gupta PK, Dube A, Chourasia MK (2015) Th-1 biased immunomodulation and synergistic antileishmanial activity of stable cationic lipid–polymer hybrid nanoparticle: biodistribution and toxicity assessment of encapsulated amphotericin B. Eur J Pharm Biopharm 89:62–73

Atta KFM, Ibrahim TM, Farahat OOM et al (2017) Synthesis, modelling and biological evaluation of hybrids from pyrazolo[1,5-*c*]pyrimidine as antileishmanial agents. Future Med Chem 9:1913–1929

Aubé J (2012) Drug repurposing and the medicinal chemist. ACS Med Chem Lett 3:442–444

Bora D (1999) Epidemiology of visceral leishmaniasis in India. Natl Med J India 12:62–68

Bruni N, Stella B, Giraudo L, Della PC, Gastaldi D, Dosio F (2017) Nanostructured delivery systems with improved leishmanicidal activity: a critical review. Int J Nanomedicine 12:5289–5311

Burza S, Mahajan R, Sanz MG et al (2014) HIV and visceral leishmaniasis coinfection in Bihar, India: an under-recognized and under-diagnosed threat against elimination. Clin Infect Dis 59:552–555

Burza S, Croft SL, Boelaert M (2018) Leishmaniasis. Lancet 392:951–970

Chakravarty J, Sundar S (2010) Drug resistance in leishmaniasis. J Global Infect Dis 2:167

Chandrasekar R, Debnath S, Sivagami B, Jayasree P, Niranjan BM (2018) Therapeutic efficacy of flavonoids and terpenoids an ongoing herbal therapy in the treatment of leishmaniasis. Nat Prod Ind J 14:124

Chauhan SS, Pandey S, Shivahare R et al (2015) Novel β-carboline–quinazolinone hybrid as an inhibitor of *Leishmania donovani* trypanothione reductase: synthesis, molecular docking and bioevaluation. Med Chem Commun 6:351–356

ClinicalTrials.gov (2013) Trial to determine efficacy of fexinidazole in visceral leihmaniasispatients in Sudan. http://clinicaltrials.gov/show/NCT01980199. Accessed 12 Nov 2013

Correia D, Macedo VO, Carvalho EM et al (1996) Comparative study of meglumine antimoniate, pentamidine isethionate and aminosidinesulfate in the treatment of primary skin lesions caused by *Leishmania* (Viannia) *braziliensis*. Rev Soc Bras Med Trop 29:447–453

Cota GF, de Sousa MR, de Mendonca AL, Patrocinio A, Assunção LS, de Faria SR, Rabello A (2014) Leishmania-HIV co-infection: clinical presentation and outcomes in an urban area in Brazil. PLoS Negl Trop Dis 8:e2816

Croft SL, Seifert K, Duchêne M (2003) Antiprotozoal activities of phospholipid analogues. Mol Biochem Parasitol 126:165–172

Dar AA, Shadab M, Khan S, Ali N, Khan AT (2016) One-pot synthesis and evaluation of antileishmanial activities of functionalized *S*-Alkyl/Aryl benzothiazole-2-carbothioate scaffold. J Organomet Chem 81:3149–3160

Dea-Ayuela P, Bilbao-Ramos MA, Bolás-Fernández F, González-Cardenete MA (2016) Synthesis and antileishmanial activity of C7- and C12-functionalized dehydroabietylamine derivatives. Eur J Med Chem 121:445–450

Desjeux P, Ghosh RS, Dhalaria P, Strub-Wourgaft N, Zijlstra EE (2013) Report of the post kala-azar dermal leishmaniasis (PKDL) consortium meeting, New Delhi, India, 27–29 June, 2012. Parasit Vectors 6:196

Diro E, Lynen L, Mohammed R, Boelaert M, Hailu A, van Griensven J (2014) High parasitological failure rate of visceral leishmaniasis to sodium stibogluconate among HIV co-infected adults in Ethiopia. PLoS Negl Trop Dis 8:e2875

Diro E, Ritmeijer K, Boelaert M et al (2015) Use of pentamidine as secondary prophylaxisto prevent visceral leishmaniasis relapse in HIV infected patients, thefirst twelve months of a prospective cohort study. PLoS Negl Trop Dis 9:e0004087

DNDI (n.d.) DNDI-5561 – DNDi. https://www.dndi.org/diseases-projects/portfolio/dndi-5561/

DNDi (2016) Fexinidazole/miltefosine combination (VL). Available from https://www.dndi.org/diseases-projects/portfolio/completed-projects/fexinidazole-vl/

Dorlo TP, Balasegaram M, Beijnen JH, de Vries PJ (2012) Miltefosine: a review ofits pharmacology and therapeutic efficacy in the treatment of leishmaniasis. J Antimicrob Chemother 67:2576–2597

Espuelas MS, Legrand P, Campanero MA, Appel M, Chéron M, Gamazo C, Barratt G, Irache JM (2003) Polymeric carriers for amphotericin B: *in vitro* activity, toxicity and therapeutic efficacy against systemic candidiasis in neutropenic mice. J Antimicrob Chemother 52:419–427

Fan Y, Lu Y, Chen X, Tekwani BL, Li XC, Shen Y (2018) Anti-leishmanial and cytotoxic activities of a series of maleimides: synthesis, biological evaluation and structure-activity relationship. Molecules 23:2878

Félix MB, de Souza ER, de Lima MCA et al (2016) Antileishmanial activity of new thiophene-indole hybrids: design, synthesis, biological and cytotoxic evaluation, and chemometric studies. Bioorg Med Chem 24:3972–3977

Ferreira LLG, Andricopulo AD (2018) Chemoinformatics strategies for leishmaniasis drug discovery. Front Pharmacol 9:1278

Ferrins L, Pollastri MP (2018) The importance of collaboration between industry, academics, and nonprofits in tropical disease drug discovery. ACS Infect Dis 4:445–448

Fersing C, Basmaciyan L, Boudot C et al (2019) Nongenotoxic 3-nitroimidazo[1,2-*a*]pyridines Are NTR1 substrates that display potent *in vitro* antileishmanial activity. ACS Med Chem Lett 10:34–39

Fonseca-Berzal C, Ibáñez-Escribano A, Vela N et al (2018) Antichagasic, leishmanicidal, and trichomonacidal activity of 2-benzyl-5-nitroindazole-derived amines. ChemMedChem 13:1246–1259

de Freitas EO, Leoratti FM, Freire-de-Lima CG, Morrot A, Feijó DF (2016) The contribution of immune evasive mechanisms to parasite persistence in visceral leishmaniasis. Front Immunol 7:153

Freitas-Junior LH, Chatelain E, Kim HA, Siqueira-Neto JL (2012) Visceral leishmaniasis treatment: what do we have, what do we need and how to deliver it? Int J Parasitol Drugs Drug Resist 2:11–19

Ghozlani ME, Bouissane L, Berkani M et al (2019) Synthesis and biological evaluation against *Leishmania donovani* of novel hybrid molecules containing indazole-based 2-pyrone scaffolds. Med Chem Commun 10:120–127

Government of Nepal - Ministry of Health and Population (2009) Kala-azar elimination program in Nepal. National strategic guideline on kala-azar elimination program in Nepal

Guo LS, Working PK (1993) Complexes of amphotericin B and cholesteryl sulfate. J Liposome Res 3:473–490

Gupta S, Yardley V, Vishwakarma P, Shivahare R, Sharma B, Launay D, Martin D, Puri SK (2015) Nitroimidazo-oxazole compound DNDI-VL-2098: an orally effective preclinical drug candidate for the treatment of visceral leishmaniasis. J Antimicrob Chemother 70:518–527

Handman E (1999) Cell biology of *Leishmania*. Adv Parasitol 44:1–39

Hendrickx S, Beyers J, Mondelaers A, Eberhardt E, Lachaud L, Delputte P, Cos P, Maes L (2016) Evidence of a drug-specific impact of experimentally selected paromomycin and miltefosine resistance on parasite fitness in *Leishmania infantum*. J Antimicrob Chemother 71:1914–1921

Hnik P (2016) Stable formulations for the oral administration of amphotericin B. WO2016112339A1. Chem Abstr 165:169462

Jacob RT, Liu Y, Sciotti RJ (2018) Novel oxaborole analogs and uses thereof. WO2018160845A1. Chem Abstr 169:347172

Jacomini AP, Silva MJV, Silva RGM et al (2016) Synthesis and evaluation against *Leishmania amazonensis* of novelpyrazolo[3,4-*d*]pyridazinone-*N*-acylhydrazone-(bi)thiophene hybrids. Eur J Med Chem 124:340–349

Jain K, Jain NK (2013) Novel therapeutic strategies for treatment of visceral leishmaniasis. Drug Discov Today 18:1272–1281

Jameson MB, Rischin D, Pegram M, Gutheil J, Patterson AV, Denny WA, Wilson WR (2010) A nitrogen mustard prodrug activated by both hypoxia andaldo-keto reductase 1C3, in patients with solid tumors. Cancer Chemother Pharmacol 65:791–801

Javed I, Hussain SZ, Ullah I et al (2015) Synthesis, characterization and evaluation of lecithin-based nanocarriers for the enhanced pharmacological and oral pharmacokinetic profile of amphotericin B. J Mater Chem B 3:8359–8365

Jha SN, Singh NK, Jha TK (1991) Changing response to diamidine compounds in cases of kala-azar unresponsive to antimonial. J Assoc Physicians India 39:314–316

Jha TK, Sundar S, Thakur CP, Felton JM, Sabin AJ, Horton J (2005) A phase II dose-ranging study of sitamaquine for the treatment of visceral leishmaniasis in India. Am J Trop Med Hyg 73:1005–1011

Kanwar A, Eduful BJ, Barbeto L et al (2017) Synthesis and activity of a new series of antileishmanial agents. ACS Med Chem Lett 8:797–801

Khare S, Nagle AS, Biggart A et al (2016) Proteasome inhibition for treatment of leishmaniasis, Chagas disease and sleeping sickness. Nature 537:229–233

Khatta SN, Haiba NS, Asal AM, Bekhit AA, Guemei AA, Amer A, El-Faham A (2017) Study of antileishmanial activity of 2-aminobenzoyl amino acid hydrazides and their quinazoline derivatives. Bioorg Med Chem Lett 27:918–921

Khattab SN, Khalil HH, Bekhit AA, El-Rahman MMA, de la Torre BG, El-Faham A, Albericio F (2018) 1,3,5-Triazino peptide derivatives: synthesis, characterization, and preliminary antileishmanial activity. ChemMedChem 13:725–735

Kima PE (2014) Leishmania molecules that mediate intracellular pathogenesis. Microbes Infect 16:721–726

Klug DM, Gelb MH, Pollastri MP (2016) Repurposing strategies for tropical disease drug discovery. Bioorg Med Chem Lett 26:2569–2576

Krause G, Kroeger A (1994) Topical treatment of American cutaneous leishmaniasis with paromomycin and methyl benzethonium chloride: a clinical study under field conditions in Ecuador. Trans R Soc Trop Med Hyg 88:92–94

Kumar R, Sahoo GC, Pandey K, Das V, Das P (2016) Development of PLGA-PEG encapsulated miltefosine based drug delivery system against visceral leishmaniasis. Mater Sci Eng C Mater Biol Appl 59:748–753

Leañez J, Nuñez J, García-Marchan Y et al (2019) Anti-leishmanial effect of spiro dihydroquinoline-oxindoles on volume regulation decrease and sterol biosynthesis of *Leishmania braziliensis*. Exp Parasitol 198:31–38

Lindoso JAL, Moreira CHV, Cunha MA, Queiroz IT (2018) Visceral leishmaniasis and HIV coinfection: current perspectives. HIV/AIDS 10:193–201

Mathur P, Samantaray JC, Vajpayee M, Samanta P (2006) Visceral Leishmaniasis/human immunodeficiency virus co-infection in India: the focus of two epidemics. J Med Microbiol 55:919–922

Mbekeani AJ, Jones RS, Llorens MB et al (2019) Mining for natural product antileishmanials in a fungal extract library. Int J Parasitol Drugs Drug Resist 11:118. https://doi.org/10.1016/j.ijpddr.2019.05.003

McGwire BS, Satoskar AR (2013) Leishmaniasis: clinical syndromes and treatment. QJM 107:7–14

Melaku Y, Collin SM, Keus K, Gatluak F, Ritmeijer K, Davidson RN (2007) Treatment of kala-azar in southern Sudan using a 17-day regimen of sodium stibogluconate combined with paromomycin: a retrospective comparison with 30-daysodium stibogluconate monotherapy. Am J Trop Med Hyg 77:89–94

Mengesha B, Endris M, Takele Y, Mekonnen K, Tadesse T, Feleke A, Diro E (2014) Prevalence of malnutrition and associated risk factors among adult visceral leishmaniasis patients in Northwest Ethiopia: a cross sectional study. BMC Res Notes 7:75

Mishra KB, Tiwari N, Bose P, Singh R, Rawat AK, Singh SK, Mishra RC, Singh RK, Tiwari VK (2019) Design, synthesis and pharmacological evaluation of noscapine glycoconjugates. Chem Select 4:2644–2648

Molyneux DH, Killick-Kendrick R (1987) Morphology, ultrastructure and life cycles. In: Peters W, Killick-Kendrick R (eds) The leishmaniases in biology and medicine, vol 1. Academic Press, London, pp 121–176

Mondal D, Singh SP, Kumar N et al (2009) Visceral leishmaniasis elimination programme in India, Bangladesh, and Nepal: reshaping the case finding/case management strategy. PLoS Negl Trop Dis 3:e355

Morizot G, Jouffroy R, Faye A et al (2016) Antimony to cure visceral leishmaniasis unresponsive to liposomal amphotericin B. PLoS Negl Trop Dis 10:e0004304

Mowbray CE, Braillard S, Speed W (2015) Novel amino-pyrazole Ureas with potent in vitro and in vivo antileishmanial activity. J Med Chem 58:9615–9624

Mukhopadhyay R, Mukherjee, S, Mukherjee B, Naskar K, Mondal D, Decuypere S, Ostyn B, Prajapati VK, Sundar S, Dujardin JC, Roy S. (2011) Characterisation of antimony-resistant Leishmania donovani isolates: Biochemical and biophysical studies and interaction with host cells. Int. J Parasitol. 41:1311–1321

Murillo JA, Gil JF, Upegui YA et al (2019) Antileishmanial activity and cytotoxicity of *ent*-beyerene diterpenoids. Bioorg Med Chem 27:3972–3977

Musa A, Khalil E, Hailu A et al (2012) Sodium stibogluconate (SSG) and paromomycin combination compared to SSG for visceral leishmaniasis in East Africa: a randomised controlled trial. PLoS Negl Trop Dis 6:e1674

Nagle AS, Khare S, Kumar AB et al (2014) Recent developments in drug discovery for leishmaniasis and human african trypanosomiasis. Chem Rev 114:11305–11347

Narayan S, Thakur CP, Bahadur S, Thakur M, Pandey SN, Thakur AK, Mitra DK, Mukherjee PK (2017) *Cedrus deodara*: *in vitro* antileishmanial efficacy and immumomodulatory activity. Indian J Med Res 146:780–787

Nocentini A, Supuran CT, Winum J-Y (2018) Benzoxaborole compounds for therapeutic uses: a patent review (2010-2018). Expert Opin Ther Patent 28:493–504

Oliveira LFG, Souza-Silva F, Côrtes LMC et al (2018) Antileishmanial activity of 2-methoxy-4*H*-spiro-[naphthalene-1,20-oxiran]-4-one(epoxymethoxy-lawsone): a promising new drug candidate for leishmaniasis treatment. Molecules 23:E864

Olliaro PL, Guerin PJ, Gerstl S, Haaskjold AA, Rottingen J-A, Sundar S (2005) Treatment options for visceral leishmaniasis: a systematic review of clinical studies done in India,1980–2004. Lancet Infect Dis 5:763–774

Ortiz D, Guiguemde WA, Hammill JT et al (2017) Discovery of novel, orally bioavailable, antileishmanial compounds using phenotypic screening. PLoS Negl Trop Dis 11:e0006157

Oryan A (2015) Plant-derived compounds in treatment of leishmaniasis. Iran J Vet Res 16:1–19

Ostyn B, Gidwani K, Khanal B, Picado A, Chappuis F, Singh SP, Rijal S, Sundar S, Boelaert M (2011) Incidence of symptomatic and asymptomatic *Leishmania donovani* infections in high-endemic foci in India and Nepal: a prospective study. PLoS Negl Trop Dis 5:e1284

Pagliero RJ, Kaiser M, Brun R, Nieto MJ, Mazzieri MR (2017) Lead selection of antiparasitic compounds from a focused library of benzenesulfonyl derivatives of heterocycles. Bioorg Med Chem Lett 27:3945–3949

Panda SK, Luyten W (2018) Antiparasitic activity in Asteraceae with special attention to ethnobotanical use by the tribes of Odisha, India. Parasite 25:10

Pandey A, Sahu D, Bakkali T et al (2012) Estimate of HIV prevalence and number of people living with HIV in India 2008–2009. BMJ Open 2:e000926

Pandey RK, Verma P, Sharma D, Bhatt TK, Sundar S, Prajapati VK (2016a) High-throughput virtual screening and quantum mechanics approach to develop imipramine analogues as leads against trypanothione reductase of leishmania. Biomed Pharmacother 83:141–152

Pandey RK, Kumbhar BV, Srivastava S, Malik R, Sundar S, Kunwar A, Prajapati VK (2016b) Febrifugine analogues as *Leishmania donovani* trypanothione reductase inhibitors: binding energy analysis assisted by molecular docking, ADMET and molecular dynamics simulation. J Biomol Struct Dyn 35:141–158

Patterson S, Wyllie S, Stojanovski L et al (2013) The R enantiomer of the antitubercular drug PA-824 as a potential oral treatment for visceral leishmaniasis. Antimicrob Agents Chemother 57:4699–4706

Penta A, Chander S, Smith TK, Sankaranarayanan M (2018) Design, synthesis and biological evaluation of piperazinyl-β-carbolinederivatives as anti-leishmanial agents. Eur J Med Chem 150:559–566

Penta A, Chander S, Smith TK, Singh RP, Jha PN, Sankaranarayanana M (2019) Biological evaluation and structure activity relationship of 9-methyl-1-phenyl-9*H*-pyrido[3,4-*b*]indole derivatives as anti-leishmanial agents. Bioorg Chem 84:98–105

Petitti M, Vanni M, Barresi AA (2008) Controlled release of drug encapsulated as a solid core: theoretical model and sensitivity analysis. Chem Eng Res Des 86:1294–1300

Petri e Silva SC, Palace-Berl F, Tavares LC, Soares SR, Lindoso JA (2016) Effects of nitro-heterocyclic derivatives against Leishmania (Leishmania) infantum promastigotes and intracellular amastigotes. Exp Parasitol 163:68–75

Pigott DM, Golding N, Messina JP et al (2014) Global database of leishmaniasis occurrence locations, 1960-2012. Sci Data 1:140036

Ponte-Sucre A, Gamarro F, Dujardin J-C et al (2017) Drug resistance and treatment failure in leishmaniasis: a 21st century challenge. PLoS Negl Trop Dis 11:e0006052

Prajapati VK, Awasthi K, Yadav TP, Rai M, Srivastava ON, Sundar S (2011) An oral formulation ofamphotericin B attached to functionalized carbon nanotubes is an effective treatment for experimental visceral leishmaniasis. J Infect Dis 205:333–336

Prajapati VK, Mehrotra S, Gautam S, Rai M, Sundar S (2012) *In vitro* antileishmanial drug susceptibility of clinical isolates from patients with Indian visceral leishmaniasis–status of newly introduced drugs. Am J Trop Med Hyg 87:655–657

Purkait B, Kumar A, Nandi N et al (2012) Mechanism of amphotericin B resistancein clinical isolates of *Leishmania donovani*. Antimicrob Agents Chemother 56:1031–1041

Purohit P, Pandey AK, Singh D et al (2017) An insight into tetrahydro-β-carboline–tetrazole hybrids: synthesis and bioevaluation as potent antileishmanial agents. Med Chem Commun 8:1824–1834

Rahman M, Ahmed B-N, Faiz MA et al (2011) Phase IV trial of miltefosinein adults and children for treatment of visceral leishmaniasis (kala-azar) in Bangladesh. Am J Trop Med Hyg 85:66–69

Ramesh V, Avishek K, Sharma V, Salotra P (2014) Combination therapy with amphotericin-B and miltefosine for post-kalaaardermal leishmaniasis: a preliminary report. Acta Derm Venereol 94:242–243

Ramu D, Garg S, Ayana R et al (2017) Novel β-carboline-quinazolinone hybrids disrupt *Leishmania donovani* redox homeostasis and show promising antileishmanial activity. Biochem Pharmacol 129:26–42

Ramu D, Jain R, Kumar RR et al (2019) Design and synthesis of imidazolidinone derivatives as potent anti-leishmanial agents by bioisosterism. Arch Pharm Chem Life Sci 352:e1800290

Ready PD (2013) Biology of phlebotomine sand flies as vectors of disease agents. Annu Rev Entomol 58:227–250

Redhu NS, Dey A, Balooni V, Singh S (2006) Leishmania-HIV co-infection: an emerging problem in India. AIDS 20:1213–1215

Reguera RM, Perez-Pertejo Y, Gutierrez-Corbo C et al (2019) Current and promising novel drug candidates against visceral leishmaniasis. Pure Appl Chem 91:1385–1404

Rijal S, Ostyn B, Uranw S et al (2013) Increasing failure of miltefosine in the treatment of Kala-azar in Nepal and the potential role of parasite drug resistance, reinfection, or noncompliance. Clin Infect Dis 56:1530–1538

Ritmeijer K, Ter Horst R, Chane S, Aderie EM, Piening T, Collin SM, Davidson RN (2011) Limited effectiveness of high-dose liposomal amphotericin B (AmBisome) for treatment of visceral leishmaniasis in an Ethiopian population with high HIV prevalence. Clin Infect Dis 53: e152–e158

Romero G, Lessa H, Macedo VO et al (1996) Open therapeutic study with aminosidine sulfate in mucosal leishmaniasis caused by Leishmania (Viannia) braziliensis. Rev Soc Bras Med Trop 29:557–565

da Rosa R, de Moraes MH, Zimmermann LA, Schenkel EP, Steindel M, Bernardes LSC (2017) Design and synthesis of a new series of 3,5-disubstituted isoxazoles active against *Trypanosoma cruzi* and *Leishmania amazonensis*. Eur J Med Chem 128:25–35

Roychoudhury J, Sinha R, Ali N (2011) Therapy with sodium stibogluconate in stearylamine-bearing liposomes confers cure against SSG-resistant *Leishmania donovani* in BALB/c mice. PLoS One 6:e17376

Ruiz-Santaquiteria M, de Castro S, Toro MA et al (2018) Trypanothione reductase inhibition and anti-leishmanial activity of all-hydrocarbon stapled a-helical peptides with improved proteolytic stability Eur. J Med Chem 149:238–247

Saccoliti F, Angiulli G, Pupo G et al (2017) Inhibition of Leishmania infantum trypanothione reductase by diaryl sulfide derivatives. J Enzyme Inhib Med Chem 32:304–310

Saha S, Acharya C, Pal U, Chowdhury SR, Sarkar K, Maiti NC, Jaisankar P, Majumder HK (2016) A novel spirooxindole derivative inhibits the growth of *Leishmania donovani* parasites both *in vitro* and *in vivo* by targeting type IB topoisomerase. Antimicrob Angents Chemother 60:6281–6293

Sangshetti JN, Khan FAK, Kulkarni AA, Arote R, Patil RH (2015) Antileishmanial drug discovery: comprehensive review of the last 10 years. RSC Adv 5:32376–32415

Saudagar P, Dubey VK (2014) Carbon nanotube based betulin formulationshows better efficacy against Leishmania parasite. Parasitol Int 63:772–776

Sen Gupta PC (1968) Leishmaniasis in India. J Indian Med Assoc 50:34–36

Sesana AM, Monti-Rocha R, Vinhas SA, Morais CG, Dietze R, Lemos EM (2011) In vitro activity of amphotericin B cochleates against *Leishmania chagasi.* Mem Inst Oswaldo Cruz 106:251–253

Shio MT, Paquet M, Martel C, Bosschaerts T, Stienstra S, Olivier M, Fortin A (2014) Drug delivery by tattooing to treat cutaneous leishmaniasis. Sci Rep 4:4156

da Silva AD, dos Santosa JA, Machadob PA et al (2018) Insights about resveratrol analogs against trypanothione reductase of *Leishmania braziliensis*: molecular modelling, computational docking and in vitro antileishmanial studies. J Biomol Struct Dyn 37:2960–2969

Srivastava P, Prajapati VK, Rai M, Sundar S (2011) Unusual case of resistance to amphotericin B in visceral leishmaniasis in a region in India where leishmaniasis isnot endemic. J Clin Microbiol 49:3088–3091

Sundar S (2001) Drug resistance in Indian visceral leishmaniasis. Tropical Med Int Health 6:849–854

Sundar S, Chakravarty J (2013) Leishmaniasis: an update of current pharmacotherapy. Expert Opin Pharmacother 14:53–63

Sundar S, Chakravarty J (2015) Investigational drugs for visceral leishmaniasis. Expert Opin Investig Drugs 24:43–59

Sundar S, Chatterjee M (2006) Visceral leishmaniasis-current therapeutic modalities. Indian J Med Res 123:345–352

Sundar S, Thakur BB, Tandon AK, Agrawal NR, Mishra CP, Mahapatra TM, Singh VP (1994) Clinico-epidemiological study of drug resistance in Indian kala-azar. Br Med J 308:307

Sundar S, Jha T, Thakur CP, Sinha PK, Bhattacharya SK (2007) Injectable paromomycin for visceral leishmaniasis in India. N Engl J Med 356:2571–2581

Sundar S, Rai M, Chakravarty J, Agarwal D, Agrawal N, Vaillant M, Olliaro P, Murray HW (2008) New treatment approach in Indian visceral leishmaniasis: single-dose liposomal amphotericinB followed by short-course oral miltefosine. Clin Infect Dis 47:1000–1006

Sundar S, Chakravarty J, Agarwal D, Rai M, Murray HW (2010) Single-dose liposomal amphotericin B for visceral leishmaniasis in India. N Engl J Med 362:504–512

Sundar S, Sinha PK, Rai M et al (2011a) Comparison of short-course multidrug treatment with standard therapy for visceral leishmaniasis in India: an open-label, non-inferiority, randomised controlled trial. Lancet 377:477–486

Sundar S, Sinha PK, Verma DK et al (2011b) Ambisome plus miltefosine for Indian patients with kala-azar. Trans R Soc Trop Med Hyg 105:115–117

Sundar S, Singh A, Rai M et al (2012) Efficacy of miltefosine in thetreatment of visceral leishmaniasis in India after a decade of use. Clin Infect Dis 55:543–550

Sundar S, Singh A, Tiwari A, Shukla S, Chakravarty J, Rai M (2014) Efficacy and safety of paromomycin in treatment of post-kala-azar dermal leishmaniasis. ISRN Parasitol 2014:548010

Sundar S, Chakravarty J, Meena LP (2019) Leishmaniasis: treatment, drug resistance and emerging therapies. Expert Opin Orphan Drugs 7:1–10

Tamiru A, Tigabu B, Yifru S, Diro E, Hailu A (2016) Safety and efficacy of liposomal amphotericin B for treatment of complicated visceral leishmaniasis in patients without HIV, North-West Ethiopia. BMC Infect Dis 16:548

Teixeira de Macedo SS, Visbal G, Lima PGJ, Urbina JA, de Souza W, Cola FRJ (2018) In vitro antileishmanial activity of ravuconazole, a triazole antifungal drug, as a potential treatment for leishmaniasis. J Antimicrob Chemother 73:2360–2373

Tejería A, Pèrez-Pertejo Y, Reguera RM et al (2019) Antileishmanial activity of new hybrid tetrahydroquinoline and quinoline derivatives with phosphorus substituents. Eur J Med Chem 162:18–31

Temraz MG, Elzahhar PA, Bekhit AE-DA, Bekhit AA, Labib HF, Belal ASF (2018) Anti-leishmanial click modifiable thiosemicarbazones: design, synthesis, biological evaluation and in silico studies. Eur J Med Chem 151:585–600

Thakur CP, Singh RK, Hassan SM, Kumar R, Narain S, Kumar A (1999) Amphotericin B deoxycholate treatment of visceral leishmaniasis with newer modes of administration and precautions: a study of 938 cases. Trans R Soc Trop Med Hyg 93:319–323

Thompson AM, O'Connor PD, Blaser A et al (2016) Repositioning antitubercular 6-nitro-2,3-dihydroimidazo[2,1-*b*][1,3]oxazoles for neglected tropical diseases: structure–activity studies on a preclinical candidate for visceral leishmaniasis. J Med Chem 59:2530–2550

Thompson AM, O'Connor PD, Marshall AJ et al (2017) 7- Substituted 2-nitro-5,6-dihydroimidazo [2,1-*b*][1,3]oxazines: novel antitubercular agents lead to a new preclinical candidate for visceral leishmaniasis. J Med Chem 60:4212–4233

Thompson AM, O'Connor PD, Marshall AJ et al (2018) Development of (6*R*)-2-nitro-6-[4-(trifluoromethoxy)phenoxy]-6,7- dihydro-5*H*-imidazo[2,1-*b*][1,3]oxazine (DNDI-8219): a new lead for visceral leishmaniasis. J Med Chem 61:2329–2352

Tiuman TS, Santosm AO, Ueda-Nakamuram T, Filho BP, Nakamura CV (2011) Recent advances inleishmaniasis treatment. Int J Infect Dis 15:e525–e532

Torrado JJ, Serrano DR, Uchegbu IF (2013) The oral delivery of amphotericin B. Ther Deliv 4:9–12

Upadhyay A, Kushwaha P, Gupta S, Dodda RP, Ramalingam K, Kant R, Goyal N, Sashidhara KV (2018) Synthesis and evaluation of novel triazolyl quinoline derivatives as potential antileishmanial agents. Eur J Med Chem 154:172–181

Upadhyay A, Chandrakar P, Gupta S et al (2019) Synthesis, biological evaluation, structure–activity relationship, and mechanism of action studies of quinoline–metronidazole derivatives against experimental visceral leishmaniasis. J Med Chem 62:5655–5671

Velásquez AMA, Ribeiro WC, Venn V et al (2017) Efficacy of a binuclear cyclopalladated compound therapy for cutaneous Leishmaniasis in the murine model of infection with *Leishmania amazonensis* and its inhibitory effect on topoisomerase 1B. Antimicrob Agents Chemother 61:e00688–e00617

Vijayakumar S, Das P (2018) Recent progress in drug targets and inhibitors towards combating leishmaniasis. Acta Trop 181:95–104

de Vries HJ, Reedijk SH, Schallig HD (2015) Cutaneous leishmaniasis: recent developments in diagnosis and management. Am J Clin Dermatol 16:99–109

Wadhone P, Maiti M, Agarwal R, Kamat V, Martin S, Saha B (2009) Miltefosine promotes IFN-c-dominated anti-leishmanial immune response. J Immunol 182:7146–7154

Walkinshaw M (2014) Multiple chemical scaffolds inhibit a promising Leishmania drug target. IUCrJ 1:202–203

Wasan EK, Gershkovich P, Zhao J et al (2010) A novel tropically stable oral amphotericin B formulation (iCo-010) exhibits efficacy against visceral leishmaniasis in a murine model. PLoS Negl Trop Dis 4:e913

WHO (2002) New therapy for visceral leishmaniasis. Wkly Epidemiol Rec 77:205–212

WHO (2010) Control of the leishmaniasis: report of a meeting of the World Health Organization Expert Committee on the control of leishmaniasis, Geneva, 2010. World Health Organ Tech Rep Ser 949:1–186

WHO (2017) WHO: weekly epidemiological record: global leishmaniasis update, 2006–2015, a turning point in leishmaniasis surveillance. World Health Organ 92:520–525

World Health Organization (1995) Report on the consultative meeting on Leishmania/HIV co-infection. Report no. WHO/LEISH/95.35. World Health Organization, Geneva, pp 1–14

World Health Organization (2004) Report of the scientific working group meeting on leishmaniasis. World Health Organization, Geneva

World Health Organization (2005) Regional strategic framework for elimination of kala-azar from the SouthEast Asia region (2005–2015). World Health Organization, New Delhi

World Health Organization (2011) Trend of reported cases of Kala-azar in SEA Region 1994–2011. WHO, Geneva, pp 12–15

World Health Organization (2012) Regional strategic framework for elimination of kala-azar from the South East Asia Region (2011–2015). WHO, Geneva

World Health Organization, South-East Asia, Government of the People's Republic of Bangladesh, Royal Government of Bhutan, Government of India, Government of Nepal, Royal Thai Government (2014) Memorandum of understanding among Bangladesh, Bhutan, India, Nepal and Thailand on the elimination of kala-Azar from the South-East Asia Region, Dhaka. WHO, Geneva

World Health Organization (2015) World health statistics, 2015. WHO, Geneva. http://www.who.int/gho/publications/world_health_statistics/2015/en/

World Health Organization (2019). https://www.who.int.Newsroom.Factsheets.Detail

World Health Organization Global Health Observatory (2017) Leishmaniasis. WHO, Geneva. http://www.who.int/gho/neglected_diseases/leishmaniasis/en/

Wu G, Zhao T, Kang D et al (2019) Overview of recent strategic advances in medicinal chemistry. J Med Chem 62:9375–9414

Wyllie S, Thomas M, Patterson S et al (2018) Cyclin-dependent kinase 12 is a drug target for visceral leishmaniasis. Nature 560:192–197

Wyllie S, Brand S, Thomas M et al (2019) Preclinical candidate for the treatment of visceralleishmaniasis that acts through proteasome inhibition. Proc Natl Acad Sci U S A 116:9318–9323

Yamthe LRT, Appiah-Opong R, Fokou PVT, Tsabang N, Boyom FF, Nyarko AK, Wilson MD (2017) Marine algae as source of novel antileishmanial drugs: a review. Mar Drugs 15:323

Yousuf M, Mukherjee D, Dey S, Pal C, Adhikari S (2016) Antileishmanial ferrocenylquinoline derivatives: synthesis and biological evaluation against Leishmania donovani. Eur J Med Chem 124:468–479

Yousuf M, Mukherjee D, Dey S, Chatterjee S, Pal A, Sarkar B, Pal C, Adhikari S (2018) Synthesis and biological evaluation of polyhydroxylated oxindole derivatives as potential antileishmanial agent. Bioorg Med Chem Lett 28:1056–1062

Zarif L (2005) Drug delivery by lipid cochleates. Methods Enzymol 391:314–329

Zhu X, Farahat M, Mattamana AA et al (2016) Synthesis and pharmacological evaluation of mono-arylimidamides as antileishmanial agents. Bioorg Med Chem Lett 26:2551–2556

Zijlstra EE, Musa AM, Khalil EA, el-Hassan IM, el-Hassan AM (2003) Post-kala-azar dermal Leishmaniasis. Lancet Infect Dis 3:87–98

**Neena Goyal** completed her Ph.D. in 1989 from the Division of Biochemistry, CSIR - Central Drug Research Institute, Lucknow and is presently working as head and chief scientist in the same department. Her research interests include identification and characterization of novel drug targets for visceral leishmaniasis where chemical intervention may prove lethal to the parasite and hence, generate new leads for the development of antileishmanial chemotherapy. With the expertise in molecular biology, proteomics and classical analytical biochemistry, her goal is to develop an understanding of clinical resistance against first-line and second-line drugs with an aim to treat resistant leishmania. She is a recipient of INSA Young Scientist Medal (1993) and CSIR Young Scientist Award (1998).

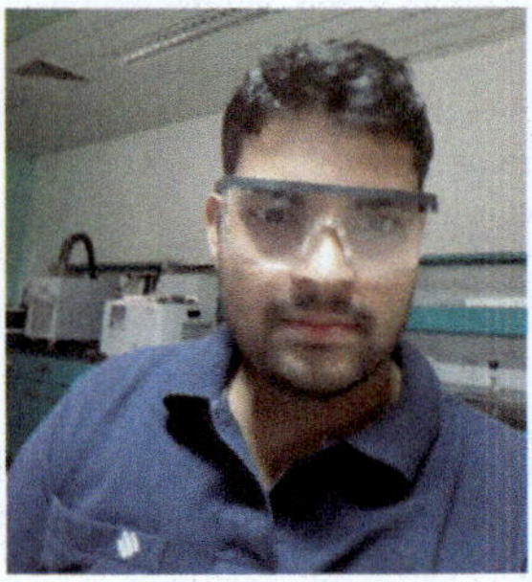

**Manveer Patel** was born in Uttar Pradesh, India, in 1994. He completed his M.Sc. in applied chemistry from Babasaheb Bhimrao Ambedkar Central University, Lucknow, India in 2016. After passing CSIR-UGC-NET-Lectureship exam, he joined the CSIR-Central Drug Research Institute in 2018, as a Project fellow under the supervision of Dr. Sanjay Batra. His current research focuses on design and synthesis of anti-infective agents.

**Sanjay Batra** completed his Ph.D. in 1993 from the Medicinal and Process Chemistry Division at Central Drug Research Institute, Lucknow and is presently working as a senior principal scientist in the same department. His research interests include development of compounds for neglected diseases and CNS disorders, chemistry associated with Morita–Baylis–Hillman adducts, and transition metal-catalyzed reactions and heterocyclic chemistry. He is a Fellow of Royal Society of Chemistry and Fellow of National Academy of Sciences, India. He is also a recipient of Bronze medal from the Chemical Research Society of India.

# Introduction to Diabetes and Prevalence in India 6

Parimal Misra and Ranjan Chakrabarti

Diabetes is a chronic progressive disease associated with different co-morbidities. The availability of insulin or effective use of insulin in the body is the major cause of this disease. Enhanced blood sugar or hyperglycemia is the manifestation of this disease. If not treated timely, diabetes can damage cardiac system, blood vessels, vision, renal function and nerves leading to increased risk of cardiac disease, stroke and may cause neuronal, retinal, and kidney disease. The overall risk of morbidity and mortality among diabetes affected people is more than double compared to non-diabetic people (WHO n.d.).

Mainly two types of diabetes exist. Type 1 diabetes is associated with deficient insulin production. It is diagnosed in the early childhood and symptoms include frequent urination, enhanced thirst, hunger, weight loss, vision changes, and fatigue. Treatment of type 1 diabetes is the daily administration of insulin.

Among diabetic population, ~90% of people are comprised of type 2 diabetes. Type 2 diabetes occurs mainly due to obesity, enhanced plasma lipid (dyslipidemia), stress, and sedentary life style (lack of enough physical activity). Symptoms are initially not that severe like type 1 diabetes but increase slowly but progressively. Generally the disease is diagnosed in the late and in the advanced stage. So, type 2 diabetes is known as a silent killer disease.

Insulin resistance occurs when body cannot use insulin effectively resulting enhanced insulin secretion from the pancreatic β-cells causing hyperinsulinemia. Overtime, enhanced insulin resistance leads to impaired fasting glucose (IFG) and impaired glucose tolerance (IGT). This is known as prediabetic stage. IFG condition is diagnosed when fasting plasma glucose level ranges from 6.1 to 6.9 mmol/L (110–

P. Misra (✉)
Dr. Reddy's Institute of Life Sciences, Hyderabad, India
e-mail: parimalm@drils.org

R. Chakrabarti
United States Pharmacopeia, Hyderabad, India
e-mail: rxc@usp.org



M. Dikshit (ed.), *Drug Discovery and Drug Development*,
https://doi.org/10.1007/978-981-15-8002-4_6

125 mg/dL) and 2 h plasma glucose is measured as >7.8 mmol/L (140 mg/dl) where as in the case of IGT, the fasting plasma glucose level is measured as <7.0 mmol/L (126 mg/dL) and 2 h plasma glucose ranges from 7.8 mmol/L (140 mg/dL) to 11.1 mmol/L (200 mg/dL). If untreated, insulin resistance increases drastically and rapid progressive degeneration of pancreatic β-cells leads to the decreased secretion of insulin resulting the transformation of prediabetic stage into full blown diabetic condition. Diabetic condition is detected when measured plasma glucose level in fasting condition is ≥7.0 mmol/L (126 mg/dL) or 2 h plasma glucose level is 11.1 mmol/dL (200 mg/dL). Ultimately, in the last stage of type 2 diabetes, the level of secreted insulin is measured below the normal level. During this stage, the patients are administered daily with exogenous insulin (Fig. 6.1).

Epidemiological and scientific data are now emerging which suggest that there may exist differences in the genesis of diabetes in Indians relative to the western population. Multiple risk factors such as genetic susceptibility, thin-fat phenotype, low birth weight, fetal programming, urbanization, sedentary life style, socioeconomic transition, smoking, and alcohol contribute to the origin of Indian insulin resistance syndrome. Indians are thinner with less muscle mass but are centrally obese with higher plasma free fatty acids, triglycerides, and insulin, characteristic of insulin resistance. Obesity and being overweight are relatively lower in Indian population than Western population but at a given Body Mass Index (BMI), Indians have a higher body fat composition (thin-fat phenotype) compared to other populations (Radha and Mohan 2007; Anjana et al. 2014).

A new susceptibility locus at 2q21 was identified in Indians in a genome-wide association study for type 2 diabetes (Tabassum et al. 2013). Study also identified that the common variants of IL6, LEPR, and PBEF1 are associated with obesity in Indian children (Tabassum et al. 2012). Several genetic variants have been detected in genome-wide association studies of single-nucleotide polymorphism (SNPs), which may be associated with β-cell dysfunction and insulin resistance (Khardori et al. n.d.). About 40 independent loci have been identified and found to be associated with increased type 2 diabetes (Wheeler and Barroso 2011). A subset of them are: TCF7L2, MTNR1B, FSADS1, PPARγ, KCNJ11 (Nielsen et al. 2003); FTO and IGF2BP2 (Ukkola et al. 2001); HHEX, SLC30A8 (Sladek et al. 2007), and WFS1 (Sandhu et al. 2007). Genetic variants in incretin hormone—gastric inhibitory polypeptide (GIPR) might also affect type 2 diabetes (Saxena et al. 2010) and has been found to be associated with reduced β-cell function. A clear association of genetic defects with some form of diabetes have been proposed. Maturity Onset Diabetes of Youth (MODY) syndrome has been deserved to be associated with β-cell dysfunction for 2–5% of type 2 diabetes. Eleven MODY subtypes have been identified to date (Winckler et al. 2007).

In 2015, worldwide prevalence of diabetes was ~422 million (8.5% of adults aged 20–79) (WHO 2016) and it is projected to reach ~642 million in 2040, which is equivalent to one in every ten adults (International Diabetes Federation 2015; Kaveeshwar and Cornwall 2014). The prevalence of diabetes in India was ~33 million in 2005, ~69 million in 2015 and predicted to reach ~98 million in 2030 (Tripathy et al. 2017; Ramachandran 2005). About 7.8% of Indians above

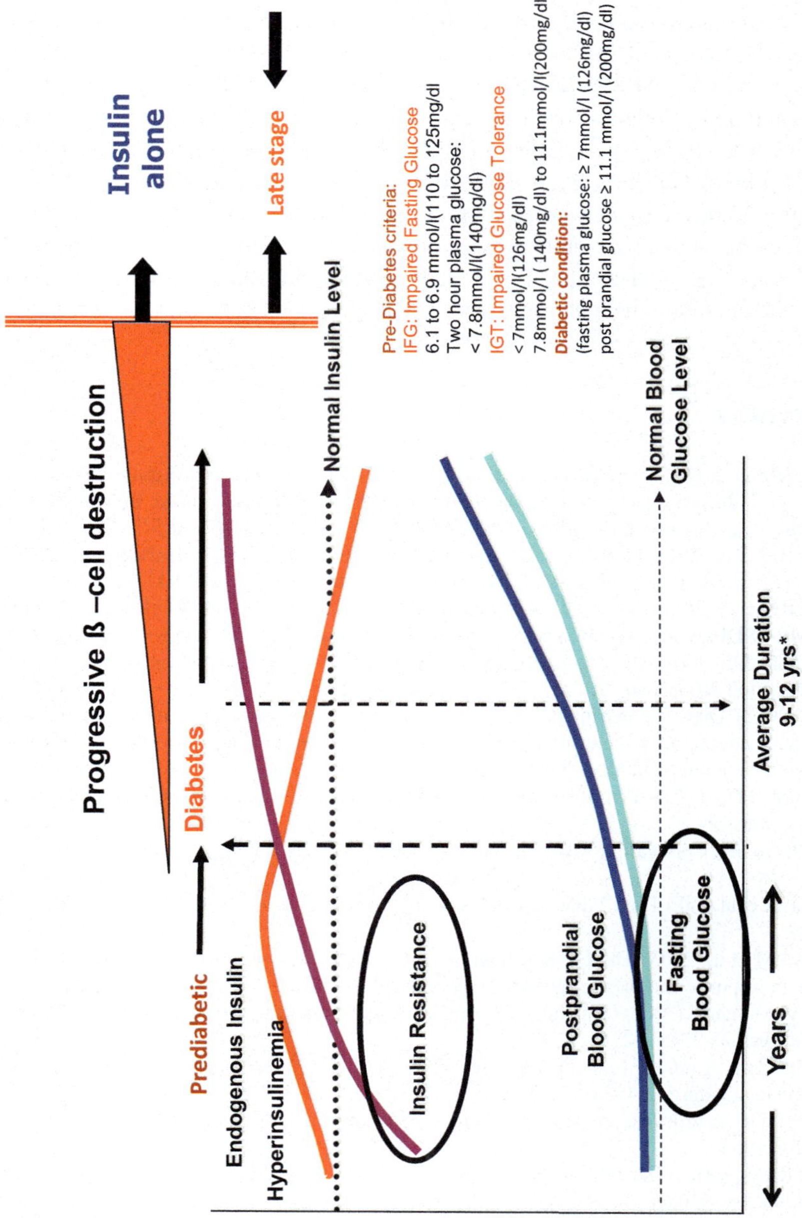


**Fig. 6.1** Progressive nature of type 2 diabetes mellitus

18 years has been found to have high blood glucose levels or are being treated for diabetes (WHO 2014). Approximately 36.5 million impaired glucose tolerant adults were reported in India in 2015 and India ranks highest in the list of top ten countries having impaired glucose tolerance (International Diabetes Federation 2015). Prevalence of impaired blood glucose in India is greater than the impaired glucose tolerance (Anjana et al. 2011). A study was conducted by Indian Council of Medical Research-India (ICMR–INDIAB) on >20-year-old adults covering 15 different states of India. It reveals that frequency of diabetes is different among states in India. It varies from 4.3% in Bihar to 13.6% in Chandigarh (Anjana et al. 2017). In 2012, National Nutrition Monitoring Bureau carried out the third repeat survey involving adult men and women of India in different states. Their report showed that in Indian adult men and women, prevalence of diabetes is 8.2% and 6.8%, respectively. States of Kerala, Tamil Nadu, and Gujarat showed higher prevalence (8.2–16.4%) among male and females (ICMR 2012).

## References

Anjana RM et al (2011) Prevalence of diabetes and prediabetes (impaired fasting glucose and/or impaired glucose tolerance) in urban and rural India: phase I results of the Indian Council of Medical Research-INdia DIABetes (ICMR-INDIAB) study. Diabetologia 54:3022

Anjana RM et al (2014) Physical activity and inactivity patterns in India – results from the ICMR-INDIAB study (Phase-1) [ICMR-INDIAB-5]. Int J Behav Nutr Phys Act 11:26

Anjana RM et al (2017) Prevalence of diabetes and prediabetes in 15 states of India: results from the ICMR-INDIAB population-based cross-sectional study. Lancet Diabetes Endocrinol 5 (8):585–596. https://doi.org/10.1016/S2213-8587(17)30174-2. Erratum in: Lancet Diabetes Endocrinol. 2017;5(8): e5

ICMR (2012) Diet and nutritional status of rural population, prevalence of hypertension and diabetes among adults and infant and young child feeding practices – report of third repeat survey. National Institute of Nutrition, ICMR, Hyderabad

International Diabetes Federation (2015) IDF diabetes atlas, 7th edn. International Diabetes Federation, Brussels

Kaveeshwar SA, Cornwall J (2014) The current state of diabetes mellitus in India. Australas Med J 7:45–48

Khardori R et al (n.d.) Type 2 diabetes mellitus. https://emedicine.medscape.com/. Accessed 4 Mar 2019

Nielsen EM et al (2003) The E23K variant of Kir6.2 associates with impaired post-OGTT serum insulin response and increased risk of type 2 diabetes. Diabetes 52(2):573–577

Radha V, Mohan V (2007) Genetic predisposition to type 2 diabetes among Asian Indians. Indian J Med Res 125(3):259–274

Ramachandran A (2005) Epidemiology of diabetes in India--three decades of research. J Assoc Physicians India 53:34–38

Sandhu MS et al (2007) Common variants in WFS1 confer risk of type 2 diabetes. Nat Genet 39 (8):951–953

Saxena R et al (2010) Genetic variation in GIPR influences the glucose and insulin responses to an oral glucose challenge. Nat Genet 42(2):142–148

Sladek R et al (2007) A genome-wide association study identifies novel risk loci for type 2 diabetes. Nature 445(7130):881–885

Tabassum R et al (2012) Common variants of IL6, LEPR, and PBEF1 are associated with obesity in Indian children. Diabetes 61:626–631

Tabassum R et al (2013) Genome-wide association study for type 2 diabetes in Indians identifies a new susceptibility locus at 2q21. Diabetes 62:977–986

Tripathy JP et al (2017) Prevalence and risk factors of diabetes in a large community-based study in North India: results from a STEPS survey in Punjab, India. Diabetol Metab Syndr 9:8. https://doi.org/10.1186/s13098-017-0207-3

Ukkola O et al (2001) Insulin like growth factor 2 and IGF-binding protein 1 (IGFBP1) gene variants are associated with overfeeding-induced metabolic changes. Diabetologia 44(12):2231–2236

Wheeler E, Barroso I (2011) Genome-wide association studies and type 2 diabetes. Brief Funct Genomics 10(2):52–60

WHO (2014) Global status report on non-communicable diseases 2014. World Health Organization, Geneva

WHO (2016) Global report on diabetes. WHO, Geneva

WHO (n.d.). http://www.who.int/mediacentre/factsheets/fs312/en/. Accessed 24 Mar 2016

Winckler W et al (2007) Evaluation of common variants in the six known maturity-onset diabetes of the young (MODY) genes for association with type 2 diabetes. Diabetes 56(3):685–693

**Parimal Misra** is Senior Professor and Chief Scientist, Center for Innovation of Molecular and Pharmaceutical Sciences, Dr. Reddy's Institute of Life Sciences, Hyderabad and worked in Dr. Reddy's Laboratories Limited in the areas of metabolic disorder, inflammation, and drug discovery. He was a visiting scientist of the Feinberg School of Medicine, Northwestern University, Chicago in the Dept of Pathology. He has published 48 articles in scientific journals, authored two book chapters and published 11 international and 12 national patents and trained more than 60 drug discovery scientists, Ph.D., and postdocs. He has been elected as a Fellow of the Royal Society of Chemistry (UK) in 2016 and a Fellow of the Royal Society of Biology (UK) in 2017.

**Ranjan Chakrabarti** is currently Vice President—Scientific Outreach, Biologics at United States Pharmacopeia.

Dr. Ranjan Chakrabarti has over 23 years of experience in Pharmaceutical and Biopharma industries. Before joining to Industry, he worked in Academics at the USA and successfully coordinated research projects in Cancer Cell Biology and Diabetes. He has guided several Ph.D. students. Before joining USP, Dr. Ranjan was leading the Biology Group at Dr. Reddy's drug discovery and also served at key management position in GVK Biosciences. He has worked with several national and international companies for discovery and development of both chemical and biological molecules.

Dr. Ranjan is the Co-Inventor of 32 US patents, published 58 papers in peer reviewed International Journals and presented 73 lectures in International and National Conferences.

# Type 2 Diabetes Mellitus: Marketed Drugs and Mechanisms

7

Parimal Misra and Ranjan Chakrabarti

Clinical studies have proven that rigorous control of hyperglycemia in type 2 diabetes patients can reduce the occurrence of chronic complications (Fig. 7.1). At present therapies of type 2 diabetes can be classified as following based on their mode of action and class:

1. Insulin secretagogues
2. Absorption inhibitors
3. Biguanides
4. Insulin sensitizers
5. Enhancing glucose-stimulated insulin secretion
6. Selective sodium-glucose transporter-2 (SGLT-2) inhibitors

## 7.1 Insulin Secretagogues: Drugs Which Trigger Insulin Release from Pancreatic β-cells

### 7.1.1 Sulfonylureas (SU)

Sulfonylureas are insulin secretagogues (e.g., glyburide, glipizide, glimepiride) which are used for treatment of type 2 diabetes for over 40 years. These agents work by inhibiting ATP-sensitive K+ Channels on pancreatic β-cells. The beneficial

P. Misra
Dr. Reddy's Institute of Life Sciences, Hyderabad, India
e-mail: parimalm@drils.org

R. Chakrabarti (✉)
United States Pharmacopeia, Hyderabad, India
e-mail: rxc@usp.org



M. Dikshit (ed.), *Drug Discovery and Drug Development*,
https://doi.org/10.1007/978-981-15-8002-4_7

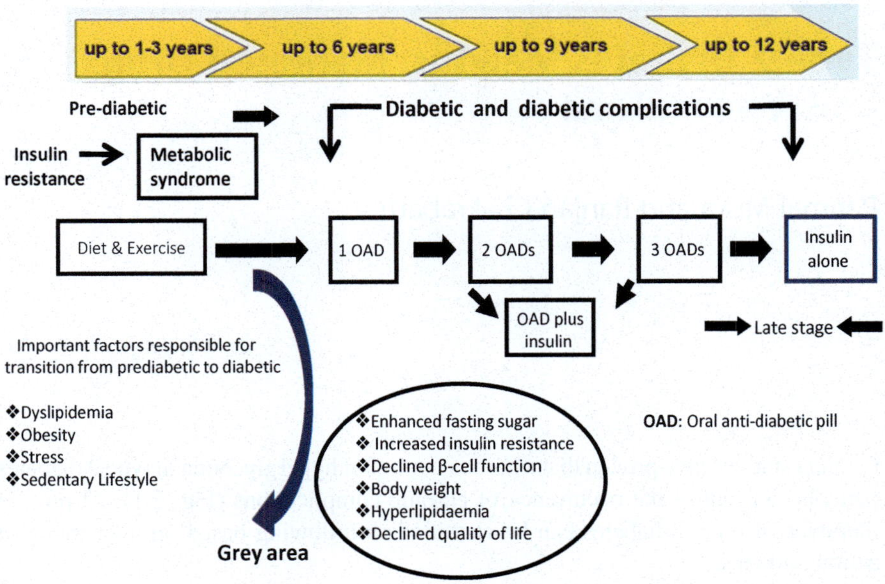


**Fig. 7.1** Progressive nature of type 2 diabetes mellitus and available treatment options at present

effects of sulfonylureas on glucose levels are well known, but it has modest effect on the progression of the disease including its micro- and macro-vascular complications. These drugs probably have greatest efficacy on glycemic control among the oral antidiabetic drugs.

Sulphonyl ureas are being used as adjunct therapy to diet and exercise in type 2 diabetes patients. They are in general well tolerated, but overt hypoglycemia is a major concern. Glyburides are metabolized to an active moiety, which is secreted through urine. Thereby it is advisable to avoid them in elderly patients having impaired renal function. Lower incidence of hypoglycemia has been observed with glipizide and glimepiride. There are reports of higher mortality with glyburide (7.5%) as compared to glimepiride (2.7%) (Zeller et al. 2010). All drugs of this class are associated with weight gain and thereby usage for obese patients is restricted.

### 7.1.2 Meglinitides

Meglinitides (repaglinide, nateglinide) are structurally different class of insulin secretagogues. These drugs bind at a different site in beta-cells as compared to SU receptors to stimulate insulin secretion (Fuhlendorff et al. 1998). These are much short acting and can reduce postprandial glucose level and risk of hypoglycemia is much less (Bellomo Damato et al. 2011). These drugs are prescribed just before meals due to their quick onset of action.

## 7.2 Absorption Inhibitors

Sugar absorption is delayed by these agents which help to prevent postprandial glucose surge. α-glucosidase inhibitors (acarbose, voglibose, miglitol) inhibit the brush border enzyme α-glucosidase (which degrade complex carbohydrate to sugar) in the small intestine, thereby decreasing the absorption of dextrin and disaccharides in intestine. Degradation of carbohydrates and the absorption of glucose into circulation after a meal are delayed by these agents. Accordingly, these drugs have maximum impact on postprandial hyperglycemia (Roger et al. 1995). No hypoglycemia and weight gain was observed with these agents alone. In contrary, these drugs have serious effects at gastrointestinal system which includes abdominal discomfort such as bloating, flatulence, and diarrhea. If titrated slowly, gastrointestinal intolerance can be reduced.

## 7.3 Biguanides

Metformin, phenformin, and buformin entered into the market short after sulfonylureas were introduced. Metformin and phenformin were widely used until 1977, when due to lactic acidosis and associated mortality phenformin was withdrawn. Buformin on the other hand was introduced in a limited manner. Metformin has been found to be safe and effective (Scarpello and Howlett 2008). Basal and postprandial plasma glucose levels were lowered by metformin by decreasing hepatic gluconeogenesis and thereby hepatic glucose output (Jackson et al. 1987). Metformin decreases intestinal glucose absorption. It also increases peripheral glucose uptake and utilization and thereby improves insulin sensitivity. In general, hypoglycemia is not observed with metformin. In contrast to other antidiabetic drugs, body weight lowering effect has been observed with metformin treatment (Bailey 1992). Twenty percent of patients suffer from diarrhea, abdominal discomfort, nausea, and anorexia during metformin treatment (Kecskemeti et al. 2002). In UKPDS, metformin reduced macrovascular disease endpoints in obese patients (Turner et al. 1999). American college of Physicians (ACP) in a guideline released in January 2017, recommended metformin as a first line of treatment for type 2 diabetes.

## 7.4 Insulin Sensitizers

Thiazolidinediones (TZDs) belong to this class and act as insulin sensitizers. Sensitivity of insulin in both fat and muscle tissues has been increased by TZDs. They also inhibit hepatic glucose production, but to a lesser extent. As a class, TZDs also decrease triglyceride and increase HDL-cholesterol level in type 2 diabetes patients. TZDs are agonist of Peroxisome Proliferator Activated Receptor ϒ (PPARϒ) (Olefsky 2000). PPARϒ is a nuclear superfamily orphan receptor that mediates adipocyte differentiation. It also regulates gene expression and thereby modulates insulin sensitivity (Olefsky 2000). Fibrates, the triglyceride lowering drugs are PPARα agonists, another isoform of same PPAR family. TZDs are used as monotherapy and also in combination with other antidiabetic agents, such as sulfonylureas, metformin, meglinitides, DPP-4 inhibitors, GLP-1 receptor agonists or insulin.

Although TZDs have several desirable effects, it has severe safety issues, specially edema (including muscular edema) and weight gain. These drugs should be used with caution for the patients having renal dysfunction. Liver dysfunction has been reported after rosiglitazone treatment (Lewis et al. 2011). Patients with normal or compromised left ventricular function may show induction or worsening heart failure with these drugs. There are few reports of increased bladder cancer when pioglitazone is used for more than 2 years (Piccinni et al. 2011; Mathis et al. 2001). The USFDA has recommended not to prescribe pioglitazone for active bladder cancer patients. It also restricts its use in patients, with a history of bladder cancer. The FDA has restricted access to this class of drugs as patients treated with rosiglitazone showed elevated risk of myocardial infarction. It can be used only to the patients who are already treated successfully with this agent or whose glycemic control could not be achieved by pioglitazone. Due to its potential risk of osteoporosis, pioglitazone is not clinically preferred in postmenopausal women with severe osteoporosis (Wu et al. 2013; Lecka-Czernik 2010; Schwartz and Sellmeyer 2008).

## 7.5 Enhancing Glucose-Stimulated Insulin Secretion

A key defect in type 2 diabetes is the inability of glucose to induce insulin secretion from pancreatic β-cells (Miyawaki et al. 1999). Due to this defect, β-cells fail to ameliorate increasing insulin resistance and ultimate development of overt hyperglycemia. Unlike sulfonylureas, which stimulate insulin independent of high glucose level, more desirable alternative approach will be to stimulate glucose-dependent insulin secretion. Glucagon like peptide 1 (GLP-1) and gastric inhibitory peptide (GIP) are two gut-derived peptide hormones, which act through their respective G-protein coupled receptors (GPCR) on pancreatic β-cells to stimulate glucose-induced insulin secretion. Incretin hormone GLP-1 promotes β-cell growth and increases glucose disposal in fat, muscle, and liver. It also delays gastric motility and promotes satiety. All these effects together leads to overall normalization of glucose (Jens and Deacon 1998).

GLP-1 is thereby considered as an obvious drug candidate for the treatment of type 2 diabetes. Continuous intravenous infusion of GLP-1 in type 2 diabetic patients shows near normalization of diurnal plasma glucose concentration (Nauck et al. 1996). Rapid metabolism by the ubiquitous enzyme dipeptidyl peptidase IV (DPP IV) inactivates GLP-1 after simple subcutaneous injection and becomes ineffective (Pencek 2012). Modified GLP-1 peptide agonists which are resistant to DPP IV and/or specific DPP IV inhibitors are being used to solve this problem.

### 7.5.1 GLP-1 Agonists

GLP-1 agonists were developed to mimic the effect of endogenous incretin GLP-1. All these agents stimulate insulin in a glucose-dependent manner. Modest loss of weight was observed with these agents in combination with metformin. Prevention of beta-cell destruction by these agents has been observed in animal studies. These studies also indicate that these agents may in time restore β-cells in these models. If proven in human, this will have a tremendous therapeutic potential.

Exenatide, a 39-aminoacid peptide derived from the saliva of the Glia monster lizard is the first in the class of these drugs approved in 2005. Exenatide requires twice daily injection, a long acting version that is given weekly was approved in 2012. It shows improvement in glycemic control along with body weight (Marso et al. 2016). Major adverse event include gastrointestinal events. Post the success of exenatide, several other GLP-1 agonists flooded the market. Liraglutide was approved in 2010, it also has an approval for reduction of major cardiovascular events in patients with type 2 diabetes and established cardiovascular disease (Douglas n.d.). In April 2014, albiglutide was approved in combination with diet and exercise. It is administered once weekly. Patients receiving this drug have fewer GI events than liraglutide, but more injection site reaction and less weight loss (Pratley et al. 2014; Tucker 2014). In September 2014, another agonist dulaglutide was approved as once a weekly subcutaneous injection dose, but it is not recommended as first line of treatment for type 2 diabetes. It is also contraindicated in patients with history of medullary thyroid carcinoma (He 2017). There are reports that this class of drugs increase risk of pancreatitis (Azoulay 2016). In 2016, lixisenatide is recommended for subcutaneous injection (once daily) within 1 hr before the first meal of the day. In 2017, semaglutide was approved. In cardiovascular outcome study, the drug was found to reduce cardiovascular risk (Aschner et al. 2010).

### 7.5.2 DPPIV Inhibitors

DPPIV inhibitors are a class of drug that extend the duration of action of incretin hormones such as GLP-1 and GIP. DPPIV inhibitors can be used as monotherapy. They can also be used in combination with metformin or TZD. Their dosage is once a day and they have no effect on body weight. In clinical studies, adverse GI effect with sitagliptin was lower than metformin (Willemen et al. 2011; Ferrannini and Solini 2012). One major side effect reported among DPPIV inhibitor users is increasing upper respiratory infection, although the exact mechanism is still not clear (Elkinson and Scott 2013).

## 7.6 Selective Sodium-Glucose Transporter-2 Inhibitors

Sodium-glucose transporter-2 (SGLT-2) inhibitors are the latest group of oral medications used for treatment of type 2 diabetes. The kidneys filter glucose out of the blood and then reabsorb it back into the blood. The proteins involved in this reabsorption are called sodium-glucose transport (SGLT) proteins. There are six identified SGLTs of which two SGLT-1 and 2 considered most important. SGLT-2 inhibitors block this protein and thereby prevent the reabsorption of glucose back to blood. Their action is dependent on blood glucose level and independent of insulin (Saeed and Narendran 2014). SGLT-2 inhibitors have been approved for use in type 2 diabetes treatment from 2013. The first SGLT-2 inhibitor approved in the USA was canagliflozin (Fala 2015). Two more SGLT-2 inhibitors were approved afterwards—dapagliflozin and empagliflozin both in 2014. Most common side effects associated with these drugs are vaginal yeast and urinary tract infections (Li et al. 2017; Liu et al. 2017). There are some concerns related to risk of fractures (Wolverton and Blair 2017) and risk of renal dysfunction in elderly patients following this drug treatment (Perlman 2018). There are also report on potential risk of ketoacidosis with these drugs (Limenta 2019).

### 7.6.1 Unmet Medical Need in Metabolic Disorders and Way Forward

There are three potential approaches for the management of chronic and multifactorial disease like metabolic disorders.

(1) Treating the disease, (2) Delaying the onslaught of the disease, and (3) Curing the disease.

As stated in Fig. 7.1, the current paradigm favors a step-by step approach to the treatment of diabetes. The treatment starts with control of diets and exercise and most of the patients fall within 12 months and mainly progress to monotherapy either with metformin or sulfonylurea. But as monotherapy generally does not produce optimal results (Table 7.1), patients are treated with either dual combination therapy such as Met/SU + TZD (Pioglitazone), Met/SU + GLP-1, Met + SGLT-2 inhibitors, Met + DPPIV inhibitors or with triple therapy consisting of Met + TZD + GLP-1, TZD + Met + Insulin, Met + DPPIV inhibitors + SU, TZD + SU + Insulin or Met + SU + Insulin, etc. Still 2/3 of the diabetic patients do not satisfy the American Diabetes Association (ADA) recommended glycemic goal. Diabetes is a multifactorial disease and often associated with hyperlipidemia, obesity, and high blood pressure. In addition to these diabetic drugs, diabetic patients are treated with fibrates (dyslipidemia), statins (hypercholesterolemia), and blood pressure reducing drugs to treat these comorbidities.

In the last three decades, scientists and clinicians developed medicines to treat the disease but metabolic disorder being a multifactorial and very complex disease no single class of present oral anti-hyperglycemic monotherapy targets all key pathophysiologies of metabolic disorders (Table 7.1). Current therapy is based on

**Table 7.1** Strengths and weaknesses of the marketed antidiabetic drugs

| | | A. Insulin secretagogues | B. Absorption Inhibitors | C. Biguanides | D. Insulin sensitizers | E. Enhancing glucose stimulated insulin secretion | F. Selective Sodium-Glucose Transporter-2 inhibitors |
|---|---|---|---|---|---|---|---|
| Strengths | Insulin sensitization | | | ✓ | ✓ | | |
| | Control of glycemic parameters | ✓ | ✓ | ✓ | ✓ | ✓ | ✓ |
| | Impact on lipids | | | | ✓ | | |
| | Weight neutral/ weight loss | | | ✓ | | ✓ | ✓ |
| | Reversal of the disease | | | | | ✓ | |
| Weaknesses | Hypoglycemia | ✓ | | | | | |
| | Weight gain | | | | ✓ | | |
| | CV side effects | | | | ✓ | | |
| | Chances of carcinogenesis | | | | ✓ | ✓ | |

No single class of oral anti-hyperglycemic monotherapy targets all key pathophysiologies of metabolic disorders

combination therapy but not that efficacious and associated with several side effects (Table 7.1) and so still a huge unmet medical need is there. In the last two decades, scientists are trying to validate some potential novel targets which can be exploited for the development of drugs and these drugs may delay the onslaught of the disease, if cannot cure.

Challenges for the drug discovery scientists are to discover new agents which should enhance insulin sensitivity, reduce glucose, lipids, body weight and should have beneficial cardiovascular profiles for the treatment of metabolic disorders. Thiazolidinedione (PPARγ agonist) and glitazar (PPARα/γ dual agonist) drugs failed due to cardiovascular side effects and this saga emphasizes that scientists should discover drugs with all these essential properties by modulating different pathways (Psaty and Furberg 2007; Nissen et al. 2005; Henry et al. 2009; Lincoff et al. 2014; Cavender and Lincoff 2010).

Several new targets are being explored for future antidiabetic drug discovery. Most of the molecules against these targets are in discovery or early preclinical stage. A brief report on these targets is given below:

#### 7.6.1.1 Adenosine Monophosphate Activated Kinase (AMPK): A Potential Target for Total Metabolic Disorders

Exercise keeps us healthy. Exercise activates adenosine monophosphate activated kinase (AMPK) (Hardie and Carling 1997). Activators of the human AMPK enzyme are considered to be a new high-value target to the treatment of type 2 diabetes which should demonstrate clear cardio metabolic benefits. In the liver, activation of AMPK results in decreased fasting sugar, hyperlipidemia but enhanced fatty acid oxidation. In the skeletal muscle, activated AMPK stimulates glucose transport and fatty acid oxidation. In adipose tissue, activation of AMPK leads to reduction of fat deposition but enhancement of breakdown and burning of stored fat. These two opposite effects result in the reduction of body weight. Thus, AMPK has been identified as a possible target for controlling metabolic syndrome for many years. Several novel AMPK activators have been tried in clinic with limited success so far (Misra et al. 2009; Bung et al. 2018; Steinberg and Carling 2019).

#### 7.6.1.2 Protein Tyrosine Phosphatase Inhibitors

Binding of insulin to its receptor induces phosphorylation of certain tyrosine moieties within the receptor and thereby activates the receptor kinase. Protein tyrosine phosphatase on the other hand dephosphorylates the activated insulin receptor and thereby makes it inactive. Inhibitors of PTP1B are expected to block deactivation of insulin receptor and enhance insulin sensitivity and thereby a potential target for the antidiabetic drug discovery (Maheswari et al. 2018). Although research is going on for last several years, most of the molecules today at discovery or preclinical stage. Developing a specific inhibitor is a major challenge as ATPase interacting residues in protein kinase are conserved, several molecules were discontinued in clinical stage.

#### 7.6.1.3 Fructose-1,6-bisphosphatase (FBPase)

The enzyme which catalyzes the conversion of fructose-1,6-bisphosphate to fructose-6-phosphate in gluconeogenic pathway is FBPase (Hers 1990). Inhibitor developed for this enzyme suppresses hepatic gluconeogenesis in hepatocytes, thereby hepatic glucose output (Vincent et al. 1991). Sankyo and Pfizer were developing few molecules, but it did not proceed further. There are still several molecules in discovery phase, we need to watch how they proceed.

#### 7.6.1.4 Phosphoenol Pyruvate Carboxykinase (PEPCK)

PEPCK is another important gluconeogenic enzyme. Hepatic parenchymal cells produce glucose from pyruvate-derived amino acid metabolism using this enzyme (Devi 2010). Insulin inhibits PEPCK enzyme expression and thereby inhibits its upregulation. It has been observed that patients with diabetes have elevated blood glucose levels as insulin is unable to regulate PEPCK expression in them and thereby liver glucose output is affected. Inhibition of expression of PEPCK is thereby a potential mean to manage treatment of diabetes. All the molecules are at early preclinical stage.

#### 7.6.1.5 Free Fatty Acid Receptors (FFAR-1)

Free fatty acid receptor 1 (FFAR 1) regulates glucose-stimulated insulin release from pancreatic β-cells is regulated by (Itoh et al. 2003). Accordingly, it is an attractive target for antidiabetic drug development and several groups are working on this target. Some of the molecules have shown impressive activity in diabetic animal models, none have advanced to clinical stage yet.

#### 7.6.1.6 G-protein Coupled Receptors (GPCR)

G-protein coupled receptors are a large family of receptors. They play a key role in cellular signal transduction for several critical molecules inside human body. Accordingly, they are highly preferred druggable targets for different therapeutic interventions.

GPR 119, GPR 142, and GPR 120, members of GPCR family have been identified as the potential targets for type 2 diabetes therapy. GPR 119 is expressed on the pancreatic β-cells, enter endocrine cells of gastrointestinal tract. GPR 119 activation increases intracellular CAMP levels. This then stimulates insulin secretion from the pancreatic β-cells. In enter endocrine cells of GI tract, secretion of incretins is stimulated by enhanced expression of GPR 119, which has a profound role in glycemic control (Kerru et al. 2012). Therefore, GPR 119 agonists are being explored as potential therapeutic tool for type 2 diabetes. Several molecules are being explored for GPR 119 agonism and most of them are in early preclinical stage. Few molecules are also being studied as GPR 142 agonist, which showed decent glucose lowering activity in animal models.

Glucagon receptor (GCGr): The 29 amino acid peptide glucagon is a memeber of secretin family. It is secreted from the pancreatic α-cells, inhibits hepatic glycogen synthesis and stimulates glycogenolysis and gluconeogenesis, which in turn converts glucagon to glucose in liver (Shah et al. 2000) and thereby contributes significantly to hyperglycemia in type 2 diabetic patients. Glucagon activity is mediated by its receptor (GCGr), a member of GPCR family. This receptor is located on the hepatocyte plasma membrane (Brubaker and Drucker 2002).

Type 2 diabetes is considered a hormonal disease caused by deregulations in insulin and glucagon levels, which is characterized by hyperglycemia in fasting and postprandial state. Therefore, GCGr antagonists which can reduce glucagon secretion are considered a potential tool for type 2 diabetes therapy (Filipski 2015).

Several small molecules as well as large molecules such as antisense oligonucleotides and peptides are being explored as potential GCGr antagonists (Bhanot et al. 2012; Freier and Bhanot 2014). One of the LY 240920 from Eli Lilly, a small molecule GCGr antagonist was tested in phase 2 clinical study and showed significant reduction in $HbA_{1c}$ and fasting glucose level (Kazda et al. 2016). Most of the other molecules are in preclinical stage.

Increase in blood pressure and alanine aminotransferase levels with these agonists are a major safety issue observed in their clinical trials. These raised a serious question on their viability for type 2 diabetes therapy.

#### 7.6.1.7 Curing of the Disease: Vaccine Approach

All the available diabetic drugs can be used only once the disease sets in and thus are not preventative. Also none of these drugs modify the actual disease; they are "symptomatic" in nature, meaning that they decrease blood glucose but do not cure the disease.

A new approach is needed towards this disease, one that may prophylactically prevent this disease as well as cure it. Prior to onset of full-fledged T2D, patients go through a phase of gestation called prediabetes. During this phase, the patient starts to present with borderline increase in glucose levels above normal level. This prediabetes stage lasts for years with continuing increase in blood glucose. There are no symptoms of diabetes during this phase. Most people with prediabetes will go on to get full blown disease unless they are able to reduce their risk factors such as diet, lifestyle, weight/BMI, and exercise.

This prediabetes phase represents an excellent window of opportunity for introduction of a preventative therapeutic such as a vaccine against a disease causing protein. This will then prevent the simmering prediabetes from bubbling over to a full blown disease. An immune approach can also be applied to treat T2D since the protein being targeted by the vaccine also plays a role in it. To date there have been no efforts to prevent the disease. So the invention should be focused on both active as well as a passive vaccination approach to this disease.

#### 7.6.1.8 Curing of the Disease: Formulated Indian Natural Products Approach

Formulated Indian natural products are used through ages to treat the prediabetic and diabetic patients. We consume most of the ingredients of the formulated natural products daily through foods. But the quantity is less. Each ingredient may not have significant efficacy to combat the disease but in combination the total efficacy may be enhanced. The mechanism of the antidiabetic effects is poorly known for most of the natural products. Most importantly, the toxicity profile of the formulated Indian natural products should be determined before use.

## References


Aschner P, Katzeff HL et al (2010) Efficacy and safety of monotherapy of sitagliptin compared with metformin in patients with type 2 diabetes. Diabetes Obes Metab 12(3):252–261

Azoulay L (2016) Association between incretin-based drugs and the risk of acute pancreatitis. JAMA Intern Med 176(10):1464–1473

Bailey CJ (1992) Biguanides and NIDDM. Diabetes Care 15(6):755–772

Bellomo Damato A, Stefanelli G et al (2011) Nateglinide provides tighter glycemic control than glyburide in patients with Type 2 diabetes with prevalent postprandial hyperglycemia. Diabet Med 28(5):560–566

Bhanot S, Geary R et al (2012) Compounds and methods for modulating expression of GCGr. US 20120208864A1

Brubaker P, Drucker D (2002) Structure-function of the glucagon receptor family of G protein-coupled receptors: the glucagon, GIP, GLP-1 and GLP-2 receptors. Recept Channels 8:179–188

Bung N, Surepalli S, Seshadri S, Patel S, Peddasomayajula S, Kummari LK, Kumar ST, Babu PP, Parsa KVL, Poondra RR, Bulusu G, Misra P (2018) 2-[2-(4-(trifluoromethyl) phenylamino) thiazol-4-yl]acetic acid (Activator-3) is a potent activator of AMPK. Sci Rep 8(1):9599

Cavender MA, Lincoff AM (2010) Therapeutic potential of aleglitazar, a new dual PPAR-α/γ agonist: implications for cardiovascular disease in patients with diabetes mellitus. Am J Cardiovasc Drugs 10:209–216

Devi SS (2010) Structure and function of hepatic parenchymal cells. Comprehens Toxicol 9:11–30

Douglas D (n.d.) Albiglutide long-acting option for diabetes control. Medscape Medical News. http://www.medscape.com/viewarticle/821700

Elkinson S, Scott LJ (2013) Canagliflozin: first global approval. Drugs 73(9):979–988

Fala L (2015) Jardiance (Empagliflozin), an SGLT2 inhibitor, receives FDA approval for the treatment of patients with type 2 diabetes. Am Health Drug Benefits 8(Spec Feature):92–95

Ferrannini E, Solini A (2012) SGLT2 inhibition in diabetes mellitus: rationale and clinical prospects. Nat Rev Endocrinol 8:495–502

Filipski KJ (2015) Small molecule glucagon receptor antagonists: a patent review (2011-2014). Expert Opin Ther Targets 25:819–830

Freier SM, Bhanot S (2014) Antisense modulation of GCGR expression. US8865674B2

Fuhlendorff J, Rorsman P et al (1998) Stimulation of insulin release by repaglinide and glibenclamide involves both common and distinct processes. Diabetes 47:345–351

Hardie DG, Carling D (1997) The AMP-activated protein kinase–fuel gauge of the mammalian cell? Eur J Biochem 246:259–273

He J (2017) Endocrinology & Metabolic Drug Advisory Committee Meeting. FDA briefing document. https://www.fda.gov/downloads/advisorycommittees/committeesmeetingmaterials/drugs/endocrinologicandmetabolicdrugsadvisorycommittee/ucm580460.pdf

Henry RR et al (2009) Effect of the dual peroxisome proliferator-activated receptor-alpha/gamma agonist aleglitazar on risk of cardiovascular disease in patients with type 2 diabetes (SYNCHRONY): a phase II, randomised, dose-ranging study. Lancet 374:126–135

Hers HG (1990) Mechanisms of blood glucose homeostasis. J Inhert Metab Dis 13(4):395–410

Itoh Y, Kawamata Y et al (2003) Free fatty acids regulate insulin secretion from pancreatic β-cells through GPR40. Nature 422:173–176

Jackson RA, Hawa MI et al (1987) Mechanism of metformin action in non-insulin-dependent diabetes. Diabetes 36(5):632–640

Jens J, Deacon CF (1998) Inhibition of the activity of dipeptidyl-peptidase IV as a treatment for type 2 diabetes. Diabetes 47(11):1663–1670

Kazda CM, Ding Y et al (2016) Evaluation of efficacy and safety of the glucagon receptor antagonist LY 2409021 in patients with type 2 diabetes: 12 and 24- week phase 2 studies. Diabetes Care 39:1241–1249

Kecskemeti V, Bagi Z et al (2002) New trends in the development of oral antidiabetic drugs. Curr Med Chem 9(1):53–71

Kerru N, Singh-Pillay A et al (2012) Current antidiabetic agents and their molecular targets: a review. Eur J Med Chem 152:436–488

Lecka-Czernik B (2010) Bone loss in diabetes: use of antidiabetic thiazolidinediones and secondary osteoporosis. Curr Osteoporos Rep 8(4):178–184

Lewis JD, Ferrara A et al (2011) Risk of bladder cancer among diabetic patients treated with pioglitazone: interim report of a longitudinal cohort study. Diabetes Care 34(4):916–922

Li D et al (2017) Urinary tract and genital infections in patients with type 2 diabetes treated with sodium-glucose co-transporter 2 inhibitors: a meta-analysis of randomized controlled trials. Diabetes Obes Metab 19(3):348–355

Limenta M (2019) Adverse drug reaction profile of SGLT2 inhibitor-associated diabetic ketosis/ketoacidosis in singapore and their precipitating factors. Clin Drug Investig 39(7):683–690

Lincoff AM et al (2014) Effect of aleglitazar on cardiovascular outcomes after acute coronary syndrome in patients with type 2 diabetes mellitus: the AleCardio randomized clinical trial. JAMA 311:1515–1525

Liu J et al (2017) Effects of SGLT2 inhibitors on UTIs and genital infections in type 2 diabetes mellitus: a systematic review and meta-analysis. Sci Rep 7(1):2824

Maheswari N, Karthikeyan C et al (2018) Recent advances in protein tyrosine phosphatase 1B targeted drug discovery for type II diabetes and obesity. Curr Drug Targets 19:551–575

Marso SP, Daniels GH et al (2016) Liraglutide and cardiovascular outcomes in Type 2 diabetes. N Engl J Med 375(4):311–322

Mathis D, Vence L et al (2001) beta-Cell death during progression to diabetes. Nature 414 (6865):792–798

Misra P et al (2009) Modulation of endogenous AMPK levels for the treatment of obesity. WO/2009/019600 A2, 12 Feb 2009

Miyawaki K et al (1999) Glucose intolerance caused by a defect in the entero-insular axis: a study in gastric inhibitory polypeptide knowkout mice. Proc Natl Acad Sci 96:14843–14847

Nauck MA, Wollschlager D et al (1996) Effects of subcutaneous glucagon-like peptide 1 (GLP-1 [7–36 amide]) in patients with NIDDM. Diabetologia 39(12):1546–1553

Nissen SE, Wolski K, Topol EJ (2005) Effect of muraglitazar on death and major adverse cardiovascular events in patients with type 2 diabetes mellitus. JAMA 294:2581–2586

Olefsky JM (2000) Treatment of insulin resistance with peroxisome proliferator–activated receptor γ agonists. J Clin Invest 106(4):467–472

Pencek R (2012) Blickensderfer a et al., Exenatide once weekly for the treatment of type 2 diabetes: effectiveness and tolerability in patient subpopulation. Int J Clin Pract 66(11):1021–1032

Perlman A (2018) Clinical spectrum and mechanism of acute kidney injury in patients with diabetes mellitus on SGLT-2 inhibitors. Isr Med Assoc J 20(8):513–516

Piccinni C, Motola D et al (2011) Assessing the association of pioglitazone use and bladder cancer through drug adverse event reporting. Diabetes 34(6):1369–1371

Pratley RE, Nauck MA et al (2014) Once-weekly albiglutide versus once-daily liraglutide in patients with type 2 diabetes inadequately controlled on oraldrugs (HARMONY 7): a randomised open-label multicentre non-inferiority phase 3 study. Lancet Diab Endocrinol 2 (4):289–297

Psaty BM, Furberg CD (2007) The record on rosiglitazone and the risk of myocardial infarction. N Engl J Med 357:67–69

Roger NW, Chiasson JL et al (1995) Clinical experience with acarbose: results of a Canadian multicentre study. Clin Invest Med 18(4):318–324

Saeed MA, Narendran P (2014) Dapagliflozin for the treatment of type 2 diabetes: a review of the literature. Drug Dis Dev Ther 8:2493–2505

Scarpello JH, Howlett HC (2008) Metformin therapy and clinical uses. Diab Vasc Dis Res 5 (3):157–167

Schwartz AV, Sellmeyer DE (2008) Effect of thiazolidinediones on skeletal health in women with Type 2 diabetes. Expert Opin Drug Saf 7(1):69–78

Shah P, Vella A et al (2000) Lack of suppression of glucagon contributes to postprandial hyperglycemia in subjects with type 2 diabetes mellitus. J Clin Encrinol Metab 85:4053–4059

Steinberg GR, Carling D (2019) AMP-activated protein kinase: the current landscape for drug development. Nat Rev Drug Discov 18(7):527–551

Tucker M (2014) FDA approves once-weekly dulaglutide for type 2 diabetes. Medscape Medical News. http://www.medscape.com/viewarticle/831969

Turner RC, Cull CA et al (1999) Glycemic control with diet, sulfonylurea, metformin or insulin in patients with type 2 diabetes mellitus: progressive requirement for multiple therapies (UKPDS49). JAMA 281(21):2005–2012

Vincent MF, Marangos PJ et al (1991) Inhibition by AICA riboside of gluconeogenesis in isolated rat hepatocytes. Diabetes 40:1259–1266

Willemen MJ, Mantel-Teeuwisse AK et al (2011) Use of dipeptidyl peptidase-4 inhibitors and the reporting of infections: a disproportionality analysis in the WHO VigiBase. Diabetes Care 34 (2):369–374

Wolverton D, Blair MM (2017) Fracture risk associated with common medications used in treating type 2 diabetes mellitus. Am J Health Syst Pharm 74(15):1143–1151

Wu H, Li L, Ma Y, Chen Y, Zhao J, Lu Y, Shen P (2013) Regulation of selective PPARγ modulators in the differentiation of osteoclasts. J Cell Biochem 114(9):1969–1977

Zeller M, Danchin N et al (2010) Impact of type of preadmission sulfonylureas on mortality and cardiovascular outcomes in diabetic patients with acute myocardial infarction. J Clin Endocrinol Metab 95(11):4993–5002

**Parimal Misra** is Senior Professor and Chief Scientist, Center for Innovation of Molecular and Pharmaceutical Sciences, Dr. Reddy's Institute of Life Sciences, Hyderabad and worked in Dr. Reddy's Laboratories Limited in the areas of metabolic disorder, inflammation, and drug discovery. He was a visiting scientist of the Feinberg School of Medicine, Northwestern University, Chicago in the Dept of Pathology. He has published 48 articles in scientific journals, authored two book chapters and published 11 international and 12 national patents and trained more than 60 drug discovery scientists, Ph.D., and postdocs. He has been elected as a Fellow of the Royal Society of Chemistry (UK) in 2016 and a Fellow of the Royal Society of Biology (UK) in 2017.

**Ranjan Chakrabarti** is currently Vice President—Scientific Outreach, Biologics at United States Pharmacopeia.

Dr. Ranjan Chakrabarti has over 23 years of experience in Pharmaceutical and Biopharma industries. Before joining to Industry, he worked in Academics at the USA and successfully coordinated research projects in Cancer Cell Biology and Diabetes. He has guided several Ph.D. students. Before joining USP, Dr. Ranjan was leading the Biology Group at Dr. Reddy's drug discovery and also served at key management position in GVK Biosciences. He has worked with several national and international companies for discovery and development of both chemical and biological molecules.

Dr. Ranjan is the Co-Inventor of 32 US patents, published 58 papers in peer reviewed International Journals and presented 73 lectures in International and National Conferences.

# Anti-diabetes Research in India: Contributions from Industrial Organizations

# 8

Sarbani Pal and Manojit Pal

*Dedicated to late Dr. K. Anji Reddy (the founder-chairman of Dr. Reddy's Laboratories Limited, Hyderabad) on the occasion of his 80th birthday (Feb 1, 1939).*

## 8.1 Anti-diabetes Drug Discovery Effort in India

Soon after independence the new drug discovery and development efforts in India were initiated by government of India via establishing several laboratories. These include a number of CSIR Labs such as CDRI-Lucknow, NCL-Pune, RRL-Jammu (currently IIIM), IICT-Hyderabad, etc. in addition to various Regional Research Laboratories (RRLs) and organizations like Indian Drugs and Pharmaceuticals Ltd (IDPL). The Council of Scientific and Industrial Research (CSIR) was responsible for running and functioning of these CSIR Labs where the research team having background and experiences in chemistry or pharmacology or interdisciplinary areas pursued the drug discovery effort. Subsequently, R&D centers were established by several multinational pharma companies in India that boosted the effort. Later in 1980, several Indian pharma companies also entered into this area that eventually strengthened India's research efforts towards discovery and development of new drugs. The drug discovery in India has been evolving and has made significant contributions in advancement of science and identification of new therapeutics.

S. Pal (✉)
Department of Chemistry, MNR Degree and PG College, Hyderabad, India

M. Pal
Dr. Reddy's Institute of Life Sciences, Hyderabad, India

Manipal Academy of Higher Education, Manipal, Karnataka, India
e-mail: manojitp@drils.org



M. Dikshit (ed.), *Drug Discovery and Drug Development*,
https://doi.org/10.1007/978-981-15-8002-4_8

Discovery model in Indian pharma industry has aimed at developing novel therapeutics for diabetes working independently or in partnership with large pharma companies. Nevertheless, this section will focus on efforts devoted by Indian Pharma companies mainly in the area of diabetes in the past decades. The contributions made by government R&D centers or academic institutes are not generally covered here except the cases where contributions are made in collaboration with Indian Pharma companies.

### 8.1.1 Contributions Made by Indian Pharma Companies

The research activities of pharma companies in the area of discovery of anti-diabetes agents described here include that of Dr. Reddy's Laboratories Limited (DRL), Zydus Cadila Healthcare (or Cadila Healthcare), Glenmark Pharmaceuticals, Orchid Pharma, and other Indian companies. The information provided here mainly include the identification number of compound or agent, mechanism of action if known, highest developmental phase, e.g., preclinical or clinical stage reached/completed, current status as per published reports or bulletins or literature, and finally the status of collaborations or licensing deals with other pharmaceutical companies (Indian or global), if any. Attempt has been made to cover information available till end of 2018.

#### 8.1.1.1 Dr. Reddy's Laboratories Limited

The drug discovery effort in India in the area of diabetes received an initial breakthrough in 1997 when Dr. Reddy's Laboratories Limited (DRL), Hyderabad, India out-licensed its preclinical-stage anti-diabetes compound DRF-2593 (balaglitazone) to Novo Nordisk, a Denmark-based multinational pharmaceutical company (Reddy 2015). Indeed, DRF-2593, a small-molecule based peroxisome proliferator-activated receptor γ (PPARγ) agonist was the first compound discovered and out-licensed by an Indian pharma company. This development was viewed as one of the key landmarks in the history of India's New Drug Discovery effort on small molecules (Fig. 8.1) and was thought to be the beginning of a new era for the influx of new drugs from India. Among the Indian pharma companies DRL was the first to launch drug discovery research in 1994 (Sridharan 1998). The company was established in 1984 with the initial goal of manufacturing of active pharmaceutical ingredients (APIs) and was slowly became one of the top pharma giants in India. In addition to diabetes, DRL also focused on conducting research for the discovery of anti-inflammatory, anticancer, and antibacterial agents as well as agents for treating cardiovascular diseases.

**DRF-2593** Being a selective partial PPARγ agonist DRF-2593 (Fig. 8.2) showed an $EC_{50}$ of 1.351 μM for human PPARγ. The PPARs or peroxisome proliferator-activated receptors are a family of nuclear proteins that work as transcription factors regulating the expression of genes. The essential roles of PPARs are known in the regulation of cellular differentiation, development, and metabolism (protein, lipid,

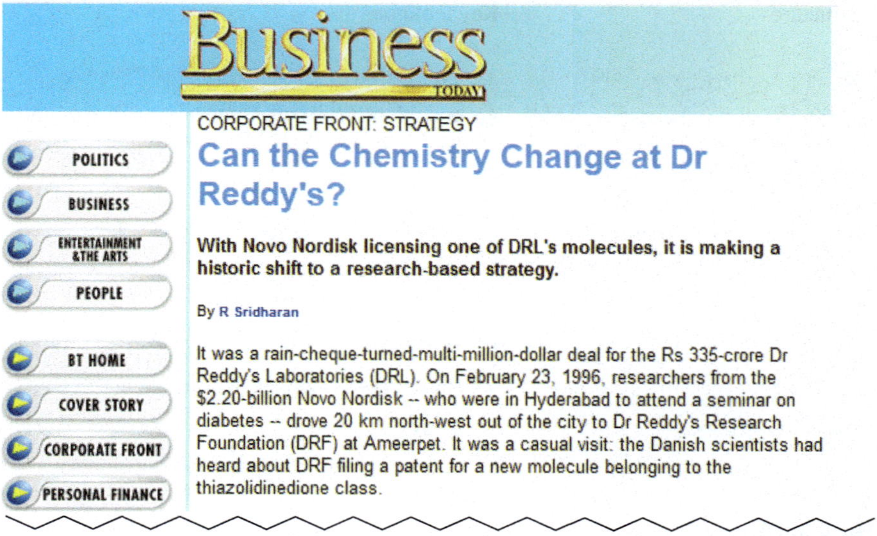

Business TODAY

POLITICS
BUSINESS
ENTERTAINMENT &THE ARTS
PEOPLE
BT HOME
COVER STORY
CORPORATE FRONT
PERSONAL FINANCE

CORPORATE FRONT: STRATEGY

**Can the Chemistry Change at Dr Reddy's?**

**With Novo Nordisk licensing one of DRL's molecules, it is making a historic shift to a research-based strategy.**

By R Sridharan

It was a rain-cheque-turned-multi-million-dollar deal for the Rs 335-crore Dr Reddy's Laboratories (DRL). On February 23, 1996, researchers from the $2.20-billion Novo Nordisk -- who were in Hyderabad to attend a seminar on diabetes -- drove 20 km north-west out of the city to Dr Reddy's Research Foundation (DRF) at Ameerpet. It was a casual visit: the Danish scientists had heard about DRF filing a patent for a new molecule belonging to the thiazolidinedione class.

**Fig. 8.1** Part of the report published in Business Today in 1998 regarding out-licensing of DRF-2593 to Novo Nordisk by DRL. (http://archives.digitaltoday.in/businesstoday/22051998/cf2.html)

| Structure | Key features |
|---|---|
| 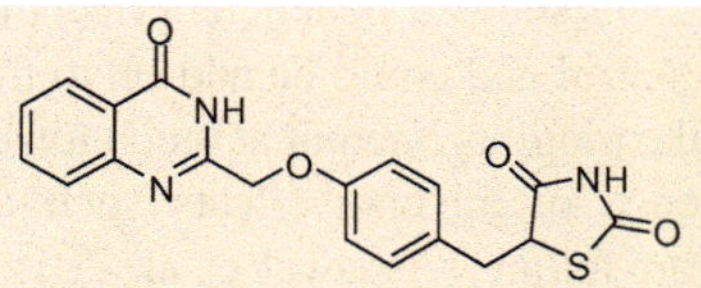 | • belongs to thiazolidinedione (TZD) class<br>• partial agonist of PPARγ ($EC_{50}$ ~ 1.351 μM for human PPARγ)<br>• impressive glucose lowering ability<br>• showed lesser body fluid as well as fat accumulation and less heart enlargement<br>• no decrease in bone formation<br>• completed phase 3 studies |

**Fig. 8.2** Chemical structure of DRF-2593 (balaglitazone) and its key features (Agrawal et al. 2012)

and carbohydrate) of higher organisms. Three distinct PPAR subtypes or isoforms have been identified in most mammalian species, e.g., PPARα, PPARγ, and PPARδ. The PPARα is expressed mostly in the metabolically active tissues such as liver, muscle, heart, kidney, and intestine. PPARα is known to be the pharmacological target for the anti-hyperlipidemic drugs such as fibrates, whereas PPARγ function as the cellular receptor of the insulin-sensitizing drugs, e.g., thiazolidinediones (TZDs). Indeed, DRF-2593 belongs to the thiazolidinedione class of compounds and showed antihyperglycemic activity (at 3 mg/kg, p.o.) in completely diabetic and insulin

| Structure | Key features |
|---|---|
| 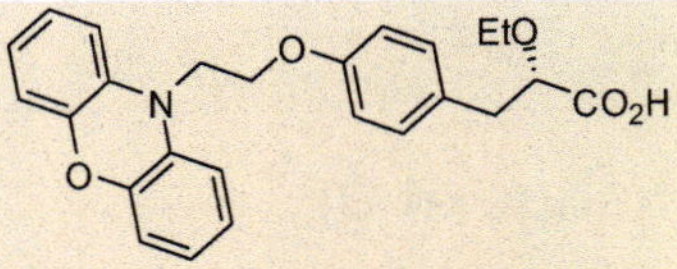 | • belongs to α-alkoxy propanoic acid class<br>• potent agonist of PPARα (hPPARα) and PPARγ (hPPARγ) with $IC_{50}$ values ~ 0.98 and 0.092 μM, respectively<br>• significant decrease in blood glucose and plasma lipid along with insulin sensitizing activities<br>• favorable pharmacokinetics in Phase 1 study<br>• HbA1c and plasma lipid lowering in Phase 2 study<br>• reached phase 3 clinical trial |

**Fig. 8.3** Chemical structure of DRF-2725 (ragaglitazar) and its key features (Chakrabarti et al. 2003; Saad et al. 2004)

resistant db/db mice (Agrawal et al. 2012). Balaglitazone (10 mg/kg, p.o.) decreased overall glucose, insulin levels, and increased bodyweight in male diet-induced obese rats, and such effects were equal to that of 30 mg/kg pioglitazone. Being more potent than the full PPARγ agonist rosiglitazone, DRF-2593 showed beneficial effects in phase 3 clinical studies including high glycemic control and could be add-on to the insulin therapy. It showed similar or same effect like pioglitazone but at much lower dose and better safety profile with less incidence of adverse effects such as myocardial infarction, peripheral edema, and heart failure (Agrawal et al. 2012). Additionally, during preclinical studies DRF-2593 showed lower effects related to the fluid retention, heart enlargement, and reduction of bone formation as observed for full PPAR gamma agonists (Agrawal et al. 2012). Although balaglitazone was returned to DRL in 2004, DRL undertook further development of this molecule after partnering with Rheosciences in 2005. As a result balaglitazone entered into Phase 3 clinical studies in 2007 and became the first Indian compound to reach this level. While the clinical data reported in 2010 were promising, the compound however was abandoned in 2011 that was mainly triggered by the fact that rosiglitazone was banned in 2010 both in India and Europe. Being another compound of the glitazone family rosiglitazone was developed and launched into the market by GlaxoSmithKline (GSK), but later the compound was linked to the apparent risks for increased heart attacks.

**DRF-2725** After discovering the thiazolidinone (or glitazone) class of anti-diabetes agent balaglitazone (Agrawal et al. 2012; Maji and Samanta 2015), another molecule, i.e., ragaglitazar (DRF-2725) (Fig. 8.3), was discovered by DRL (Chakrabarti et al. 2003; Saad et al. 2004). Ragaglitazar that belonged to the α-alkoxy propanoic

| Structure | Key features |
|---|---|
| 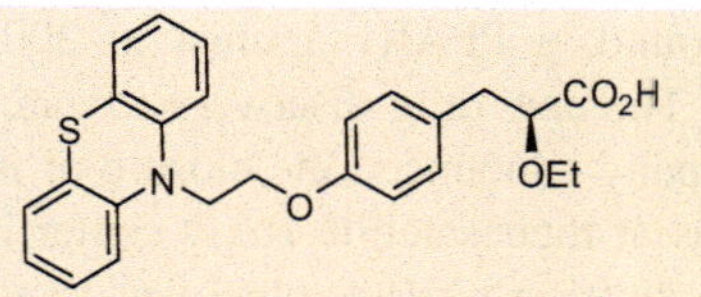 | • belongs to α-alkoxy propanoic acid class<br>• a dual PPARα/γ agonist<br>• Preclinical studies |

**Fig. 8.4** Chemical structure of DRF-4158 and its key features

acid class was a potent and efficacious dual PPAR α/γ agonist [$IC_{50}$ ~ 0.98 (PPAR α) and 0.092 μM (PPAR γ)]. Restoring insulin sensitivity and correcting dyslipidemia was the main goal for the use of this compound. In phase 2 clinical studies, DRF-2725 showed significant decrease of plasma triglycerides, total cholesterol, blood glucose, and HbA1C. The increase of HDL cholesterol was also observed at pharmacological relevant doses. During a 12-week study in patients with type 2 diabetes, DRF-2725 showed glycemic control comparable to that of pioglitazone. Moreover, improvement in the lipid profile was significant. Indeed, with such a profile of effects DRF-2725 was thought to be highly promising in reducing the morbidity and mortality related to the long-term cardiovascular complications. This molecule was also out-licensed to Novo Nordisk in 1998 and was undergoing phase 3 clinical trials with 2500 T2D patients. However, further progress of this molecule was halted by Novo Nordisk in 2002 due to the bladder tumors identified in rats during toxicological studies.

**DRF-4158** DRL discovered another α-alkoxy propanoic acid based anti-diabetes molecule DRF-4158 (or DRF 554158 or DRF-MDX8 or LBL-752) (Fig. 8.4) that was a dual PPARα/γ agonist and inhibitor of hydroxymethyl-glutaryl-coenzyme A (HMG-CoA) reductase (Lohray et al. 1999). In 2001, DRL joined with Novartis for the advancement of this molecule which however was reverted back to DRL and was stopped in 2003.

**Other Molecules** To develop its four other compounds including three for the treatment of metabolic disorder, DRL participated in a joint venture that caused the creation of Perlecan Pharma, the first company in India to focus on integrated drug development (Dr. Reddy's Laboratories BusinessWire 2005). The molecules that were taken for clinical trials include DRL-16536, an AMP-activated protein kinase (AMPK) modulator to treat metabolic disorders; DRL-11605 a pan-PPARα/δ/γ agonist to treat obesity and dyslipidemia; and DRF-10945, a non-fibrate predominantly PPARα agonist for the treatment of metabolic disorders along with RUS-3108 for the treatment of atherosclerosis (Dr. Reddy's Laboratories BusinessWire 2005). However, none of these molecules could make further progress because of their failure in preclinical or early clinical studies. Indeed, the joint venture was ended in 2008. DRL also collaborated with Argenta Discovery, an

UK-based company with the aim of developing new treatments for chronic obstructive pulmonary disease (COPD) (Argenta Discovery 2007). This collaboration facilitated the clinical development of a compound, a PPARγ agonist in 2007 which however was abandoned in the next year. The new drug discovery research effort at DRL was halted in 2009 when the company announced the closure of its research activities at Hyderabad, India as well as at their satellite R&D center in Atlanta, USA (Businessworld 2014). Its research division in Hyderabad was transferred to a Bangalore-based subsidiary company called Aurigene that offers research services to Pharma firms.

#### 8.1.1.2 Zydus Cadila

In 2000, Zydus Cadila initiated its new drug discovery research program at Zydus Research Center, Ahmedabad with a major focus on metabolic disorders. Accordingly, a PPARα/γ agonist, i.e., ZYH1 (INN saroglitazar, trademark Lipaglyn) (Fig. 8.5), was approved in 2013 in India. This drug was intended to cure patients having diabetic dyslipidemia and hypertriglyceridemia with type 2 diabetes mellitus that were not controlled by statin treatment (Bronson et al. 2014). Indeed, this drug was not only discovered but developed first time by an Indian company (The Times of India 2013).

**ZYH1** The drug ZYH1 belongs to α-alkoxy propanoic acid class of compound. It is a PPAR agonist with potent PPARα ($EC_{50}$ ~ 0.65 pM in HepG2 cells) and moderate PPARγ activity ($EC_{50}$ ~ 3 nM in HepG2 cells). The drug's agonist action at PPARα helped in lowering of high blood triglycerides, whereas its agonist action on PPARγ improved insulin resistance and consequently lowering of blood sugar. In db/db

| Structure | Key features |
|---|---|
| 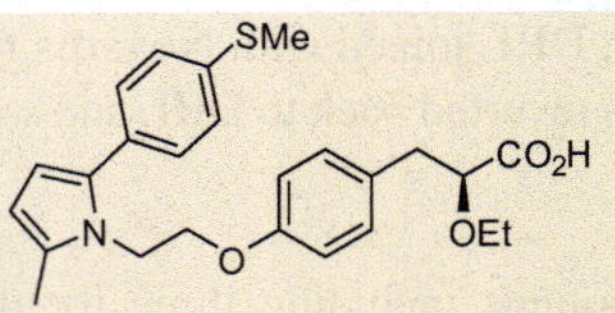 | • a PPARα/γ agonist<br>• $EC_{50}$ values 0.65 pM (hPPARα, in HepG2 cells), 3 nM (hPPARγ, in HepG2 cells)<br>• decreased serum TG, FFA, and glucose in db/db mice in dose-dependent manner<br>• reduced plasma triglyceride and HbA1c in clinical trial<br>• improved glycaemic and lipid parameters in Indian T2D patients<br>• showed good safety and tolerance in patients<br>• launched in India |

**Fig. 8.5** Chemical structure of ZYH1 (saroglitazar) and its key features (Jani et al. 2014; Chatterjee et al. 2015)

mice, 12-day oral dosing with Saroglitazar (0.01–3 mg/kg/day) caused decreases in serum triglycerides (TG), free fatty acids (FFA), and glucose in a dose-dependent manner ($ED_{50} \sim 0.05$ mg/kg, 0.19 mg/kg, and 0.19 mg/kg, respectively). Additionally, decrease in serum insulin (91%) and AUC-glucose (59%) was observed following an oral glucose administration at 1 mg/kg dose. Further, the efficacy (TG lowering) potential of Saroglitazar was confirmed via a 90-day repeated dose comparative study in Wistar rats and marmosets (small, squirrel-like monkeys). This study also showed low risk of PPAR-associated side effects in humans. In the clinical trial, ZYH1 (4 mg tablets) decreased the mean plasma triglyceride levels by $-46.7 \pm 3.02\%$ (mean ± SE) at week 12. A mean HbA1c (hemoglobin A1c) reduction of 0.3% was also observed. Overall, the study indicated that ZYH1 was safe and well tolerated by patients (Jani et al. 2014). In Indian patients with Type 2 diabetes, ZYH1 improved both glycemic and lipid parameters (Chatterjee et al. 2015). Indeed significant mean reductions were observed in fasting plasma glucose, postprandial plasma glucose, glycosylated hemoglobin, total cholesterol, low-density lipoprotein cholesterol, triglyceride, non-high-density lipoprotein cholesterol, and the ratio of triglyceride and high-density lipoprotein cholesterol. Moreover, no significant change observed in body weight, blood pressure, high-density lipoprotein cholesterol, and serum creatinine.

**ZYH7** Zydus Cadila developed another drug, i.e., ZYH7, a PPARα agonist to reduce triglyceride level and address dyslipidemia. The molecule entered into the Phase 2 clinical trials in India (ClinicalTrials.gov 2012a) in 2011 but further details are not known.

**ZYDPLA1** Zydus Cadila was also developing a competitive and long acting and selective DPP-4 (dipeptidyl peptidase 4) inhibitor, i.e., ZYDPLA1 (DPP4 Ki 0.0027 μM) (Fig. 8.6), to provide once a week therapy for the treatment of type 2 diabetes (Jain et al. 2015; Desai et al. 2014; Giri et al. 2017; Patel et al. 2017). This molecule, a derivative of substituted tetrahydro-2*H*-pyran-3-amine, showed selectivity (>7000-fold) towards DPP-4 over DPP-8 and DPP-9. It also showed selectivity (>60,000-fold) towards DPP-4 over fibroblast activation protein (FAP) in vitro. In mice and rats, it showed antihyperglycemic effect and elevation of circulating GLP-1 as well as insulin levels in a dose-dependent manner. ZYDPLA1 showed favorable pharmacokinetics with excellent oral bioavailability in all species. It showed >48 h of plasma DPP-4 inhibition in mice and rats, and for up to 168 h in dogs and non-human primates (Jain et al. 2015). In a separate study, ZYDPLA1 showed full DPP-4 inhibition in rats irrespective of its route of administration (oral or intravenous). While IND approval from the USFDA was received by Zydus in October 2013 to initiate a Phase I trial in type II diabetes however, the phase 1 study in the USA was discontinued according to the report of Zydus Cadila pipeline in August 2017.

**Other Molecules** The other compounds that were under development include ZYH2, a dual PPARα/γ agonist (Cadila Healthcare 2000–2016; Zydus Cadila

| Structure | Key features |
|---|---|
| 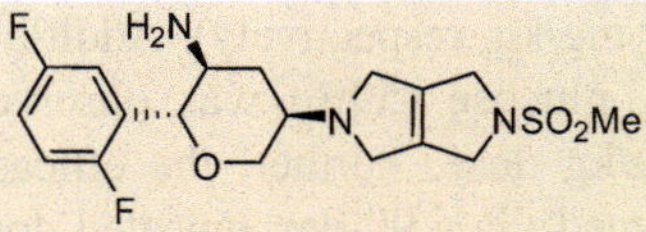 | • a DPP-4 inhibitor ($K_i$ 0.0027 μM; Koff $2.3 \times 10^{-4}$/s)<br>• > 7000 fold selectivity for DPP-4 over DPP-8 and -9<br>• raised circulating GLP-1 and insulin levels (mice and rats) and showed antihyperglycaemic effect in a dose-dependent manner.<br>• showed antihyperglycaemic effects in *db/db* mice and Zucker fatty rats<br>• entered in phase 1 trial |

**Fig. 8.6** Chemical structure of ZYDPLA1 and its key features (Jain et al. 2015; Desai et al. 2014; Giri et al. 2017; Patel et al. 2017)

2011), ZYD1, a peptidic GLP-1 agonist (Bahekar et al. 2008), ZYOG1, an oral peptidomimetic GLP-1 agonist (Bahekar et al. 2011), ZYGK1, a glucokinase activator (Kharul et al. 2011), and ZYG19, a G-protein-coupled receptor 119 (GPR119) agonist (Pingali et al. 2010). According to the press note of Zydus Cadila published in April 2011, the phase 1 study was completed for the molecule ZYH2 (Zydus Cadila 2011). Among other molecules the ZYD1, an agonist of GLP-1 receptor showed enhanced stability to proteolytic breakdown, especially against DPP-4 and did not show gastrointestinal side effects. Initiation of a Phase I study intending to evaluate the various effects of ZYD1 in healthy volunteers was reported in 2011. Similarly, phase 1 clinical trial was initiated for the GLP-1 agonist ZYOG1 that unlike previous injectable agonists was an oral peptidomimetics based agent. When administered via the oral route ZYOG1 showed favorable effects on glucose and HbA1c reduction with an added advantage of weight loss in preclinical models. Additionally, a differentiated safety profile with no nausea-like symptoms was observed in the preclinical studies. The molecule completed phase 1 trial according to the press release in 2011 (Zydus Cadila 2011). In the same year, the IND application of Zydus Cadila was approved by the USFDA for initiating the phase 1 clinical trials of glucokinase activator ZYGK1 (Zydus Cadila 2011). A substituted benzamide derivative ZYGK1 showed beneficial effects in regulating blood glucose (fasting and non-fasting) in several preclinical models of Type 2 diabetes. Additionally, ZYGK1 showed promising safety profile in preclinical studies. The Phase 1 clinical studies was also initiated for the potent and small-molecule based GPR119 agonist ZYG19 that showed in vitro activation of GPR119 receptor in cell-based assays and antihyperglycemic effects in animal models. ZYG19 was extensively tested to evaluate its acute and repeat dose toxicity, genotoxicity, and male reproductive toxicity. It appeared to have an acceptable safety profile for initiating the clinical trials below the declared NOAEL levels. In spite of showing promising results in the preclinical stage, most of these compounds are now abandoned.

#### 8.1.1.3 Glenmark Pharmaceuticals

The Mumbai-based Glenmark Pharmaceuticals initiated its new drug discovery research efforts in 2001. The company delivered a number of developmental compounds of which several entered into the clinical trials. Most of these compounds were inhibitors of PDE4 (phosphodiesterase 4) (Swamy 2012). In the area of anti-diabetes research, the company successfully developed and out-licensed one of its DPP-4 inhibitors, e.g., melogliptin (GRC-8200), to Merck KGaA in 2006 (Thomas et al. 2006; Gupta et al. 2009; Kushwaha et al. 2014).

**Melogliptin** A member of 4-fluoropyrrolidine-2-carbonitrile class of compounds melogliptin (Thomas et al. 2006) was found to be a potent inhibitor ($IC_{50}$ ~ 1.61 nM) of DPP-4 with a 10,000-fold selectivity over DPP2, PPCE, and other proteases tested. At 5 mg/kg dose it showed good pharmacokinetic profile (half-lives ~1.28 h, 4.31 h, and 2.15 h) with the oral bioavailability of 60%, 90%, and 94% in rat, dog, and cynomolgus monkeys, respectively. In db/db mice melogliptin (3 mg/kg/day p.o.) showed 30% reduction of AUC in OGTT and increased the insulin levels by twofolds. In beagle dogs (overnight fasted) melogliptin (5 mg/kg p. o.) resulted in an acceptable peak concentration (2.15 µg/mL) and plasma clearance (1.17 L/h/kg). Additionally, inhibition of plasma DPP-4 (>90%) was observed after 6 h post administration of melogliptin. The drug was well tolerated both at single ascending and multiple ascending doses in phase 1 studies, thereby favoring the once-daily dosing option. Within an hour of dosing, the drug showed >90% inhibition of DPP-4. The molecule entered into phase 2 clinical trials but was abandoned in 2011 (Fig. 8.7).

| Structure | Key features |
|---|---|
| 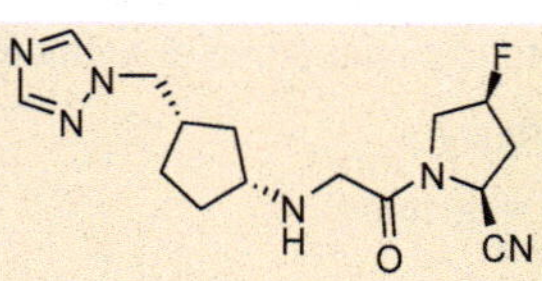 | • a DPP-4 inhibitor ($IC_{50}$ 1.61 nM against hDPP-4)<br>• selective for DPP-4 (10,000 fold over DPP2, PPCE and other proteases)<br>• excellent PK profile (oral bioavailability 50-95%)<br>• showed dose-dependent antihyperglycaemic effect.<br>• showed antihyperglycaemic effects in *db/db* mice and Zucker fatty rats<br>• entered phase 2 clinical trials |

**Fig. 8.7** Chemical structure of melogliptin (GRC-8200) and its key features (Thomas et al. 2006)

#### 8.1.1.4 Orchid Chemicals & Pharmaceuticals

In 2002 the Chennai based Orchid Chemicals & Pharmaceuticals, or Orchid Pharma engaged in a joint venture to develop BLX-1002 of Bexel Biotechnology (a US-based firm) for the treatment of diabetes.

**BLX-1002** BLX-1002 (Fig. 8.8) belongs to the amino acid phenoxy ether class of compounds (structure described in U.S. Patent 6794401) (Zhang et al. 2009). It is a water-soluble thiazolidinedione derivative (no structural resemblance with any PPAR class of TZDs) possessing no PPAR affinity. The molecule (and its major metabolite) inhibited aldose reductase, an important enzyme for polyol pathway that plays a key role in diabetic retinopathy and neuropathy. BLX-1002 stimulated AMP-activated protein kinase activity in human liver cells like metformin, raised cytosolic $Ca^{2+}$, and enhanced glucose-aided insulin release in a PI3K (phosphatidylinositol 3-kinase)-dependent manner (Nag et al. 2004). Indeed, the antihyperglycemic effects of this molecule in diabetic animal models seemed to be not associated with the body weight gain as commonly observed in case of PPAR agonists. The molecule entered into the Phase II trial that was completed in January 2008. However, no further progress of this molecule is known (Adis Insight n.d.-a).

#### 8.1.1.5 Panacea Biotec

Panacea Biotec established its Mohali R&D Centre in Punjab, in 2005 to devote research efforts in the area of metabolic disorders and diabetes in addition to infectious diseases. The company mainly focused on inhibitors of DPP-4 and of sodium-glucose co-transporter-2 (SGLT2) that are known to be beneficial for curing diabetes. Consequently, a β-amino acid derivative PBL-1427 was identified as a reversible, competitive and second-generation inhibitor of DPP-4 that reached Phase 1 clinical trials in India in 2012 (Jain et al. 2009; ClinicalTrials.gov 2012b). The molecule by inhibiting DPP-4 ($IC_{50} = 12$ nM) prolonged the half-life of incretins, potentiated GSIS and thus reduced glucose excursion without provoking hypoglycemia. It showed considerable selectivity (>15,000-fold) towards DPP-4 over

| Structure | Key features |
|---|---|
| 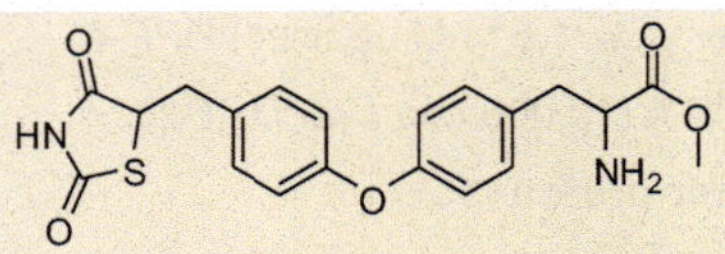 | • a water-soluble thiazolidinedione derivative with no PPAR affinity<br>• inhibited aldose reductase<br>• stimulated AMP-activated protein kinase activity<br>• activates PI 3-kinase pathway<br>• improved hyperglycemia without body weight gain |

**Fig. 8.8** Chemical structure of BLX-1002 and its key features (Zhang et al. 2009; Nag et al. 2004)

DPP-8 and 9. PBL 1427 is still in the company's pipeline (Panacea Biotec 2003–2016).

#### 8.1.1.6 Matrix Laboratories

A Hyderabad-based company Matrix Laboratories was conducting research on discovery of DPP-4 inhibitors (Gopalan et al. 2007; Kodimuthali et al. 2010; Singh et al. 2008) that led to the identification of the cyanopyrrolidine-based lead molecule MX-6001 (Gopalan et al. 2007). The molecule was a competitive and slow binding inhibitor of DPP-4 [$IC_{50} = 4$ nM (recombinant enzyme), 5 nM (rat plasma) and 7 nM (human plasma)] and showed selectivity >10,000-fold over DPP-II, 1315-fold over DPP-8, and 154-fold over DPP-9. With 67% oral bioavailability ($T_{1/2} = 2.12$ h) MX-6001 showed OGTT profile better than Vildagliptin and GLP-1 elevation as well as insulin-level increase in C57BL6 mice. However, no further progress of this molecule is reported.

#### 8.1.1.7 Piramal Enterprises Ltd

As part of their drug discovery program on metabolic disorder Piramal Enterprises Ltd with its R&D center at Mumbai collaborated with Eli Lilly & Co in 2007 to develop preclinical and clinical candidates. Accordingly, the company in-licensed two candidates from Eli Lilly. The first one, i.e., P1201 (of which the mode of action was not disclosed), entered Phase 1 trial in Europe but was stopped in 2013 (Piramal 1997–2016). The second one, i.e., P2202, was an 11β-hydroxysteroid dehydrogenase type 1 (β-HSD1) inhibitor that entered Phase 1 in Europe in 2009 and reached Phase 2 in India and Canada. However, this compound was also abandoned (Piramal 1997–2016). A non-thiazolidinedione insulin sensitizer P1736 (Fig. 8.9) was

| Structure | Key features |
|---|---|
| 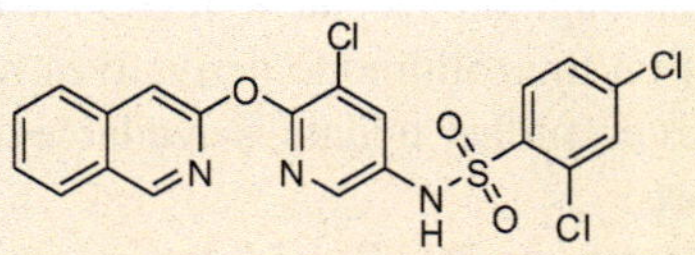 | • a non-TZD and non-PPAR insulin sensitizer<br>• causing GLUT-4 translocation and inhibition of h11$\beta$-HSD1as likely mode of action<br>• decreased plasma glucose and triglyceride levels in diabetic *db/db* mice<br>• decreased plasma glucose in the obese and diabetic *ob/ob* mice<br>• enhanced glucose uptake in the muscles and decreased plasma insulin in *ob/ob* mice<br>• did not show adverse effects on liver safety and was weight neutral<br>• completed phase 1 trial |

**Fig. 8.9** Chemical structure of P1736 and its key features (Anthony et al. 2013)

discovered using fat cell-based phenotypic screening for the treatment of diabetes (Anthony et al. 2013). P1736 did not activate human PPAR (PPARg) receptors and increased glucose uptake ($EC_{50} \sim 400$ nM) in the insulin resistant 3T3 adipocytes in a dose-dependent manner. Indeed, translocation of GLUT-4 (Glucose Transporter type 4) transporters was observed in these adipocytes when this compound was used at 10 μM. The compound showed better effects in lowering plasma glucose and triglyceride levels in diabetic *db/db* mice when compared with metformin. However, no adverse effects on body weight or liver function were observed in the same model. In *ob/ob* mice P1736 reduced glucose by 30–35% and improved hyperinsulinemia significantly. Additionally, it increased glucose uptake in soleus muscles. Nevertheless, the safety profile of P1736 (that was found to be weight neutral) was excellent in all preclinical models of diabetes. Study indicated that P1736 might have elicited its antidiabetic action partly by inhibiting 11β-HSD1 in the liver or adipose tissues of diabetic animal models. The molecule though completed Phase 1 clinical trial but was discontinued in 2014 (Cola 2015).

Piramal was developing a G-protein-coupled receptor 40 (GPR40) agonist P11187 for metabolic disorders/type 2 diabetes that also entered phase 1 clinical trial in the USA (Cola 2015). The potent and oral agent P11187 was a selective and partial agonist of human ($EC_{50} \sim 14.4$ nM), rat ($EC_{50} \sim 2.53$ nM), and mouse GPR40 ($EC_{50} \sim 4.25$ nM). It showed desirable effects in glucose tolerance test in Sprague-Dawley (SD) rats ($ED_{50} \sim 0.28$ mg/kg) and C57BL/6J mice ($ED_{50} \sim 0.62$ mg/kg). There was a considerable increase in glucose infusion rates following chronic administration of P11187 in DIO mice (1–10 mg/kg, orally, 3 weeks) and in a hyperglycemic clamp study in SD rats. However, this compound was finally stopped in 2014 (Cola 2015).

#### 8.1.1.8 Other Companies

A Mumbai-based company Elder Pharmaceuticals though filed a patent in 2008 with Poona College of Pharmacy (Pune, Maharashtra) on thiazolidinone derivatives for the treatment of diabetes (Sakhardande et al. 2010) no further update is available on their internal drug discovery or development efforts.

Pune-based pharma company Lupin Limited identified a PPAR modulator (Lupin Ltd 2001–2016) LL-6531 for the treatment of metabolic diseases which however was stopped at the preclinical stage in 2009. The new drug discovery effort of the Ahmedabad-based company Torrent Pharmaceuticals Ltd in the area of anti-diabetes agents led to the identification of a cyanopyrrolidine-based competitive and reversible DPP-4 inhibitor (Gupta et al. n.d.) TRC-8156 ($IC_{50} = 88$ nM). The compound however failed at the preclinical stage. With the goal of out-licensing preclinical drug candidates and collaboration, Advinus Therapeutics focused on metabolic diseases in addition to inflammatory and neglected diseases at their R&D center in Pune. In the area of anti-diabetes agent Advinus developed a glucokinase activator GKM-001 that finished a 14-day Phase 2 proof-of-concept study (Mookhtiar 2015; Filipski et al. 2012; The Pharmatimes 2012) and a backup compound GKM-002 that was undergoing preclinical studies (Adis Insight n.d.-b). However, further progress of these compounds became uncertain due to the change in company's policies and operations (Dandekar 2016).

## 8.2 Perspective of Current Anti-diabetes Research

This section will provide an overview on current trend and strategies being pursued in the area of anti-diabetes research worldwide followed by the current situation in India.

### 8.2.1 Need for New Drugs Based on Existing/Novel Targets

The class of antidiabetic drugs that are currently in clinical practice includes sulfonylureas, α-glucosidase inhibitors, biguanides, meglitinides, DPP-4 inhibitors, thiazolidinediones, SGLT-2 inhibitors, dopamine agonists, bile acid sequestrants, insulin and its analogues, amylin agonist, and GLP 1 analogues. At a first glance it may appear that adequate numbers of agents or drugs are available to treat T2D, with each differing in mechanism of action, efficacy and safety profile, and the route of administration. However, despite the access to all these anti-diabetes agents with the currently used drugs an appropriate glycemic control is not achieved or maintained for more than a third of diabetic patients (De Pablos-Velasco et al. 2014). Indeed, the poor response to multiple oral antidiabetic drugs has been observed due to the progressive β-cell failure over time (Khan et al. 2006). Additionally, the frequent multiple chronic complications of diabetes itself (e.g., comorbidities especially cardiovascular disease and renal impairment) and potential drug interactions with other drugs prescribed for comorbidities have made the drug therapy more challenging and complex (Tahrani et al. 2011). While concerns/drawbacks of some of the new anti-diabetes drugs are already highlighted in the earlier section, a brief overview on issues with existing and commonly used drugs is presented here (Table 8.1).

#### 8.2.1.1 Drawbacks of Common Anti-diabetes Drugs

Metformin, a well-known first-line remedy in multiple treatment algorithms, is reported to work chiefly by restricting the hepatic glucose release, one of the main concerns of T2D (Inzucchi et al. 2012; Garber et al. 2013; Qaseem et al. 2012; Mancini and Poitout 2015). In addition to its long track record of efficacy, metformin showed nominal weight gain and decreased risk of hypoglycemia. In spite of favorable benefits of metformin, its gastrointestinal (GI) side effects are not tolerated in some patients (Inzucchi et al. 2012). The uses of insulin secretagogues, e.g., sulfonylureas and meglitinides, are common and early in the treatment of a diabetic patient because of their clinical efficacy in the short term. However, it has been observed that these drugs did not maintain efficacy over time. Moreover, their uses are often accompanying with hypoglycemia and weight gain. These side effects frequently restrict the further uses of these drugs. The thiazolidinedione class of compounds addresses a third aspect of the physiologic defects of diabetic patients

**Table 8.1** Concerns of commonly used anti-diabetes agents

| Therapy | Concerns |
|---|---|
| Insulin:<br>• Basal insulin: glargine, detemir, NPH<br>• Parndial (bolus) insulin: aspart, lispro, glulisine, human regular insulin | • Increased risk of hypoglycemia<br>• Common source of hospital errors<br>• Subcutaneous injections<br>• Need for glucose monitoring |
| GLP-1 receptor agonists:<br>• exenatide<br>• liraglutide | • Subcutaneous injections<br>• Gastrointestinal side effects<br>• Decreased appetite and weight loss<br>• Concern regarding acute pancreatitis |
| Metformin | • Risk of lactic acidosis in patients with impaired kidney function, heart failure, hypoxemia, alcoholism, cirrhosis, contrast exposure, sepsis, and shock<br>• Gastrointestinal side effects |
| Insulin secretagogues:<br>• Sulfonylureas: glyburide, glibenclamide, glipizide, gliclazide, and glimepiride<br>• Glinides: repaglinide and nateglinide | • Risk of hypoglycemia<br>• Significant drug-to-drug interactions<br>• Risk of cardiovascular events |
| α-Glucosidase inhibitors:<br>• acarbose<br>• miglitol | • Gastrointestinal side effects<br>• Contraindicated in patients with inflammatory bowel disease, partial bowel obstruction, or severe renal or hepatic disease |
| TZDs:<br>• pioglitazone | • Slow onset of action<br>• Contraindicated in patients with heart failure, hemodynamic instability, and hepatic dysfunction |
| DPP-4 inhibitors:<br>• Sitagliptin<br>• Saxagliptin<br>• Linagliptin<br>• Alogliptin | • Concern regarding acute pancreatitis |
| SGLT-2 inhibitors:<br>• Canagliflozin<br>• Dapagliflozin | • Increased risk of urinary and genital tract infections<br>• Risk of dehydration |

via enhancing the insulin sensitivity in the periphery. These agents have shown continued durability of effect and lesser occurrences of hypoglycemia. However, these agents have shown other side effects including edema and weight gain that possibly restrict their uses. It is widely accepted that insulin therapies that target insulin deficiency/insufficiency directly are better agents for glycemic control. The problem of dose limitations or other side effects (as seen with non-insulin therapies) is not associated with the uses of insulin. However, considerable risks of hypoglycemia and weight gain are concerns for insulin therapies. Moreover, patients normally show less enthusiasm in adopting the insulin therapy at the initial phase of their disease (Inzucchi et al. 2012).

#### 8.2.1.2 Newer Goals of Anti-diabetes Research

While different combinations of drugs have been tested and used for glycemic control but that too did not achieve the desired outcome for many patients and enforced them to depend on insulin for optimal glycemic control. Moreover, as mentioned earlier the risk of hypoglycemia posed a particular clinical challenge during intensified application of antidiabetic drugs especially insulin and secretagogues. Thus the unmet medical need for optimizing the glycemic control but with minimal hypoglycemia and weight gain appeared to be persistent in spite of availability of a range of anti-diabetes drugs or agents. And this has been reflected in the increase of number of patients suffering from diabetes in every year. Presently the newer goal of current research has been to devote efforts towards addressing these issues, i.e., discovery and development of new, efficacious, and safer drugs that can take care of these complications of T2D especially comorbidities, hypoglycemia, weight gain, and drug interactions. Needless to say that most of these approaches ideally could be based on novel targets that are known to be dysregulated in T2D and were not addressed previously. Notably, development of new formulations as well as delivery methods for known drugs and other incretin-based approaches are also being pursued but are not included here.

### 8.2.2 Targets and Agents Currently Being Pursued Worldwide

In order to address the β-cell dysfunction in T2D, a number of investigational agents are being developed that are designed to target novel sites in pancreatic β-cell, thereby increasing the insulin secretion. Some of these are presented here.

#### 8.2.2.1 G-Protein-Coupled Receptors

Activation of certain G-protein-coupled receptors (GPRs that are exceedingly expressed in β-cells and intestinal enteroendocrine L cells) causes GLP-1 release and GLP-1 activation, thereby stimulating glucose-dependent insulin secretion (GSIS) in β-cells. Thus agonists of GPR 40 and GPR 119 have been tested that caused enhanced insulin secretion without actually entering into the β-cells (Mancini and Poitout 2015; Kahn et al. 2014). Notably, activation of GPR 119 in β-cells can stimulate GSIS directly (Ohishi and Yoshida 2012). Recently, a dual-target compound, i.e., HBK001 (Fig 8.10), has been reported to inhibit DPP4 and activate GPR119 ex and in vivo (Huan et al. 2017). Indeed, this compound promoted glucose-stimulated insulin secretion in primary islets of mouse and human. HBK001, on a single administration in ICR mice increased plasma incretins levels better than the known DPP4 inhibitor linagliptin. Moreover, long-term treatment of

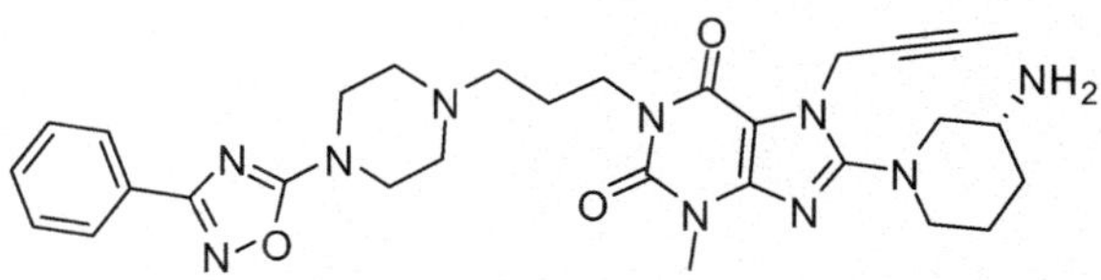

**Fig. 8.10** Chemical structure of HBK001

HBK001 in KKAy mice ameliorated hyperglycemia and improved glucose tolerance more effectively than linagliptin. The direct effect of HBK001 on islet β-cells via GPR119 activation was indicated by the increase of first-phase insulin secretion caused by HBK001 in KKAy mice. The other effects of HBK001, i.e., improvement of islet morphology, increase of β-cell proliferation, and upregulation of genes involved in improving β-cell function suggested that this agent could be beneficial for patients who are insensitive to the existing DPP-4 inhibitors.

#### 8.2.2.2 Glucokinase Activators

Glucokinase, a member of the hexokinase family of enzymes is an important enzyme that plays the role of a glucose sensor in β-cell. It functions as a monomer and facilitates phosphorylation of glucose (at C-6) using magnesium adenosine triphosphate as the second substrate. This results in the formation of glucose-6-phosphate (G6P). Glucose appeared to be the preferred substrate of glucokinase under the physiological conditions. Expression of glucokinase is generally occurred in organs like pancreas, liver, brain, and the gastrointestinal tract. These organs are believed to have a distinct role in glucose sensing (Matschinsky et al. 2006). The glycogen synthesis is promoted in the liver as a result of phosphorylation of glucose by glucokinase. However, it results in insulin release in the pancreatic β-cells. Thus activation of glucokinase enzyme results in increase of insulin secretion as well as hepatic glucose metabolism simultaneously. Several glucokinase activators were developed as potential anti-diabetes agents (Pal 2009a, b; Matschinsky 2013, 2009) that showed their blood glucose lowering ability in a number of animal models of T2D (Bailey et al. 2016). Also, studies in humans (single- and multiple-dose placebo-controlled study) have revealed the effect of these activators in decreasing the glucose levels (fasting and postprandial) in patients with T2D as well as in healthy adults (Nakamura and Terauchi 2015; Pirags et al. 2012). However, for one of these activators, i.e., MK-0941, the loss of efficacy over time and rise of the serum triglyceride levels were noted during its trial that halted this particular trial (Meininger et al. 2011). While liver-specific glucokinase activators, e.g., PF-04991532 (a Phase 2 clinical candidate) (Fig. 8.11), were developed to offset hyperglycemic side effect of these drugs (Erion et al. 2014), the non-alcoholic fatty

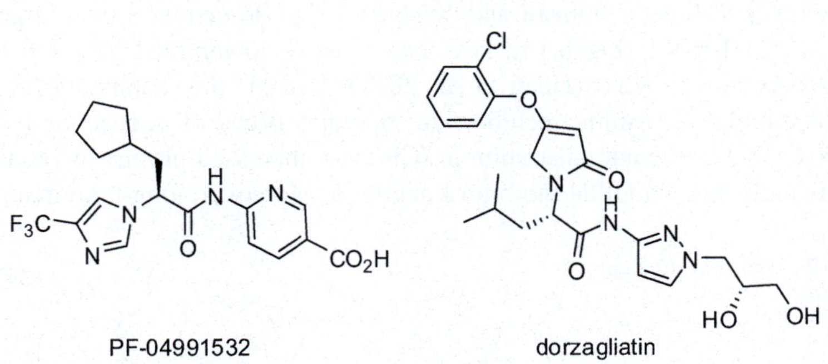


**Fig. 8.11** Chemical structure of PF-04991532 and dorzagliatin

liver disease appeared to be a concern (De Ceuninck et al. 2013) (though PF-04991532 did not show changes in hepatic triglyceride in rats and the specific mechanism of non-alcoholic fatty liver disease was not clear). Nevertheless, the allosteric dual activation of glucokinase by dorzagliatin (HMS5552) (Fig. 8.11) was entirely different from glucokinase activation due to disruption of glucokinase–glucokinase regulatory protein (GKRP) complex. This was evident from the observation that the serum triglyceride concentrations were not enhanced in patients with T2D who received dorzagliatin in phase 2 study (Zhu et al. 2018a, b). Moreover, no dyslipidemia or abnormality of liver was observed during drug safety studies in rats or dogs.

#### 8.2.2.3 Triazine-Based Agent Imeglimin

A novel first-in-class oral antidiabetic agent, i.e., imeglimin (a triazine derivative) (Fig. 8.12), not only enhanced insulin secretion in response to glucose but also modulated mitochondrial energetics to increase insulin sensitivity and suppress gluconeogenesis. A phase 3 ready molecule imeglimin has shown the potential to prevent endothelial and diastolic dysfunction and provided protective effects on β-cell survival and function. Earlier the molecule was studied in rodent. It showed reduction of fasting plasma glucose (FPG) and hemoglobin A1c (A1c) and inhibited hepatic glucose production comparable to metformin. Imeglimin also caused the stimulation of skeletal muscle glucose uptake and showed protection against beta-cell apoptosis. Besides, the decrease in liver steatosis was also observed. Notably, imeglimin was found to be comparable to DPP-4 inhibitors in terms of glucose-dependent insulin secretion (Vuylsteke et al. 2015). The Phase 3 trial of this molecule is expected to commence soon.

#### 8.2.2.4 New Incretin Mimetics

The incretin mimetics (a new class of drugs emerged during the past decade) have become safer and efficacious alternatives to other oral anti-diabetes drugs because of their glucose-dependent action. These include various GLP-1 analogues and DPP-4 inhibitors that are currently available for patients use for last few years. Recently, a new approach involving fixed dose combination of a GLP-1 analogue and basal insulin has been introduced. Indeed, better glycemic control with favorable effects on weight change and hypoglycemia incidence have been observed with the use of these novel combinations, e.g., liraglutide with insulin degludec or lixisenatide with insulin glargine (Rosenstock et al. 2016; Linjawi et al. 2017; Valentine et al. 2017). These effects were observed in adults with T2D for whom the disease was ineffectively controlled with a GLP-1 receptor agonist or oral agents. The combination liraglutide/insulin degludec (Xultophy) is approved by USFDA as an adjunct to diet

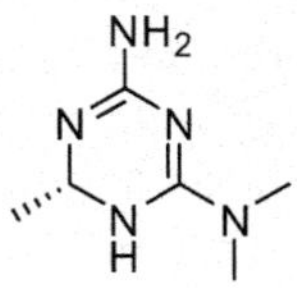


**Fig. 8.12** Chemical structure of imeglimin

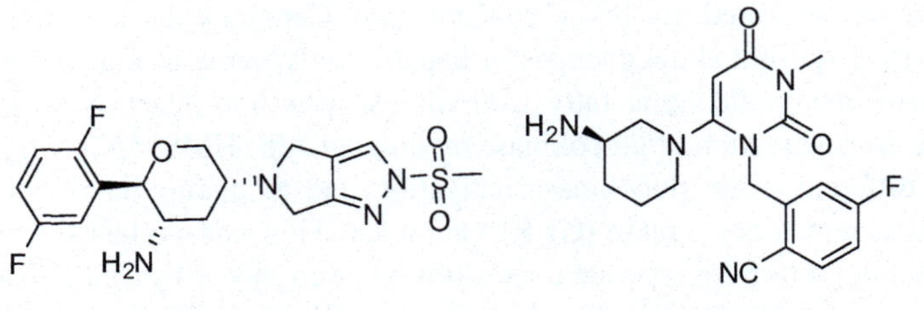


**Fig. 8.13** Chemical structure of omarigliptin and trelagliptin

and exercise. The combination is intended to improve glycemic control in T2D patients (adults) ineffectively controlled on basal insulin (<50 units daily) or liraglutide ($\leq$1.8 mg daily). Similarly, in 2016 the USFDA approved Soliqua 100/33 (insulin glargine [Lantus] 100 units/mL and lixisenatide [Adlyxin] 33 mcg/mL; sanofi-aventis) for T2D patients (adults) whose disease was ineffectively controlled with basal insulin (<60 units daily) or lixisenatide.

Nevertheless, efforts have been devoted to develop orally active GLP1 analogue as all of currently available GLP1 analogues are either once-daily or once-weekly injectable preparations. Thus, semaglutide an orally active GLP1 analogue has been evaluated for efficacy and long-term safety in subjects with Type 2 Diabetes along with DPP-4 inhibitor sitagliptin (Novo Nordisk A/S 2015). Chemically, it has structural similarity with GLP-1 with differences in two amino acid substations at position 8 and 34 (where 2-aminoisobutyric acid and arginine are present, respectively) and lysine at position 26 is acylated with stearic diacid (Lau et al. 2015). The phase 3 clinical trials of this molecule (trade name Ozempic) was completed in 2016 and the drug is being marketed in the USA and Europe which can be used as both injectable and oral drug. The other innovations in the domain of incretin mimetics include development of an implanted subcutaneous mini-pump (ITCA 650) which delivers up to 80 μg of exenatide daily and an exenatide once-monthly suspension (Henry et al. 2014; Wysham et al. 2016). In phase 3 study, ITCA 650 showed considerable decrease of $HbA_{1c}$ and weight compared to placebo. It was well tolerated in patients with uncontrolled T2D on oral anti-diabetes medications. Nevertheless, the other major group of incretin mimetics developed is the oral DPP-4 inhibitors which could be taken once in a week instead of daily dose as needed for existing inhibitors (e.g., gliptins are in the market since 2006 currently with 8 in number). These include omarigliptin (MK-3102) (Goldenberg et al. 2017) and trelagliptin (Zafatek) (Inagaki et al. 2015) (Fig. 8.13) that have shown to be efficacious in clinical studies. Omarigliptin with a prolonged half-life allowing for once-weekly dosing was approved and launched in Japan. However, Merck & Co that developed this drug has decided not to pursue marketing authorization of omarigliptin in Europe and the United States because of commercial reasons. The other drug trelagliptin in the form of its salt succinate has been approved for use in Japan in March 2015. However, Takeda, the company that developed trelagliptin after purchasing it from Furiex Pharmaceuticals, chose not to get approval for the drug in the USA and EU due to the high cost quoted by Furiex.

#### 8.2.2.5 Inhibitors of SGLT2

While concerns has been raised regarding the therapeutic usage of SGLT2 inhibitors as mentioned in the earlier section, some recent inhibitors have shown considerable promise. For example, empagliflozin (Fig. 8.14) a potent and selective inhibitor was approved for medical use in the United States in 2014 and subsequently in the EU, Japan, and other parts of the world (Frampton 2018). Apart from being an active and well-tolerated antihyperglycemic agent empagliflozin showed low risk of hypoglycemia and exerted a favorable effect on several nonglycemic consequences, including modest decrease in bodyweight and blood pressure. It also showed cardioprotective and renoprotective effects mostly independent of glycemic control in patients with T2D and cardiovascular disease (CVD) in a mandated cardiovascular (CV) outcomes trial (EMPA-REG OUTCOME). Notably, canagliflozin was the only other inhibitor of SGLT2 that showed renal and CV benefits. However, unlike canagliflozin the inhibition of empagliflozin has not been associated with an enhanced risk of bone fractures or amputation. Overall, the drug is useful particularly for patients who are at high CV risk but in need of an (additional) antidiabetic treatment in order to reach their glycemic goal. Considering the fact that dual inhibition of SGLT1 and SGLT2 by a single molecule would offer complementary insulin-independent mechanisms to cure diabetes sotagliflozin (LX4211, Zynquista) (Fig. 8.14) has been discovered and developed. Indeed, it has been developed as a dual inhibitor of SGLT1 (facilitates glucose absorption in the gastrointestinal tract) and SGLT2 (facilitates for glucose reabsorption by the kidney) (Lampueta et al. 2015). The molecule, a first-in-class was developed jointly by Lexicon Pharmaceuticals and Sanofi as an orally administered drug for the cure of type 1 and type 2 diabetes mellitus. The USFDA accepted the NDA for sotagliflozin on May 22, 2018. The NDA contains data from the Tandem clinical trial program that includes three Phase 3 clinical trials evaluating the safety and effectiveness of sotagliflozin in nearly 3000 adults with ineffectively controlled type 1 diabetes (Sanofi 2018). The target date for an FDA action regarding sotagliflozin under the Prescription Drug User Fee Act (PDUFA) has been expected to be March 22, 2019.

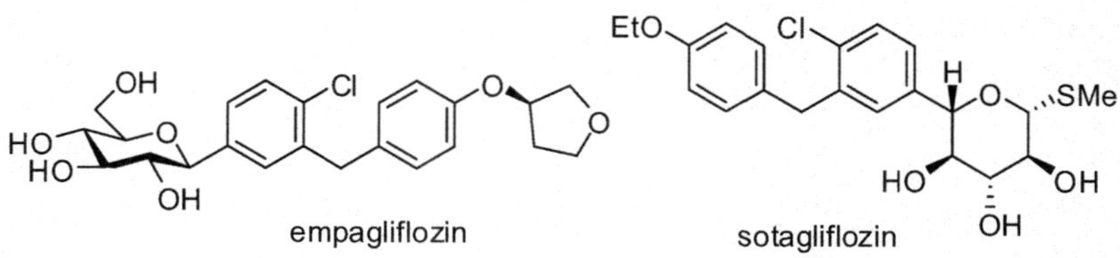


**Fig. 8.14** Chemical structure of empagliflozin and sotagliflozin

#### 8.2.2.6 Selective PPARγ Modulators

While the PPARγ activation is known to improve markers of inflammation, hyperglycemia, insulin resistance, and endothelial function, however, clinical applications of thiazolidinediones (TZDs) have been restricted by their side effects such as congestive heart failure, adipogenic weight increase, fluid retention, and a reduction in bone mineral density linked to fractures. For example pioglitazone, a full PPARγ agonist belonging to the thiazolidinedione class has shown most of these adverse effects. However, a selective PPARγ modulator (SPPARM), e.g., INT131 (T0903131 or T131 or AMG131), has shown to improve glucose metabolism while minimizing the side effects of pioglitazone (DePaoli et al. 2014). With respect to the chemical structure, INT131 is different from TZDs and belongs a new group of non-TZD PPARγ ligands (Fig. 8.15). The besylate salt of INT131 was well tolerated and considerably improved $HbA_{1c}$ compared with placebo in subjects with T2D not effectively controlled on sulfonylureas and metformin or sulfonylurea alone. The compound also showed dose-dependent decreases in $HbA_{1c}$, equivalent to 45 mg pioglitazone, but with less fluid buildup and weight increase, consistent with its SPPARM design. The compound completed phase 2 *clinical* trials and moved into Phase 3 trials.

### 8.2.3 Current Scenario in India

As discussed earlier that efforts are currently being devoted worldwide towards newer innovations in the area of discovery and development of anti-diabetes agents or drugs. However, the situation is different in India at present especially in industrial organizations where the activities towards these directions are either slowed down or halted (Differding 2017). For example, in-house drug discovery research activities were halted by Dr. Reddy's Lab in 2009 and Piramal Enterprises in 2014. Prior to this similar step was taken by Ranbaxy Lab in 2008. Several other pharma companies in India have suspended or abandoned their drug discovery activities. Thus the expectation that India could become one of the major hubs for the discovery and development of new drugs received a big blow. Nevertheless, a brief description on current situation in India in the area of anti-diabetes research is presented in the following sections.

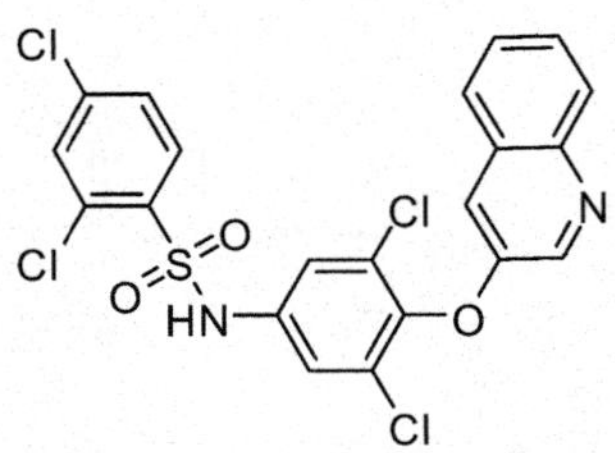


**Fig. 8.15** Chemical structure of INT131

#### 8.2.3.1 Compounds Being Pursued

While most of the anti-diabetes agents being pursued by various Indian pharmaceutical companies are either discontinued or abandoned, some are still in the pipeline. Brief particulars of these compounds along with their status if known and information about the corresponding companies are presented in Table 8.2.

#### 8.2.3.2 Quest for Global Partners

With these compounds in their pipeline one of the key issues that Indian companies encountered with is the lack of interest shown by potential global partner who could offer necessary support for further development of their molecules either via collaboration or via in-licensing these molecules. In fact, several molecules listed in Table 8.2 are waiting for such events. However, the global pharma companies seem to be more interested in out-sourcing their drug discovery related activities to Indian CROs (contract research organizations) rather than looking into the assets, i.e., leads/preclinical candidates already generated independently by Indian Pharma companies. Though it is not always true that these global pharma giants are not interested in in-licensing/partnering suitable compounds from Indian companies (indeed, Dr. Reddy's Lab, Glenmark, and other companies successfully out-licensed several of their compounds in the past) but in certain cases their focus seems to be on candidates that completed most of the clinical studies such as Phase 2 or more. This strategy perhaps offer lower risk of attrition (as the higher rate of attrition was observed in the past for compounds in-licensed from Indian companies) compared to that involved in in-licensing preclinical candidates for which efficacy/safety data in human are yet to be generated. However, unfortunately except few most of the Indian Pharma companies are financially not so strong enough that they alone can push their molecules to clinical trials. The other reason that disallowed Indian companies from finding global partner could be the lack of novelty or lack of differentiation of their compounds from the existing products already in the development elsewhere. This is supported by the fact that past success of Dr. Reddy's in out-licensing their molecules was due to its early entry into the area of glitazars. In fact, Dr. Reddy's had been among the first companies to work on glitazars. Nevertheless, this does not mean that all compounds listed in Table 8.2 lack novelty or differentiation. For example, Advinus's glucokinase activator has been the prospective first-in-class drug with the benefits over earlier discontinued developmental candidates of being liver-selective with a lower hypoglycemic risk (Filipski and Pfefferkorn 2014). Shantani's compound has been reported to possess a new mechanism of action for the treatment of T2D (Saxena 2015b). It has now become important that more such drug discovery efforts should be directed towards similar direction to a greater extent by Indian companies.

#### 8.2.3.3 Role of Academic Institutions/National Laboratories

While anti-diabetes research is witnessing a slow progress in Indian industrial organizations, it did not impact ongoing activities in the same area in academic research institutes, universities, or national laboratories that are generally funded by government agencies such as Council of Scientific & Industrial Research (CSIR),

**Table 8.2** Anti-diabetes agents being pursued by Indian pharmaceutical companies

| Compound | Mechanism of action | Status | Company | Remarks | References |
|---|---|---|---|---|---|
| GKM-001 | Glucokinase activators | Completed a 14-day Phase 2 proof-of-concept study | Advinus Therapeutics, Pune | Established in 2005 and backed by Tata group. | Mookhtiar (2015), Filipski et al. (2012), The Pharmatimes (2012), Adis Insight (n.d.-b) |
| GKM-002 | | Preclinical development | | | |
| KU-5039 | Possibly a fatty acid analogue activator of AMPK | Preclinical development | Kareus Therapeutics, Headquarter in Switzerland | A virtual company established in 2007 by earlier members of Dr. Reddy's Lab | Khanna and Pillarisetti (2014) |
| CNX-012-570 | AMPK activator | Out-licensed to Boehringer Ingelheim | Connexios Life Sciences, Bangaluru | Established in 2003 and supported by the venture investing arm of a co-founder of Indian IT service company Infosys | Rao et al. (2014a, b) |
| CNX-011-67 | GPR40 agonist | | | | Rao et al. (2012) |
| CNX-013-B2 | Activator of RXR (Retinoid X Receptor) α,β,γ | | | | Ranga et al. (2015) |
| CNX-010-49 | βHSD inhibitor | | | | Rao et al. (2013) |
| NDS100179 | Not known | Preclinical development | Shantani Proteome Analytics | A Pune-based company established in 2010 | Saxena (2015a), Reddy et al. (2015) |
| KBR2001 (or KBRPL2001) | GPR120 agonist and first in class to have shown preclinical proof of concept in a chronic efficacy DIO mouse model | In-licensed from Piramal and currently in preclinical development | Krish Biotech | A Kalyani (near Kolkata) based company established in 2009 | Anupindi et al. (2016) |

Department of Biotechnology (DBT), *Indian Council of Medical Research* (ICMR), Department of Science & Technology (DST), etc. For example, Lucknow-based CSIR-Central Drug Research Institute (CSIR-CDRI) is one of the leading national laboratories possessing state-of-the-art facilities necessary for drug discovery where research is being performed in the area of diabetes and subsequently discovery of new anti-diabetes drugs. Similarly, Indian institute of Chemical Technology (IICT) in Hyderabad, Indian Institute of Integrated Medicine (IIIM) in Jammu are the other institutes capable of pursuing anti-diabetes research. While teaching and training along with academic research are the major focus of National Institute of Pharmaceutical Education and Research (NIPER) in Mohali, the institute also focuses on drug discovery research activities as evident from its research publications especially in the area of diabetes. Indeed, the newly created several NIPERs throughout the country could play important role if these institutes are supported and empowered properly. Notably, over the years a non-governmental organization known as Dr. Mohan's Diabetes Specialities Centre with its headquarter in Chennai, Tamil Nadu is highly active not only in providing medical services to patients suffering from diabetes but also conducting research in various areas of diabetes and spreading awareness about this disease throughout the country (Dr. Mohan's Diabetes Specialities Centre n.d.). Nevertheless, details of anti-diabetes research activities being pursued in CSIR-CDRI along with other academic research institutes/ universities in India are covered separately in another part of this chapter.

It is worthy to mention that on May 30, 2016 CSIR launched an anti-diabetes ayurvedic drug/formulation BGR-34 in the Indian market that has been reported to possess the DPP-4 inhibitory activity (Times of India 2016). The drug developed by National Botanical Research Institute (NBRI) and Central Institute for Medicinal and Aromatic Plants (CIMAP) (both belong to CSIR) has been found to be effective in lowering blood sugar levels and increasing immunity as well. Indeed, trials conducted for 18 months scientifically validated the efficacy and safety of BGR-34 that is currently available at much cheaper cost than existing DPP-4 inhibitors (Gupta et al. 2017, 2018). The drug was reported to be composed of six Indian medicinal herbs, e.g., *Tinospora cordifolia*, *Berberis aristata*, *Gymnema sylvestre*, *Pterocarpus marsupium*, *Trigonella foenumgraecum,* and *Rubia cordifolia*, that were optimized for their antidiabetic activity. The research knowledge and technology was transferred to Aimil Pharmaceuticals (India) Ltd, New Delhi, India for production and marketing of this drug.

## 8.3 Future of Anti-diabetes Research in India and Way Forward

It is clear and evident from the previous sections that most of the Indian Pharma majors devoted their efforts in the area of anti-diabetes research in the past with the goal of discovery and development of new molecules useful for curing T2D. As a result, a number of promising and potent candidates were discovered of which several entered into Phase 1 trials. However, not many of them subsequently entered in the next phase of clinical trials because of a high attrition rate for various reasons and factors some of which are not disclosed or explained properly. Till date just one compound was able to make it to the market successfully and several were discontinued at Phase 2 or 3 clinical trials stage. Notably, this trend is not just common in India but very much witnessed in other parts of the world though success rate of launching molecules in the market is higher in the USA and European countries.

### 8.3.1 Refocusing and Restructuring Drug Discovery Efforts

While the reasons for discontinuing molecules were either not known or reported to be the result of business considerations in some cases it is important to evolve a strategy to counter this problem. Since the safety issues emerged as one of the major hurdles on several occasions, it is essential to address the potential for safety issues with new molecules at every stage of the development. However, in spite of this strategy the complete picture may only be realized after longer testing in the clinic. Besides this safety concern it is also important to realize that the efficacy observed in preclinical models may not always be repeated in clinic or efficacy observed in clinic may not always translate into the same clinical benefits. The development process is more complex and time consuming for molecules with unproven mode of action and the question of safety remains as a bigger concern throughout the development program. The drug development project may also be at high risk if it is based on a wrong hypothesis or if a precedence molecule or agent failed during clinical studies. In view of these concerns, it is therefore obvious that a number of drug development projects might fail as it was observed in case of Indian Pharma companies in the recent past. And this number is considerably high especially in case of development of anti-diabetes molecules. It is also important to note that regulatory considerations have driven up the cost of drug development particularly in the area of diabetes. This has impacted the anti-diabetes drug discovery research in India as several industrial R&Ds have diverted their focus and efforts from diabetes to other areas or scaled down/halted their overall drug discovery efforts. However, as far as India's drug discovery effort is concerned in spite of large number of failures the limited success that has been achieved so far is in the area of diabetes. Moreover, India is one of the top listed countries having a large number of diabetic populations in the world. Thus defocusing or halting research efforts in the area of diabetes may not be an appropriate option for India. Rather, the efforts should be devoted for development of programs that would directly impact the pathophysiology of diabetes including both T1D and T2D. In other words, proven approach(s) should be established for finding molecules or combinations of molecules that actually impact the pathophysiology and can be explained as well as clearly demonstrated. If successful then significant increase in number of successful drug development programs is expected along with the necessary support that would facilitate the delivery of new treatments for diabetes. This may also help in addressing the issue of lack of novelty or lack of differentiation from existing development compounds that created hurdles for Indian companies from tracing global partners in a number of cases. Overall, with the number of newer innovations either has been reported or being pursued in the area of discovery and development of anti-diabetes agents or drugs, the future of anti-diabetes drugs research is undoubtedly exciting and promising. Moreover, with the remarkable buildup of R&D capability and of know-how during past two decades Indian pharma companies should not ignore or overlook their potential and opportunities but strengthen and consolidate their past initiatives.

### 8.3.2 Overcoming Other Barriers

During 2016–2017 India at 0.83% of GDP (gross domestic product) was among the countries that have made the lowest investment in the area of scientific research. This was the reflection of scenario in industrial especially pharmaceutical research too. While the situation has improved marginally, e.g., investment in R&D as a percentage of sales has been increased to more than 6% for some Indian companies however the Indian industry has a long way to go to match the figure of Western Pharma companies. India also ranked low in a survey concerning the attractiveness of biomedical investment in countries when factors like infrastructure and scientific capabilities, effectiveness of intellectual property protections, clinical research settings and framework, regulatory framework, market access and financing, etc. were assessed. Additionally, weakness of current scientific education system that lacks emphasis on practical aspects of drug development and applications has worsened the situation. Thus a major effort and a long-term commitment by industry and Government in addressing these issues especially finances, human resources, infrastructure, and legal and regulatory framework is necessary. However, numerous government-supported initiatives such as Make in India, Skill India, Atal Innovation Mission, etc. could be useful in addressing some of these issues. Nevertheless, most of these and other general issues and concerns related to India's overall drug discovery approach along with the potential solutions are described in details earlier (Differding 2017; Dikshit and Dikshit 2016). Needless to say that these are very much relevant to the strategies and approaches followed in the area of anti-diabetes research in India too.

Thus some of the key way forward may be summarized as follows:

1. **Strategy and focus**: Choosing projects that can deliver best-in-class compounds if not first-in-class with clear focus on research goals, deliverables, and timeliness; this should also include India's own diseases problems that are of least interest to the western world.
2. **Challenges**: Realizing challenges, complexities, and difficulties involved in the new drug discovery research and developments, accepting failure as a rule in drug discovery and not giving up early.
3. **Issues:** Fixing issues related to inadequate idea of IP protection, regulatory uncertainties concerning clinical trials, unethical or wrong practices, pricing uncertainties, etc.

4. **R&D investment:** Increasing overall investment in R&Ds especially in new drug discovery research programs with endurance, India's current R&D investment in NCE research is far below than the international standard especially like the USA, Japan, UK, and other European countries.
5. **Clinical trials:** Participating actively in the clinical trials as a partner with global pharma giants (rather than out-licensing molecules to them) having knowledge, expertise, and experiences for over several decades not only for the smooth progress of molecules but also for gaining valuable exposure and experiences for the future.
6. **Skill gap:** Eliminating the skill gap by addressing the weakness of scientific education system and training scientists in the area of new drug discovery and development with the focus on understanding the difference between academic and applied research (e.g., Chemical Biology vs. Medicinal Chemistry, Basic Biology vs. Disease Biology, etc.).
7. **Brain drain:** Taking necessary step to stop brain drain and creating vibrant as well as exciting environment for retaining, encouraging, and nurturing the necessary, especially young talent.
8. **Leadership:** Building appropriate leadership via attracting professionals with relevant track record and appropriate experiences from abroad, this is extremely important for achieving the desired success instead of making mere progress (for example, unlike process research and development India's traditional skill and talent in Organic Synthesis still lack guidance to redirect the effort towards Med Chem/NCE research).
9. **CDRI like institutes**: Creating more national laboratories possessing state-of-the-art facilities like CDRI and providing necessary support for clear strategy and goals (similar like creating additional IITs, IISERs, and NISERs throughout the country).
10. **Collaborations:** Increasing interactions/collaborations between industry and academia especially to enhance/spread awareness about the need for new drug discovery research effort in the country. Institutes following PPP (public-private partnership) model like DRILS, Hyderabad could play a central role in this effort.

Overall, the need for encouragement of funding options (private and public), collaborations within India (pharma–pharma or academia—pharma with complementary strengths), exposure to new technologies, access to global domain experts, etc. are the key way forward. Needless to say that there is a need for visionary scientists to lead and take up the challenges and be in the area of drug discovery for the long haul. Nevertheless, the views and opinions suggested by others as well as summarized above may help in bringing the required changes and renewed interest as well as enthusiasm in anti-diabetes research in India. Or else diabetes will likely be remained as one of the major undertreated medical issues in India.

## 8.4 Conclusions

In conclusion, various aspects of past, present, and future of anti-diabetes research in India are presented with particular details of several historical events. The enormous contributions made by various industrial R&Ds and organizations are mainly highlighted with the specific focus on organic molecules that were discovered and developed as potential anti-diabetes agents or drugs. While other agents or formulations or repositioned existing agents, etc. are generally not covered, the recently launched ayurvedic drug BGR-34 is mentioned in an appropriate section. A brief overview of leading molecules that either launched or reached to various stages of clinical trials is presented along with their chemical structure or class and mechanism of action wherever possible, pharmacological effects overserved in various studies including preclinical/clinical phases if reported and the current status if known. These include balaglitazone (DRF-2593) and ragaglitazar (DRF-2725) of Dr. Reddy's Lab Ltd, saroglitazar (ZYH1) of Zydus Cadila, melogliptin (GRC-8200) of Glenmark Pharmaceuticals, BLX-1002 of Orchid Pharma, etc. Attempt was also made to provide the current status of anti-diabetes research in India with the commentary on future perspectives. Overall, India has long been a formidable player in pharmaceutical manufacturing but currently lagging behind in the area of new drug discovery. While a number of factors contributed to this cause (the major one being the high attrition rate), there are several encouraging facts that can boost and revamp India's drug discovery efforts in future especially in the area of diabetes. It's growing economy, coupled with its growing population of diabetes, cardiovascular and other diseases, poising India a strong and potential hub to become a future powerhouse of R&D in addition to the manufacturing in pharmaceuticals.

It's increasing scientific expertise along with a large pool of highly educated, English speaking scientists are becoming more conducive for undertaking drug discovery research and conducting trials more cheaply and in some cases faster than their Western counterparts. These can be viewed as major advantages in a world where drug discovery and development costs are soaring high and quick entry to the market is of tremendous importance. However, it remains to be seen that apart from countering various challenges how quickly India can bring back the same enthusiasm and passion among its drug discovery researchers that was observed in the past in the area of anti-diabetes research at the initial stage.

**Acknowledgement** MP thanks management of Dr. Reddy's Institute of Life Sciences, Hyderabad for encouragement and support.

## References

Adis Insight (n.d.-a) Drug profile BLX 1002. https://adisinsight.springer.com/drugs/800019061. Accessed 24 Jan 2019

Adis Insight (n.d.-b) Drug profile GKM 002. https://adisinsight.springer.com/drugs/800038106. Accessed 24 Jan 2019

Agrawal R, Jain P, Dikshit SN (2012) Balaglitazone: a second generation peroxisome proliferator-activated receptor (PPAR) gamma (γ) agonist. Mini-Rev Med Chem 12:87–97

Anthony J, Kelkar A, Wilankar C, Ranjith V, Bhumra SK, Mutt S, Deka N, Sivaramakrishnan H, Sharma S, Marita AR (2013) Discovery of p1736, a novel antidiabetic compound that improves peripheral insulin sensitivity in mice models. PLoS One 8:e77946

Anupindi R, Sharma R, Husain R (2016) KBRPL2001, a novel GPR120 agonist, improves insulin sensitivity, glucose tolerance and decreases hepatic steatosis in rodent models of type 2 diabetes. In: ADA 2016—American Diabetes Association 76th Scientific Sessions, New Orleans, LA, USA, 10–14 June 2016

Argenta Discovery (2007) Argenta Discovery and Dr. Reddy's progress pre-clinical anti-inflammatory candidate to treat chronic respiratory disease. http://www.mvm.com/upload/news/downloads/26112007Argenta%20Press%20Release.pdf. Accessed 23 Jan 2019

Bahekar RH, Lohray BB, Lohray VB, Jain MR, Banerjee KM, Patel PR (2008) Antidiabetic compounds. Cadila Healthcare Ltd., Ahmedabad, Gujarat, India, PCT Pat. Appl. WO2008/062457 A2, 29 May 2008

Bahekar R, Jain MR, Patel PR (2011) Short chain peptidomimetics based orally active GLP-1 agonist and glucagon receptor antagonist. Cadila Healthcare Ltd., Ahmedabad, India, PCT Pat. Appl. WO2011/048614A2, 28 Apr 2011

Bailey CJ, Tahrani AA, Barnett AH (2016) Future glucose-lowering drugs for type 2 diabetes. Lancet Diabetes Endocrinol 4:350–359

Bronson J, Black A, Dhar M, Ellsworth B, Merritt JR (2014) Chapter Twenty Seven-To market, to market – 2013. In: Desai MC (ed) Annual reports in medicinal chemistry, vol 49. Elsevier, Amsterdam, pp 437–508

Businessworld (2014) Death of a dream. http://www.businessworld.in/article/Death-Of-A-Dream/08-11-2014-65256/. Accessed 24 Jan 2019

Cadila Healthcare (2000–2016) Annual reports, Corporate presentations. http://zyduscadila.com/

Chakrabarti R, Vikramadithyan RK, Misra P, Hiriyan J, Raichur S, Damarla RK, Gershome C, Suresh J, Rajagopalan R (2003) Ragaglitazar: a novel PPAR alpha PPAR gamma agonist with potent lipid-lowering and insulin-sensitizing efficacy in animal models. Br J Pharmacol 140:527–537

Chatterjee S, Majumder A, Roy S (2015) Observational study of effects of saroglitazar on glycaemic and lipid parameters on Indian patients with type 2 diabetes. Sci Rep 5:7706. https://doi.org/10.1038/srep07706

ClinicalTrials.gov (2012a) A clinical trial to evaluate the safety and efficacy of ZYH7 compared to fenofibrate in patients with dyslipidemia. US National Library of Medicine, Bethesda, MD. NLM Identifier: NCT01539616. https://clinicaltrials.gov/ct2/show/record/NCT01539616. Accessed 23 Jan 2019

ClinicalTrials.gov (2012b) Study of single ascending doses of PBL-1427 in healthy volunteers. US National Library of Medicine, Bethesda, MD. NLM Identifier: NCT01554293. https://clinicaltrials.gov/ct2/show/NCT01554293. Accessed 24 Jan 2019

Cola JR (2015) Discontinued drug therapies to treat diabetes in 2014. Expert Opin Investig Drugs 24:1241–1245

Dandekar V (2016) Amidst crisis, Tata Group's pharma unit advinus therapeutics lays off 50 people. The Economic Times. http://economictimes.indiatimes.com/industry/healthcare/biotech/pharmaceuticals/amidst-crisis-tata-groupspharma-unit-advinus-therapeutics-lays-off-50-people/articleshow/55424515.cms. Accessed 11 Dec 2016

De Ceuninck F, Kargar C, Ilic C, Caliez A, Rolin JO, Umbdenstock T, Vinson C, Combettes M, de Fanti B, Harley E, Sadlo M, Lefèvre AL, Broux O, Wierzbicki M, Fourquez JM, Perron-Sierra F, Kotschy A, Ktorza A (2013) Small molecule glucokinase activators disturb lipid homeostasis and induce fatty liver in rodents: a warning for therapeutic applications in humans. Br J Pharmacol 168:339–353

De Pablos-Velasco P, Parhofer KG, Bradley C, Eschwege E, Gönder-Frederick L, Maheux P, Wood I, Simon D (2014) Current level of glycaemic control and its associated factors in patients with type 2 diabetes across Europe: data from the PANORAMA study. Clin Endocrinol 80:47–56

DePaoli AM, Higgins LS, Henry RR, Mantzoros C, Dunn FL, INT131-007 Study Group (2014) Can a selective PPARγ modulator improve glycemic control in patients with type 2 diabetes with fewer side effects compared with pioglitazone? Diabetes Care 37:1918–1923

Desai RC, Bahekar R, Jadav P, Goswami A, Patel P (2014) 2-Phenyl-5-heterocyclyl-tetrahydro-2H-pyran-3-amine compounds for use in the treatment of diabetes and its associated disorders. Cadila Healthcare Ltd., Ahmedabad, Gujarat, India, PCT Pat. Appl. WO2014/061031 A1

Differding E (2017) The drug discovery and development industry in India-two decades of proprietary small-molecule R&D. Chem Med Chem 12:786–818. See this reference: For an in-depth analysis and comprehensive overview

Dikshit M, Dikshit DK (2016) Drug discovery research in India. Curr Sci 111:252–255. See this reference: For a commentary

Dr. Mohan's Diabetes Specialities Centre (n.d.). https://drmohans.com/. Accessed 25 Jan 2019

Dr. Reddy's Laboratories BusinessWire (2005) Dr. Reddy's announces the formation of perlecan pharma. http://www.businesswire.com/news/home/20050927006290/en/Dr.-Reddys-Announces-Formation-Perlecan-Pharma-Perlecan#.VFofh00tC70. Accessed 23 Jan 2019

Erion DM, Lapworth A, Amor PA, Bai G, Vera NB, Clark RW, Yan Q, Zhu Y, Ross TT, Purkal J, Gorgoglione M, Zhang G, Bonato V, Baker L, Barucci N, D'Aquila T, Robertson A, Aiello RJ, Yan J, Trimmer J, Rolph TP, Pfefferkorn JA (2014) The hepatoselective glucokinase activator PF-04991532 ameliorates hyperglycemia without causing hepatic steatosis in diabetic rats. PLoS One 9(5):e97139. https://doi.org/10.1371/journal.pone.0097139

Filipski KJ, Pfefferkorn JA (2014) A patent review of glucokinase activators and disruptors of the glucokinase--glucokinase regulatory protein interaction: 2011-2014. Expert Opin Ther Pat 24:875–891

Filipski KJ, Stevens BD, Pfefferkorn JA (2012) Glucokinase activators in development. In: Jones RM (ed) New therapeutic strategies for type 2 diabetes: small-molecule approaches. RSC Publishing, Cambridge, pp 88–108

Frampton JE (2018) Empagliflozin: a review in type 2 diabetes. Drugs 78:1037–1048. https://doi.org/10.1007/s40265-018-0937-z

Garber AJ, Abrahamson MJ, Barzilay J, Blonde L, Bloomgarden Z, Bush M, Dagogo-Jack S, Davidson M, Einhorn D, Garvey W, Grunberger G, Handelsman Y, Hirsch I, Jellinger P, McGill J, Mechanick J, Rosenblit P, Umpierrez G, Davidson M (2013) AACE comprehensive diabetes management algorithm 2013 consensus statement. Endocr Pract 19(Suppl 2):1–48. https://doi.org/10.4158/EP13176.CSUPPL

Giri P, Patel N, Patel B, Patel H, Bahekar R, Srinivas NR, Patel PR, Desai R (2017) Sensitive and specific LC-ESI-MS/MS method for determination of ZYDPLA1, a novel long-acting dipeptidyl peptidase 4 inhibitor in rat plasma: an application for toxicokinetic study in rats. Biomed Chromatogr 31:e3984. https://doi.org/10.1002/bmc.3984

Goldenberg R, Gantz I, Andryuk PJ, O'Neill EA, Kaufman KD, Lai E, Wang YN, Suryawanshi S, Engel SS (2017) Randomized clinical trial comparing the efficacy and safety of treatment with the once-weekly dipeptidyl peptidase-4 (DPP-4) inhibitor omarigliptin or the once-daily DPP-4 inhibitor sitagliptin in patients with type 2 diabetes inadequately controlled on metformin monotherapy. Diabetes Obes Metab 19:394–400

Gopalan B, Ravi D, Rasheed M, Sreedhara SKH, Ishtiyaque A (2007) Novel dipeptidyl peptidase IV inhibitors and processes for their preparation and pharmaceutical compositions containing them. Matrix Laboratories Ltd., Secunderabad, India, PCT Pat. Appl. WO2007/113634 A1, 11 Oct 2007

Gupta S, Walunj SS, Tokala RK, Parsa KVL, Singhand SK, Pal M (2009) Emerging drug candidates of dipeptidyl peptidase iv (DPP IV) Inhibitor class for the treatment of type 2 diabetes. Curr Drug Targets 10:71–87

Gupta BP, Sharma I, Kohli N, Sharma S, Rathi A, Sharma AK (2017) Preliminary clinical assessment and non-toxicity evaluation of an Ayurvedic formulation BGR-34 in NIDDM. Euro J Pharm Med Res 4:302–311

Gupta BP, Sharma I, Kohli N, Sharma S, Rathi A, Sharma AK (2018) Preliminary clinical assessment and non- toxicity evaluation of an ayurvedic formulation BGR-34 in NIDDM. J Tradit Complement Med 8:506–514. https://doi.org/10.1016/j.jtcme.2017.11.00

Gupta RC, Chhipa L, Mandhare AB, Nadkarni SS, Joshi D, Zambad S, Pathak P, Chauthaiwale V, Dutt C (n.d.) Design and synthesis of novel thiazolidine and pyrrolidine derivatives as DPP-IV inhibitors. https://www.technologynetworks.com/drug-discovery/posters/design-and-synthesis-of-novel-thiazolidine-and-pyrrolidine-derivatives-as-dppiv-inhibitors-230478. Accessed 24 Jan 2019

Henry RR, Rosenstock J, Logan D, Alessi T, Luskey K, Baron MA (2014) Continuous subcutaneous delivery of exenatide via ITCA 650 leads to sustained glycemic control and weight loss for 48 weeks in metformin-treated subjects with type 2 diabetes. J Diabetes Complicat 28:393–398

Huan Y, Jiang Q, Li G, Bai G, Zhou T, Liu S, Li C, Liu Q, Sun S, Yang M, Guo N, Wang X, Wang S, Liu Y, Wang G, Huang H, Shen Z (2017) The dual DPP4 inhibitor and GPR119 agonist HBK001 regulates glycemic control and beta cell function ex and in vivo. Sci Rep 7:4351. https://doi.org/10.1038/s41598-017-04633-5

Inagaki N, Onouchi H, Maezawa H, Kuroda S, Kaku K (2015) Once-weekly trelagliptin versus daily alogliptin in Japanese patients with type 2 diabetes: a randomised, double-blind, phase 3, non-inferiority study. Lancet Diabetes Endocrinol 3:191–197

Inzucchi SE, Bergenstal RM, Buse JB, Diamant M, Ferrannini E, Nauck M, Peters AL, Tsapas A, Wender R, Matthews DR, American Diabetes Association (ADA), European Association for the Study of Diabetes (EASD) (2012) Management of hyperglycemia in type 2 diabetes: a patient-centered approach: position statement of the American Diabetes Association (ADA) and the European Association for the Study of Diabetes (EASD). Diabetes Care 35:1364–1379. https://doi.org/10.2337/dc12-0413

Jain R, Trehan S, Das J, Singh N, Nanda GK, Magadi SK, Sharma SK (2009) Novel heterocyclic compounds. Panacea Biotec Ltd., New Delhi, India, PCT Pat. Appl. WO2009/093269 A1, 30 Jul 2009

Jain MR, Joharapurkar AA, Bahekar RH, Patel H, Jadav P, Kshirsagar SG, Patel VJ, Patel KN, Ramanathan VK, Patel PR, Desai RC (2015) Pharmacological characterization of ZYDPLA1, a novel long-acting dipeptidyl peptidase-4 inhibitor. J Diabetes 7:708–717. https://doi.org/10.1111/1753-0407.12233

Jani RH, Pai V, Jha P, Jariwala G, Mukhopadhyay S, Bhansali A, Joshi S (2014) A multicenter, prospective, randomized, double-blind study to evaluate the safety and efficacy of Saroglitazar 2 and 4 mg compared with placebo in type 2 diabetes mellitus patients having hypertriglyceridemia not controlled with atorvastatin therapy (PRESS VI). Diabetes Technol Ther 16:63–71

Kahn SE, Cooper ME, Del Prato S (2014) Pathophysiology and treatment of type 2 diabetes: perspectives on the past, present, and future. Lancet 383:1068–1083

Khan SE, Haffner SM, Heise MA, Herman WH, Holman RR, Jones NP, Kravitz BG, Lachin JM, O'Neill MC, Zinman B, Viberti G, ADOPT Study Group (2006) Glycemic durability of rosiglitazone, metformin, or glyburide monotherapy. N Engl J Med 355:2427–2443

Khanna I, Pillarisetti S (2014) Methods and compositions for treatment of diabetes and dyslipidemia. Kareus Therapeutics S.A., La Chaux-de-Fonds, Switzerland, US Pat. US8.623.897 B2, 7 Jan 2014

Kharul R, Jain MR, Patel PR (2011) Substituted benzamide derivatives as glucokinase (GK) activators. Cadila Healthcare Ltd., Ahmedabad, India, PCT Pat. Appl. WO2011/013141 A2, 3 Feb 2011

Kodimuthali A, Prasunamba PL, Pal M (2010) Synthesis of a novel analogue of DPP-4 inhibitor Alogliptin: introduction of a spirocyclic moiety on the piperidine ring. Beilstein J Org Chem 6:71. https://doi.org/10.3762/bjoc.6.71

Kushwaha RN, Haq W, Katti SB (2014) Sixteen-years of clinically relevant dipeptidyl peptidase-IV (DPP-IV) inhibitors for treatment of type-2 diabetes: a perspective. Curr Med Chem 21:4013–4045

Lampueta P, Zambrowicz B, Strumph P, Sands A (2015) Development of sotagliflozin, a dual sodium-dependent glucose transporter ½ inhibitor. Diabetes Vas Dis Res 12:101–110

Lau J, Bloch P, Schäffer L, Pettersson I, Spetzler J, Kofoed J, Madsen K, Knudsen LB, McGuire J, Steensgaard DB, Strauss HM, Gram DX, Knudsen SM, Nielsen FS, Thygesen P, Reedtz-Runge S, Kruse T (2015) Discovery of the once-weekly glucagon-like peptide-1 (GLP-1) analogue semaglutide. J Med Chem 58:7370–7380

Linjawi S, Bode BW, Chaykin LB, Courrèges JP, Handelsman Y, Lehmann LM, Mishra A, Simpson RW (2017) The efficacy of IDegLira (insulin degludec/liraglutide combination) in adults with type 2 diabetes inadequately controlled with a GLP-1 receptor agonist and oral therapy: DUAL III randomized clinical trial. Diabetes Ther 8:101–114

Lohray VB, Lohray BB, Paraselli RB, Ramanujam R, Chakrabarti R (1999) New heterocyclic compounds and their use in medicine, process for their preparation and pharmaceutical compositions containing them. Dr. Reddy's Research Foundation, Hyderabad, India, PCT Pat. Appl. WO1999/008501 A2

Lupin Ltd (2001–2016) Annual reports. http://www.lupin.com/archives-annual-report.php

Maji D, Samanta S (2015) A review on the role of PPARÎ³ agonist and hybrids in type 2 diabetes and cardiomyopathy. Asian J Pharm Clin Res 8:26–31. https://innovareacademics.in/journals/index.php/ajpcr/article/view/3993

Mancini AD, Poitout V (2015) GPR40 agonists for the treatment of type 2 diabetes: life after 'TAKing' a hit. Diabetes Obes Metab 17:622–629

Matschinsky FM (2009) Assessing the potential of glucokinase activators in diabetes therapy. Nat Rev Drug Discov 8:399–416

Matschinsky FM (2013) GKAs for diabetes therapy: why no clinically useful drug after two decades of trying? Trends Pharmacol Sci 34:90–99

Matschinsky FM, Magnuson MA, Zelent D, Jetton TL, Doliba N, Han Y, Taub R, Grimsby J (2006) The network of glucokinase-expressing cells in glucose homeostasis and the potential of glucokinase activators for diabetes therapy. Diabetes 55:1–12

Meininger GE, Scott R, Alba M, Shentu Y, Luo E, Amin H, Davies MJ, Kaufman KD, Goldstein BJ (2011) Effects of MK-0941, a novel glucokinase activator, on glycemic control in insulin-treated patients with type 2 diabetes. Diabetes Care 34:2560–2566

Mookhtiar KA (2015) Liver-selective glucokinase activation for treating type 2 diabetes: translation from mouse to man. In: Ramanbhai Foundation 7th International Symposium on Current Trends in Pharmaceutical Sciences, Ahmedabad, India, 2015

Nag B, Nag A, Dey D, Agarwal SK (2004) Amino acid phenoxy ethers. US Patent Application No: 6794401, 21 Sep 2004

Nakamura A, Terauchi Y (2015) Present status of clinical deployment of glucokinase activators. J Diabetes Investig 6:124–132

Novo Nordisk A/S (2015) Efficacy and long-term safety of oral semaglutide versus sitagliptin in subjects with type 2 diabetes (PIONEER 3). https://clinicaltrials.gov/ct2/show/NCT02607865. Accessed 24 Jan 2019

Ohishi T, Yoshida S (2012) The therapeutic potential of GPR119 agonists for type 2 diabetes. Expert Opin Investig Drugs 21:321–328

Pal M (2009a) Recent advances in glucokinase activators for the treatment of type 2 diabetes. Drug Discov Today 14:784–792

Pal M (2009b) Medicinal chemistry approaches for glucokinase activation to treat type 2 diabetes. Curr Med Chem 16:3858–3874

Panacea Biotec (2003–2016) Annual reports. http://www.panacea-biotec.com/

Patel H, Joharapurkar AA, Bahekar R, Patel P, Kshirsagar SG, Modi N, Ghoghari A, Patel VJ, Jain MR, Srinivas NR, Patel PR, Desai RC (2017) Is the inhibition of dipeptidyl peptidase-4 (DDP-4) enzyme route dependent and/or driven by high peak concentration? Seeking answers with ZYDPLA1, a novel long acting DPP-4 inhibitor, in a rodent model. Drug Res (Stuttg) 67:223–227. https://doi.org/10.1055/s-0042-122776

Pingali H, Zaware P, Jain M (2010) Novel GPR119 agonists. Cadila Healthcare Ltd., Ahmedabad, India, PCT Pat. Appl. WO2010/146605 A1, 23 Dec 2010

Pirags V, Lebovitz H, Fouqueray P (2012) Imeglimin, a novel oral antidiabetic, exhibits a good efficacy and safety profile in type 2 diabetic patients. Diabetes Obes Metab 14:852–858

Piramal (1997–2016) Annual reports. http://www.piramal.com/investors/financial-reports#parentVertical-Tab2

Qaseem A, Humphrey L, Sweet D, Starkey M, Shekelle P, Clinical Guidelines Committee of the American College of Physicians (2012) Oral pharmacologic treatment of type 2 diabetes mellitus: a clinical practice guideline from the American College of Physicians. Ann Intern Med 156:218–231. https://doi.org/10.7326/0003-4819-156-3-201202070-00011

Ranga MG, Rao JMR, Gudla CS (2015) Substituted bicyclic heteroaryl compounds as RXR agonists. Connexios Life Sciences Pvt. Ltd., Bangalore, India, PCT Pat. Appl. WO2015/107549 A1, 23 Jul 2015

Rao JMR, Arumugam N, Ansari MM, Gudla C, Pachiyappan S, Ramalingam M, George J, Arul GF, Bommegowda YK, Angupillai SK, Kottamalai R, Jidugu P, Rao DS (2012) Agonists of GPR40. Connexios Life Sciences Pvt. Ltd., Bangalore, India, PCT Pat. Appl. WO2012/011125 A1, 26 Jan 2012

Rao JMR, Venkatesham U, George J, Fernand G, Doppalapudi SR, Madhavan GR, Arumugam N, Ansari M, Murugavel K, Pradeep J, Allavuddeen S, Vijayaramalingam K, Prasad HS, Raj AM, Gnanavel S, Kottamalai R, Babu NMPS, Kenchegowda BY (2013) Cyclic amide derivatives as inhibitors of 11b-hydroxysteroid dehydrogenase and uses thereof. Connexios Life Sciences Pvt. Ltd., Bangalore, India, PCT Pat. Appl. WO2013/128465 A1, 6 Sep 2013

Rao JMR, Ranga MG, Pachiyappan S (2014a) Olefin substituted oxindoles having AMPK activity. Boehringer Ingelheim International GmbH, Ingelheim am Rhein, Germany, PCT Pat. Appl. WO2014/202528 A1, 24 Dec 2014

Rao JMR, Ranga MG, Pachiyappan S (2014b) Spiro-substituted oxindole derivatives having AMPK activity. Boehringer Ingelheim International GmbH, Ingelheim am Rhein, Germany, PCT Pat. Appl. WO2014/202580 A1, 24 Dec 2014

Reddy KA (2015) An unfinished agenda: my life in the pharmaceutical industry. Penguin Books India Pvt. Ltd., Gurgaon. http://www.drreddys.com/our-story/our-heritage/an-unfinished-agenda/

Reddy DS, Saxena C, Komirishetty K (2015) Novel indazole compounds and a process for the preparation thereof. Council of Scientific & Industrial Research, New Delhi, India, Shantani Proteome Analytics Pvt. Ltd., Pune, India, PCT Pat. Appl. WO2015/015519 A1, 5 Feb 2015

Rosenstock J, Aronson R, Grunberger G, Hanefeld M, Piatti P, Serusclat P, Cheng X, Zhou T, Niemoeller E, Souhami E, Davies M, LixiLan-O Trial Investigators (2016) Benefits of LixiLan, a titratable fixed-ratio combination of insulin glargine plus Lixisenatide, versus insulin glargine and lixisenatide monocomponents in type 2 diabetes inadequately controlled on oral agents: the LixiLan-O randomized trial. Diabetes Care 39:2026–2035

Saad MF, Greco S, Osei K, Lewin AJ, Edwards C, Nunez M, Reinhardt RR (2004) Ragaglitazar improves glycemic control and lipid profile in type 2 diabetic subjects: a 12-week, double-blind, placebo-controlled dose-ranging study with an open pioglitazone arm. Diabetes Care 27:1324–1329. https://doi.org/10.2337/diacare.27.6.1324

Sakhardande R, Kulkarni V, Sachan N, Nimbalkar M (2010) N-biphenylacyl thiazolidine-2,4-dione derivatives, their synthesis and uses. Elder Pharmaceuticals Ltd., Mumbai, India, PCT Pat. Appl. WO2010/082212 A2, 22 Jul 2010

Sanofi (2018) FDA to review Zynquista™ (sotagliflozin) as potential treatment for type 1 diabetes. http://www.news.sanofi.us/2018-05-22-FDA-to-review-Zynquista-TM-sotagliflozin-as-potential-treatment-for-type-1-diabetes. Accessed 24 Jan 2019

Saxena C (2015a) A "first-in-class" pre-IND compound for management of type-2-diabetes. In: 2015 BIO International Convention, Philadelphia, 15–18 June 2015

Saxena C (2015b) Shantani corporate overview 2014–2015

Singh SK, Manne N, Pal M (2008) Synthesis of (S)-1-(2-chloroacetyl)pyrrolidine-2-carbonitrile: a key intermediate for dipeptidyl peptidase IV inhibitors. Beilstein J Org Chem 4:20. https://doi.org/10.3762/bjoc.4.20

Sridharan R (1998) Can the chemistry change at Dr. Reddy's? Business Today – corporate front: strategy. http://archives.digitaltoday.in/businesstoday/22051998/cf2.html. Accessed 24 Jan 2019

Swamy D (2012) Glenmark: will innovation pay?. http://forum.valuepickr.com/t/glenmark-will-innovation-pay/286. Accessed 23 Jan 2019

Tahrani AA, Bailey CJ, Del Prato S, Barnett AH (2011) Management of type 2 diabetes: new and future developments in treatment. Lancet 378:182–197

The Pharmatimes (2012) Advinus' GK-activator achieves early POC for diabetes. http://thepharmatimes.in/advinus-gk-activator-achieves-early-poc-for-diabetes/. Accessed 24 Jan 2019

The Times of India (2013) Zydus Group launches new diabetic drug. https://timesofindia.indiatimes.com/business/india-business/Zydus-Group-launches-new-diabetic-drug/articleshow/20459662.cms. Accessed 23 Jan 2019

Thomas A, Gopalan B, Lingam PRVS, Shah DM (2006) Novel dipeptidyl peptidase IV inhibitors, pharmaceutical compositions containing them, and process for their preparation. Glenmark Pharmaceuticals S.A., Neuchatel, Switzerland, PCT Pat. Appl. WO2006/040625 A1, 20 Apr 2006

Times of India (2016) CSIR launches Ayurveda based anti diabetic drug for Rs 5 per tablet. http://timesofindia.indiatimes.com/articleshow/52503738.cms?utm_source=contentofinterest&utm_medium=text&utm_campaign=cppst

Valentine V, Goldman J, Shubrook JH (2017) Rationale for, initiation and titration of the basal insulin/GLP-1RA fixed-ratio combination products, IDegLira and IGlarLixi, for the management of type 2 diabetes. Diabetes Ther 8:739–752

Vuylsteke V, Chastain LM, Maggu GA, Brown C (2015) Imeglimin: a potential new multi-target drug for type 2 diabetes. Drugs R&D 15:227–232. https://doi.org/10.1007/s40268-015-0099-3

Wysham CH, MacConell L, Hardy E (2016) Efficacy and safety of multiple doses of exenatide once-monthly suspension in patients with type 2 diabetes: a phase II randomized clinical trial. Diabetes Care 39:1768–1776

Zhang F, Dey D, Bränström R, Forsberg L, Lu M, Zhang Q, Sjöholm Å (2009) BLX-1002, a novel thiazolidinedione with no PPAR affinity, stimulates AMP-activated protein kinase activity, raises cytosolic $Ca^{2+}$, and enhances glucose-stimulated insulin secretion in a PI3K-dependent manner. Am J Phys Cell Phys 296:C346–C354

Zhu D, Gan S, Liu Y, Ma J, Dong X, Song W, Zeng J, Wang G, Zhao W, Zhang Q, Li Y, Fang H, Lv X, Shi Y, Tian H, Ji L, Gao X, Zhang L, Bao Y, Lei M, Li L, Zeng L, Li X, Hou X, Zhao Y, Hu T, Ge X, Zhao G, Li Y, Zhang Y, Chen L (2018a) Dorzagliatin monotherapy in Chinese patients with type 2 diabetes: a dose-ranging, randomised, double-blind, placebo-controlled, phase 2 study. Lancet Diabetes Endocrinol 6:627–636

Zhu XX, Zhu DL, Li XY, Li YL, Jin XW, Hu TX, Zhao Y, Li YG, Zhao GY, Ren S, Zhang Y, Ding YH, Chen L (2018b) Dorzagliatin (HMS5552), a novel dual-acting glucokinase activator, improves glycaemic control and pancreatic β-cell function in patients with type 2 diabetes: a 28-day treatment study using biomarker-guided patient selection. Diabetes Obes Metab 20:2113–2120. https://doi.org/10.1111/dom.13338

Zydus Cadila (2011) Zydus' IND application of ZYGK1 for treating diabetes receives USFDA approval (Press Note). http://www.zydususa.com/Press%20Releases/2011/April%207%20-%20IND%20Application%20of%20ZYGK1%20Receives%20USFDA%20Approval.pdf

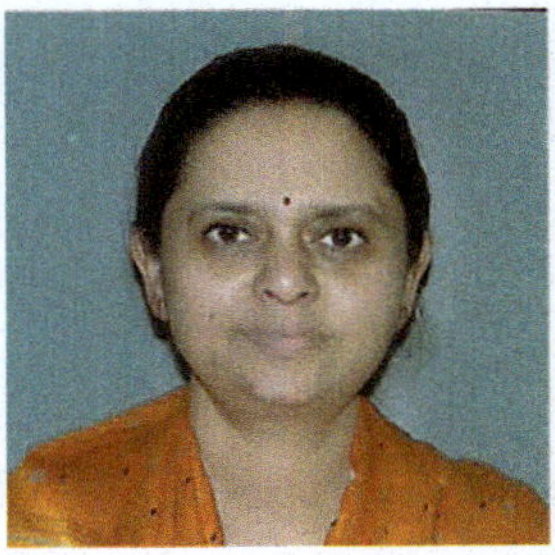

**Sarbani Pal** (nee Das) is a faculty and Head of the Chemistry Department, M.N.R. Degree and Post-Graduate College, Hyderabad, India. She received her MSc degree in Chemistry from University of Calcutta (India) in 1990 and PhD degree from IACS/Jadavpur University, Kolkata (India) in 1996. She then moved to the Department of Chemistry, M.S. University, Baroda, Gujarat (India), where she worked as a Lecturer in Chemistry (1996–1997). Subsequently, she moved to Hyderabad (India), where she joined the M.N.R. College (affiliated to Osmania University, Hyderabad) and continuing since then. Her research interests include the total synthesis of bicyclic and tricyclic sesquiterpenes, synthesis of organo-ruthenium complexes, design and synthesis of anti-inflammatory agents and macrolide antibiotics. She has supervised several PhD students and published more than 65 research papers and review articles in various international journals. She was awarded Bioorganic and Medicinal Chemistry Letters Top 10 Cited Author for 2011–2012.

**Manojit Pal** (MSc, IIT, Kharagpur and PhD, IACS/Jadavpur University, Kolkata, India) worked in industrial R&Ds, e.g., Alembic, Sun Pharma, Dr. Reddy's, and Mylan-Matrix. He then joined with Dr. Reddy's Institute of Life Science, Hyderabad and presently continuing as a Professor-Med. Chem. He received CSIR Certificate of Merit 1989, BMCL Most Cited Paper 2003–2006 Award, BMCL Top 10 Cited Author certificate 2011–2012, Publons Peer Review Awards 2017, 2018 and 2019, Bentham Certificate of Recognition 2019, and others. He was one of the most cited organic chemists in India (Current Science 2015, 109, 869) and featured in Stanford's list of 2% scientists in the world 2020. He became Associate Editor-RSC Advances in 2015, FRSC in 2016, Adjunct Faculty-Manipal University in 2018, Editorial Board member-Bioorganic Chemistry in 2019, and invited member of ACS in 2019. His research interests include pursuing Med Chem in areas like tuberculosis, inflammation, obesity, psoriasis, and cancer. He has authored/co-authored ~250 papers, 17 review articles, patents, and a book (Elsevier).

# 9 Indian Academia Tryst with Cardio-metabolic Drug Discovery and Development

Manoj K. Barthwal

## 9.1 Introduction

A 2017 WHO report on non-communicable diseases for India indicates highest percentage of deaths (around 30%) due to cardiovascular and metabolic diseases. Therefore, in spite of available medications there is a huge unmet need in this area.

Metabolic and cardiovascular drug discovery and development is an expensive and long drawn affair where failures far exceed the triumphs. Since the drugs approved in this area are often taken for life, the regulatory studies associated with such type of drugs are extensive and stringent, thus increasing the overall pre-clinical and clinical research program budget which runs into billions of dollars. However, the fruits of a single blockbuster drug can easily overcome all the failures and make unprecedented changes to the lifestyle of an individual. This hope is presently driving the remnant drug discovery efforts in the sub-continent. Indian academia has contributed in its own way to the cardio-metabolic drug discovery and development program, with major efforts from few institutes like Council of Scientific and Industrial Research-Central Drug Research Institute (CSIR-CDRI), Central Institute of Medicinal and Aromatic Plants (CIMAP), National Botanical Research Institute (NBRI) Lucknow, Indian Institute of Integrative Medicine (IIIM), Jammu and Indian Institute of Microbial Technology (IIMT), Chandigarh. From the ongoing and previous efforts, it looks like that the Indian institutes and universities have some appetite for the cardio-metabolic drug discovery and development. However, whether this is sufficient for the launch of a successful blockbuster in a time bound manner can always be debated. Majority of the products developed by the academia and research institutes are either herbals or derived from the natural

M. K. Barthwal (✉)
Pharmacology Division, CSIR-Central Drug Research Institute, Lucknow, UP, India

Academy of Scientific and Innovative Research (AcSIR), Ghaziabad, UP, India
e-mail: manojbarthwal@cdri.res.in



M. Dikshit (ed.), *Drug Discovery and Drug Development*,
https://doi.org/10.1007/978-981-15-8002-4_9

sources. The launch of such products in the Ayush mode is often faster when compared to small molecules due to less regulatory work required for their launch often coinciding with the limited resources available. This also leads to the overall cost reduction and makes the product affordable. Also there is often traditional knowledge available around these products and addition of scientific proof and package makes them attractive for the end users. However, the cost and effort that goes in the discovery and development of the small molecules which often takes the drug controller general of India (DCGI) route for their launch, is quite huge. Therefore, as mentioned above only few premiere institutes have ventured into this arena. Subsequently we discuss here the available products and those in the advanced pipeline for the treatment of cardio-metabolic disorders. Few products where preliminary information is available is also discussed.

### 9.1.1 Marketed Products

#### 9.1.1.1 AYUSH-82/IME-9

Central Council for Research in Ayurvedic Sciences (CCRAS) developed this anti-diabetic ayurvedic formulation in the name of AYUSH-82 and later sold the distribution and manufacturing rights to Kudos laboratories, who later repackaged it as IME9 and commercialized with the help of Council of Scientific & Industrial Research-National Research Development Corporation (CSIR-NRDC).

This was the most common tested herbal formulation. In around 22 studies carried out, diet was used as an intervention or co-intervention in 15 (Shekelle et al. 2005). Fasting and postprandial glucose was monitored in these studies from 6 to 12 weeks.

AYUSH-82 is synergistic mixture of five essential herbs, viz. Karela (Momordica charantia), Jamun (Syzygium cumini), Amra (Mangifera indica), Gudmar (Gymnema sylvestre), and Shilajeet (Asphaltum). Mechanistic studies suggest that AYUSH-82 induced beta cell activation and insulin production which may explain the anti-diabetic effect of this formulation. AYUSH-82 induced decrease in peripheral insulin resistance and slow absorption of glucose from the intestines also explains the anti-diabetic effect (Shekelle et al. 2005; Kumar et al. 1999; Chowdhary et al. 1998; Pandey et al. 1995).

#### 9.1.1.2 BGR-34

This drug has been collectively developed by Council of Scientific & Industrial Research (CSIR)-National Botanical Research Institute (NBRI) and CSIR-Central Institute for Medicinal and Aromatic Plant (CIMAP).

BGR-34 is the combination of Daruharidar (Berberis aristata, Stem), Vijaysar (Pterocarpus marsupium, Heartwood), Gudmar (Gymnema sylvestre, Leaf), Manjeeth (Rubia cordifolia, Root), Methika (Trigonella foenum graceum, Seed), and Giloy (Tinospora cordifolia, Stem) which are previously reported in literature to exert an anti-diabetic effect and their combination was decided on the basis of their traditional use (Chakrabarti et al. 2011; Sangeetha et al. 2011; Ahmad et al. 1989;

Aralelimath and Bhise 2012; Rani et al. 2013; Patel et al. 2012). Aimil Pharmaceuticals (India) Ltd is licensee for the production and worldwide marketing of BGR-34. The pre-clinical studies carried out in diabetic rats showed beneficial effects of BGR-34 and no adverse effect was observed on liver, heart, and kidney (unpublished data CSIR).

BGR-34 is an option for adults with type 2 diabetes who are naïve or on common anti-diabetic medications like metformin, sulfonylurea, or glitazones. BGR-34 is an alternative and add-on option to consider for managing type 2 diabetes. A clinical trial was planned and registered in the clinical trial registry-India. In a study conducted at the Aggarwal hospital, New Delhi, India, 64 NIDDM outpatients were recruited after carefully considering the inclusion and exclusion criteria. Preliminary assessment of clinical and toxicity parameters indicated that treatment with BGR-34 significantly improved fasting plasma glucose, postprandial blood glucose and HbA1c when compared with placebo (Triphala, comparator agent) (Gupta et al. 2018). Glycemic parameters measured after BGR-34 treatment in diabetic patients showed significant improvement with a 34.3% decrease in fasting blood sugar, 35.5% decrease in postprandial blood sugar, and improvement in glycosylated hemoglobin levels by 20.31% as compared to placebo group which showed reduction by 13.2%, 10.9%, and 10.87%, respectively (Gupta et al. 2018).

#### 9.1.1.3 D5 Chooranam

This drug is developed by the Central Council for Research in Siddha (CCRS). D5 chooranam is one of the many Siddha medicine formulations that is indicated for diabetes. In an e Book on Siddha, Clinical Research by central council for Research in siddha ministry of ayush government of India, it is indicated that a multicentric clinical trial conducted with the coded polyherbal formulation D5 chooranam found beneficial results in the drug naïve diabetic population [CTRI/2013/12/004231] [Registered on: 20/12/2013] (Central Council for Research in Siddha n.d.). CCRS is conducting evidence-based research for validating the efficacy of such formulations. The peripheral institutes involved in the CCRS multicentric open clinical trial on diabetes (Neerizhivu) with D5 chooranam were SCRI, SRRI Puducherry, and SCRU Palayamkottai. The preliminary analysis shows promising anti-diabetic effect of D5 chooranam since it significantly reduced blood sugar. The trial indicates promising results and suggests that D5 chooranam can be an effective anti-diabetic agent (IEC Approval No: IEC No: CCRS/SCRI-1/2011-12/04CTRI Registration No/Ref no: 2013/08/005563).

In the patent filed for D5 chooranam, claim has been made for the development of an economical, effective, and safe process for the preparation of a synergistic compound D5 for the management of Type II Diabetes (Ramaswamy et al. 2018). In this invention, the herbal constituents, namely Cassia auriculata, Cassia fistula, Syzygium cumini, Salada oblonga, Cyperus rotundus, Costus speciosus, and Cinnamomum zeylanicum are isolated and purified by a novel and distinct process. Subsequently drying is done in the sunshade and the material is pulverized into micro-fine powder which can be converted as per requirement into a synergistic compound D5 capsule, granules, powder, caplet, tablet, or in any other nutraceutical form for oral consumption during the management of Type II Diabetes and related disorders. According to this process, the yielded D5 formulation demonstrates high efficacy in managing hyperglycemia and dyslipidemia in diabetes patients and is found to be hepato-protective and generally safe.

#### 9.1.1.4 Picroliv

Picroliv, an extract from the roots and rhizomes of Picrorhiza kurroa, is a herbal hepato-protective agent developed by CSIR-Central Drug Research Institute (CDRI 2017a). A Phase 1 clinical trial was carried out in healthy individuals, where candidate drug was given orally. Initially two groups were made where volunteers received either placebo or a single dosage of 300 mg of picroliv. Subsequently, in a 4-week multiple dose study 100 mg (low dose) and 200 mg (high dose) of picroliv was administered orally. The drug was well tolerated in 10 volunteers receiving the low dose and no side effect was observed. However, out of the 10 volunteers who received the higher dose, one incidence of heavy abdomen and another of mild constipation was reported. Irrespective of the above, the drug in both cases continued for 4 weeks. In both single and multiple dose studies, no marked change was observed in the vitals, hematological and biochemical parameters including electrocardiogram. Subsequently, in patients with acute severe viral hepatitis, a bi-phasic phase II clinical trial was conducted. In the first phase open trials were conducted at Seth GSMC, Mumbai and KGMC, Lucknow. Out of total 35 patients that completed the acute severe viral hepatitis 12-week trial, 21 were recruited at the center at Mumbai and 14 at the center at Lucknow. Progressive improvements in the biochemical and clinical parameters were observed in all the patients. As a consequence of this study, a preliminary safety and efficacy profile of picroliv was established in select groups of severe viral hepatitis patients. Therefore, from the trial studies it was concluded that when 100 mg of picroliv was administered twice daily and orally for 12 weeks, it was well tolerated and found safe with some therapeutic benefit was seen in acute viral hepatitis patients.

Subsequently, a double-blind, placebo control multicentric clinical trial of picroliv involving 102 patients of severe acute viral hepatitis was conducted at Seth GSMC, Mumbai, IMS, B.H.U. Varanasi and S.M.S. Medical college, Jaipur. Out of 102 patients that completed the trial, 55 patients were in picroliv group while 47 formed the placebo. Patients administered with picroliv showed much more rapid clinical improvement as assessed by symptomatology and liver function test scores when compared to placebo-treated patients. The test medicines were well tolerated and no significant adverse effect was observed in the study (Mehrotra et al. 1990; Dwivedi et al. 1991, 1990). Recently, CSIR-CDRI has undertaken experiments to evaluate the hepato-protective effect of Picrorhiza kurroa derived fractions and picroliv on metabolic liver diseases like Non-alcoholic Fatty Liver Disease (NAFLD) and nonalcoholic steatohepatitis (NASH).

#### 9.1.1.5 PMZ-2010 (Centhaquin)

PMZ-2010 is being developed and repositioned as a novel resuscitative agent for hypovolemic shock.

PMZ-2010 (Centhaquin) was originally discovered at CDRI, Lucknow (CDRI-7173) as an anti-hypertensive agent. However, drawbacks like lack of blood pressure lowering activity for longer times reduced the interest of the researchers and this drug was discontinued. However, in later studies it was discovered that it had sustained lactate lowering effect which led to the identification of its resuscitative activity. In the animal models of hemorrhagic shock, PMZ-2010 was found to be a highly effective resuscitative agent and is currently in clinical development (CTRI/2017/03/008184) (Gulati et al. 2018). After an initial analysis of 23 subjects, 19 subjects meeting the inclusion and exclusion criteria were included in the study. All participants received standard treatment for shock. Patients participating in the study were randomly divided into either control cohort ($N = 7$) that received standard treatment along with normal saline or PMZ-2010 cohort ($N = 12$) that received standard treatment along with PMZ-2010. The results indicate that PMZ-2010 (Centhaquin) is an effective resuscitative agent and may improve the outcome of patients of hypovolemic shock (Hegde et al. 1997).

#### 9.1.1.6 Gugulipid

This is a hypolipidemic agent developed by CSIR-CDRI (Nityanand and Kapoor 1973). Taking leads from the ancient ayurvedic medicine system of India, gugulipid was developed from the plant Commiphora mukul. From the ancient times ayurvedic practitioners have been prescribing "gugglu" or the tree-derived gum for the treatment of lipid disorders, rheumatism, arthritis, and gout. Guggulsterones was identified as an active constituent of the non-toxic fraction showing hypolipidemic activity. Since the fraction was equally potent as the individual components, it was subsequently developed as a phytopharmaceutical preparation called gugulipid. Gugulipid successfully passed all the three phases of clinical trial and its effect was comparable to the drug clofibrate. During the clinical trials, gugulipid was found superior to the other drugs due to the absence of obvious side effects. Out of 330 patients recruited in the clinical trial, 80% responded to the treatment of gugulipid. There was a significant lowering of 24% and 22% in the total cholesterol and triglycerides, respectively after gugulipid treatment (Agarwal et al. 1986; Gopal et al. 1986; Nityanand et al. 1989). Gugulipid was approved for marketing in 1986 by the Drug Controller General of India after the successful completion of three phases of clinical trials. Gugulipid was manufactured and marketed by Cipla Ltd, Mumbai under the brand name Guglip.

### 9.1.2 IND Stage and Pre-launch Molecules

#### 9.1.2.1 S007-867

S007-867 is a safe, small molecule anti-thrombotic drug being developed at CDRI. All pre-clinical and IND enabling studies are completed and IND is filed (CDRI 2017b). According to a recent WHO report on the status of non-communicable diseases (NCD) in India, cardiovascular disease accounts for the largest number of deaths (26% of all NCD deaths). Damage to the vessel wall exposes the vascular collagen leading to platelet adhesion and activation. This can lead to series of events culminating into myocardial infarction or stroke. CDRI has developed a small molecule anti-platelet molecule that has profound anti-thrombotic effect and inhibits only collagen-induced platelet activation. Acute coronary syndrome patients taking dual anti-platelet therapy may benefit from such kind of molecule due to inhibition of thrombus formation at the injury site. Salient features of the technology are that S007-867 prevents only collagen-induced platelet activation and therefore is a first in class small molecule inhibitor. Its efficacy is better than aspirin and shows reduced bleeding risk in the animal models. Therefore, given the present pre-clinical profile of S007-867 it can be said that this molecule may have a better profile in clinic when compared to other standard drug like aspirin or clopidogrel. One more advantage with the molecule is its route of administration which is oral. It reversibly inhibits the collagen-induced platelet aggregation and mechanism of action suggests that early thrombus formation during vessel damage can be prevented by this molecule. It has decent oral bioavailability as assessed in rat and rabbit. It is easy to synthesize and scale-up. The molecule has minimal drug interaction/CYP inhibition profile. The molecule can be positioned for conditions like myocardial infarction and stroke. To the best of our knowledge only one collagen specific molecule, Revacept (PR-15) is in development mode. This is a GPVI bivalent soluble form of humanized Fc fusion of GPVI ectodomain. Phase I clinical trial is already completed. Recruitment for Phase II clinical trial is in progress. The molecule will be administered through I.V route for the treatment of ischemic attacks and associated pathologies like stroke (Misra et al. 2018; Patelis et al. 2018; Chandasana et al. 2015, 2016). S007-867 is open for licensing.

#### 9.1.2.2 Curcuma Extract (HM)

CDRI, Lucknow developed a standardized curcuma extract (HM) for treating neuro-cerebrovascular disorders (CDRI 2017c). It is a well-characterized hexane soluble fraction of curcuma longa which shows beneficial effects in animal models of stroke, myocardial ischemia reperfusion injury, metabolic dyslipidemia, and atherosclerotic cardiovascular disease (Dohare et al. 2008a, b).

Interruption of blood supply to the brain leads to neuro-cerebrovascular complications like cerebrovascular infarction, stroke, and ischemic attacks. Temporary interruptions of the blood supply to the brain lead to transient ischemic attacks (TIAs) and an ischemic stroke calls for an immediate attention for preventing any mortality. Usually surgical or medical interventions are done in case of TIA or stroke. To restore blood supply after a blockage in the brain, a surgical intervention may be required. Anti-thrombotics may help in preventing the formation of clots and occlusion of the blood vessel. No efficient remedy is available for the brain swelling which commonly accompanies brain infarction or hemorrhage. The use of anti-platelet drug ticlopidine for managing stroke is not encouraged due to long-term adverse side effects associated with this drug. Tissue plasminogen activator (t-PA) can be effective in removing clots from the blocked arteries. Though t-PA which is naturally occurring in the body can restore the blood supply in the blocked vessel by dissolving the clot, many other aspects of stroke need to be taken care of if permanent damage has to be prevented. Most strokes lead to an infarction due to death in the core area of the affected brain regions and the reduction in the blood flow is so huge that the damage is beyond repair and the cells usually cannot recover (CDRI 2017b).

The CDRI standardized fraction is prepared from the Zingiberaceae family Curcuma longa L rhizomes. Extraction is done in a lipid affinity solvent. The extract is effective in the rodent stroke model of middle cerebral artery occlusion. This extract is standardized and besides others has fixed amount of ar-Turmerone along with zedoarondiol, isozedoarondiol, turmerone, zingiberine, curcumerone, curcumene, curcumenone, zedoarone, and curlone. Most of the studies for IND application are done. Required toxicity and safety pharmacology studies with the fraction are complete and it is found quite safe. Since the fraction confers protection in animal models of cerebral stroke (Ray et al. 2002), it can be positioned for such clinical conditions. Some studies carried out with the hexane soluble fraction also indicate a hypolipidemic, anti-atherosclerotic and anti-inflammatory effect of this standardized fraction (Singh et al. 2015, 2013).

#### 9.1.2.3 CDR-267-F018

This salubrious mangrove plant (*xylocarpus moluccensis*) extract developed by CDRI, Lucknow has immense medicinal value for treating cardio-metabolic complications (CDRI 2017d). CDR-267-F018 is a standardized extract obtained from the fruits (renewable source) of mangrove plant (CDRI 2017b). The extract demonstrates anti-diabetic and anti-dyslipidemic activities. Most of the pre-clinical studies with the extract are over. The safety pharmacology and toxicity studies are also done. In an animal model of metabolic dyslipidemia, CDR-267-F018 improves vascular function and demonstrates triglyceride and cholesterol lowering effects comparable to fenofibrate. In an animal model of isoproterenol-induced cardiac hypertrophy in mice, CDR-267-F018 provides cardio-protection better than the standard comparator propranolol as assessed by echocardiography. The extract shows anti-inflammatory, anti-arthritic, and anti-atherosclerotic effects in the animal models. In an atherosclerosis ApoE knockout mouse model, the plant extract reversed dyslipidemia, atherosclerosis and improved plaque stability and was also compared with atorvastatin (Central Council for Research in Siddha n.d.; Kanshana et al. 2018).

#### 9.1.2.4 Novel Streptokinase

This anti-thrombotic drug was developed at CSIR - Institute of Microbial Technology, Chandigarh. The research team developed a novel process for the preparation of novel streptokinase proteins which are clot specific and possess altered plasminogen properties, including enhanced fibrin selectivity. The unique feature of these proteins is that the kinetics of plasminogen activation induced by these proteins are quite different from natural streptokinase. It was observed that there is a temporary lag or delay in the initial rate of reaction that catalytically converts plasminogen to plasmin. Hybrid proteins developed show high fibrin specificity and display “delayed-action” thrombolysis. Therefore, these new hybrid proteins can be of added advantage in thrombolytic therapy of various cardiovascular disorders (Sahni et al. 1998).

### 9.1.3 Molecules with Preliminary Studies

#### 9.1.3.1 Dialyzed Fenugreek Seed Extract

In this study, the researchers developed a novel method for preparing a unique dialyzed fenugreek seed extract which exerts anti-diabetic effect. The hypoglycemic effect induced by the extract was shown to be partially mediated by the insulin signalling pathway (PCT/IN2006/000217). The process of preparing the dialyzed aqueous extract of fenugreek seeds involved washing the fenugreek seeds in distilled water followed by sterilization and grinding. The ground powder is suspended in phosphate buffered saline (PBS), and subsequently the filtrate obtained after the filtration process was treated with charcoal to obtain a clear supernatant. The supernatant was lyophilized and subsequently dissolved in PBS. Finally dialysis was done to obtain the dialyzed aqueous extract of fenugreek seeds (FSE), which was aliquoted and stored till further use (Bhat and Vijayakumar 2006).

#### 9.1.3.2 Anti-diabetic Natural Product

In a patent filed by the council of scientific and industrial research, department of biotechnology, a novel process involving natural products from Perna viridis was developed for the cure and control of diabetes mellitus. Extract of this Indian green mussel can be used for the cure of diabetes mellitus since it possesses hypoglycemic activity and pancreatic regeneration properties. When tested on isolated islets of Langerhans from mouse pancreas, the natural product was found to be non-toxic and cyto-protective. Since the extract reversed experimental diabetes in rats, it can be speculated that it possesses islet neogenesis activity (Bhonde and Chatterji 2002).

#### 9.1.3.3 Anti-atherosclerotic Polyherbal Formulation

Researchers at the Banaras Hindu University, Department of Biotechnology developed a polyherbal preparation for the prevention of atherosclerosis and hyperlipidemia comprising a mixture of Commiphora mukul, Boswellia serrata, Semecarpus anacardium Strych nox nux vomica, Terminalia arjuna, and Shankha Bhasma. This polyherbal composition may further include Rubia cordifolia, Bacopa monnieri, Triphala, and Trikatu. The efficacy of the formulation was established in the animal models of hyperlipidemia and atherosclerosis. This polyherbal formulation displays anti-inflammatory and anti-oxidant effects. Mechanistically, it has been shown to inhibit LOX-15, Cox-2 and Ca-deposition in the plaques. The circulating HDL and plaque collagen contents are increased after the treatment with the polyherbal formulation. Significant reduction in serum TG is observed (Tripathi 2003).

## 9.2 Challenges in Developing Drugs from Natural Sources

There are several challenges while performing natural product drug discovery. While the collection of the plant material is usually done from its natural habitat, a rigorous characterizing exercise involving anatomical, morphological, genetic, and chemical characterization of the plant should be performed before proceeding further. This will help to reduce variability in plant material during repeat collections as continuous ongoing plant taxonomy modifications and synonymy cannot be avoided (Sanjay et al. 2007; Atanasov et al. 2015). Moreover, the collection of the material, its identification and preparation of the herbarium vouchers are usually a tedious and manual process, carried out by rare specialized persons and cannot be automated.

Once the pure compound or the natural product is isolated, its quantity is usually low and obtained after several tedious process of isolation. Though the availability of the product may not be a problem, its amount may be an issue when carrying out detailed pharmacological activities including animal studies where quantity of the compound required is quite high. Also the isolated natural products usually have unique features like presence of several chiral centers and oxygen-containing substituents and their laboratory synthesis may require several lengthy steps. Also compatibility of natural products and their source with the high throughput screening system also remains a challenge since several of the products may directly interfere with the assay (Sanjay et al. 2007; Atanasov et al. 2015; David et al. 2015). Besides this the clinical trials with natural resources may be quite expensive and sometime not possible without the involvement of the industry, the IPR issues with the isolated products may be a further dampener for industry participation.

Also when a successful natural product is identified there is a continuous pressure on its cultivation and harvesting which often becomes unsustainable leading to supply crisis. Also there may be local, legal, and IPR issues which may restrict the utilization of the plant material. After the issue of new guidelines on 4th of March 2014 by the United States Patent and Trademark Office ("Guidance for Determining Subject Matter Eligibility of Claims Reciting or Involving Laws of Nature, Natural Phenomena, and Natural Products"), patenting natural products will be more difficult. According to the new guidelines a patent claim should include "marked difference" from a known natural law, material, or phenomenon (Atanasov et al. 2015). There are several factors that may govern the availability of a natural product from a particular plant resource. Besides the species and the harvest time, soil quality, altitude, actual climate and processing, storage and isolation procedures may greatly vary the type and amount of natural product being isolated. These all factors may affect the type and amount of the natural product isolated often leading to its degradation or chemical modification which may give a totally different pharmacological profile. Besides these plants are also inhabited with endophytic organisms like bacteria and fungi. Therefore, it has to be made sure that the isolated natural product is not a metabolite of these organisms (Atanasov et al. 2015; David et al. 2015).

Developing drugs from plant sources is a huge challenge and recently CDSCO has also come up with phytopharmaceutical guidelines that will encourage, simplify, and channelize the drug discovery and development in this area. Mentioned in the gazette of India and ministry of health and family welfare (Department of Health and Family Welfare) notification, New Delhi, on the 19th March 2019 "phytopharmaceutical drug" means a drug of purified and standardized fraction, assessed qualitatively and quantitatively with defined minimum four bioactive or phytochemical compounds of an extract of a medicinal plant or its part, for internal or external use on human beings or animals, for diagnosis, treatment, mitigation, or prevention of any disease or disorder but does not include drug administered through parenteral route (Bhatt 2016).

The activity should be confirmed in plant material preferably grown in captive cultivation and batch and season variations need to be ascertained. Also plant should not be in endangered category and the source of the drug should be preferably from renewable parts like leaves or flowers.

Besides other things, there are several important information to be generated for this class of drugs. The plants source has to be identified and properly authenticated. Proper fractionation, activity guided sub-fractionation needs to be done. Purification of the marker compounds has to be carried out and proper manufacturing process for the formulation should be in place. In addition to this stability studies and in vitro and in vivo pharmacokinetic studies need to be performed. Also proper toxicity and safety pharmacology studies need to be carried out as per norms. The product should have been through proper human studies and clinical trials as per norms. Also there should be a clear information regarding regulatory status, marketing information, and post-marketing surveillance plan. The important points covered in these guidelines also address microbial and ash contamination in plant materials. Besides this the regular safety and toxicological studies need to be completed. Therefore, the amount of work to be carried out for developing a phytopharmaceutical is nowhere less than what is required for an NCE. Besides this during the initial stage proper plant collection and its authentication also remains a challenge. Establishment of high throughput bioassay and scale-up of bioactive lead compounds also remains an issue. Therefore, developing a natural product based drug or phytopharmaceutical may be more challenging than ever thought.

## 9.3 Future Perspective

World and India's cardio-metabolic burden is increasing with time and needs proper attention. Efforts from Indian academia are less though sincere wherever observed. Besides serious efforts from the premiere drug institutes like CDRI, Lucknow and IMTech, Chandigarh in developing single molecules, novel chemical entities, and phytopharmaceuticals, most of the minor research has been carried out by universities and the products dealt with are naturals or herbals. Government of India has taken the challenge of drug discovery very seriously and has launched several programs towards this which include but not limited to phytopharmaceutical mission, nutraceutical mission, and several others.

Recently department of biotechnology, India has come out with a program to fund drug discovery and development in India through specific programs which will also involve DBT and BIRAC. The aim of this program is to identify important hits and leads in the country in the area of cardio-metabolic and other disorders and fund the real exciting ones so that the drug discovery program gets an impetus.

Therefore, there is no doubt that there will be ample opportunities available to the Indian institutes and universities to tap the government fund for doing drug discovery and development. However, how much they are ready for this will only be known with time. The chances of getting a block buster will significantly increase if the stakeholders like Indian industry are involved during the inception of the project. Therefore, some commitment from the Industry is also required for bearing the fruits of the overall efforts. Besides this with changing times it is also important that the academia does lot of studies in its internationally accredited facilities so that there is global acceptance of the launched product. In this direction recently CDRI, Lucknow got the National Good Laboratory Practice (GLP) Compliance Monitoring Authority (NGCMA) approved GLP facility for pre-clinical studies involving toxicity and safety pharmacology. GLP is a system evolved by Organisation for Economic Co-operation and Development (OECD). IIM, Jammu and Indian Institute of Toxicology Research (IITR), are the other two accredited national facilities for carrying out toxicity studies and production of GMP grade materials, respectively. Also due to the emergence of new category of drugs like biologics, the academia has to build and blend with the changing landscape of drug discovery and development.

**Acknowledgement** Help in referencing and patent search from Mr. Sukka Santosh Reddy (SRF, ICMR) and Mr. Shivnath Banerjee (PA, MOES) is highly acknowledged. This is a CDRI communication no. 10139

## References

Agarwal RC et al (1986) Clinical trial of gugulipid--a new hypolipidemic agent of plant origin in primary hyperlipidemia. Indian J Med Res 84:626–634

Ahmad F et al (1989) Insulin like activity in (-) epicatechin. Acta Diabetol Lat 26(4):291–300

Aralelimath VR, Bhise SB (2012) Anti-diabetic effects of gymnema sylvester extract on streptozotocin induced diabetic rats and possible Î²-cell protective and regenerative evaluations. Dig J Nanomater Biostruct 7(1):135

Atanasov AG, Waltenberger B, Pferschy-Wenzig EM, Linder T, Wawrosch C, Uhrin P, Temml V, Wang L, Schwaiger S, Heiss EH, Rollinger JM, Schuster D, Breuss JM, Bochkov V, Mihovilovic MD, Kopp B, Bauer R, Dirsch VM, Stuppner H (2015) Discovery and resupply of pharmacologically active plant-derived natural products: a review. Biotechnol Adv 33 (8):1582–1614. https://doi.org/10.1016/j.biotechadv.2015.08.001. Review

Bhat MK, Vijayakumar MV (2006). https://patents.google.com/patent/WO2007138609A1/en?oq=WO+2007%2f138609+A1

Bhatt A (2016) Phytopharmaceuticals: a new drug class regulated in India. Perspect Clin Res 7 (2):59–61

Bhonde RR, Chatterji A (2002). https://patents.google.com/patent/US7074433B2/en?oq=US+7%2c074%2c433+B2

CDRI (2017a). https://www.cdri.res.in/pdf/Research/Compendium2017.pdf. Page 5

CDRI (2017b). https://www.cdri.res.in/pdf/Research/Compendium2017.pdf. Page 8
CDRI (2017c). https://www.cdri.res.in/pdf/Research/Compendium 2017.pdf. Page 11
CDRI (2017d). https://www.cdri.res.in/pdf/Research/Compendium2017.pdf. Page 13
Central Council for Research in Siddha (n.d.). http://siddhacouncil.com/downloads/ccrs-ebook.pdf/ . Page 19
Chakrabarti R et al (2011) Dipeptidyl peptidase-IV inhibitory activity of Berberis aristata. J Nat Prod 4:158–163
Chandasana H et al (2015) Pharmacokinetics and tissue distribution study of novel potent antiplatelet agent S007-867 in mice using HPLC-MS/MS. Xenobiotica 45(6):530–537
Chandasana H et al (2016) Assessment of the pharmacokinetics, tissue distribution and excretion studies of a novel antiplatelet agent S007–867, following administration to rats. Drug Test Anal 8(7):723–729
Chowdhary DP et al (1998) Hypoglycaemic effect of a coded formulation: Aysuh-82. J Res Ayurveda Siddha 19(3-4):107–115
David B, Wolfender J-L, Dias DA (2015) The pharmaceutical industry and natural products: historical status and new trends. Phytochem Rev 14:299–315
Dohare P et al (2008a) Neuroprotective efficacy and therapeutic window of curcuma oil: in rat embolic stroke model. BMC Complement Altern Med 8(1):55
Dohare P, Varma S, Ray M (2008b) Curcuma oil modulates the nitric oxide system response to cerebral ischemia/reperfusion injury. Nitric Oxide 19(1):1–11
Dwivedi Y et al (1990) Hepatoprotective activity of picroliv against carbon tetrachloride-induced liver damage in rats. Indian J Med Res 92:195–200
Dwivedi Y et al (1991) Picroliv affords protection against thioacetamide-induced hepatic damage in rats1. Planta Med 57(01):25–28
Gopal K et al (1986) Clinical trial of ethyl acetate extract of gum gugulu (gugulipid) in primary hyperlipidemia. J Assoc Physicians India 34(4):249–251
Gulati A et al (2018) Clinical results indicating effectiveness of PMZ-2010 (Centhaquin) as a novel resuscitative agent for hypovolemic shock. Circulation 138(Suppl_1):A13092
Gupta BP et al (2018) Preliminary clinical assessment and non-toxicity evaluation of an ayurvedic formulation BGR-34 in NIDDM. J Tradit Complement Med 8(4):506–514
Hegde LG et al (1997) Attenuation in rat brain nitric oxide synthase activity in the coarctation model of hypertension. Pharmacol Res 36(2):109–114
Kanshana JS et al (2018) Standardized fraction of Xylocarpus moluccensis fruits improve vascular relaxation and plaque stability in dyslipidemic models of atherosclerosis. J Ethnopharmacol 213:81–91
Kumar N, Kumar A, Sharma ML (1999) Clinical evaluation of single and herbo-mineral compound drugs in the management of madhumeha. J Res Ayurveda Siddha 20:1–9
Mehrotra R et al (1990) In vitro studies on the effect of certain natural products against hepatitis B virus. Indian J Med Res 92:133–138
Misra A et al (2018) Anti-thrombotic efficacy of S007-867: pre-clinical evaluation in experimental models of thrombosis in vivo and in vitro. Biochem Pharmacol 148:288–297
Nityanand S, Kapoor NK (1973) Cholesterol lowering activity of the various fractions of the guggal. Indian J Exp Biol 11(5):395–396
Nityanand S, Srivastava JS, Asthana OP (1989) Clinical trials with gugulipid. A new hypolipidaemic agent. J Assoc Physicians India 37(5):323–328
Pandey VN, Rajagopalan SS, Chowdhary DP (1995) An effective Ayurvedic hypoglycemic formulation. J Res Ayurveda Siddha 16(1-2):1–14
Patel DK et al (2012) An overview on antidiabetic medicinal plants having insulin mimetic property. Asian Pac J Trop Biomed 2(4):320–330
Patelis N et al (2018) An update on novel antiplatelets in vascular patients. Curr Pharm Des 24 (38):4558–4563
Ramaswamy RS et al (2018) Process for the preparation of synergistic compound D5 for the management of diabetes mellitus and related disorders. https://patents.google.com/patent/

IN2015CH02578A/en?q=process-for-the-preparation-of-synergistic-compound-d5-for-the-management-of-diabetes-mellitus-and-related-disorders

Rani S et al (2013) Research article Antiglycation, antioxidant and antidiabetic activity of traditional medicinal plant: Rubia cordifolia linn. For management of hyperglycemia. Int J Plant Anim Environ Sci 3:42

Ray M et al (2002). https://patents.google.com/patent/US6991814B2/en?oq=US+6%2c991%2c814+B2

Sahni G et al (1998). https://patents.google.com/patent/US20030059921A1/en?oq=US+2003%2f0059921A1

Sangeetha MK et al (2011) Tinospora cordifolia attenuates oxidative stress and distorted carbohydrate metabolism in experimentally induced type 2 diabetes in rats. J Nat Med 65(3-4):544

Sanjay MJ et al (2007) Challenges and opportunities in drug discovery from plants. J Curr Sci 92 (9):1251

Shekelle PG et al (2005) Are Ayurvedic herbs for diabetes effective? J Fam Pract 54(10):876–886

Singh V et al (2013) Curcuma oil ameliorates hyperlipidaemia and associated deleterious effects in golden Syrian hamsters. Br J Nutr 110(3):437–446

Singh V et al (2015) Curcuma oil attenuates accelerated atherosclerosis and macrophage foam-cell formation by modulating genes involved in plaque stability, lipid homeostasis and inflammation. Br J Nutr 113(1):100–113

Tripathi YB (2003). https://patents.google.com/patent/US7416743B2/en?oq=US+7.416%2c743+B2

**Manoj K. Barthwal** is Senior Principal Scientist at the division of Pharmacology, CSIR-Central Drug Research Institute, Lucknow. He has over 20 years of rich international and national experience in the cardiovascular field. His laboratory is involved in the understanding of disease mechanism and development of novel strategies for the management of cardiovascular disorders. His pioneering work has shed light on the role of macrophages in the progression of cardio-metabolic diseases. He has identified unique macrophage signalling mechanisms that can be targeted for therapeutic intervention. His research efforts have led to the identification of novel small molecules for the treatment of atherosclerotic cardiovascular diseases.

# Modern Drug Discovery and Development in the Area of Cancer: Indian Context 10

Tapas K. Kundu, Anand P. Kulkarni, Dipak Datta, and Sweta Sikder

## 10.1 What Is Cancer?

Cancer is a perturbation of intricate balance between "cell division" and "cell death." Normal cells divide up to certain extent or limit which is called "Hayflick" limit and then die to maintain normal body homeostasis. Cancer cells do not obey "Hayflick" limit and proliferate uncontrollably to form a tumor. In a simplistic description, cancer is a deregulated cell division due to changes in the genetic material of a cell. According to WHO, cancer is a commonly used name for a large group of diseases with similar characteristic features. Some other commonly used terms to describe cancer include malignant tumors and neoplasms. The most important characteristic feature of cancer is rapid multiplication of cancerous cells and spreading to different parts of the body is known as metastasis. Metastatic cancer is a leading cause of death among the cancer patients across the world (WHO 2018). Some other major

T. K. Kundu (✉)
CSIR-Central Drug Research institute, Lucknow, India

Transcription and Disease Laboratory, Molecular Biology and Genetics Unit, Jawaharlal Nehru Centre for Advanced Scientific Research, Bangalore, India
e-mail: director@cdri.res.in

A. P. Kulkarni
Scientific Directorate, CSIR - Central Drug Research Institute, Lucknow, India
e-mail: anand_kulkarni@cdri.res.in

D. Datta
Division of Cancer Biology, CSIR - Central Drug Research Institute, Lucknow, India
e-mail: dipak.datta@cdri.res.in

S. Sikder
Transcription and Disease Laboratory, Molecular Biology and Genetics Unit, Jawaharlal Nehru Centre for Advanced Scientific Research, Bangalore, India

Division of Neuroscience & Ageing Biology, CSIR - Central Drug Research Institute, Lucknow, India

 
M. Dikshit (ed.), *Drug Discovery and Drug Development*,
https://doi.org/10.1007/978-981-15-8002-4_10

characteristic features of cancer cells include the capability to acquire autonomous growth signals, evasion of growth inhibitory signals, angiogenesis, and unlimited replication.

**Cancer Key Facts Worldwide**

- According to WHO, cancer is the second leading cause of death due to diseases in the world.
- In 2018, an estimated 9.6 million have died from cancer.
- One sixth of deaths worldwide are due to different types of cancer.
- Lung and breast cancers are most common type of cancer by incidence.
- Lung and colorectal cancer are leading causes of death.
- Tobacco, alcohol, higher BMI, low consumption of fruit and vegetables, and sedentary life style are the leading causes of mortality in cancer patients worldwide.

Source: WHO (2018)

**Cancer Statistics of India**

- From 1990 to 2016, death due to cancer is doubled from 0.38 to 0.81 million deaths.
- About 1.15 million estimated incidence of new cancer in 2018 and is expected to be doubled by year 2040.
- About 2.25 million people are living with cancer disease.
- Breast and oral cancers are predominant types.

Source: Cancerindia.Org (2020)

### 10.1.1 Whether Cancer Is a Disease of Humans?

Cancer disease is not limited to human beings. Incidence of cancer is reported in many of the animal species as well as plants. It is reported that in over 140 species of eudicots plants, *Agrobacterium tumefaciens* causes a disease named as "Crown gall," which is characterized by the uncontrolled division of cells around the infection, forming a tumor (Helen and Stafford 2000). Unlike cancer in humans, metastasis does not occur in plants as the plant cells are anchored in place by the cell walls.

In the animal kingdom, cancer is a much more widespread disease. Death due to cancer has been reported in wild animals, marine animals as well as domestic animals. The mortality rate due to cancer in many of the animal species is similar to that of human beings. Recent news coverage by National media regarding cancer death in animals is that of an eleven-year-old Royal Bengal Tiger named Yash, which was suffering from a rare form of cancer and died at Sanjay Gandhi National Park (SGNP) in Mumbai on 28 May 2019. A point of scientific interest in this news coverage is "rare form of cancer." It indicates the incidence of different types of cancer in animals similar to humans.

**Peto's Paradox**

Cancer is basically a consequence of mutation in the somatic cells. By virtue of cause of cancer, it should have been the fact that more the number of somatic cells, higher the chances of cancer. Statistically, it was not found true. This is called as "Peto's paradox," after the scientist Richard Peto, who noticed that cancer prevalence is not correlated with body size. For example, blue whales have thousand times more somatic cells than human; however, cancer incidence is much lower in blue whales compared to humans. Similarly, elephants have trillions more cells than us and live long time, yet they have lower cancer rates.

Wide variety of cancer types, their causes and characteristic features have thrown a mammoth challenge to the researchers to find out the effective cure for cancer. Though many of the cancer types, particularly in early stages, are being cured through variety of therapeutics available, the later stage of cancer is a huge challenge for researchers. Many of the researchers are focusing their research on prophylactics—a medication used to prevent the occurrence of disease. Providentially, Indian system of traditional medicine, Ayurveda, has given new hopes to discover prophylactics.

### 10.1.2 How Old Is Cancer?

The fossil remains of duck-billed dinosaurs, which lived 7 million years ago, is the oldest evidence of cancer incidence shown by Rothschild BM of the North-eastern Ohio Universities College of Medicine (Rothschild et al. 2003) while evidence of cancer incidence in human beings has been witnessed throughout recorded history. Fossils, mummies in Egypt, and manuscripts of ancient history are valuable evidence of cancer incidence during ancient history. The "Edwin Smith Papyrus," an ancient Egyptian medical text on "Trauma surgery" which dates back to about 3000 BC is the available oldest description of cancer disease. In this textbook, there is a mention about cauterization to remove eight cases of tumors using a tool called fire drill. The textbook describes that there is no treatment for this disease (The American Cancer Society Medical and Editorial Content Team 2014).

However, in a study published in "Nature Reviews Cancer" in 2010 by researchers from University of Manchester (David and Zimmerman 2010), it has been debated that cancer is a modern, man-made disease caused by external factors such as pollution and diet. They have cited the fact that evidence of cancer in fossils and early humans is very limited. Many types of cancer commonly identified in adult human are not seen in non-human primates. It is evidently true that modern world is having higher exposure for carcinogens.

In the literature, Hippocrates (460–370 BC), an ancient Greek physician, used the term "Karkinos" (Greek word for crab) to describe cancer. Around 47 AD, the Grecko-Roman philosopher Celsus translated the Greek word karkinos to the Latin word for crab: Cancer. Interestingly, the term "Oncology"—a branch of medicine that deals with the cancer, is also derived from ancient Greek. Aelius Galenus (130–200 AD), another famous Greek physician used the term Oncos (swelling) to describe cancer.

The invention of compound microscope in 1590 and its subsequent use for scientific investigations facilitated major breakthroughs in the area of cancer. Using microscope, pathologists described specific cancers, documented the incidences and types of cancer. Today, many advancements have been made in research in the area of cancer. More than 100 types of cancers have been reported worldwide so far. Researchers have succeeded in developing treatments for some of the cancer types already including radiation, chemo-, immuno-, hormone, and targeted therapy. However, still we are far away from finding cure to many types of cancer which are affecting human beings irrespective of their class or lifestyle, including celebrities, world leaders to poor people.

## 10.2 Cancer Types and Incidence

International Classification of Diseases for Oncology has grouped cancer into six major categories based on histology (National Cancer Institute n.d.).

**Carcinoma** is a malignant neoplasm of epithelial origin. It accounts for 80–90% of all cancer cases. Two major subtypes of carcinoma are: (1) adenocarcinoma and (2) squamous cell carcinoma. Examples include skin, breast, lung, prostate cancer, etc.

**Sarcoma** is a rare type of cancer which originates from connective tissues like bone, cartilage, and muscle. Examples includes osteosarcoma, glioma, chondrosarcoma, etc.

**Myeloma** is a cancer of plasma cells, a type of white blood cells of blood.

**Leukemia** also called as blood cancer refers to the cancers of the bone marrow, which leads to overproduction of abnormal blood cells.

**Lymphoma** are solid cancers and usually originates from the glands or nodes of the lymphatic system.

**Mixed Type** of cancer may have two or more components of cancer. Example, adenosquamous carcinoma, carcinosarcoma, etc.

Another type of classification of cancer is based on the site of cancer and is more popular. Examples include breast, oral, lung, cervical, stomach, colon, and prostate cancer.

## 10.3 Cancer in India

There are no reports available regarding incidence of cancer in ancient Indian civilizations. However, there is a mention of cancer like disease, particularly Granthi (small in size) or Arbuda (large in size) and remedies for the same in ancient manuscripts of Ayurveda. According to available reports, cancer cases in India are recorded from seventeenth century onwards. During later part of the nineteenth century, doctors of Indian Medical Services started publishing cancer case series (Smith and Mallath 2019). The efforts at policy level was seen in 1946, in which a national committee recommended setting up of facilities to diagnose and manage cancer burden in Indian states. Soon after independence, in 1950, the Cancer Research Institute was established by Prof. VR Khanolkar as a dedicated center to understand intricacies of disease to evolve treatment regime. Currently, CRI is part of the Advanced Centre for Treatment, Research and Education in Cancer (ACTREC), which is mandated to function as a national center for treatment, research, and education in cancer. Indian Cancer Society, established in 1951 by Dr. DJ Jussawalla and Naval Tata, is a first non-government, non-profit organization in India with a mission to create awareness about cancer, facilitate early detection, and providing support for treatment of cancer for poor cancer patients. Subsequently, Dr. Jussawalla established the first Population Based Cancer Registry (PBCR) in Mumbai in 1963 with financial support from Indian Cancer Society. India's National Cancer Registry Program started in 1982 and over the years, several population based cancer registries were added. At present, a total of 31 population based cancer registries and 29 hospital based cancer registries are functional across India. These cancer registries are playing imperative role in controlling the impact of cancer on the community by providing a framework for assessment.

It is pertinent to mention that the Lady Tata Memorial Trust, established by Sir Dorabji Tata in 1932 in memory of his wife, Lady Meherbai, who died of leukemia, started financial support for R&D in the area of cancer. Subsequently, considering the incidences of cancer and its impact on Indian population, CSIR, DAE, ICMR, and DBT started support for the cancer research in India. According to bibliometric reports, number of publications in the area of cancer in 1990 was about 300. By 2010, the number was almost 1500 publications per year (Sullivan et al. 2014).

### 10.3.1 Role of Council of Scientific & Industrial Research, India

Council of Scientific & Industrial Research, India was formally established in 1942 for development of natural resources and industries in India by setting up of research laboratories to assist the nation building. The health of the increasing population was then a major area of concern. Life expectancy was very low. The Central Drug Research Institute (CDRI) at Lucknow was the first laboratory of CSIR established in 1951 exclusively for drug research. In due course of time, other research laboratories to advance the biomedical research were established in India. National Chemical Laboratory, Pune, and Indian Institute of Chemical Technology were focusing on development of indigenous process technologies for lifesaving drugs to support Indian pharma industries. CSIR-CDRI, apart from new drugs, also significantly contributed to the development of indigenous process technologies for more than 80 drugs. During the last more than 75 years of CSIR's existence, a large number of newer laboratories were added to its fold. With advancement in biomedical sciences, newer aspects of drug research were initiated. Due to intermixing of various disciplines many of the laboratories, not directly working in healthcare sector, also made valuable contributions to this area. Among the CSIR Institutes, Central Drug Research Institute, Lucknow; Indian Institute of Chemical Biology, Kolkata; Centre for Cellular & Molecular Biology, Hyderabad; Indian Institute of Integrative Medicine, Jammu; and Indian Genomics and Integrative Biology are currently pursuing fundamental as well as translational research programs in the area of cancer.

### 10.3.2 Role of Other Government Organizations

Several institutes established by Department of Biotechnology, Department of Science & Technology, Indian Council for Medical Research also significantly contributing to the cancer research in Indian context. Contributions from the research teams at Indian Institute of Science, Bengaluru and Jawaharlal Nehru Centre for Advanced Scientific Research, Bengaluru in advancing knowledge frontiers in the area of cancer is noteworthy. ICMR played a major role in setting up cancer therapy regimen across India. They set up indigenous treatment protocols for adjuvant and neo-adjuvant therapy in Indian patients. ACTREC and AIIMS also developed their indigenous therapy protocol by modifying international guidelines which is very useful in Indian context.

Significant contributions of Indian Research Organizations in Cancer research are discussed in the subsequent sections of this of chapter.

### 10.3.3 Predominant Cancer Types in India

Available reports indicate that incidence of most of the types of cancers that affect world population is prevalent in Indian population as well. During last two and a half decades, the cancer burden has actually doubled from 0.38 million in 1991 to 0.81 million deaths in 2016. About 2.25 million people in India are living with cancer. Predominant types of cancer in 1990 and 2016 are given in Table 10.1. Various studies indicated that use of tobacco products is the single most preventable cause of cancer.

Among the cancer types, breast cancer is predominant in Indian women in terms of morbidity and mortality, followed by cervical, ovarian, and uterine cancer. Interestingly, in urban women population, breast cancer is leading whereas, in rural women population, cervical cancer is the leading cause of morbidity and mortality. Death due to cervical cancer is almost one women per 8 min in India

**Table 10.1** Top ten cancer types in India in 1990 and 2016 based on incidence rate

| | 1990 | 2016 |
|---|---|---|
| 1 | Cervical cancer | Breast cancer |
| 2 | Stomach cancer | Lip and oral cavity cancer |
| 3 | Lip and oral cavity cancer | Cervical cancer |
| 4 | Breast cancer | Stomach cancer |
| 5 | Pharynx cancer other than nasopharynx | Lung cancer |
| 6 | Lung cancer | Pharynx cancer other than nasopharynx |
| 7 | Colon and rectum cancer | Colon and rectum cancer |
| 8 | Esophageal cancer | Esophageal cancer |
| 9 | Leukemia | Leukemia |
| 10 | Larynx cancer | Prostate cancer |

*Source: The burden of cancers and their variations across the states of India: the Global Burden of Disease Study 1990–2016* (India State-Level Disease Burden Initiative Cancer Collaborations 2018)

(NCDIR 2012–2014). Currently, the estimated rate of incidence of breast cancer is 25.8 per one lakh women, which is expected to become 35 per one lakh women by 2026 if adequate measures are not taken. Mortality rate among the breast cancer patients is nearly 50% (Fig. 10.1).

Among the Indian men, the predominant cancer types include oral cancer followed by lung, esophagus, and stomach. Analysis of incidence of cancer in different states of India provided very interesting information regarding lifestyle and predominant cancer types. Figure 10.2 gives a pictorial representation of cancer incidence and risk factors in different states of India. Though different geographical regions of India have different predominant types of cancer, consumption of tobacco and pollution are reported to be major factors causing cancer in the Indian population in several regions.

According to 3-year report of population based cancer registries 2012–2014 published by National Centre for Disease Informatics and Research, ICMR, Aizawl district in Mizoram reported highest cancer incidence among males. In females, the Papumpare district in Arunachal Pradesh reported highest incidence of cancer (NCDIR 2012–2014).

According to the aforesaid report, lung cancer, mouth cancer, esophagus cancer, and stomach cancer are predominant types in males across the Population Based Cancer Registries (PBCRs) in India. Lung cancer is predominant type in Delhi, Mumbai, Chennai, Kolkata, Bangalore, Thiruvananthapuram, etc. Oral cancer is the leading type of cancer in Bhopal, Ahmedabad Urban, Nagpur, Pune, Barshi, etc. The PBCRs in the state of Meghalaya, Assam, and Patiala reported highest incidence of cancer in esophagus while breast cancer is predominant type of cancer in 19 registry areas and Cervix uteri is the leading site in six registry areas among females.

Data from various PBCRs and HBCRs indicate that cancer incidence is highest in North-East region of Indian subcontinent. Major factors attributed for high incidence include lack of awareness, socioeconomic conditions, and difficulty of access to the facilities.

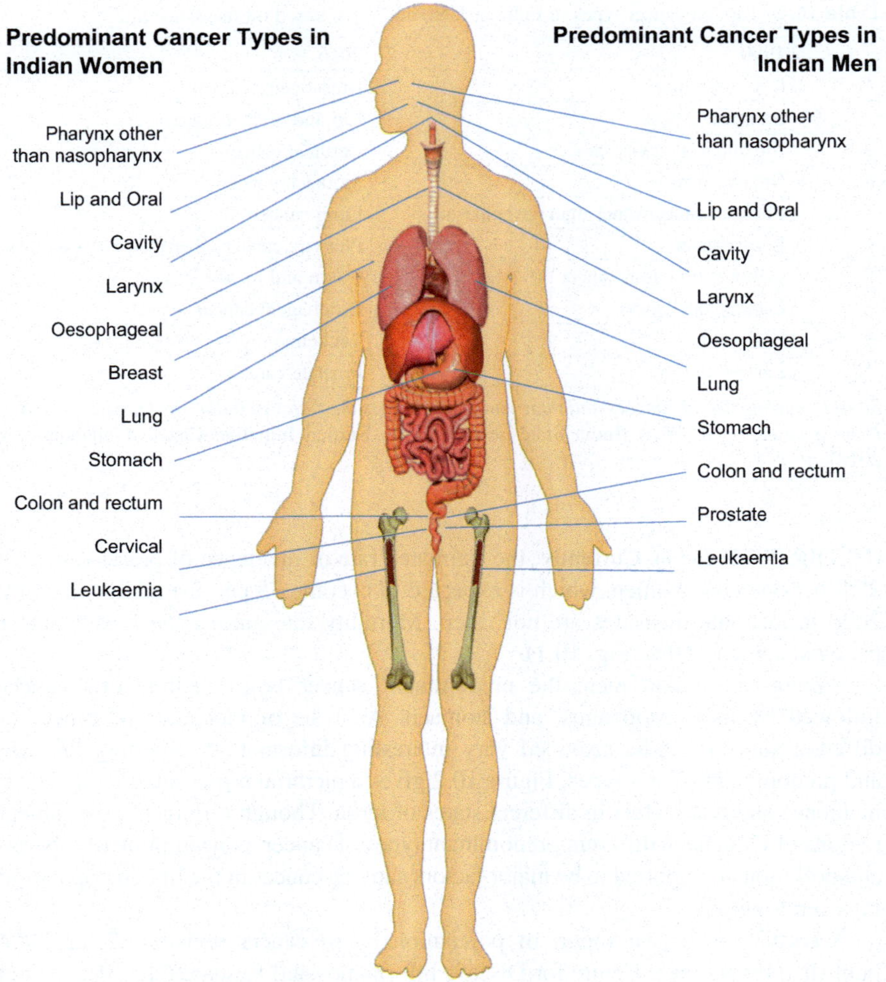


**Fig. 10.1** Different types of cancer predominant in Indian population

## 10.4 Cancer Therapy

Advancement of Science & Technology led to development of multiple strategies and several therapeutics to treat cancer. Type of treatment to be given to a cancer patient is being decided based on the cancer types and other associated factors. Some of the popular treatment methods and therapies are briefly described below:

**Surgery** is a procedure to remove cancer from patient body. Ways include cryosurgery, precise surgery through laser beams, hyperthermia, etc.

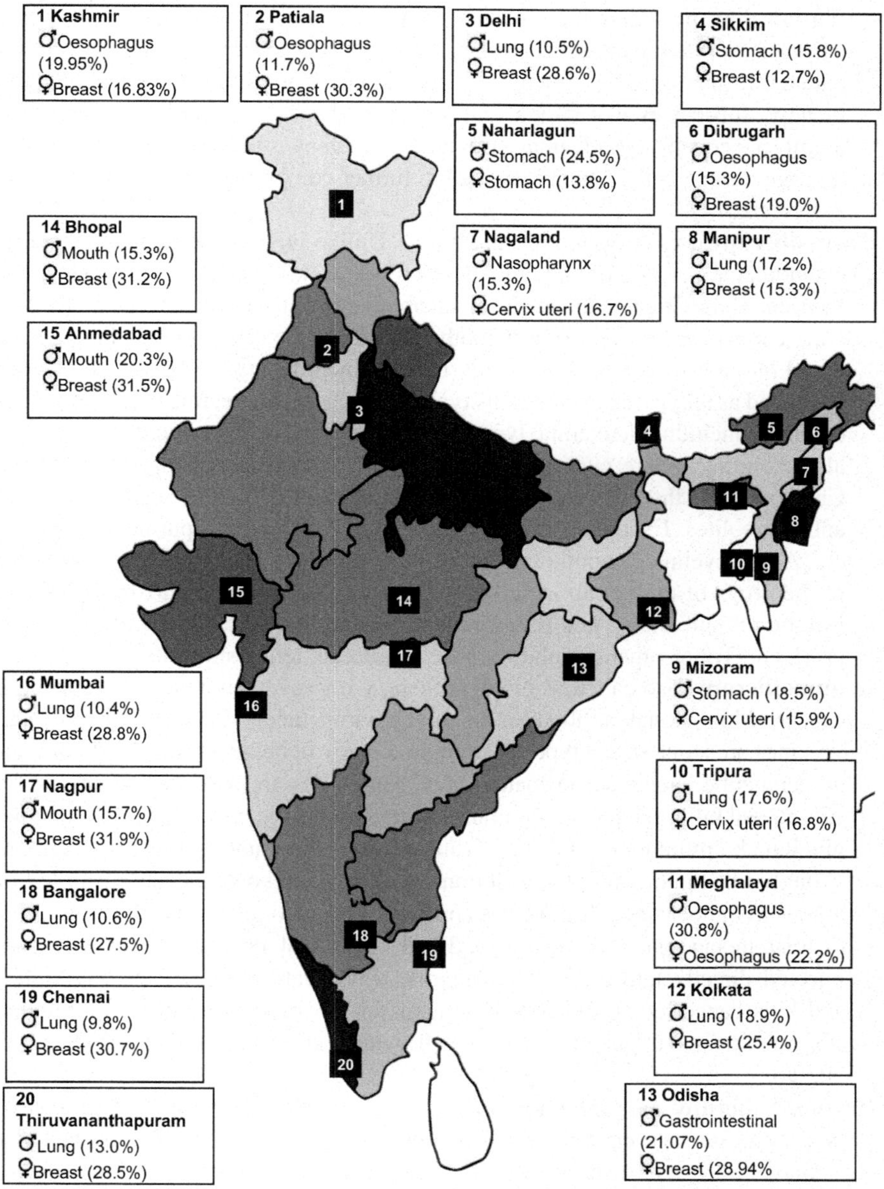


**Fig. 10.2** Leading sites of cancer in male and female as reported in population based cancer registry across India. Leading site based on the proportion relative to all sites of cancer incidence. Percentage value in parenthesis indicates the relative proportions of cancer incidence at leading site relative to all sites of cancer

**Radiation therapy** uses high dose of radiation (X-rays, gamma rays, and charged particles, etc.) to kill cancer cells and shrink tumor. Radiation kills cancer cells by damaging their genetic structure or functional proteins. In systemic radiation therapy, substances like radioactive iodine used, which traverse through blood to kill cancer cells. Disadvantage of radiation therapy is that the radiation can also damage normal cells, which may lead to further complications (Oncology Nurse Advisor 2014).

**Chemotherapy** uses drugs to kill cancer cells. Unlike surgery and radiation therapy, chemotherapy acts throughout the body and kill cancer cells that have metastasized. Three main goals of chemotherapy in cancer treatment include cure, control, and palliation. Chemotherapy drugs are often divided into groups based on their mechanism of action. Drugs which damage DNA of cancer cells are called as alkylating agents. This type of drugs used in treatment of many types of cancer including leukemia, lymphoma, myeloma, sarcoma, etc. Some popular alkylating agents are cyclophosphamide, melphalan, carboplatin, cisplatin, etc. Chemo drug that interferes with normal metabolism of cells is called antimetabolites. Examples include 5-fluorouracil, 6-mercaptopurine, cytarabine, etc. Anthracycline chemotherapy acts on the enzymes inside cancer cells. Some of the drugs of this group include actinomycin-D, bleomycin, doxorubicin, etc. Mitotic inhibitors stop cell division. Mitotic inhibitors include docetaxel, paclitaxel, etc. Topoisomerase inhibitors are irinotecan, teniposide, etc.

**Immunotherapy** also called as biologic therapy improves immune system to fight cancer. Monoclonal antibodies, oncolytic virus therapy, T-cell therapy, and vaccines are some of the types of immunotherapy being used worldwide. Immunotherapeutic agents act in many ways. Antibodies are used to flag cancer for subsequent killing of them by immune system. Immune checkpoint inhibitors like nivolumab, ipilimumab, and atezolizumab help the immune system respond more strongly to a tumor. Adoptive cell transfer attempts to boost the natural ability of T cells to fight cancer. USFDA has approved first oncolytic virus therapy in 2015 to treat melanoma. Genetically modified version of herpes simplex virus are injected directly into areas of melanoma, which enters cancer cells, multiplies and kills cancer cells. Release of antigens in this process triggers the immune system of host to target all the cancer cells which have same antigens (Cancer.Net 2020a).

**Hormone therapy** uses inhibitors of hormone and slows or stops the growth of breast and prostate cancers, which are dependent of hormones for growth. For example, estrogen causes growth of breast cancer in some women. Clinicians use medicine like tamoxifen to block the effect of estrogen on the growth of cancer cells in breast tissue. Aromatase inhibitors like anastrozole and letrozole are used to prevent estrogen production in postmenopausal women to prevent the recurrence of breast cancer. Fulvestrant binds with the estrogen receptor and eliminates it rather than just blocking it. Similarly, variety of medicines are used to control testosterone, which causes growth of prostate cancer in men (University of Rochester Medical Centre n.d.) (Table 10.2).

**Table 10.2** Top 20 blockbuster cancer drugs 2017

| | Drug name | Manufacturer | Condition or diseases treated | Generic name |
|---|---|---|---|---|
| 1 | Revlimid | Celgene | Multiple myeloma | Lenalidomide |
| 2 | Avastin | Roche | Breast, colorectal, lung, kidney, ovarian cancers | Bevacizumab |
| 3 | Herceptin | Roche | HER2+ breast cancer | Trastuzumab |
| 4 | Rituxan | Roche | Non-Hodgkin's lymphoma, chronic lymphocytic leukemia | Rituximab |
| 5 | Opdivo | Bristol-Myers Squibb; Ono Pharmaceutical | Metastatic melanoma, colon cancer, lung cancer, renal cell carcinoma, head and neck cancer, Hodgkin lymphoma, and liver cancer | Nivolumab |
| 6 | Gleevec | Novartis | Chronic myeloid leukemia, gastrointestinal stromal tumors | Imatinib |
| 7 | Imbruvica | Johnson & Johnson/ Pharmacyclics | Mantel cell lymphoma, chronic lymphocytic leukemia | Ibrutinib |
| 8 | Velcade | Johnson & Johnson/ Takeda | Multiple myeloma, mantle cell lymphoma | Bortezomib |
| 9 | Zytiga | Johnson & Johnson | Prostate cancer | Abiraterone acetate |
| 10 | Xtandi | Astellas Pharma/Pfizer | Prostate cancer | Enzalutamide |
| 11 | Alimta | Eli Lilly | Non-small cell lung cancer | Pemetrexed |
| 12 | Gardasil | Merck & Co. | Cervical cancer | Human papillomavirus quadrivalent (Types 6, 11, 16, and 18) Vaccine, Recombinant |
| 13 | Ibrance | Pfizer | Breast cancer | Palbociclib |
| 14 | Perjeta | Roche | HER2-positive breast cancer | Pertuzumab |
| 15 | Tasigna | Novartis | Chronic myeloid leukemia | Nilotinib |
| 16 | Xgeva | Amgen | Bone metastases | Denosumab |
| 17 | Afinitor | Novartis | Breast cancer | Everolimus |
| 18 | Jakafi | Incyte/ Novartis | Polycythemia vera; myelofibrosis | Ruxolitinib |
| 19 | Tarceva | Roche | Non-small cell lung, pancreatic cancers | Erlotinib |
| 20 | Keytruda | Merck & Co. | Advanced melanoma; non-small cell lung cancer; head and neck squamous cell cancer | Pembrolizumab |

*Source: The balance. Top 20 Blockbuster Cancer Drugs. Available from URL: https://www.thebalance.com/top-cancer-drugs-2663234*

**Targeted therapy** deploys targeted delivery techniques for delivery of specific drugs to target the specific genes, proteins, or the tissue that contributes to growth and survival of cancer. Differential biochemical/physiological features of cancer cells compared to normal cell are targeted. For example, targeted drugs can work to block or turn off chemical signals that control growth and division of cancer cells. They may stop angiogenesis to kill cancer cells or may carry specific toxins to kill cancer cells, but nor normal cells. Clinicians are using several types of agents in targeted therapy including monoclonal antibodies and small molecule drugs depending upon the need with due diligence. Diligence is important considering the fact that all cancers do not have same targets.

For example, about 20–25% cases of breast cancers are marked with high concentration of Human Epidermal Growth Factor Receptor 2 (HER2). HER2 has been indicated for its role in the growth of cancer cells. Trastuzumab, a monoclonal antibody, is used to treat non-metastatic HER2-positive breast cancer. Pertuzumab, a recombinant humanized monoclonal antibody is used to treat stage II and III breast cancer in combination with trastuzumab and chemotherapy.

Epidermal Growth Factor Receptor (EGFR) content is high in colorectal cancers, which play important role in growth of colorectal cancer. Panitumumab, a fully human monoclonal antibody, is used in targeted therapy for colorectal cancer. Similarly, angiogenesis in colorectal cancer is also targeted using antibodies as well as small molecule drugs. EGFR is also targeted in treatment for lung cancer.

For treating melanoma, BRAF gene is targeted on case to case basis. Dabrafenib and vemurafenib are FDA approved targeted therapies for stage III and IV melanoma (Cancer.Net 2020b).

The advancement of knowledge frontiers has opened up new methodologies for targeted therapy for cancer. Recently, a team of researchers led by Dr. Prabhat Mishra has developed a platform technology of nanocarriers for personalized oncotherapy. In this platform technology, they have used inverse hexagonal nano liquid crystalline particles to host anticancer drug formononetin, an optical beacon tetraphenylethene, and a tumor targeting ligand anisamide to ensemble drug targeting, imaging, and non-invasive therapeutic properties. In vitro and in vivo studies confirmed the enhanced efficacy of targeted. In the coming years, it is anticipated that targeted therapy will be a major type of cancer therapy (Urandur et al. 2018) (Fig. 10.3).

### 10.4.1 Drug Resistance in Cancer

Resistance to chemotherapy and targeted therapies at molecular level has posed a greater challenge to researchers and clinicians in finding a cure to cancer. Drug resistance has rendered the arsenal of anticancer drugs to remain on shelves. In-depth studies on cancer drug resistance mechanism have revealed complexity of resistance process in cancer cells. Individual's genetic differences, multi-drug resistance, epigenetic factors, altered drug metabolism, and gene amplification are some of

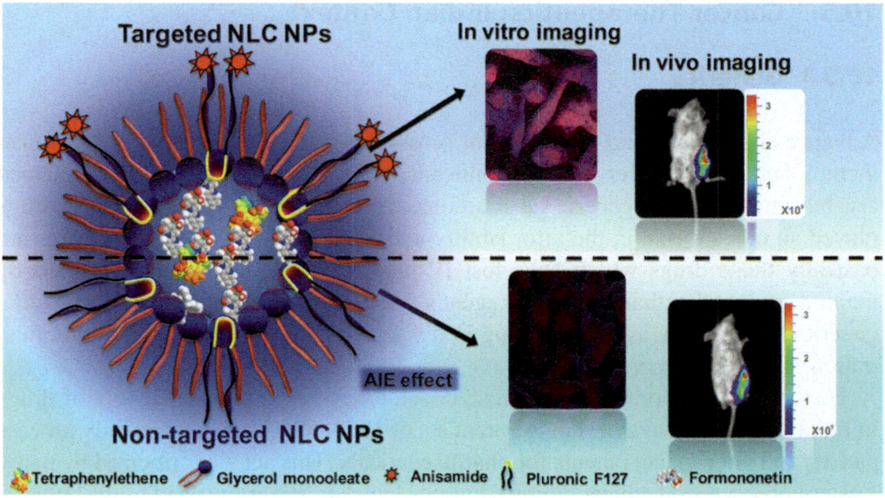


**Fig. 10.3** Anisamide anchored lyotropic nano liquid crystalline particles with aggregation-induced-emission effect—a smart optical beacon for tumor imaging and therapy

the major reasons of drug resistance. Drug resistance to tyrosine kinase inhibitors is one of the very well-studied drug resistance mechanism in cancer. Tyrosine kinases play important role at cellular level, including cell cycle control. Several drugs have been discovered targeting tyrosine kinase inhibitors for treatment of various cancers and were found successful in treating cancer. Imatinib, gefitinib, erlotinib, sorafenib, sunitinib, and dasatinib are some of the widely used TK inhibitors, which have same mechanism of action—competitive ATP inhibition at the catalytic binding site. Studies have revealed that cancer cells have acquired resistance post-treatment with above TK inhibitors. Studies also revealed that apart from mutation, alternative splicing, alternative signaling pathways, and epigenetic factors contributed to the resistance to TK in patients. Following the drug resistance cases, researchers and clinicians are now focusing on alternative approaches for cancer treatment using the available drugs with a minimum scope for resistance and also alternate drug targets.

Similar to drug resistance, cancer cells have also exhibited resistance to radiation therapy. Several recent breakthrough discoveries have advanced the knowledge frontiers in cancer cell resistance to radiation therapy. Recently, a team of researchers from JNCASR, led by Prof Tapas K. Kundu has shown that downregulation of Positive Co-factor (PC4) leads to enhanced autophagy. The enhanced level of autophagy allows cancer cells to withstand the stress caused by radiation and cells become resistant to radiation (Sikder et al. 2019). This finding will have implications on radiation therapy of cancer.

## 10.5 Cancer Therapeutics: Indian Context

### 10.5.1 Generics

Advances in research and development across the globe had augmented the cancer therapy options. However, access to anticancer drugs, due to high cost has remained as a burning issue. The generic substitutions of commonly used chemotherapy drugs played significant role in the affordability and accessibility. Generic medicines are basically those drugs which have lost IP protection, and are being produced by manufacturers other than the firm or agency which innovated the drug. By definition, generic drug is same as original drug in dosage, safety, strength, route of administration, quality, and performance.

Historically important and fortunate development towards affordability of drugs in India was the passage of Indian Patent act of 1970, which recognized only process patents and not product patents for pharmaceuticals. This act was criticized by many western observers on ethical grounds but this legislation was the need of the hour for the Nation. The architect of the patent law of 1970, S. Vedaraman, the then Director of the Indian Patent Office defended the act by stating "We are not against patents. We are prepared to pay decent license fees. But we in India cannot afford monopolies." This act provided impetus to the Indian pharmaceutical industries. In the subsequent decades, driven by the abundant skilled human and material resource, low costs, and enhanced demand (domestic and international), the Indian pharmaceuticals sector witnessed a tremendous growth.

### 10.5.2 Contribution of Indian Pharmaceutical Sector

Indian pharmaceutical industry is playing instrumental role in world pharma sector owing to its outstanding competency and capabilities, particularly, reverse engineering and manufacture of generic medicines at significantly lower costs. Indian industries export drugs to more than 200 countries, which comprises of 20% of generic drugs. In 2018, the Indian pharmaceutical sector was valued at US$ 36.7 billion. Market share of generic drugs estimated to be around 71%. Affordability and accessibility of drugs worldwide is attributed to the low cost supply of quality drugs from India. As per the available reports, global healthcare programs like "Global Fund to Fight AIDS, Tuberculosis and Malaria" and UNICEF rely on generics manufactured in India. India is popularly known as Pharmacy of world.

Among the Indian pharma companies, CIPLA has played important role in bringing down the cost of cancer therapy in India. In 2012, Cipla cut prices of key cancer drugs by nearly 75%. Price of the kidney cancer drug sorafenib was reduced to Rs. 6840 for a month's supply, down from 28,000. Lung cancer drug gestinib was reduced to Rs. 4250 down from 10,000. Price of temozolomide, used to treat brain tumor, reduced to Rs. 5000 from Rs. 20,000. According to Dr. Y. K. Hamied, former Chairman and Managing Director, "Cipla had brought down the cost of treatment of AIDS and malaria worldwide. Continuing its contribution towards affordable and accessible treatment for patients, they have included cancer, not only in India but globally. It was one of the bold step taken by Indian pharma, which was followed by other pharma companies in due course of time." The price cut triggered price war among the pharmaceutical companies and many other firms followed the footsteps of Cipla and reduced the cost of cancer drugs in India subsequently.

### 10.5.3 Contribution of Council of Scientific & Industrial Research, India

CSIR labs like CDRI, IICT, NCL, IIIM, IGIB, IICB, and CCMB have played significant role in two major aspects—development of process technology for manufacture of generics and development of scientific and technical workforce to cater to the needs of pharma companies. CSIR-CDRI developed process technology for cyclophosphamide and transferred to Sarabhai Research Centre, Baroda in 1978. Cyclophosphamide is used as chemotherapy to treat lymphoma, multiple myeloma, leukemia, ovarian cancer, and breast cancer. In 1985, CSIR-CDRI developed indigenous technology for tamoxifen citrate (anti-breast cancer) and licensed to Cipla, Mumbai. Similarly, CSIR-IICT was involved in development of cheaper alternative chemical routes for several lifesaving anticancer drugs such as etoposide, vinblastine, vincristine, and mitoxantrone, which were licensed to industries for commercial production.

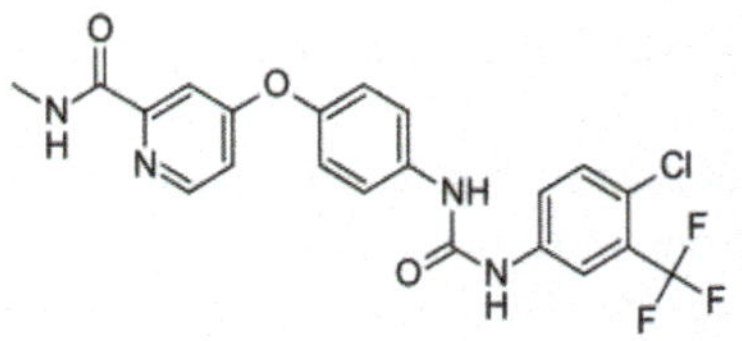


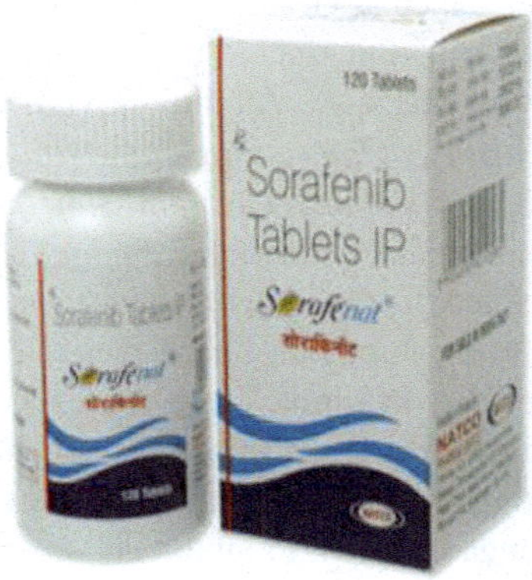


A noteworthy contribution from CSIR-IICT, Hyderabad has been development of technology for Bayer's Nexavar (Sorafenib)—a lifesaving anticancer drug, priced at Rs. 280,000 for a month's dosage. IICT licensed the technology to the Indian company NATCO which is selling the same anticancer drug sorafenib tosylate at Rs. 8800 per month's dosage at a whopping 97% cost reduction. This anticancer drug is the first example of "compulsory license" in India even when the patent on product was valid to mandate a generic drug maker to produce an inexpensive medicine in public interest.

### 10.5.4 Biosimilars

Indian pharmaceutical companies have grown up as the global market leaders in biosimilars which is also true in case of cancer therapeutics. Notably, Herceptin (trastuzumab), used in certain breast and stomach cancer patients was the first biologic to be approved by FDA, was also the first similar biologics manufactured by an Indian company, which received USFDA approval to market in the United States. Indian guidelines for approval of biologics is even stringent than generics. Recently, Mylan and Biocon launched a bevacizumab biosimilar in India which treats several types of cancers.

At the moment, there are more than 100 Indian biopharmaceutical companies busy in manufacturing and marketing of biosimilars in India which is called as "similar biologics" by Indian regulatory agencies. Biosimilar market in India was approximately US$ 300 million in 2015. Indian sales are close to US$ 250 million and has CAGR of about 14%.

### 10.5.5 Anticancer Drugs from Traditional System of Medicine

Traditional Indian system of medicine has always been a source of healthcare need for Indian population from time immemorial. It has also inspired modern drug discovery for several diseases. According to Ayurveda, cancer is inflammatory or non-inflammatory swelling called as Granthi (small in size) or Arbuda (large in size). Researchers have provided scientific data to prove the anticancer potential of many of the herbs used in Ayurveda, e.g., Allium sativum, *Curcuma longa*, *Annona atemoya*, *Phyllanthus niruri*, *Piper longum*, *Withania somnifera*, *Emblica officinalis*, *Andrographis paniculata*, *Ocimum sanctum*, *Tinospora cordifolia*, *Semecarpus anacardium*, *Ziziphus mauritiana*, *Podophyllum hexandrum*, etc.

Paclitaxel (Taxol) isolated from *Taxus brevifolia* is used in cancer treatment clinically. It acts as mitotic inhibitor. Vincristine and vinblastine isolated from *Catharanthus roseus* are also in clinical use for cancer treatment. They act as microtubule inhibitors and arrest cell cycle. The anticancer properties of several plants is still under investigation and some have shown promising results. Curcumin isolated from Curcuma longa is shown to inhibit cell proliferation in wide variety of cell limes. It has also shown to reduce VGEF and bFGF mediated angiogenesis. Withaferin A isolated from *Withania somnifera* is shown to induce apoptosis in variety of cancer cells. Triterpenic acids isolated from *Boswellia serrata* are shown to inhibit topoisomerase I and II. It is found to be effective against brain tumor.

Literature discloses strong evidence from traditional medicinal plants based cancer medicines with relatively few side effects.

### 10.5.6 Contribution of CSIR Institute

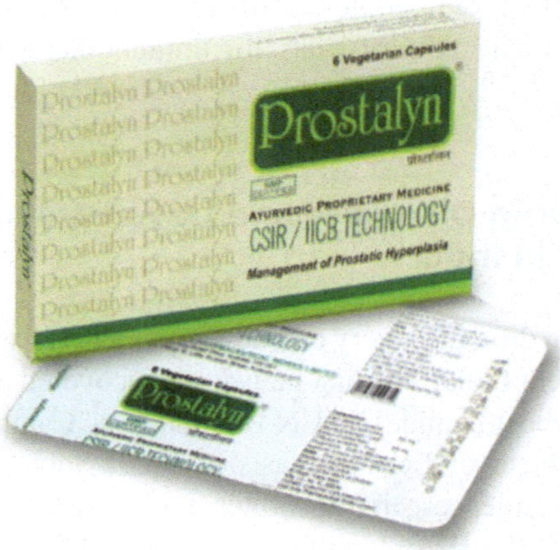


Another success story of CSIR in the area of cancer is development of herbal medicine by CSIR-IICB, Kolkata. The invention relates to the use of extract from leaves of *Murraya koenigii* and *Tribulus terrestris* for the treatment of benign prostate hyperplasia. This herbal formulation for prostate diseases has been licensed to M/s East India Pharmaceuticals Works (L), Kolkata. Product is in market with Brand Name: Prostalyn.

### 10.5.7 Advances in Comprehensive Cancer Care in India

Though cancer therapy regimen advanced lot in terms of targeted therapy applications, in major Indian hospitals, classical therapy belongs to surgery, radiation, and chemotherapy. Hormone therapy is mostly used in endocrine cancers. Many times, the treatment also include combination of two or more therapy from the above, the common ones are radiation therapy, chemotherapy, and surgery. The treatment course depends upon the diagnosis of the condition and its current stage. Cyber Knife Surgery, a non-invasive, pain free robotic radiation therapy is the latest in cancer treatment in India claiming no side effects as such.

Other advanced therapies like stem cell transplant therapy and immunotherapy are still very costly and mostly unaffordable to common Indian patients. At present, three immunotherapy drugs are available in India—Pembrolizumab (Keytruda®), Nivolumab (Opdivo®, Opdyta®), and Atezolizumab (Tecentriq®)—which are approved by DCGI (Drug Controller General of India) for stage-IV lung cancer patients. Recently, another ICI or checkpoint inhibitor—Durvalumab (Imfinzi®)—showed efficacy in stage 3 inoperable lung cancer. The drug will be launched in India soon. When immunotherapy suits for a patient, it works really well. But there are lot of cases, it does not even work and ends up with its side effects. Therefore, we have to be always cautious in applying to each case if we can manage its affordability issues.

Recently, AIIMS took an initiative and established 182 bed specialized cancer hospital (Dr. BR Ambedkar Institute Rotary Hospital) where they are planning to start advanced cancer care clinical trials for immunotherapy and biosimilars and other mAbs.

## 10.6 Modern Drug Discovery and Development in the Area of Cancer in India

In general, the modern drug discovery and development efforts in India can be traced back to pre-independent India. The first new drug discovered and developed in India is Urea stibamine (carbostibamide) by UN Brahmachari, Campbell Medical College, Kolkata for the treatment of Leishmaniasis in 1922. Soon after Independence of India, setting up of premier research institutes, particularly CSIR-CDRI, Lucknow; CSIR-NCL, Pune; CSIR-IIIM, Jammu, and CSIR-IICT, Hyderabad, under the umbrella of Council of Scientific & Industrial Research, ushered a new era for drug discovery for diseases of national priorities. These laboratories successfully discovered and developed several new drugs. Out of the 20 new drugs discovered and developed in independent India, 13 are from CSIR.

Though there are no success story of new cancer drug discovery and development in India, it is worth to mention the efforts of Indian Pharma in finding cure to different types of cancer. Table 10.3 enlists the efforts of Indian companies and organizations towards discovery and development of cancer drugs.

In 2002, CSIR-CDRI collaborated with Dabur Research Foundation in an ambitious program on discovery and development of novel anticancer agents. The program was extended further for lead optimization and drug candidate selection up to 2009. However, the efforts didn't yield the desired level. Couple of leads identified under this program was dropped during preclinical development stage.

Among the Pharma companies, Dr. Reddy's Laboratories, Hyderabad, an Indian multinational Pharmaceutical company had been the first Indian company to launch drug discovery research in India. It launched drug discovery program in 1994, followed by Ranbaxy, Torrent Pharmaceuticals Ltd., Wockhardt Ltd., Piramal Enterprises Ltd., Dabur Research Foundation, etc.

Dr. Reddy's laboratory developed topoisomerase inhibitor DRF-1042 up to Phase I clinical trial. Further development abandoned as licensee "Clintech" could not raise funds for Phase II clinical trial. Piramal Life Sciences is developing P276 and P1446 targeting cyclin-dependent kinases. P276 is in Phase II clinical trials in the USA, whereas P1446 is in Phase I clinical trial in Canada and India.

**Table 10.3** Status of modern drug discovery and development efforts by Indian pharmaceutical companies

| Name of the firm | Candidate drug | IUPAC name | Structure | Drug target | Latest stage of development as per available reports |
|---|---|---|---|---|---|
| Dr. Reddy's Laboratories | DRF 1042 | (19*S*)-19-ethyl-19-hydroxy-12-(2-hydroxyethoxy)-17-oxa-3,13-diazapentacyclo [11.8.0.0$^{2,11}$.0$^{4,9}$.0$^{15,20}$] henicosa-1 (21),2,4,6,8,10,15(20)-heptaene-14,18-dione | 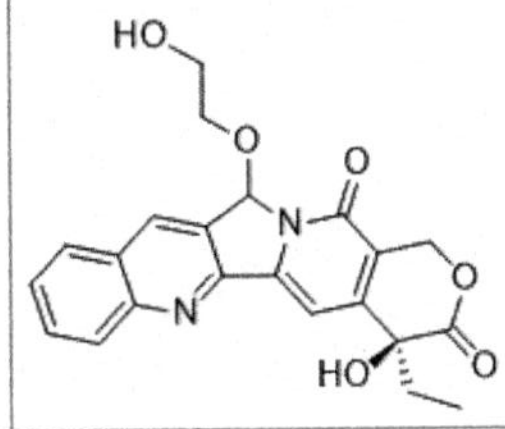 | Topoisomerase-I | Abandoned after partner Clintech couldn't raise funds for Phase II |
| | DRF 1644 | Unknown | Unknown | Topoisomerase-I | Phase I clinical trial |
| Piramal Life Sciences | P276 | 2-(2-chlorophenyl)-5,7-dihydroxy-8-[(2*S*,3*R*)-2-(hydroxymethyl)-1-methylpyrrolidin-3-yl] chromen-4-one | 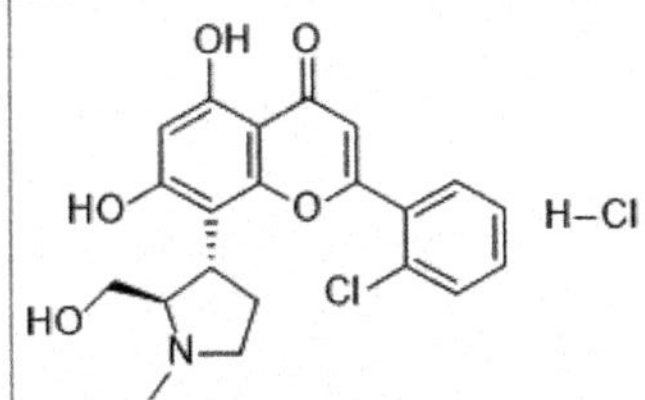 | Cyclin-dependent kinase | Received IND status from USFDA for mantle cell lymphoma and currently in Phase II clinical trials in the USA |
| | P1446 | 4H-1-Benzopyran-4-one, 2-[2-chloro-4-(trifluoromethyl)phenyl]-5,7-dihydroxy-8-[(2R,3S)-2-(hydroxymethyl)-1-methyl-3-pyrrolidinyl]-, hydrochloride (1:1) | 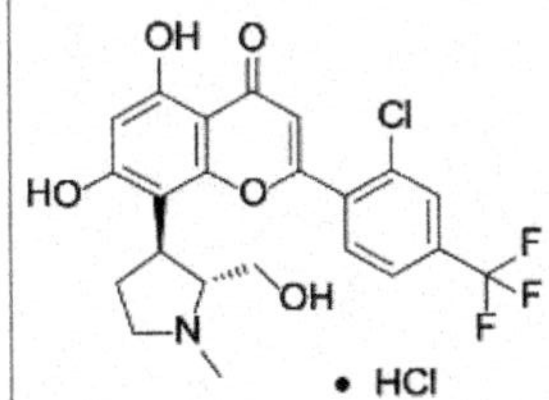 | Cyclin-dependent kinases | Phase I in Canada and India. Does not have IND status from USFDA |

(continued)

**Table 10.3** (continued)

| Name of the firm | Candidate drug | IUPAC name | Structure | Drug target | Latest stage of development as per available reports |
|---|---|---|---|---|---|
| Aurigene Discovery Technologies | CA-170 (AUPM-170) | (((S)-3-amino-1-(3-((S)-1-amino-2-hydroxyethyl)-1,2,4-oxadiazol-5-yl)-3-oxopropyl)carbamoyl)-L-allothreonine | 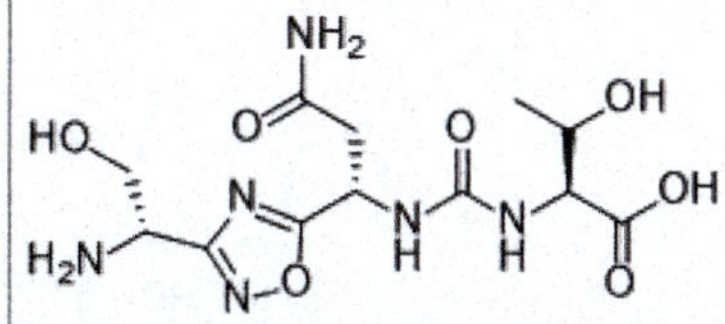 | Programmed cell death protein-1 and V-domain Ig suppressor of T-cell activation | Licensed to Curis (2015); Under Phase I clinical trial in patients with advanced tumors and lymphomas |
| | CA-4948 (AU-4948) | 6-(6-aminopyridin-3-yl)-*N*-(2-morpholin-4-yl-1,3-benzothiazol-6-yl)pyridine-2-carboxamide | 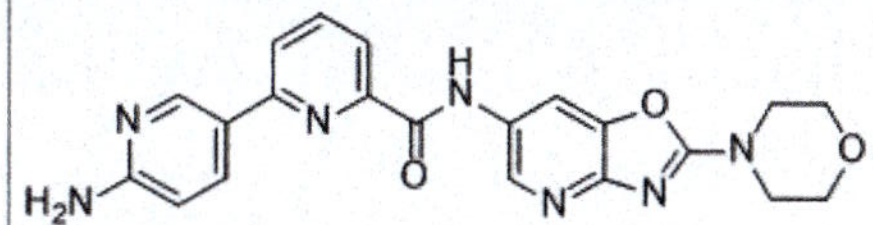 | IRAK4 kinase | Licensed to Curis (2015); Phase 1 clinical trial in patients with non-Hodgkin's lymphoma |
| | Debio-1142 | *N*-[4-methyl-3-[2-[4-(4-methylpiperazin-1-yl)anilino]-5-oxo-7,8-dihydropyrido[4,3-d]pyrimidin-6-yl]phenyl]-3-(trifluoromethyl)benzamide | 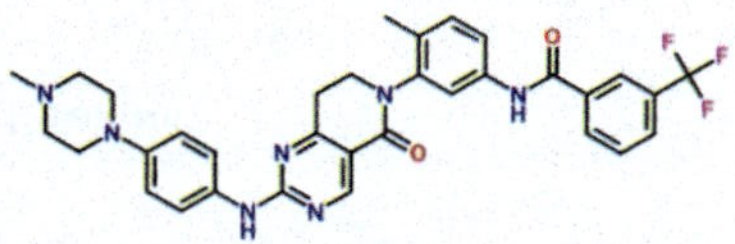 | Jak2 tyrosine kinase and Src tyrosine kinase | Licensed to Debiopharm (2011), preclinical studies carried out in 2013. No further updates since then |
| | AUNP-012 | CAS Number 1353563-85-5 | SNTSESF-NH<br>SNTSESFKFRVTQLAPKAQIKE-$NH_2$ | Immune checkpoint modulator | Licensed to Pierre Fabre (2014)—preclinical stop 2015 |

| | | | | | |
|---|---|---|---|---|---|
| Jubilant | CK-103 (TG-1601) | Unknown | Unknown | BRD4 protein and Bromodomain and extra-terminal domain protein | Licensed to Checkpoint Therapeutics; Phase I clinical trial to be initiated |
| Curadev | RG70099 (CRD1152) | Unknown | Unknown | Indolamine 2,3-dioxygenase and Trp 2,3-dioxygenase | Licensed to Roche (2015), preclinical—ongoing |

*Source: Websites of the concerned pharmaceutical companies and information available in public domain including Clinical Trial Registries*

Aurigene Discovery Technologies has discovered multiple anticancer candidate drugs. It has licensed AUPM-170 (CA-170) and AU-4948 (CA-4948) to Curis in 2015. CA-170 is being developed as dual PD-L1/Vista inhibitor and is under Phase I clinical trial in patients with advanced tumors and lymphomas. CA-4948 is being developed as inhibitor of IRAK4 kinase and is currently under Phase 1 clinical trial in patients with non-Hodgkin's lymphoma. Two more leads discovered by Aurigene, Debio-1142 (Jak2 tyrosine kinase inhibitor; Src tyrosine kinase inhibitor) and AUNP-012 (immune checkpoint modulator) have reached preclinical stage.

Another lead compound CK-103 (TG-1601) discovered by Jubilant Biosys as BET (bromodomain and extra-terminal) inhibitor has been licensed to Checkpoint Therapeutics; Phase I Clinical Trial to be initiated. RG70099 (CRD1152) discovered by Curadev as Dual IDO1/TDO inhibitor has been licensed to Roche (2015). As per the latest reports available, preclinical studies are ongoing.

Few more leads discovered by Indian pharma companies and government organizations are currently under preclinical developmental studies. Few of them are potential reach to clinical trial stage shortly.

## 10.7 Novel target for Cancer therapy

Targeted therapy revolutionized the cancer treatment by minimizing the side effects in patients over conventional chemotherapeutics. However, there are still unmet medical needs in cancer, especially for patients with advanced metastatic disease where drug resistance plays a pivotal role for disease recurrence. Currently, monoclonal antibodies (MAbs) against different modern day drug targets are of better choice than their small molecule inhibitors as off-target effects or toxicity of small molecule inhibitors are way too high than MAbs. Immune Checkpoint Inhibitors (ICI) like PD1, PDL1, and CTLA4 are in the hit list of current novel targets and are the major stakeholder of coming 5 years of cancer therapy.

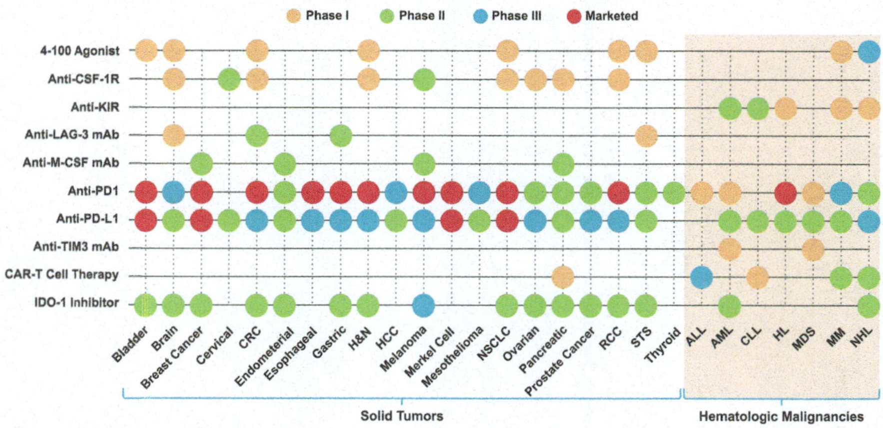


Source: Clinicaltrials.gov, 2020 & US FDA

Here are some basics of immune checkpoint blockade therapy. The fundamental function of immune system is to discriminate between self and foreign molecule.

This led immune system to attack the harmful foreign substances and rescue its own healthy cells. To perform this, immune system uses the checkpoint mechanism (molecules on certain immune cells like T cells that need to be activated or inactivated to start an immune response). Under normal conditions, the checkpoint proteins like PD-1 (Programmed Death-1 and its ligand PDL-1) and CTLA-4 (cytotoxic T-lymphocyte-associated protein 4) are expressed in T-cell surface. They perform a critical role in downregulating the immune response against the self cells and promoting the self-tolerance by suppressing T-cell inflammatory activity. However, the evolution in cancer leads to the overexpression of these checkpoints proteins in the surface of cancer cells which help them to rescue from the attack of immune system. Recently in the field of immunotherapy, the Nobel prize in medicine was awarded to James P. Allison and Tasuku Honjo for their immense contribution in immune checkpoint blockade therapy against PDL-1 and CTLA-4 immune proteins, respectively. There are several checkpoints inhibitors which are approved by FDA and some drugs are in clinical trials in all forms of cancer per se. The diagram below indicates the predominance of anti-PD1/PDL1 in overall cancer therapy trends for last 2 years.

From January to June 2019, FDA released 23 approvals for different oncology indications and closer look of the list clearly emphasizes two facts; one is how important is proper drug combinations and another is the recent surge in ICI based immunotherapy combinations with targeted small molecule drugs.

Considering the disease heterogeneity even in a single type of cancer, modern day cancer therapy is like a personalized medicine. Examples are many, such as melanoma with and without B-Raf mutations are having different treatment regimen, lung cancer treatment depends on EGFR mutation status. In this context, discovery of new drugs targeting novel proteins are of paramount importance. Here are some examples of clinically validated novel cancer drug targets:

**PI3K-mTOR pathway inhibitor**: Phosphatidylinositol-3-kinase (PI3K)/AKT/mammalian target of rapamycin (mTOR) signaling cascade is one of the most major intracellular pathways that are being known as a master regulator for different types of cancer. Herculean efforts have been made to the development of drugs targeting PI3K-mTOR signaling pathways, some of them are currently being pursued in active clinical trials, and it is becoming increasingly evident that PI3K-mTOR inhibitors are shown to be promising agents as anticancer therapeutics. PI3K inhibitors are categorized into dual PI3K-mTOR inhibitors, pan-PI3K inhibitors and isoform-specific inhibitors, on the other hand, mTOR inhibitor like rapamycin are either being used as combination therapy or in the active clinical trial against solid tumors in combination with some chemotherapeutic drugs.

**Smac Mimetics as Cancer Therapeutics**: One of the hallmarks of cancer is to bypass the programmed cell death. The well-known IAP (inhibitor of apoptosis) proteins are overexpressed in numerous cancer types, thus making them promising target for cancer therapeutics. There are several Smac mimetics small molecule inhibitors that mimic Smac, an endogenous antagonist of IAP proteins. In recent years, the Smac mimetics inhibitors show favorable response in preclinical trials. The potential of these mimetics are to directly trigger cancer cell death by alone and more importantly, synergistic response with the conventional therapeutics, radiotherapy, and immunotherapy. Presently, numerous Smac mimetics are under evaluation in early clinical trials as monotherapy and in combination for several cancers (i.e., GDC-0917/CUDC-427, LCL161, AT-406/Debio1143, HGS1029, and TL32711/birinapant) (Fulda 2015).

**Hedgehog pathway Inhibitor:** The hedgehog (Hh) pathway plays a critical role in cancer development and around 25% of all cancers have aberrant Hh pathway activation. Vismodegib (Erivedge) is the first hedgehog pathway inhibitor approved by FDA in 2012. It is used in some situations to treat people with metastatic basal cell skin carcinoma.

**Proteasome Inhibitor:** Proteasome helps to break down proteins into smaller parts that the cell doesn't need. Drug that blocks proteasomes from working are called proteasome inhibitors. They result in a load of unwanted proteins in the cell, which forces cancer cells to die. Bortezomib (Velcade) is the first FDA approved proteasome inhibitor drug used to treat myeloma and melanoma. Currently, multiple clinical trials are ongoing using proteasome inhibitor as a combination with other drugs or MAbs.

**Epigenetic regulation in Cancer manifestation:**

Chromosomal abnormalities like aneuploidy and translocations are regarded as hallmarks of malignancy. The abnormal expression of the non-histone proteins, which are also involved in the maintenance of the chromatin structure, might result in these chromosomal abnormalities as well as alteration in gene regulation during oncogenesis. Recent advances in the field of epigenetics suggest that oncogenic development could be closely associated with the altered epigenetic state of the genome. Such epigenetic changes involve aberrant DNA methylation, the alteration of chromatin components in DNA packaging (Ellis et al. 2009; Sharma et al. 2009), altered posttranslational modifications of histones, and also anomalous expression of noncoding RNAs like miRNAs. Chromatin alterations and DNA methylation cumulatively alter the epigenetic regulation of gene transcription in all the stages of tumor progression. Histone acetyltransferases and deacetylases are known to aberrantly express in cancers, resulting in anomalous acetylation of histones as well as non-histone proteins. Thus both expression as well as the posttranslational modifications of non-histone proteins might play determining roles in the onset of cancer and its progression. Microarray analysis from various tumor tissues has revealed the importance of miRNAs in the prediction, diagnosis, and prognosis of tumor formation. **Oncogenic miRNAs (oncomiRs)** are usually overexpressed in cancers while tumor suppressive miRNAs are downregulated quite similar to their mRNA counterparts. When these oncomiRs or tumor suppressor miRNAs are repressed or stimulated, respectively, the oncogenic properties of a tumor cell is significantly reduced. Certain cancers become addicted to these oncomiR to such an extent that suppression of the oncomiR results in complete reduction of the tumor (Reddy 2015; Block et al. 2017).

Apart from the DNA methylation status, the most studied histone modification studied in the context of Cancer is histone methylation. From studies of various cancer genomes and epigenomes through the high throughput sequencing and mass spectrometric analysis, we find recurrent translocation and mutations in a variety of lysine methyltransferases. Various studies have established the dichotomous role of EZH2 in human cancer cells. EZH2 was found to be overexpressed in prostrate and breast cancer which resulted in poor prognosis (Margueron and Reinberg 2011). On the contrary, recent studies have revealed coding mutations in the EZH2 gene in various lymphoid and myeloid neoplasms establishing it as a tumor suppressive gene. However, EZH2 inhibition for cancer treatment holds a great promise and also disrupts the quiescence phase of cancer stem cells. There are several EZH2 inhibitors like EPZ6438, CPI-1205, GSK126, CPI-0610, etc. that have been under consideration, gaining potential success in hematologic malignancies along with solid tumors.

**Perspective in the Indian Context:** In India, the number of cancer patients is dramatically increasing. The reason could be due to the gradual increase in life expectancy and also rapidly changing life style of Indian population. Although over the period large number of cancer clinics (most of these are private hospital/clinics) have emerged in different parts of the country, the fundamental drugs have hardly changed for the middle class and lower middle class Indian (in terms of annual income). Most of the drugs used are broad based cytotoxic agents, such as doxorubicin, cisplatin, and 5FU. The cancer specific or target specific drugs are not reachable for the common Indian, due to the cost of these drugs. So need of the hour would be to produce less toxic targeted drugs in cheaper rate. Although the world is moving towards the combination of targeted chemotherapy and immunotherapy, for most of the Indian, it is impossible to pay for these treatments. India need to invest more effort to generate immunotherapeutics and antineoplastic biologicals indigenously. The traditional Indian medicine, in combination with cheaper chemotherapeutics, is experimentally shown to be more effective and less toxic. We need to explore this possibility with mega consortium.

**Acknowledgement** We acknowledge CSIR-Central Drug Research institute, Lucknow. TKK is a recipient of Sir J.C. Bose National Fellowship.

## References

Block TM, Rawat S, Brosgart CL, Francisco S (2017) HHS Public Access 76:69–81

Cancer.Net (2020a) Understanding immunotherapy. https://www.cancer.net/navigating-cancer-care/how-cancer-treated/immunotherapy-and-vaccines/understanding-immunotherapy

Cancer.Net (2020b) Understanding targeted therapy. https://www.cancer.net/navigating-cancer-care/how-cancer-treated/personalized-and-targeted-therapies/understanding-targeted-therapy

David AR, Zimmerman MR (2010) Cancer: an old disease, a new disease or something in between? Nat Rev Cancer 10(10):728–733. https://doi.org/10.1038/nrc2914

Ellis L, Atadja PW, Johnstone RW (2009) Epigenetics in cancer: targeting chromatin modifications. Mol Cancer Ther 8:1409–1420

Fulda S (2015) Promises and challenges of smac mimetics as cancer therapeutics. Clin Cancer Res 21(22): 5030–5036

Helen A, Stafford (2000) Crown Gall disease and Agrobacterium Tumefaciens: a study of the history, present knowledge, missing information, and impact on molecular genetics. Bot Rev 66 (1):99–118

India State-Level Disease Burden Initiative Cancer Collaborators (2018) The burden of cancers and their variations across the states of India: the global burden of disease study 1990–2016. Lancet Oncol 19:1289–1306. https://doi.org/10.1016/S1470-2045(18)30447-9

Margueron R, Reinberg D (2011) The Polycomb complex PRC2 and its mark in life. Nature 469 (7330):343–349. https://doi.org/10.1038/nature09784

National Cancer Institute (n.d.) SEER training modules, Cancer registration & surveillance modules: cancer classification. https://training.seer.cancer.gov/disease/categories/classification.html

NCDIR (2012–2014) Three-year report of population based cancer registries, Bangalore: National Cancer Registry Programme, National Centre for Disease Informatics & Research, Indian Council of Medical Research. http://ncdirindia.org/All_Reports/PBCR_REPORT_2012_2014/ALL_CONTENT/PDF_Printed_Version/Chapter2_Printed.pdf

Oncology Nurse Advisor (2014) Radiation therapy for cancer (fact sheet). https://www.oncologynurseadvisor.com/home/for-patients/fact-sheets/radiation-therapy-for-cancer-fact-sheet/

Reddy KB (2015) MicroRNA (miRNA) in cancer. Cancer Cell Int 15:4–9

Rothschild BM, Tanke DH, Helbling M et al (2003) Epidemiologic study of tumors in dinosaurs. Naturwissenschaften 90:495. https://doi.org/10.1007/s00114-003-0473-9

Sharma S, Kelly TK, Jones PA (2009) Epigenetics in cancer. Carcinogenesis 31:27–36

Sikder S, Kumari S, Mustafi P, Ramdas N, Padhi S, Saha A, Bhaduri U, Banerjee B, Manjithaya R, Kundu TK (2019) Non-histone human chromatin protein PC4 is critical for genomic integrity and negatively regulates autophagy. FEBS J 286:4422. https://doi.org/10.1111/febs.14952

Smith RD, Mallath MK (2019) History of the growing burden of cancer in India: from antiquity to the 21st century. J Glob Oncol 5:1–15. https://doi.org/10.1200/JGO.19.00048

Sullivan R, Badwe RA, Rath GK, Pramesh CS, Shanta V, Digumarti R, D'Cruz A, Sharma SC, Viswanath L, Shet A, Vijayakumar M, Lewison G, Chandy M, Kulkarni P, Bardia MR, Kumar S, Sarin R, Sebastian P, Dhillon PK, Rajaraman P, Trimble EL, Aggarwal A, Vijaykumar DK, Purushotham AD (2014) Cancer research in India: national priorities, global results. Lancet Oncol 15(6):e213–e222

The American Cancer Society Medical and Editorial Content Team (2014) Early history of cancer; Defining cancer. https://www.cancer.org/cancer/cancer-basics/history-of-cancer/what-is-cancer.html

The Balance (2018) Top 20 blockbuster cancer drugs. https://www.thebalance.com/top-cancer-drugs-2663234

University of Rochester Medical Centre (n.d.) Health encyclopedia; Hormone therapy for cancer treatment. https://www.urmc.rochester.edu/encyclopedia/content.aspx?contenttypeid=85&contentid=p07236

Urandur S, Banala VT, Shukla RP, Mittapelly N, Pandey G, Kalleti N, Mitra K, Rath SK, Trivedi R, Ramarao P, Mishra PR (2018) Anisamide anchored lyotropic nano liquid crystalline particles with AIE effect - a smart optical beacon for tumor imaging and therapy. ACS Appl Mater Interfaces 10:12960. https://doi.org/10.1021/acsami.7b19109

WHO (2018) Cancer, Key facts. https://www.who.int/news-room/fact-sheets/detail/cancer

Cancerindia. Org (2020) Cancer statistics. http://cancerindia.org.in/cancer-statistics

**Tapas Kumar Kundu**, Ph.D. has created an excellent blend of Epigenetics and Chemical Biology, in context of cancer diagnostics and therapeutics. He has not only discovered several therapeutically relevant inhibitors of acetyltransferases, but also has shown that these inhibitors are highly effective to reduce the cancer (head and neck, liver) growth efficiently using various model systems. Prof. Kundu has done his PhD from the IISc, Bengaluru in 1995 and worked as Post-Doctoral Fellow at Rockefeller University, USA from 1996 to 1999. He joined JNCASR, Bengaluru in 1999. Currently, he is serving as the Director, CSIR-CDRI, Lucknow since 2018.

**Anand P. Kulkarni**, Ph.D. did his Master's Degree in Botany in 2001 from Karnataka University, Dharwad, Karnataka and Ph.D. in 2005 from CSIR-Central Food Technological Research Institute, Mysore with CSIR-NET Junior & Senior Research Fellowship (2001–2005). He has published several research publications on bioactive principles from Pomegranate, which are cited widely. He joined CSIR-CDRI in 2005 as Scientist—Planning. Currently, he is looking after the Scientific Directorate and Program Planning at CSIR-CDRI.

**Dipak Datta**, Ph.D. is currently Chairperson of the Division of Cancer Biology at CSIR-CDRI. In his doctoral work, he studied the immune-regulatory role of chemokines in cancer. He did his post-doctoral studies in Children's Hospital Boston and Harvard Medical School till 2011. Being a Principal Investigator in the current position, his lab research focuses on the understanding of genetic and epigenetic bases of solid tumor heterogeneity, Cancer Stem Cell (CSC) plasticity, and regulation of their key features in relation to therapy resistance and cancer metastasis.

**Sweta Sikder**, Ph.D. completed her Ph.D. in 2019 from Transcription and Disease Laboratory, JNCASR, Bangalore. She has published a few innovative research papers in the area of autophagy and cancer. She is recipient of Indian Academy of Sciences Summer Research Fellowship, CSIR-NET Junior & also Senior Research Fellowship for Ph.D. During the time of writting this book chapter, she was working as a Visiting Postdoctoral fellow in National Cancer Institute, NIH, USA.

# 11 Biopharmaceutical Development in India: Recommendations on Collaboration and Innovation to Enable Affordable Healthcare

Narendra Chirmule, Shilpa Bhat, and Sabyasachi Mondal

In India, the pharmaceutical industry has excelled in the commercialization of generics; a few biopharmaceutical companies are now obtaining global approvals for biosimilars. However, there is a lack-of-drive to develop novel biopharmaceuticals, which may be attributed to limited understanding of the complexities of drug development, low investor confidence, and an absence of a robust culture of innovation.

Institutions across India notably Department of Biotechnology (DBT), Indian Council of Medical Research (ICMR), Council for Scientific and Industrial Research (CSIR), and Department of Atomic Energy (DAE) support basic and applied research. Most government and private academic institutions have curriculums in biotechnology. The Indian biopharmaceutical industry also invests ~5–10% of revenues in R&D. India can therefore experience exponential growth in novel drug development provided that government, academics, and industry collaborate effectively.

The goal of this review is to address the problem statement: "the biopharmaceutical industry in India faces challenges in: (1) scientific inclination and depth for novel drug discovery and clinical development, (2) effective collaboration of academia and industry and (3) a culture of innovation."

N. Chirmule (✉)
Biocon Research Laboratories, SymphonyTech Biologics, Pune, India
e-mail: Narendra.Chirmule@symphonytech.com

S. Bhat · S. Mondal
BIOCON, Bangalore, Karnataka, India



M. Dikshit (ed.), *Drug Discovery and Drug Development*,
https://doi.org/10.1007/978-981-15-8002-4_11

## 11.1 The Biopharmaceutical Industry Landscape

Development and commercialization of biopharmaceuticals requires a tight interweaving between science and business (Pisano 2006). For the last four decades, inventions in the field of molecular biology ranging from recombinant DNA to gene editing have changed the course of drug discovery and development. Biopharmaceutical companies took advantage of these molecular biology platforms to develop drugs that could fight illnesses. They were successful in doing so by cleverly navigating the risk and uncertainty involved in this business. This risk mitigation can be seen in partnerships of small with large companies, such as Genentech with Eli Lilly (Hughes 2013) and Amgen with Johnson and Johnson (Binder and Bashe 2008), paved the way for successful commercialization of recombinant insulin and erythropoietin, respectively. Thus, investment decisions based on scientific discoveries, inventions, and technologies have become the anatomy of the biopharmaceutical industry (Pisano 2006), which has led to unprecedented success of business model (Fisken and Rutherford 2001).

### 11.1.1 The Science of Biopharmaceuticals

Emerging technology platforms such as recombinant DNA, genome sequencing and informatics, gene therapy, and gene editing tools have led to several discoveries and inventions in biological sciences. These in turn have helped identify novel mechanisms of treating diseases. For example, the sequencing of the human genome has unlocked the potential for personalized medicine and ushered in a new era in healthcare. It is interesting to note that only ~266 of the ~30,000 genes have been genome-based targets for approved drugs (Overington 2006; Alex et al. 2015) allowing for additional targets to be discovered. In addition, synthesis of synthetic nucleotides X and Y (Zhang et al. 2017) and discovery of non-coding RNA in gene regulation (Jackson et al. 2018) further expand the possible targets that can be identified for developing novel drugs. These discoveries are being enabled by technology platforms such as recombinant DNA, genome sequencing and informatics, gene therapy, and gene editing tools.

Biopharmaceuticals supplements the small molecule modality for the discovery of new molecular entities (NME). While small molecules are amenable to interacting with intracellular components such as nuclear proteins and those that cross the blood–brain barrier, biopharmaceuticals have higher specificity and less off-target effects on cell surface receptors, soluble proteins, and RNA-DNA targets. However, the major difference between biopharmaceuticals and small molecules is in process-development and manufacturing (Truscheim et al. 2001). In this respect, the cell and molecular biology technologies used to express proteins, design vectors, transduce genes and synthesize nucleic acids, require novel manufacturing capabilities that are different from chemistry-based processes. The processes of discovery, target validation [which involves understanding disease biology, pharmacology (efficacy), toxicology (safety)], and clinical development are generally similar. Elegant reviews have described these differences in drug development processes (Munos 2009; Kneller 2010). This review focuses on the advances in biopharmaceutical development.

Before the 1990s, most large pharmaceutical companies relied on internal R&D capability to generate NMEs using proprietary drug candidate-libraries and internal scientific knowledge. Since the 1990s, however, companies such as Genentech, Amgen, Regeneron, Celgene, Genzyme, and others focused their efforts by targeting protein sequences from the human genome (Ratti and Trist 2001; Edler and Herring 2016). These discovery platforms, such as phage display libraries, rabbit hybridomas, yeast surface display and xenogeneic mice, have enabled identifying novel targets using the molecular biology-based technology, combined with humanization and affinity maturation (An 2009). Biopharmaceuticals developed using these processes have succeeded in gaining approval of first-in-class therapies and are major contributor for the growth of the industry.

### 11.1.2 The Business of Novel Biopharmaceuticals

According to the analyst report (Evaluate Pharma 2018), the market share of biopharmaceutical products has increased from 10% in 2010 to 25% in 2017 and expected to represent 31% of total prescription drug and over-the-counter sales by 2024 (Fig. 11.1a). This paradigm shift is indicative of an increased reliance on biopharmaceuticals for novel therapies. The best-selling drug Humira (adalimumab, AbbVie) enjoys $ 19 billion (2018) revenue and a cumulative sale of approximately $ 115 billion since its launch 18 years ago. Three out of top six drugs that have more than $ 100 billion cumulative sales are biopharmaceuticals (Fig. 11.1b).

The advantages of biopharmaceuticals are in their higher specificity, better safety profiles, and longer duration of action. The mechanism of longer duration of action is attributed to the neonatal FcRn receptor mediated transcytosis. However, over the last four decades, the promise of biopharmaceuticals has been tempered by several failures resulting in loss of investments. Some examples of the unanticipated challenges include increased suicides with anti-IL17 antibody therapy for psoriasis, hypersensitivity reactions to anti-TNF antibody therapeutics (adalimumab and infliximab), and on-target JC virus activation resulting in progressive multifocal leukoencephalopathy (PML) (Schmidt 2015; Susan Lee 2005; Kleinschmidt-DeMasters and Tyler 2005). Despite these failures, the attrition rates of biopharmaceuticals during development are lower than small molecules, thereby enabling higher risk-taking in investments (DiMasi and Grabowski 2007). More recently, the expiry of patents of these biopharmaceuticals has led to the development of biosimilars, which are approved versions of original products. In the last decade, biosimilars have had a major impact on the economics of biopharmaceuticals (Blackstone and Joseph 2013).

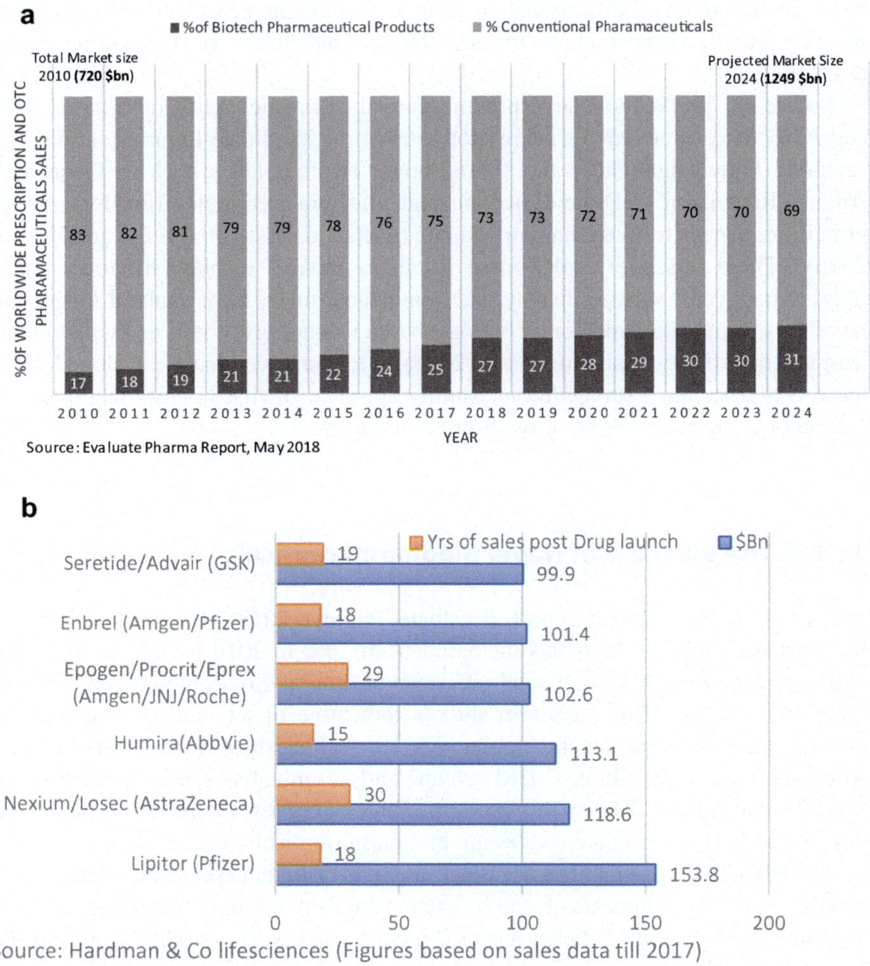


**Fig. 11.1** (**a**) Worldwide prescription drug and over-the-counter pharmaceutical sales in $ billion. There is a trend of increased reliance on biopharmaceuticals for novel therapies. Total market size 2010 was $ 720 billion; projected market size in 2024 is $ 1249 billion (source: Evaluate Pharma Report, May 2018). (**b**) $ 100 billion club. Six drugs that have cumulatively made $ 100 billion or more; three of them are biopharmaceuticals. (Source: Hardman & Co life sciences. Figures based on sales data till 2017)

### 11.1.3 The Science and Business of Biosimilars

Biopharmaceuticals account for only 1% of the total prescription drugs in the USA but 28% of total drug expenditure (Hung 2017; Scott Morton 2018). Introduction of biosimilars has resulted in an increased competition which has reduced healthcare costs (Mehr and Brook 2017). The European Medicines Agency (EMA) and the US FDA have approved several biosimilars since 2009. Several emerging markets such as Brazil, Russia, India, China, and South Korea are following suit.

Complex market drivers such as regulatory-uncertainty, production-complexity, interchangeability-rules, payer-environment, therapeutic-area, physician-support, and patient-awareness are involved in the use of biosimilar products; and each of these factors have different levels of importance in different geographies. The factors that influence the approval and use of biosimilars in each region are: (1) the biosimilar approval regulatory process, (2) economic condition of patient population (per capita income, GDP of the country, payment mechanism—out of pocket payment vs. payer driven), (3) market size, and (4) market access. There is an enormous potential for reducing costs and increasing access to biopharmaceuticals in emerging markets. The healthcare industry in India has introduced biosimilar guidance in 2012 (CDSCO 2011a, b, Guidance for New Drug Approval 2011), with a major upgrade in 2016 (CDSCO 2016, Guidance for New Drug Approval 2016), resulting in improved access to these classes of drugs.

## 11.2 The Advent of Indian Biopharmaceutical Industry

Indian pharmaceutical companies have excelled in manufacturing generics for global markets and currently supply ~40% of the drugs worldwide (Mukherjee 2018). A large part of the accomplishment can be attributed to the economic policies created by pre-TRIPS (Trade-Related Aspects of Intellectual Property Rights of the World Trade Organization 2005) laws such as the India Patent Act (IPA) and Drug Price Control Order (DPCO). Indian companies could manufacture products at a fraction of the cost required overseas. In addition, US FDA regulations for generics do not require extensive clinical studies under the 505(j) regulatory pathways. Thus, the Indian pharmaceutical industry has enjoyed extraordinary success in the generics space. For patients, the competition in the marketplace has resulted in an exponential reduction in the cost of these drugs.

Several Indian biopharmaceutical companies have forayed into developing biosimilars gaining from their experience in generics manufacturing and commercialization. The biopharmaceutical product pipeline of major Indian companies is listed in Table 11.1. However, the complex manufacturing processes of this class of drugs, the high investment costs both in manufacturing and clinical trials have resulted in companies not being able to sustain their business models. A few companies like Biocon Ltd (NSE: BIOCON), Intas Private Limited, and Lupin Ltd (NSE: LUPIN) have emerged as outliers in this context with molecules such as glargine, trastuzumab, pegfilgrastim, and etanercept (Nawrat 2018). With the enhanced efficiency and attention to compliance, it remains to be seen how Indian biopharmaceutical companies will develop biosimilars for the global markets.

Novel biopharmaceuticals, on the other hand, require significantly more investments than biosimilars. A handful of Indian companies such as Sun Pharma, Intas, Reliance, and Biocon have developed novels and obtained approval in India (Das 2017). In this review, we discuss that one of the critical requirements to advance the field of novel biopharmaceuticals in India is creation of a collaborative

**Table 11.1** Indian companies developing biopharmaceuticals

| Company name | Product name (approved in India) | Active substance | Disease area |
|---|---|---|---|
| Biocon Ltd | Basalog | *Insulin glargine* | Diabetes |
| | CanMab | *Trastuzumab* | Breast cancer |
| | Erypro | *Epoetin alfa* | Anemia, cancer, chronic kidney failure |
| | Insugen | *Human insulin* | Diabetes mellitus |
| | Krabeva | *Bevacizumab* | Metastatic colorectal cancer, lung cancer, kidney cancer, cervical cancer, ovarian cancer, brain cancers |
| Dr. Reddy's Laboratory | Cresp | *Darbepoetin alfa* | Anemia, cancer, chronic kidney failure |
| | Grafeel | *Filgrastim* | Neutropenia, hematopoietic stem cell transplantation, cancer |
| | Peg-grafeel | *Pegfilgrastim* | Cancer, neutropenia |
| | Reditux | *Rituximab* | Leukemia, lymphoma, rheumatoid arthritis |
| Intas Pharmaceuticals | Epofit/ Erykine | *Epoetin alfa* | Anemia, cancer, chronic kidney failure |
| | Folisurge | *Follitropin alfa (follicle stimulating hormone)* | Female infertility, spermatogenesis in men |
| | Intacept | *Etanercept* | Ankylosing spondylitis, juvenile idiopathic arthritis, psoriasis, psoriatic arthritis, rheumatoid arthritis |
| | Intalfa | *Interferon alfa-2b* | Carcinoid tumor, chronic hepatitis B, chronic hepatitis C, hairy cell leukemia, chronic myelogenous leukemia, BCR-ABL positive, follicular lymphoma, malignant melanoma, multiple myeloma |
| | MabTas | *Rituximab* | Lymphoma, non-Hodgkin's lymphoma |
| | Neukine | *Filgrastim* | Neutropenia, hematopoietic stem cell transplantation, cancer |
| | Neupeg | *Pegfilgrastim* | Cancer, neutropenia |
| | Peg-interferon alfa 2b | *Pegylated recombinant human interferon alfa 2b* | Chronic hepatitis B |
| | Razumab | *Ranibizumab* | Wet macular degeneration, macular edema, degenerative myopia, diabetes complications |

(continued)

**Table 11.1** (continued)

| Company name | Product name (approved in India) | Active substance | Disease area |
|---|---|---|---|
| | Terifrac | *Teriparatide (parathyroid hormone)* | Postmenopausal women with osteoporosis who are at high risk for fracture |
| Reliance Life Sciences | AbcixiRel | Abciximab | Angina, cardiac ischemia |
| | AdaliRel | Adalimumab | Rheumatoid arthritis, juvenile idiopathic arthritis, psoriatic arthritis, ankylosing spondylitis, Crohn's disease, ulcerative colitis, plaque psoriasis, hidradenitis suppurativa, uveitis |
| | BevaciRel | Bevacizumab | Metastatic colorectal cancer (mCRC), non-Squamous non–Small cell lung cancer (NSCLC), recurrent glioblastoma, metastatic renal cell carcinoma (mRCC), persistent, recurrent, or metastatic carcinoma of the cervix, recurrent epithelial ovarian, fallopian tube, or primary peritoneal cancer |
| | ChorioRel | *Chorionic gonadotrophin hormone r-hCG* | Female infertility |
| | DarbeRel | Darbepoetin | Anemia due to chronic kidney disease, anemia due to chemotherapy in patient with cancer |
| | EtanaRel | Etanercept | Rheumatoid arthritis, polyarticular juvenile idiopathic arthritis, psoriatic arthritis, ankylosing spondylitis, plaque psoriasis |
| | FostiRel | | |
| | Infimab | Infliximab | Rheumatoid arthritis, psoriatic arthritis, ankylosing spondylitis, ulcerative colitis, plaque psoriasis |
| | MiRel | | |
| | OmaliRel | Omalizumab | Asthma, chronic idiopathic urticaria (CIU) |
| | PegReliGrast | Pegfilgrastim | Patients with cancer receiving myelosuppressive chemotherapy, patients with hematopoietic subsyndrome of acute radiation syndrome |
| | Relibeta | *Interferon beta-1a* | Multiple sclerosis |
| | ReliFeron | *Interferon alfa-2b* | BCR-ABL positive, carcinoid tumor, chronic hepatitis B, chronic hepatitis |

(continued)

**Table 11.1** (continued)

| Company name | Product name (approved in India) | Active substance | Disease area |
|---|---|---|---|
| | | | C, chronic myelogenous leukemia, follicular lymphoma, hairy cell leukemia, melanoma, multiple myeloma |
| | ReliGrast | *Filgrastim* | Neutropenia |
| | Relipoietin | *Epoetin alpha* | Anemia, autologous blood transfusion, chronic kidney failure, HIV |
| | RituxiRel | *Rituximab* | Non-Hodgkin's lymphoma, rheumatoid arthritis |
| | SomatoRel | Somatotropin | For the treatment of pediatric patients with growth failure due to inadequate secretion of endogenous growth hormone, pediatric patients with short stature associated with Noonan syndrome, pediatric patients with short stature associated with Turner syndrome, pediatrics patients with short stature born small for gestational age (SGA) with no catch-up growth by age 2–4 years |
| | TenecteRel | Tenecteplase | In adults for thrombolytic treatment of suspected myocardial infarction with persistent ST segment elevation or recent Left Bundle Branch Block within 6 h after the onset of acute myocardial infarction (AMI) symptoms |
| | TrastuRel | Trastuzumab | Adjuvant breast cancer, metastatic breast cancer, metastatic gastric cancer |
| Lupin | Filgrastim | *Filgrastim* | Neutropenia |
| | Peg-filgrastim | *Pegfilgrastim* | Cancer, neutropenia |
| Wockhardt | Biovac-B | *Hepatitis B vaccine* | Hepatitis B |
| | Glaritus | *Insulin glargine* | Diabetes mellitus |
| | Wepox | *Epoetin alfa* | Anemia, cancer, chronic kidney failure |
| | Wosulin | *Human insulin* | Diabetes mellitus |
| Cipla | Etacept | *Etanercept* | Ankylosing spondylitis, rheumatoid arthritis, psoriatic arthritis, psoriasis, juvenile rheumatoid arthritis |
| Cadila Pharmaceuticals | Filgrastim | *Filgrastim* | Neutropenia |

(continued)

**Table 11.1** (continued)

| Company name | Product name (approved in India) | Active substance | Disease area |
|---|---|---|---|
| Hetero Group | Cizumab | *Bevacizumab* | Colorectal cancer |
| | Maball | *Rituximab* | Lymphoma, non-Hodgkin's lymphoma |
| USV | Filgrastim | *Filgrastim* | Neutropenia |
| | Teriparatide | *Teriparatide (parathyroid hormone)* | Postmenopausal women with osteoporosis who are at high risk for fracture |

environment. Academia, industry, government institutions and regulators need to work closely together and build an infrastructure that enables innovation.

## 11.3 A Culture of Collaboration Is Essential for Innovation

It is increasingly evident that no major accomplishments can be achieved by a single person or institution, collaboration is essential for success. Development of biopharmaceuticals involves generation and analysis of massive amounts of data that requires interaction between scientists in academia and industry. Effective collaboration enables sharing of knowledge, instrumentation, finances, and governance processes, which enhances the innovation potential of each party, far greater than each one alone (Tallman and Phene 2007).

### 11.3.1 Measuring Effective Collaboration

At the academic level, the quality of collaboration is measured by the number of co-authored articles published in a given time period. Nature index (Nature Index 2019) provides a database for measuring collaborations at institutional, corporate, national, and international levels, based on research articles published in a selected group of journals. Table 11.2 shows the collaboration indices of top ten global institutes and companies. At the industry level, Moorkens et al. (2017) have compiled a list of international collaboration between biopharmaceutical companies. Such quantitation of the nature of interactions between institutions will enable development of processes to improve the quality of collaborations (Bennett and Gadlin 2012; Ankrah and Omar 2015).

### 11.3.2 Factors for Successful Collaboration

In order for collaboration to be successful, the interacting teams must (1) build trust within team, (2) respect dissimilarities and embrace diversity, (3) list and act on

**Table 11.2** Nature-Index collaboration indices of Indian and global companies as of March 2019

| No. | Top ten institutes in India | FC | AC | No. | Top ten companies in India | FC | AC |
|---|---|---|---|---|---|---|---|
| 1 | Council of Scientific and Industrial Research (CSIR), India | 17 | 29 | 1 | Glenmark Pharmaceuticals Limited, India | 1 | 1 |
| 2 | Indian Institute of Science (IISc), India | 11 | 18 | 2 | Strand Life Sciences Pvt. Ltd., India | 0.08 | 1 |
| 3 | Tata Institute of Fundamental Research (TIFR), India | 10 | 28 | 3 | Serum Institute of India Ltd. (SII), India | 0.08 | 8 |
| 4 | Indian Institute of Science Education and Research (IISER), India | 9 | 22 | 4 | Aurigene Discovery Technologies Limited, India | 0.05 | 1 |
| 5 | National Centre for Biological Sciences (NCBS), TIFR, India | 8 | 21 | 5 | Tata Consultancy Services Limited (TCS), India | 0.03 | 1 |
| 6 | India Ministry of Science and Technology, India | 7 | 14 | | | | |
| 7 | Indian Institutes of Technology (IITs), India | 4 | 16 | | | | |
| 8 | Indian Institute of Science Education and Research Mohali (IISER Mohali), India | 4 | 6 | | – | | |
| 9 | Centre for Cellular and Molecular Biology (CCMB), CSIR, India | 3 | 6 | | | | |
| 10 | Centre for DNA Fingerprinting and Diagnostics (CDFD), India | 3 | 4 | | | | |
| | **Top ten institutes - Global** | | | | **Top ten companies - Global** | | |
| 1 | Harvard University, United States of America (USA) | 612 | 1488 | 1 | F. Hoffmann-La Roche AG, Switzerland | 35.56 | 74 |
| 2 | Stanford University, United States of America (USA) | 272 | 631 | 2 | Novartis International AG, Switzerland | 29.14 | 99 |
| 3 | Massachusetts Institute of Technology (MIT), United States of America (USA) | 200 | 740 | 3 | AstraZeneca plc, United Kingdom (UK) | 16.93 | 73 |
| 4 | Yale University, United States of America (USA) | 175 | 402 | 4 | Merck & Co., Inc., United States of America (USA) | 12.39 | 38 |
| 5 | University of California, San Diego (UC San Diego), | 173 | 480 | 5 | Amgen Inc., United States of America (USA) | 10.66 | 46 |

(continued)

**Table 11.2** (continued)

| | | | | | | | |
|---|---|---|---|---|---|---|---|
| | United States of America (USA) | | | | | | |
| 6 | University of California San Francisco (UCSF), United States of America (USA) | 172 | 491 | 6 | Pfizer Inc., United States of America (USA) | 10.54 | 40 |
| 7 | University of Pennsylvania (Penn), United States of America (USA) | 169 | 436 | 7 | GlaxoSmithKline plc. (GSK), United Kingdom (UK) | 9.51 | 42 |
| 8 | Columbia University in the City of New York (CU), United States of America (USA) | 168 | 412 | 8 | Johnson & Johnson, United States of America (USA) | 5.54 | 26 |
| 9 | University of Oxford, United Kingdom (UK) | 166 | 503 | 9 | Formosa Plastics Group, Taiwan | 5.27 | 16 |
| 10 | Johns Hopkins University (JHU), United States of America (USA) | 153 | 420 | 10 | AbbVie Inc., United States of America (USA) | 5.01 | 12 |

strengths and weaknesses of the team, (4) share data, (5) define goals and expectations, (6) articulate criteria for rewards and recognition, and (7) develop effective communication mechanisms. Leadership and program management groups should take accountability of understanding the team dynamics and ensure implementation of governance processes to manage the successful performance of teams.

### 11.3.3 Benefits of Collaboration

The rewards of a successful academia-industry collaboration are that: (1) industries gain a lucrative novel drug pipeline; (2) academic institution commercializes their research and discoveries; (3) students and faculty learn the drug-development processes; (4) the community benefits by having access to novel and affordable healthcare; (5) employees in industries nurture talent and enhance innovation; (6) partners gain better funding opportunities for innovation; and (7) regulators develop policies for innovation. These benefits are evident in the case of the top four blockbuster biopharmaceuticals—Humira, Herceptin, Enbrel, and Rituxan. Abbott's Humira employed Cambridge Antibody Technology's phage display platform. Amgen's Enbrel was developed as a chimeric molecule by Bruce Beutler and his colleagues at the University of Texas Southwestern Medical Center. The early research on Herceptin to treat breast cancer patients was done by Denis Slamon in UCLA. Rituximab was developed after Lee Nadler from the Dana Farber Cancer Institute discovered CD20 overexpression on malignant B cells (Storz 2014; Slamon et al. 1987; McCafferty 2010). Such collaborative ecosystem for innovation can transform biopharmaceutical industry in India.

### 11.3.4 The Collaboration Report Card for India

According to the international collaborations captured by Nature Index, the scores for the top ten countries are compared with India in Table 11.3. The low score for India strongly suggests the need for a concerted effort in improving the culture of collaboration. In order to enhance score of India, there is an urgency to intervene in the following three areas: (1) biotechnology education and training system to develop the relevant skills required for the biopharmaceutical industry, (2) research system to encourage new science and technology in the field of biological sciences, (3) processes of communication within and between institutes and industry, and (4) multi-disciplinary approaches of education and research, e.g., integration of quantum physics and artificial intelligence concepts in the life sciences. These recommendations are discussed further with examples:

#### 11.3.4.1 Education and Training

In the past decade, there has been a major emphasis on biotechnology in all academic institutions in India. However, the expansion of curriculum at the undergraduate and graduate levels has not been aligned with the current practices in industry. While subjects that are relevant in the biopharmaceutical industry—molecular biology, bioinformatics, cancer biology, immunology, biophysical and biochemical techniques—are a part of the curriculum, they must be taught using real-life examples and case studies by industry experts. Understanding the process of drug development requires an interdisciplinary approach that spans basic sciences of biology-chemistry-physics, chemical engineering, pharmacology, veterinary sciences, clinical studies, as well as economics and law. Consequently, the emphasis on collaboration for enabling diverse thought, and for merging interdisciplinary fields will prepare students and faculty to contribute towards progress in innovation in biopharmaceuticals.

**Table 11.3** Nature-Index collaboration score and percent contribution to collaboration in comparison with top ten nations, as of March 2019

| Country | Overall collaboration score | % Contribution | Top collaborator |
|---|---|---|---|
| USA | 4628.4 | 61.6 | China |
| UK | 1703.3 | 53.5 | USA |
| Germany | 1477.0 | 52.7 | USA |
| China | 970.0 | 48.4 | USA |
| France | 804.2 | 49.4 | USA |
| Canada | 513.8 | 44.0 | USA |
| Australia | 506.5 | 49.4 | USA |
| Switzerland | 486.2 | 45.1 | USA |
| Netherlands | 435.6 | 46.2 | USA |
| Spain | 394.78 | 47.8 | USA |
| India | 62.45 | 39.0 | USA |

#### 11.3.4.2 Interdisciplinary Science and Technology

A culture of collaboration fuels innovation by keeping pace with the advances in science and technology. The field of biological science has become highly interdisciplinary which has produced radical new discoveries such as delineating of the mechanism of action of non-coding RNA in regulation of protein synthesis (Jackson et al. 2018), and inventions of two new nucleotides to design new molecular structures (Zhang et al. 2017) both of which have the potential to transform drug development. Technological advances such as high-resolution imaging, high-order-structures analysis, high-throughput automation, and new-material science discoveries also require the understanding of diverse fields of biology, chemistry, physics, and engineering. Since academia has access to such advanced equipment and talent in an environment that enables these diverse interactions, it plays a crucial role in the innovative processes in development of new science and technology that can be utilized by industry.

#### 11.3.4.3 Micro-Environment

Inter-departmental collaborations within institutions are dependent on access to resources, capabilities, and expertise of individuals in diverse fields (principal investigators, students, and fellows). The cost of instrumentation runs in several thousand dollars. (Note, as an example, a university may have 2–5 HPLCs; there will be more than a 100 in an RND facility alone in a medium-sized firm.) Recent advancement of analytical technologies such as cryo-electron microscopy, two-dimensional nuclear magnetic resonance and CyTOF also requires significant investment not just in acquisition, but also in maintenance and enhancements.

A distinct example of the culture of the micro-environment within an institution is described by Venki Ramakrishnan in his recent book “The Gene Machine” (Ramakrishnan 2018). It was not until his move to the Medical Research Council laboratory in the University of Cambridge, that his research could make the seminal discoveries in understanding the structure and function of the ribosome. His research required use of molecular biology techniques and X-ray crystallography, combined with computing and modeling capabilities. Similarly, the discovery of the structure of DNA by Watson and Crick revolutionized the field of biotechnology by combining biology, chemistry, and X-ray diffraction, incidentally in the same lab in University of Cambridge. Thus, it is crucial that presidents, vice-chancellors, and faculty in institutions empower scientists to create a culture of active collaboration within an institution. These leadership roles should have a long enough tenure to have an impact.

#### 11.3.4.4 Macro-Environment

Inter-institutional collaboration is one of the central tenets of creating an environment of innovation. The co-locations of institutions in a geography enable such collaborations (Tallman and Phene 2007). Thirty years ago, several cities in the USA became centers of learning, as exemplified by Boston (Harvard, Boston College, Tufts University), Philadelphia (University of Pennsylvania, Temple, Jefferson) and San Francisco (University of California, Stanford). In the last decade, there has been

**Fig. 11.2** The biopharmaceutical map of India. We have pinned the website of all the major industry, government, academic and industries engaged in research and development of biopharmaceutical companies. It is a live list in Google Maps, which can be accessed through the following link. http://bit.ly/2YrvTVE

a significant investment by industry to co-locate in these regions. In India, however, despite co-location of government institutes, academic centers of excellence, and industry, illustrated in an interactive online-map in Fig. 11.2, the level of collaboration is low. A reason for this low collaboration could be attributed to the limited nature of innovation in novel drug discovery, possibly due to low risk appetite of the investors. Finally, other macro-environmental factors of collaboration include the political agenda of the country on healthcare, the health of economy (percent of GDP investment), and education policies, all of which together influence the ability of institutions to work in partnerships.

#### 11.3.4.5 Translational Science

Translational science involves the concept of developing drugs from bench to bedside. It requires highly cross-functional collaborations with interdisciplinary approaches. The activities in development of drugs spans discovery of disease biology and mechanism of action studies (primarily conducted in academic institutions), to process and product development, pharmacology, toxicology, and clinical trials (done in industry). The process requires both deep expertise in the specific fields of science (and economics and culture) and a breadth of knowledge of the entire steps of drug development (Fig. 11.3). Most academic institutions are not familiar with the complexity of drug development. A collaboration in translational science will result in the development of novel medicines to treat diseases. In India, the Translational Health Science and Technology Institute, in Faridabad, and the Rajiv Gandhi Center for Biotechnology Government Institutes and academic institutes such as the National Institute for Pharmaceutical and Educational Research (NIPER) could enhance their interactions with industries.

#### 11.3.4.6 Regulatory Governance

Regulatory agencies play a major role in the governance of the drug development process. The Central Drugs Standard Control Organization (CDSCO), headed by the Drug Controller General of India (DGCI), regulates pharmaceutical and medical devices, under the gamut of Ministry of Health and Family Welfare. The DCGI is advised by the Drug Technical Advisory Board (DTAB) and the Drug Consultative Committee (DCC). The regulation of safety under preclinical toxicology is governed by Review Committee of Genetic Manipulation (RCGM) under the Department of Biotechnology.

This complete process of governance includes a three-tier review committee comprising Institutional Biosafety Committees (IBSC) at the Institute/company; the Review Committee on Genetic Manipulation (RCGM) in the Department of Biotechnology; and the Genetic Engineering Approval Committee (GEAC) in the Ministry of Environment & Forests (MoE&F) for monitoring, evaluation and granting approval for research and development activities on recombinant DNA products. This exercise requires efficient collaboration initiatives of the various governance bodies providing a framework of establishing checks and balances in the approval of medicines (CDSCO 2007).

## 11.4 Innovation Is Central to Transforming the Biopharmaceutical Industry

A culture of innovation requires an entire society's involvement: basic and applied research, venture capital system, and internal and external motivating forces (academic, regulatory, economic, and cultural). Such an environment is required for problem-solving and inducing positive change.

**Is There a Process for Innovation?**

The "wheel" was never born through evolution; there are no animals with wheels. Humans invented it and have insisted not to re-invent it. There are two approaches for innovation—the top-down approach and the bottom-up approach.

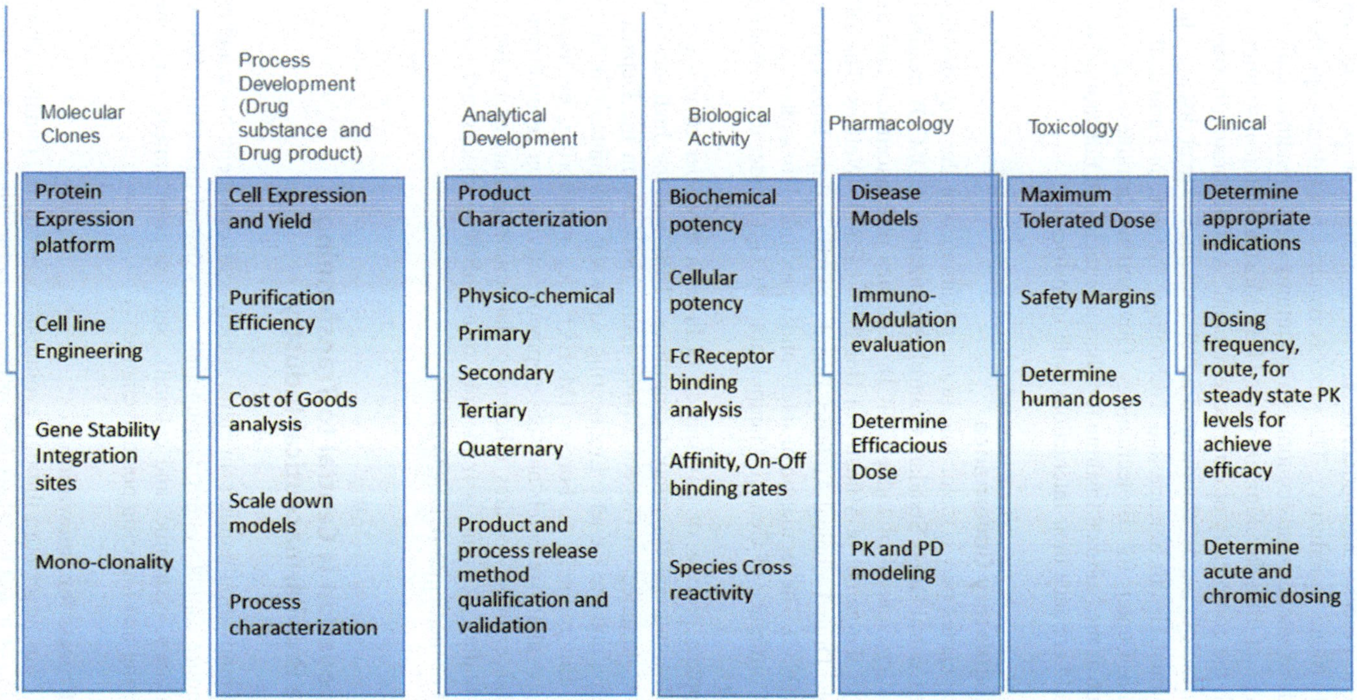


**Fig. 11.3** Key milestones in the drug development process for biopharmaceuticals. The key activities required for interactions (in the IND or BLA) with the regulatory agency in each function are listed

***The Top-Down Approach*** This process involves deconstructing the issue-at-hand, unpacking small chunks at a time, and folding it back once it is completed. Examples of this approach include solving structures of DNA, ribosomes, mitochondria and sequencing the human genome.

***The Bottom-Up Approach*** This process is nature's way of innovation (evolution) where new ingredients are added to the existing situation which results in formation of a completely new entity. The addition of ingredients is a random process and requires the survival of the fittest. Though this path is more labor and resource-intensive, the outcome shaped by the innovation is extremely robust and endures the test of time. An example is how a tadpole evolves into a frog in order to adapt to its external environment.

The following section focuses on areas of innovation that are being considered in the development of biopharmaceuticals:

### 11.4.1 Innovation in Biopharmaceuticals

Process development by design (PDbD) enables a process of innovation based on the requirement of the target product/process profile (TPP). The TPP includes parameters such as product quality, cost of goods, process scalability, sustainability, and flexibility to future change. In this review, we will outline the areas of research that will enhance the understanding of each major function in drug development.

1. **Molecular Clones**
   The efficacy (yield) of cell lines to secrete recombinant proteins has consistently increased over the past decade. From a typical minimal expectation of 1 g/L, the work-horse Chinese Hamster Ovary (CHO) cell line is recently reported to secrete 51 g/L of recombinant protein in a continuous cell culture process (Hu et al. 2007). Such high yields can be enabled by technological advances in molecular and cell biology. (1) Studying molecular patterns by "omics" analysis can provide an understanding of the signaling pathways for protein secretion. Transcriptomic analysis of lactating mammary glands of various mammals has indicated that caseins and whey proteins (i.e., alpha-lactalbumin, beta-lactoglobulin, whey acidic protein, albumin, and immunoglobulin) are secreted "naturally" with yields ranging up to 130 g/L/day (Bisana 2014) (2) Post-translational modification of proteins can also be systematically analyzed by developing in silico digital signaling and enzymatic pathways and modeling environmental conditions to mimic a physical cell. Finally, (3) new gene editing technologies (CRISP-Cas9, Talen, Zn-finger nucleases) are being used to modify genes which interfere with apoptotic, proteolytic and related intrinsic pathways (Ronda et al. 2014).

   Cell-free recombinant protein expression can enable an as-yet-undefined upper limit. The feasibility of cell-free systems has been established using cell lysates combined with the reaction mixture, vector DNA, and either heavy or light stable isotope-labeled amino acids to express recombinant proteins (Pieper et al. 1995). Expressed proteins are then purified and digested into peptides for LC-MS analysis. Optimization of these systems and developing manufacturing scale processes will require significant innovation initiatives to realize the opportunities.

2. **Process Development**
   (a) **Upstream Process Development (USP)**. The consistent improvement over the past decade in product yield and reduced impurity profile has been primarily due to advances in optimization of the upstream processes of cell growth and product yield (Gronemeyer et al. 2014). USP unit-operations encompass (1) clone selection, (2) media and feed development, (3) bioprocess development, (4) product quality development, (5) scale up, (6) cell harvesting, (7) process control, and the corresponding (8) analytics. Optimizing these processes is an intrinsic art. Multi-factorial components including media nutrients, environmental conditions of pH, temperature, and physical parameters such as sparging, fermenter and impeller dimensions have been studied using statistical approaches of design-of-experiment and principles of quality-by-design. Process controls and modular approaches for unit-operations have also yielded in development of platform technologies, specific for various cell lines and manufacturing facility designs. High-throughput and automated systems such as Ambr® bioreactor technologies enable assessing full-factorial design for optimal conditions (Velez-Suberbie et al. 2017). These processes have increased efficiencies by applying principles of six sigma.

   A trend in single-use, disposable bioreactor systems has emerged due to their promise of lower capital investment, flexibility, and potentially higher productivity. Various designs of these fermentation units, such as wave bioreactors, orbital shakers, pneumatic or stirred bioreactors, eliminate the need for cleaning and sterilization time. A deep understanding of fluid mechanics, media components, physical and environmental conditions through modeling has enabled extremely efficient processes for USP.

   (b) **Downstream Process Development (DSP)**. With a significant increase in feed volumes from USP, protein concentration, and solid contents that enter downstream purification workflows, there is a requirement for exponential enhancement of the DSP capabilities. The equipment reaches its physical capacity limit and the process also requires adding additional unit-operations which increase cost of goods and raw materials, notwithstanding time. Recent development in tangential-flow filtration processes has provided a platform to address the high-yield USP feed.

   The separation technologies involved in purification of the product include columns which utilize various chemistries in classical protein purification, such as charge (ion-exchange), affinity, hydrophobicity (reverse phase), and size. Each of these unit-operations requires significant optimization which utilize statistical modeling, automated identification and integration of impurity profiles; and continuous processing. Development of platform technologies enables increase in efficiency.

Other DSP unit-operations such as enzyme reactions, crystallization, centrifugation, lyophilization, precipitation, and also filtration for virus inactivation have also made significant advances. The use of quantum mechanics equations to understand kinetics of substrate–enzyme–cofactors–product relationships has resulted in an insight into these complex biological reactions, at scale. New analytical technologies, such as high-resolution nuclear magnetic resonance, and cryo-electron microscopy have provided feedback (process analytical tools, PAT) in real time to optimize crystal space-group dimensions, specific mechanics of controlled lyophilization and precipitation. Collectively, the improvements in efficiency and consistent performance of the DSP are necessary for meeting the requirements of a cost-competitive target product profile.

(c) **Drug product, formulation development and devices.** Distribution of biopharmaceuticals through circulatory and lymphatic systems requires innovation in formulation, molecule design, and delivery systems. Subcutaneous delivery is the preferred route (from a patient perspective) for administering these drugs. However, a major challenge in high concentration formulations is the inverse relationship of ease of delivery to viscosity and aggregation. Use of a combination of novel and generally-accepted-as-safe (GRAS) excipients that influence protein–protein interactions could enable subcutaneous injections of up to 50 mL. Devices (pens, auto-injectors, etc.) to self-administer at-home therapies can provide innovative, cost-effective tools for user acceptance. In this respect, on-body devices for continuous delivery, delivery-on-demand (based on automated diagnostic measurements) can also significantly improve adherence for chronic diseases, and maintenance therapies. Recent advances in this field include study of protein–protein interactions in order to predict protein behavior at high concentrations, computational models, reduction in material requirement, costs and time for development.

3. **Biological Activity**. Structure–function relationship studies are core components of target identification and validation. In order to identify targets that bind their ligands and influence the biological functions at cellular, organelle, and organism levels, in vitro and in vivo biological activity assays are required that faithfully represent the disease pathogenesis. A focus on "Biology-First" is one strategic initiative undertaken by the biopharmaceutical industry towards achieving this goal. Defining the target product profile and the biological activity of a drug are crucial objectives that enable development of relevant methods for assessment in these areas. The use of sophisticated, cutting edge science and technologies can enable unique biological assays that can enhance the probability of success in drugs achieving their efficacy and safety outcomes.
4. **Analytical Development**. The analytical tools used during product development are the "eyes and ears" of the process. The methodologies for characterizing the product have focused on analyzing primary, secondary, and tertiary structures of

biopharmaceuticals (Berkowitz et al. 2012). High-resolution mass spectrometry has become the primary tool for structural characterization of biopharmaceuticals. Such advanced techniques have very high sensitivity to measure primary to tertiary structures. Characterization of minor post-translational modifications of proteins (e.g., isomerization, glycosylation, deamidation, etc.) has enabled a deeper understanding of the impurity profile of biopharmaceuticals during drug development. Cryo-electron microscopy technologies have enabled understanding the structures of proteins at the single-digit Angstrom resolution. Due to the complexities of the instrumentation, these analytical methods require reliable and efficient analytical strategies. Minimization of artifact formation during sample preparation, quantitative data analyses, and multi-attribute methodology, miniaturization of the analytical techniques, automation of the instrumentation to minimize bias, and software drive data analysis can have an exponential impact on increasing the efficiency and reliability of the analytical methods.

5. **Pharmacology (Pharmacodynamics, Pharmacokinetics, Immunogenicity, Biomarkers, and companion diagnostics).** Analysis of the relationship between pharmacokinetics (PK) and pharmacodynamics (PD) is one of key principles of drug development (Meibohm 2011; Chirmule 2012). With the advent of precision medicine, precise understanding of dose and effect in each patient is becoming critical. Development of omics technologies to measure a wide range of biomarkers will help in detailed characterization of the disease pathogenesis. A critical and in-depth characterization of the methods used for biopharmaceutical assessment can be significantly enhanced by use of sophisticated data management and quantitative modeling analysis. This understanding will enable statistical analysis and interpretation of the relationship between PK and PD in the presence of anti-drug antibody (ADA), the latter being an integral component of response to biologics. Big-data collection processes involving omics (genomics, proteomics, metabolomics, etc.) technologies has become crucial for determining biomarkers for surrogates for clinical responses of safety and efficacy. The determination of biomarkers enables development of companion diagnostics to stratify patients within indications that determine responders. Finally, the bioanalytical methods for measuring these diverse responses in patients require advanced statistical tools for development, qualification, and validation for fit-for-purpose use.
6. **Predictive Toxicology**. Data from three decades of biopharmaceutical development has indicated that the attrition rates are significantly higher (ranges from 40% to 60%; personal data collection by NC from this period of analysis) than originally hypothesized (Kizhedath et al. 2017). The unexpected events that terminate molecules surface during toxicological studies. Unlike small molecules, where many off-targets effects are observed in liver and cardiac tissue, the toxicities of biopharmaceuticals range from effects on hematopoietic cells (platelets, macrophages, lymphocyte subsets) and brain (reactivation of latent virus PML); lung inflammation, cytokine storm, allergic and anaphylactic reactions.

In order to predict the probabilities of adverse reactions, in silico and in vitro/ in vivo experimental approaches have been utilized. For this purpose, a battery of analytical and bioanalytical methods needs to be created to address all anticipated biological on- or off-targeted effects. A majority of the toxicity is associated with increased immune responses or immune suppression, requiring the development of an immuno-toxicology strategy. The assays to measure hypersensitivity reactions Type I–IV can provide an understanding of the mechanism of toxicity, and develop novel genome-assisted (MHC, immunomics) predictive tools to design molecules that avert such adverse events. Thus, with major advances in new technologies of measuring analytes, as well as data analysis and modeling algorithms, safety of molecules can be predicted, rather than experienced.

7. **Clinical**. Clinical trials comprise of >60% of investment in the drug development process. Failure to meet pre-defined primary and secondary endpoints has a major impact on costs, notwithstanding approvability. Enhancing the processes of conducting clinical trials (clinical operations) can substantially reduce these costs. In this respect, adaptive clinical trials have been recognized by the FDA (Bothwell et al. 2018; Monticello 2017). These trials have processes to modify the design and endpoints after initiation without affecting the statistical validity of the outcomes. In addition, identifying biomarkers of efficacy and safety to inform association with clinical results will also have a significant impact on the probability of success and potential reduction in costs.

   To summarize, innovation in developing high quality biopharmaceuticals will be critical in making this class of drugs affordable. Improvements in yield and quality, with reduced timelines will become the norm. Personalized medicine based on individual genotype and phenotype will require precision in target product profile designs. Advances in technology will also enable higher quality to be compliant with evolving regulatory requirements. With such demands for affordable healthcare, innovation is not just required, it is essential. "Innovate or die" is the mantra.

## 11.5 Recommendation for Policies in Biopharmaceuticals

Government policies can have a major impact on the culture of innovation, enabling efficiency, compliance, and reduced costs. Before embarking on policy decisions, it is critical to (1) collect data on quantitative aspects of innovation, (2) evaluate the data for opportunities, and (3) establish high-level goals through government initiatives. The following specific steps can be taken systematically:

### 11.5.1 Collection of Relevant Information

Innovation is measured using several methods, including leading and lagging indicators. In this respect, metrics of (1) previous ideas that resulted in major impact to science and/or business, (2) number of new ideas, (3) ideas that lead to new initiatives, (4) percentage time spent on innovative concepts, (5) innovative products, (6) number of open-ended meetings that discuss and promote ideation, and (7) paradigm shifting transformative projects are examples of *leading indicators*. *Lagging indicators* can be measures that result in consequences, such as (1) innovative solutions to problems, (2) employee focus and satisfaction, (3) larger investment due to a higher risk-taking culture, (4) number of new ideas that failed, (5) adoption of new technologies, (6) number of opportunities for training programs for talent management, and (7) number of effective collaborations. Collecting specific information in this manner will enable analyzing the data to assist policy decisions with a purpose of improving innovation in developing novel biopharmaceuticals.

In order to create a culture of innovation, there needs to be a central focus on improvement. The need for innovation is to create new medicines that impact the overall health of society. While this proposition is a tall order, this section of the review makes a recommendation on step-wise approaches that can be considered for initiating policies that promote innovation in the biopharmaceuticals sector.

### 11.5.2 Evaluation of Opportunities

The Indian education system has had an emphasis on STEM (Science Technology Education Mathematics) for several decades. The current success of the education curriculums in universities in biotechnology in India can be attributed to several initiatives taken by the government and public sector, at least a decade earlier. The education systems have provided the trained scientists required for industry. India may enjoy the status of being among the top ten countries with respect to government investments in R&D in academia and institutions (Annual Report 17–18, DBT 2018), but has a very few high-impact publications (Nature Index 2019). There is a need for a consolidated budget and central funding body for public research. Although there are advances in numbers of institutions offering these courses in biotechnology, very few of them provide training on readiness-for-industry. A collaborative effort by departments of biotechnology institutions and local biotechnology industries can be extremely symbiotic. The curriculum for such courses can yield innovative ideas that could include the application of scientific concepts, and a better understanding of processes and products.

### 11.5.3 Establish Goals Through Government Initiatives

The role of the government in enhancing the innovation culture requires a strategy that can support and enhance the investment in innovation in India. An inter-ministerial cooperation at the highest level mediated through strategic program initiatives are required to increase the biotechnology science. These initiatives should have specific goals along with accountability in order to enhance the development of novel biopharmaceuticals. The outcomes of such initiatives should

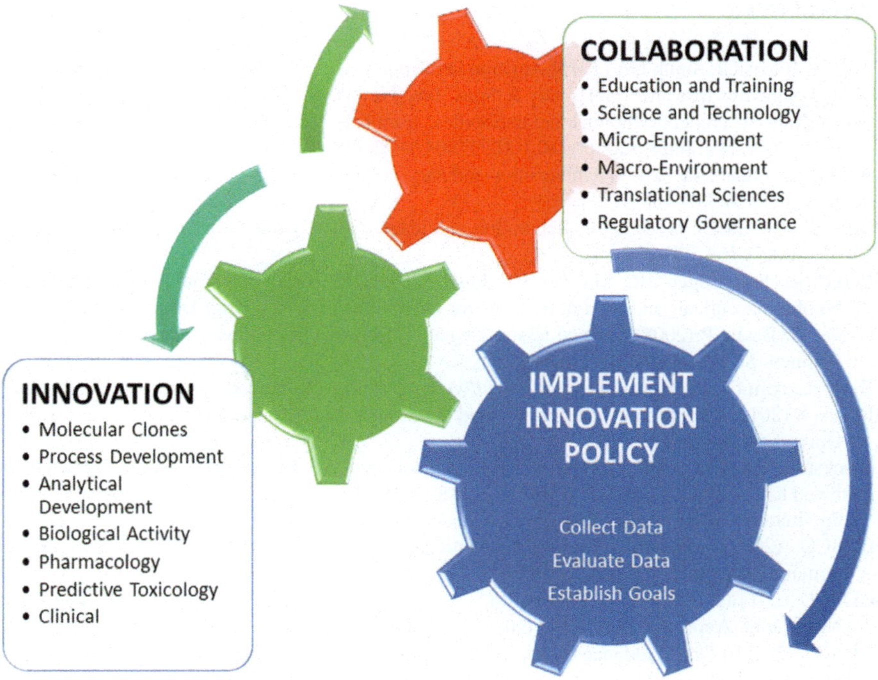


**Fig. 11.4** Collaboration, innovation, and policies for biopharmaceutical research and development. The list includes activities that link the impact of collaboration and innovation that can be utilized to developing government policies the enable affordable healthcare

support the entire process of developing biopharmaceuticals, which can find new treatments of diseases and improve the overall healthcare processes.

#### 11.5.3.1 The Road Ahead

It is estimated that the greater global investment in biotechnology particularly in emerging economies will largely drive growth in the global biotech revenue that is estimated to rise to >$ 300 billion. A report from the Department of Biotechnology estimates that the Indian biotech industry holds about 3% share of the global biotechnology industry (NC; personal communication). Increasing this share could be one policy decision. The path for success in the biopharmaceutical industry landscape in India depends on collaboration and innovation supported by new policies (Fig. 11.4). With the country offering numerous competitive advantages of R&D facilities, knowledge, skills, and cost effectiveness, the biotechnology industry in India has immense potential to emerge as a global key player.

**Acknowledgements** We wish to thank Dipti Rao (ArtX, Bangalore) for the Google map of biotechnology in India; William Cushley (University of Glasgow), Manohar Rao (Pharos Consulting, Bangalore), Dinesh Bhaskar, and Parthasarthy Sampathkumar (Biocon, Bangalore) for the critical review of the manuscript.

## References

Alex A, Harris CJ, Smith DA (2015) Attrition in the pharmaceutical industry: reasons, implications, and pathways forward. John Wiley & Sons, Inc., New York, NY

An Z (2009) Therapeutic monoclonal antibodies: from bench to clinic. John Wiley & Sons, Inc., New York, NY. https://doi.org/10.1002/9780470485408

Ankrah S, Omar AT (2015) Universities–industry collaboration: a systematic review. Scand J Manag 31(3):387–408

Bennett LM, Gadlin H (2012) Collaboration and team science: from theory to practice. J Investig Med 60(5):768–775

Berkowitz SA, Engen JR, Mazzeo JR, Jones GB (2012) Analytical tools for characterizing biopharmaceuticals and the implications for biosimilars. Nat Rev Drug Discov 11:527–540

Binder G, Bashe P (2008) Science lessons: what biotech taught me about management. Harvard Business School Press, Boston, MA

Biospace.com (2019) Hotbed regions https://www.biospace.com/hotbeds. Accessed 29 Mar 2019

Bisana S (2014) Evolution of mammary gland function: a study using monotreme models. Deakin University, Melbourne, VIC

Blackstone EA, Joseph PF (2013) Economics of biosmilars. Am Health Drug Benefit 6(8):469–478

Bothwell LE, Avorn J, Khan NF, Kesselheim AS (2018) Adaptive design clinical trials: a review of the literature and ClinicalTrials.gov. Br Med J 8:1–11

CDSCO (2007) Guidance document document no. IMP/REG/200711. http://cdsco.nic.in/writereaddata/Guidance documents.pdf

CDSCO (2011a). http://www.cdsco.nic.in, http://www.cdsco.nic.in/writereaddata/Guidance_for_New_Drug_Approval-23.07.2011.pdf. Accessed 31 Mar 2019

CDSCO (2011b) Draft guidance on approval of clinical trials & new drugs. http://www.cdsco.nic.in/writereaddata/Guidance_for_New_Drug_Approval-23.07.2011.pdf. Accessed 31 Mar 2019

CDSCO (2016). http://www.cdsco.nic.in, http://www.cdsco.nic.in/writereaddata/CDSCO-GuidanceForIndustry.pdf. Accessed 31 Mar 2019

Chirmule NJ (2012) Immunogenicity to therapeutic proteins: impact on PK/PD and efficacy. AAPS J 14:296–302

Das S (2017). https://www.business-standard.com/article/companies/indian-pharma-sees-traction-in-novel-biologics-117122701169_1.html. Accessed 30 Mar 2019

DBT (2018) Annual report 17–18. http://www.dbtindia.nic.in/wp-content/uploads/English-Annual-Report-2017-18.pdf. Accessed 31 Mar 2019

DiMasi JA, Grabowski HG (2007) The cost of biopharamceutical R&D: is biotech different. Manag Decis Econ 28:469–479

Edler J, Herring PL (2016) Trends in modern drug discovery. Handb Exp Pharmacol 232:3–22

Evaluate Pharma (2018). http://info.evaluategroup.com/rs/607-YGS-364/images/WP2018.pdf. Accessed Jan 2019

Fisken J, Rutherford J (2001) Business models and investment trends in the biotechnology industry in Europe. J Commer Biotechnol 8:191–199

Gronemeyer P, Ditz R, Strube J (2014) Trends in upstream and downstream process development for antibody manufacturing. Bioengineering 1:188–212

Hu W-S, Jayapal KP, Wlaschin KF, Hu W-S, Yap MGS (2007) Recombinant protein therapeutics from CHO cells; 20 years and counting. In: Report of Society for Biological Engineering. SBE, New York, NY. https://pdfs.semanticscholar.org/5f96/12ce9170571f296b75246e80cb671bbb886c.pdf. Accessed 31 Mar 2019

Hughes SS (2013) Genentech: the beginings of biotech. The University of Chicago Press, Chicago, IL

Hung A (2017) A systematic review of U.S. biosimilar approvals: what evidencedoes the FDA require and how are manufacturers responding? J Manag Care Spec Pharm 23(12):234–244

Jackson R, Kroehling L, Khitun A, Bailis W, Jarret A, York AG, Khan OM, Brewer JR, Skadow MH, Duizer C, Harman CD, Chang L, Bielecki P., Solis AG, Steach HR, Slavoff S, & Flavell1, RA (2018). The translation of non-canonical open reading frames controls mucosal immunity. Nature. 208, 564: 435–340.

Kizhedath A, Wilkinson S, Glassey J (2017) Applicability of predictive toxicology methods for monoclonal antibody therapeutics: status Quo and scope. Arch Toxicol 91(4):1595–1612
Kleinschmidt-DeMasters B, Tyler KL (2005) Progressive multifocal leukoencephalopathy complicating treatment with natalizumab and interferon beta-1a for multiple sclerosis. N Engl J Med 353:369–374
Kneller (2010) The Importance of new companies for drug discovery: origins of decade of new drugs. Nat Rev Drug Discov 9:955
McCafferty J (2010) The long and winding road to antibody therapeutics. MAbs 2:459–460
Mehr SR, Brook RA (2017) Factors infuencing the economics of Biosimilars in the US. J Med Econ 20(12):1268–1271
Meibohm B (2011) Population pharmacokinetic/pharmacodynamic analyses as the basis for dosing of therapeutic monoclonal antibodies. Clin Pharmacokinet 50:823–824
Monticello TM (2017) Current nonclinical testing paradigm enables safe entry to First-In-Human clinical trials. Toxicol Appl Pharamacol 334:100–109
Moorkens E, Meuwissen N, Huys I, Declerck P, Vulto AG, Simoens S (2017) The market of biopharmaceutical medicines: a snapshot of a diverse industrial landscape. Front Pharmacol 8:314
Mukherjee R (2018). https://timesofindia.indiatimes.com/business/india-business/pharma-cos-bag-40-of-us-generics/articleshow/65329865.cms. Accessed 2019
Munos B (2009) Lessons from 60 years of pharmaceutical innovation. Nat Rev Drug Discov 8:959–968
Nature Index (2019). https://www.natureindex.com/country-outputs/india. Accessed 31 Mar 2019
Nawrat A (2018). https://www.pharmaceutical-technology.com/features/expanding-generics-biosimilars-in-india/. Accessed 30 Mar 2019
Overington JP (2006) How many drug targets are there? Nat Rev Drug Discov 5:993–996
Pieper R, Luo G, Cane DE, Khosla C (1995) Cell-free synthesis of polyketides by recombinant erythromycin polyketide synthases. Nature 378:263–266
Pisano GP (2006) Science business: the promise, the reality, and the future of biotech. Harvard Business Press, Brington, MA
Ramakrishnan V (2018) Gene machine: the race to decipher the secrets of ribosomes. Hachette Book Group, Inc., New York, NY
Ratti E, Trist D (2001) The continuing evolution of drug discovery process in the pharamaceutical industry. Farmaco 56:13–19
Ronda C, Pedersen LE, Hansen HG, Thomas Beuchert Kallehauge MJ, Nielsen AT, Kildegaard HF (2014) Accelerating genome editing in CHO cells using CRISPR Cas9 and CRISPy, a web-based target finding tool. Biotechnol Bioeng 111:1604–1616
Schmidt C (2015) Suicidal thoughts end Amgen's blockbuster aspirations for psoriasis drug. Nat Biotechnol 33:894–895
Scott Morton FM (2018) The impact of the entry of biosimilars: evidence from Europe. Rev Ind Organ 53:173–210
Slamon DJ, Clark GM, Wong SG, Levin WJ, Ullrich A, McGuire WL (1987) Human breast cancer: correlation of relapse and survival with amplification of the HER-2/neu oncogene. Science 235:177–182
Storz U (2014) Rituximab: how approval history is reflected by a corresponding patent filing strategy. MAbs 4:820–837
Susan Lee AK (2005) Adverse reactions to biologic agents: focus on autoimmune disease therapies. J Allergy Clin Immunol 116:900–905
Tallman S, Phene A (2007) Leveraging knowledge across geographic boundaries. Organ Sci 18 (2):165–335
The Economist (2016) Cluster luck. https://www.economist.com/business/2016/01/16/clusterluck. Accessed 29 Mar 2019
Truscheim M, Aitken ML, Berndt ER (2001) Biopharmaceutical innovations: do biologics differ from small molecules? Forum Health Econ Pol 13:1558–9544

Velez-Suberbie ML, Betts JP, Walker KL, Robinson C, Zoro B, Keshavarz-Moore E (2017) High throughput automated microbial bioreactor system used for clone selection and rapid scale-down process optimization. Biotechnol Prog 34:58–68

Zhang, Lamba BM, Feldman AW, Zhoua AX, Lavergneb T, Lic L, and Romesberg FE. A semisynthetic organism engineered for the stable expansion of the genetic alphabet. PNAS 2017 114 (6) 1317–132

**Narendra Chirmule** is the co-founder and CEO of SymphonyTech Biologics, a data analytics company focused on engineering solutions to biology. As former Head of R&D at Biocon (Bangalore), and leadership positions in Amgen (Thousand Oaks, CA) and Merck Vaccines (West Point, PA), he has contributed to clinical development of vaccines and biopharmaceuticals. The drug development experiences include landmark *vaccines* for cervical cancer [HPV], shingles [Varicella zoster], and childhood diarrhea [Rotavirus]; *biologics* for osteoporosis (Prolia), rheumatoid arthritis (Enbrel), platelet loss (NPlate), and breast cancer (Ogrivi); and *gene therapy* using viral vectors. He also worked on the development of the adenovirus-vector based HIV vaccine, which was tested in the seminal STEP trial. The results of the trial were instrumental in informing the field of HIV vaccine on the importance of both cell and humoral protective immune responses. He has published extensively and presented seminars on subjects of basic immunology and development of biologics and vaccines. The topics of his publications include white-papers and regulatory guidances on novel methodologies of measuring immune responses to biologics and vaccines, utilization of quality-by-design approaches of molecule design, and statistical methods for analyses of pharmacokinetic and pharmacodynamics. The subject of his PhD was on development of a leprosy vaccine, from Cancer Research Institute, Mumbai; post-doctoral studies on pathogenesis of AIDS from Cornell University Medical College-North Shore Hospital, New York; teaching and research as assistant professor in gene therapy at University of Pennsylvania, Philadelphia. Dr. Chirmule is on the NIH advisory committee for HIV vaccines.

**Shilpa Bhat** is a bio-pharmaceutical R&D scientist with 10+ years of experience in early and late stage development of biotherapeutics including CMC development of large molecules, scaling up of manufacturing processes, and control of critical quality attributes of antibody therapeutics.

She is currently working as a decision scientist at Mu Sigma helping the largest Fortune 500 pharmaceutical/medical companies in going from data to decisions across their product life cycles using advanced data analytics.

**Sabyasachi Mondal** is currently working as Associate Scientific Manager at Biocon's R&D division. He has a decade of experience in bio-pharmaceutical industry. His key areas of expertise are development of bio-physical methods for biologics and analytics.

He obtained his masters degree in Biotechnology from Vellore Institute of Technology (VIT), and holds an MBA from Indian Institute of Management, Bangalore (IIMB). Before joining Biocon he had a brief stint at Saha Institute of Nuclear Physics (SINP) Kolkata, as project research associate.

Outside of his work Sabyasachi loves to take part in Urban sketching community activities and Kaggle competitions.

# Regulatory Requirements and Quality Standards in India's Clinical Trials Journey

# 12

Bobby George, Shrinivas Krishnarao Kulkarni,
and Nilima A. Kshirsagar

## 12.1 Background

The empirical use of remedies is known to man since time immemorial. The concept of evidence-based use of medicine is more recent and has evolved due to weakness of empirical clinical practice and its impact on quality and cost of patient care. Randomized controlled trials and systematic review-based meta-analysis (though with well-recognized limitations) are gold standards for evidence-based medicine. The history of clinical trials (CTs) dates back to 1537, when the first CT of a novel therapy (egg yolk, turpentine, rose oil!) was serendipitously conducted by French surgeon *Ambroise Pare* while treating wounded soldiers. Controlled CT was first conducted about 200 years later, when Scottish physician *James Lind* treated scurvy among the sailors by assigning them to subgroups for different treatments. Subsequent to the randomized controlled trial of streptomycin in tuberculosis in the UK in 1946, Harry Gold and Bradford Hill laid down the foundation of randomized, double-blind controlled clinical trials.

The Nuremberg Code framed in 1947 was essentially the first international guidance on the ethics of medical research, highlighting the essentiality of voluntariness of the informed consent for participation in research. The Helsinki Declaration (initially formulated in 1964 by the World medical association) led to the establishment of general principles and specific guidelines on use of human subjects in medical research. The Belmont Report issued in 1979 further emphasized the need

B. George (✉)
Reliance Life Sciences Pvt. Ltd., Navi Mumbai, Maharashtra, India
e-mail: bobby.george@relbio.com

S. K. Kulkarni
DUI, Panjab University, Chandigarh, India

N. A. Kshirsagar
Clinical Pharmacology, ICMR, Government of India, Delhi, India


M. Dikshit (ed.), *Drug Discovery and Drug Development*,
https://doi.org/10.1007/978-981-15-8002-4_12

for review of all clinical research by Ethics Committees (ECs). Other fundamental ethical principles and globally followed guidelines for conduct of clinical research include the WHO's International Ethical Guidelines for Biomedical Research involving human subjects [1994] and the International Conference on Harmonization (ICH) Good Clinical Practice (GCP) guideline ICH E6 R(1) [1996].

This chapter is an attempt to unfold the key regulatory developments over the course of evolution of clinical research in India. The origin of clinical research in India dates back to 1945 when the first clinical research unit of the Indian Research Fund Association (IRFA) attached to a medical institution was established at the Indian Cancer Research Centre, Bombay. It was in 1949, when IRFA was re-designated as the Indian Council of Medical Research (ICMR), which serves as the premier body responsible for the formulation, coordination, and promotion of biomedical research in the country. Apart from ICMR, several governmental organizations including Department of Biotechnology (DBT), Department of Science and Technology (DST), and the Council for Scientific and Industrial Research (CSIR) fund academic clinical research. The Government of India (GoI) passed the Drugs and Cosmetics (D&C) Act in 1940 and Rules in 1945 to regulate the import, manufacture, distribution and sale of drugs and cosmetics. The D&C Act has been divided into chapters, rules, and schedules which is amended from time to time to control the safety, efficacy, and quality of the drugs and cosmetics. The Central Drugs Standard Control Organization (CDSCO) headed by the Drugs Controller General of India (DCGI) is the national/central level licensing authority (CLA) under the ministry of health and family welfare (MoHFW) which regulates CTs in the country. Its mission is to safeguard and enhance public health by assuring the safety, efficacy and quality of drugs, cosmetics, and medical devices. Schedule Y was incorporated in 1988 under the D&C Act and rules (initially primarily keeping the Indian generic industry in mind) specifying the regulatory requirements and guidelines to seek permission to import and/or manufacture of "new drugs" for sale or to undertake CTs in India.

## 12.2 Trigger for Growth of Clinical Research

A "new drug" is typically protected by patents that provide a certain period of exclusivity in the market. Hence, during this period pharmaceutical companies usually do not face a competition from generic drugs. After attaining independence from British rule, India opted to continue with the "British Patent and Design Act of 1911" until 1972. The British law permitted patenting both processes and products for 14 years. India adopted a different patent law in 1972 to facilitate acquisition of indigenous industrial capability. Under this law, the life of a patent was limited to between 5 and 7 years and only patenting of processes was permitted. This new law was tilted towards the Indian pharmaceutical sector, keeping accessibility to affordable medicines for patients, and in the trade-off, it weakened the Intellectual Property Regime (IPR). Indian companies were able to copy still-patented drugs by slightly altering a process of production resulting in robust growth of domestic pharmaceutical industry with booming generic market. As a fallout of all this, Western multinational firms had apprehensions in introducing their new drugs in Indian markets.

India became party to the Uruguay round of General Agreement on Trade and Tariffs (GATT) negotiations which culminated in an agreement between 75 countries and the European Community leading to the formation of WTO (World Trade Organization) in 1995. WTO obliged the respective signatory countries to harmonize their intellectual property rights with the provisions of Trade Related Intellectual Property Rights (TRIPS). In context of pharmaceutical products, this meant that the countries had to recognize product patents protection. It took India about 10 years to establish a patent law that was in line with the WTO mandate. India amended the Patent (Amendment) Bill before 2005 and extended its weak process patent to strong TRIPS compliant "Product" patent system for pharmaceutical products. Indian manufacturers with reverse engineering capabilities were forced to commercialize generic versions of patented drugs only after expiry of patent. It also propelled Indian manufacturers to focus more on R&D and innovation. The harmonization of patent laws in the post-WTO scenario resulted in a steady drift of CTs to many developing countries, including India.

Realizing the potential of clinical research for newer therapies, and the stricter patent rule post GATT, the Indian government took several steps. ICMR released ethical guideline titled, "Ethical Guidelines for Biomedical Research on Human Subjects" initially in 2000 and amended in 2006 and later in 2017 addressing general and ethical issues involved in clinical evaluation of drugs/devices/diagnostic vaccines and herbal remedies. In a few years after the ICH GCP E6 Guideline (ICH GCP) was published in 1996, CDSCO released the "Good Clinical Practices (GCP) guidelines for clinical research" in 2001. It was developed with consideration of the WHO, ICH, U.S. FDA and European GCP guidelines as well as the Ethical Guidelines for Biomedical Research on Human Subjects issued by ICMR.

Major amendments were also brought about in Schedule Y in 2005 to keep pace with the requirements:

- Local CTs were made mandatory in the year 1988 under Schedule Y, but with a phase-lag, that the CT was allowed to be conducted for new drugs in India with one phase behind those being conducted in other countries. For example, companies could conduct a Phase II CT in India, only if a Phase III CT was ongoing elsewhere in the world. The 2005 amendment of Schedule Y did away with the phase-lag, such that India could participate in CTs concurrently with other nations for the same phase of the CT. However, first in man study was not allowed for a drug discovered outside of India.
- Pragmatic definitions were now provided for Phase I to Phase IV CTs.
- The 1988 version of Schedule Y had not mandated an EC review for regulated CTs with new drugs. It mentioned "It is desirable that protocols for CTs be reviewed and approved by the Institution's EC. .... In case none of the trial centres/institutions has an EC, the acceptance of the protocol by the investigator and its approval by the Drugs Controller (India) or any officer as authorized by

him to do so will be adequate to initiate the trials." The 2005 amendment mandated EC review and approval of protocol, prior to CT initiation.
- The amendment detailed out responsibilities of investigators and sponsors apart from making it mandatory to submit as part of CT application draft, an informed consent form, list of proposed investigators, and background information about the drug.
- The limitations in terms of number of patients and study sites in early phases as stipulated in former Schedule Y of 1988 were removed, thereby allowing sponsors to decide these in relation to protocol requirements.
- Greater emphasis was laid on reporting protocol amendments to CDSCO.

With the amendment of Schedule Y, the statutory board constituted under the D&C Act—The Drugs Technical Advisory Board (DTAB) endorsed the implementation of Indian GCP guideline for reforming the clinical studies in India and made adherence to it mandatory. In 2007, the GoI also removed the 12% service tax which used to be there on CTs, giving a boost to the pharmaceutical industry. In 2008, ICMR released guidelines for Good Clinical Laboratory Practices (GCLP) outlining the principles and procedures to be followed by medical laboratories involved in clinical research and/or patient care so as to provide quality data which can be used for health research and patient treatment. To facilitate fast tracking of regulatory review and approval of CTs, CDSCO classified CT applications (CTAs) into two categories in 2006.

- Category A: CTs that are taking place in countries with competent, fully fledged regulatory systems, e.g., CTs that have received approval in the USA, Japan, Canada, Europe, Australia, or Switzerland. CTAs falling in this category were committed to be reviewed by CDSCO within 2–4 weeks.
- Category B: For all the other CTAs, with a review timeline of 12 weeks.

In July 2007, to bring in more transparency, the Clinical Trial Registry-India (CTRI) was set up at the National Institute of Medical Statistics, ICMR, New Delhi, as a free, online system for registration of all CTs being conducted in India (http://ctri.nic.in). Trial registration involves public declaration and identification of trial investigators, sponsors, interventions, and patient population, target sample size, estimated date of enrollment, etc. apart from submission of EC approvals and CDSCO approval (if applicable). Since June 15, 2009, CDSCO made registration in the CTRI mandatory for CTs with new drugs. During the same period, CDSCO implemented common technical document (CTD) format for submission of dossiers for biologicals and new chemical entities. In March 2011, the CDSCO constituted 12 New Drug Advisory Committees (NDACs), comprising of experts from government medical colleges and distinguished institutions within India for review of regulatory applications including new drugs, fixed-dose combinations, additional indications, and CTs by undertaking an comprehensive assessment of non-clinical data (including pharmacological and toxicological) and clinical data furnished by the applicant. The review included face-to-face presentations by applicants to NDAC members in the presence of CDSCO officials. The intent of an additional review by NDAC was to safeguard the safety and well-being of Indian patients while ensuring strengthened regulatory environment in India. NDAC also had a key role to play in drafting guidelines for clinical research industry in evolving acceptance standards for marketing approval of new drugs.

India started emerging as one of the preferred destination for drug trials by multinational pharmaceutical companies. During the period 2003 and 2010, the CT industry witnessed a growth phase. Some of the factors apart from those listed above, which contributed towards drawing more companies to undertake global CTs in India were:

- U.S. FDA eased its regulations allowing firms to submit results of foreign CTs in applications for new drugs to be marketed in the USA. The U.S.-based drug makers started conducting about one-third of their Phase III CTs outside of the USA and India was one of the beneficiaries.
- India accounts for close to 17% of global population, has an ethnically diverse pool of potential test subjects, with a sizeable number of them being treatment "naïve" patients.
- Several government medical colleges, institutions, and laboratories having good facilities for clinical testing exist in India.
- Availability of clinical research organizations (CROs), hospitals with good infrastructure to do CTs including a large number of GCP trained doctors and associated staff and the workforce being reasonably proficient in English.
- Lower cost of conducting CT. According to Confederation of Indian Industry (CII), the overall development costs can come down by nearly 60% if the Phase 2 and 3 CTs are undertaken in India. As per a report which featured in Harvard Business in 2008, tracking Indian trial subject's costs between $1500 and $2000 while in the USA, the same would cost $20,000. Multinational companies were enticed by the potential to minimize cost and maximize efficiency by extending their drug development program to India.

## 12.3 Trigger for Revisiting Clinical Trial Regulations

However, some of these CTs were not being done as per GCP guidelines. There were lapses in the functioning of ECs and CROs; and in some cases regulatory requirements were being flouted. With the laxity of regulatory and EC oversight, the regulatory and ethical compliance by the sponsors, CROs, as well as the sanctity of the data generated through these CTs started getting compromised and questioned by media and activists alike. Some of the issues were picked up by whistle-blower(s) and non-governmental organizations (NGOs) like Swasthya Adhikar Manch, which filed a public interest lawsuit (PIL) in January 2012 complaining about ethical violations, inadequate government oversight, and unregulated practices being followed by multinational pharmaceutical companies in India. As per the PIL filed by this Indore-based health pressure group, CTs of new chemical entities (NCEs) were being conducted without following appropriate protocol, and firms were taking advantage of underprivileged people. Another PIL filed by the Women's health activists challenged the unethical advertisement and inappropriate handling of the human papilloma virus (HPV) vaccine trial funded by Bill & Melinda Gates Foundation and "Rotavirus" trial funded by multiple private and government sources. The NGO alleged violation of informed consent rules, demanding investigation of the deaths and adverse events after the immunization.

A parliamentary standing committee was set up by GoI to look into the matter. The parliamentary panel in its May 2012 report estimated that during the period 2009–2012, there were over 3.52 lakh people in India who got enrolled in various pharmaceutical studies. The panel found discrepancies in the functioning of the CDSCSO as regards the review and approval of applications and at times waiving off the requirement of CTs without seeking expert medical opinion. Besides, CDSCO had approved 33 drugs, out of a randomly selected sample of 42 by the panel, without CTs on Indian patients. Apprehensions over the conduct of CTs prompted the parliamentary panel to look into the rapidly expanding industry, and the international and domestic pharmaceutical companies sponsoring them. The Wall Street Journal was one of the first to report in international media the dark side of Indian CT business in October 2012. The Supreme Court of India heard the PIL and criticized the Union government "*for being negligent in curbing illegal CTs despite the deaths of at least 2374 persons who had undergone the dubious testing for unregistered drugs between 2007 and 2012.*"

## 12.4 Repercussions of Apex Court's Directives

In compliance with the Court's order, and with an overall objective of protecting the rights of the subjects/patients participating in CTs, the MoHFW came up with robust confidence building measures by bringing the next wave of amendments to D&C rules. Three back to back amendments were notified in January to February 2013 to protect the rights of the subjects.

1. Rule 122 DAB described the procedure for addressing and reporting serious adverse events (SAEs) and methodology for remittance for injury or death of subjects related to CTs through free medical management and financial compensation.
2. Rule 122 DAC described the conditions for seeking permission to conduct CTs. The licensing authority was given the legal right to investigate the sponsor (including their staff, subsidiaries, and contractors), investigators, and study sites for verifying regulatory compliance.
3. Rule 122 DD detailed the process for registration of ECs with the licensing authority. EC registration became a pre-requisite for review and approval of any CT protocol.

Close on the heels of these three amendments, the MoHFW released orders to set up high powered committees to build in an oversight mechanism for CTs with new drugs.

- Apex committee (under the chairmanship of Secretary, MoHFW), to take stock of new approvals which are granted.
- Technical review committee (TRC, comprising of experts), to provide inputs/recommendations related to CTs to the apex committee for overseeing and monitoring the conduct of CTs.

Some of the associated changes related to clinical research which were brought in, included:

- In 2013, MoHFW established an expert panel/committee led by Professor Ranjit Roy Choudhry to develop policies and guidelines for approval of new medicines and CTs. Some of the recommendations of this committee released in its report of July 2013 were accepted and implemented.
- Audio visual recording of the informed consent was made compulsory for each subject participating in any CT.
- Bioequivalence/bioavailability (BA/BE) studies only could be reviewed by an independent EC, while regulated CTs were to be reviewed and approved by institutional ECs.
- Investigators were not allowed to undertake more than three CTs concurrently with a view to achieve better quality of CTs being conducted.
- CDSCO set up an independent expert committee to examine the SAEs of death or injury occurring during CTs and for determining the quantum of compensation, if any, to be paid as per pre-defined formulae.
- CDSCO undertook capacity building exercise both in terms of manpower, up-gradation, and setting up new offices and testing labs. CDSCO also created a separate division to oversee post-marketing surveillance of drugs.

## 12.5 Impact of the Three Major Regulatory Amendments

While these back to back sweeping notifications were undoubtedly steps in the right direction to build up the public trust in CTs, these raised the bar for the drug industry/CROs/academic investigators, who had to go back to the drawing board and realign themselves with the new set of rules. These elicited amendments to informed consent documents and their submission to ECs and CDSCO. It also led to a pileup of project proposals (both for CT approvals and NDAs) yet to be reviewed by NDAC. Changes in SAE reporting requirements forced investigators and ECs apart from the sponsors/CROs to be more vigilant and sensitive to the issue. Though India had been a party to only $<1.4\%$ of the CTs conducted globally, it was no longer being perceived as one of the preferred destinations to do CTs. The U.S. National Institutes of Health (NIH) decided to pull out its CTs with drugs and medical devices, out of the country. Even few Indian companies took the painful decision to move some of their projects offshore to the USA, Europe, and other regions even though it meant increase in development cost and additional time and resources to be deployed. The clinical research industry got a setback from the kind of knee jerk reaction of the MoHFW. Companies and CROs engaged in clinical research and institutes conducting these programs started facing stagnation at least for an interim period. The number of study feasibilities being either requested/conducted in the Phase II and III CT settings started coming down. Many CROs started liquidating these divisions or redeploying their employees to different job profiles or in other regions. It also had an indirect impact on non-industry sponsored academic research, which also slowed down. The number of approved CTs registered in CTRI portal also went on a decline.

## 12.6 Regulatory Reforms Undertaken to Resurrect the Clinical Trial Sector

As a supportive cum corrective regulatory measure, the Indian MoHFW started amending, upgrading, and optimizing the regulatory requirements and policies with a view to encourage companies undertaking basic research, preclinical studies, CTs, BA/BE studies and associated import and/or manufacturing of drugs. CDSCO rationalized and simplified various formats of applications and made quite a few submissions and it's processing online. Some of the key reforms undertaken in the CT space are enlisted here.

- In August 2013, GoI brought in D&C Amendment Bill, 2013 to address perceived gaps in Indian CT regulations. Among the various provisions, the Bill had set forth criminal penalties for those conducting CTs without proper authorization and violation of existing regulations. The Bill though introduced in the Parliament did not get enacted.
- In 2014, it was proposed that ECs and investigator sites should undergo accreditation under the auspices of the Quality Council of India (under the National Accreditation Board of Hospitals and Healthcare providers—NABHHP). In consultation with various stakeholders, NABHHP formulated draft accreditation standards for CT sites, ECs, and investigators. Though MoHFW released order in November 2016 making accreditation of ECs mandatory from January 2018, it has not yet been made mandatory.
- The system of NDAC review had several challenges which led CDSCO to rechristen all the 12 NDACs as Subject Expert Committees (SECs) in July 2014. Each SEC comprises of about eight medical experts including pharmacologists/clinical pharmacologists and medical specialists. The applicants are called for making presentation of their CT proposal, to the respective SEC in presence of CDSCO officials. Based on the deliberations, the recommendations are passed on to TRC. The TRCs while evaluating proposal looks at apart from other aspects—the major efficacy and safety parameters proposed to be evaluated;

the risk vis-a-vis benefit ratio for the patients; innovation vis-a-vis existing therapeutic options available; and if there is an unmet medical need in the country for the study indication. The recommendations of the committee are passed on to CDSCO, which sends out formal approval or query letter to the applicant. Similar review process is adopted for new drug applications (NDA) as part of marketing authorization.

- In July 2015, MoHFW modified the requirement of making audio visual recording of the informed consent mandatory only for "vulnerable" population and wherein a "new chemical entity" or "new molecular entity" was being tried in a CT, factoring in the host of practical, cultural, and privacy concerns which were raised. The new rules permitted audio recording, but without video recording to satisfy the requirement of CTs related to HIV and leprosy, presumably with a view to reduce the possibility of confidentiality breaches with regard to such sensitive conditions.
- From November 2015, CDSCO allowed ECs to approve request for additional CT site or investigator based on ECs due diligence, without the necessity of taking CDSCO's prior approval.
- The requirement of the drug to have at least 60% shelf life at the time of import was waived (from December 2015) for the purpose of R&D and CTs as long as the drug was to be used within its shelf life. This came as a big relief for sponsors conducting CTs, as sourcing desired quantity of reference drug with 60% or more shelf life was a challenge.
- From August 2016, studies done on a marketed drug for a new indication purely for "academic/research purposes that are non-regulatory in nature (thereby meaning the data generated from the study is not intended for submission to licensing authority) were no longer required to have CDSCO permission, but were subject to EC approval. The ECs in turn were required to inform CDSCO about the proposals approved by it and also about the proposals where there could possibly be an overlap between the CTs for academic and regulatory purpose. If the CDSCO does not reply within 30 days from receipt of the application from EC, it was to be presumed that no permission was required from CDSCO.
- With effect from August 2016, export of human biological samples by the Indian diagnostic laboratories/Indian Clinical Research Centres for lab analysis/R&D or CT testing to foreign laboratories was allowed, without having to take export permit from any other government agency—like Directorate General of Foreign Trade (DGFT) or ICMR.
- In August 2016, CDSCO revoked the limitation of the maximum number of CTs concurrently allowed for any investigator. The pre-condition for CT sites to have over 50 beds to conduct any trial was also revoked. The responsibility to assess the same and take a call was left to the concerned institutional EC.
- CDSCO and DBT jointly released in August 2016, the updated version of guidelines for development of similar biologics in an attempt to streamline the regulatory process for authorization of biosimilars in India, bringing it considerably at par with global requirement.

- In September 2016, CDSCO extended the validity period of approval of BA/BE study centers and bio-analytical laboratories from 1 to 3 years giving an administrative respite.
- From September 2016, CDSCOs "SUGAM" portal started accepting online CTAs reducing the administrative burden and brought in additional transparency in the entire process.
- In January 2017, CDSCO in collaboration with the ICMR published a handbook for applicants and reviewers of CTs of new drugs. It was meant to facilitate understanding of the review process by applicants and reviewers with an intention to fasten approval timelines. It also aims to standardize and streamline the process of evaluation and subsequent decision-making of SEC. Code of conduct for the SEC members has been aptly defined in the handbook.
- Specific regulations for medical devices and in-vitro diagnostic devices were brought in through Medical device Rules, 2017 which came into effect from January 2018. The rules mandated CTs involving investigational medical devices to be approved by CDSCO and institutional EC.
- Validity of import test license for sourcing investigational product for CT or BA/BE study was extended from 1 to 2 years.
- In 2017, ICMR issued national guidelines for stem cell research which emphasized upon the fact that investigational use of therapeutic stem cells should be done only by conducting CTs with requisite regulatory approvals.
- With a view to promote CT activity in India, the Technical Committee in September 2017 announced that companies would be required to include Indian patients in global CTs in order to market in India, a new drug developed outside of the country.

These reforms were well received by the pharmaceutical industry, academia, and other stakeholders. They have further empowered the ECs in discharging their duties towards CTs. It did instill some level of confidence in the Indian as well as global pharmaceutical and biotech companies engaged in CTs. The number of CT approvals started showing a steady increase, starting from 2015. The number of global CTs being conducted picked up from 56 international studies and 37 Indian studies (in 2016) to 97 international and 67 Indian studies, respectively, in 2017 and went up further in subsequent years.

## 12.7 New Drugs and Clinical Trials Rules, 2019

With a view to further promote clinical research in India, have predictable, transparent and effective regulations for CTs and also make faster accessibility of new drugs for the Indian population, the MoHFW released the draft "New Drugs and Clinical Trials Rules" (NDCTR), in February 2018. The NDCTR were finalized after considering all objections and suggestions from stakeholders through Gazette notification GSR 227(E), in March 2019. With the new rules coming into force, Part XA and Schedule Y are no longer applicable in respect of new drugs and investigational new drugs (INDs) for human use. However, the approvals or permissions or licenses or certificates issued in respect of new drugs and INDs prior to commencement of these new rules shall be deemed to be valid till its expiry. NDCTR has been created as a separate set of rules under the D&C Act, just like (1) Drugs and Cosmetics rules and (2) Medical Device rules. NDCTR has all together 13 chapters (encompassing 107 rules) and 8 schedules. The new rules have ushered in new set of forms for different kinds of applications made to the Central Licensing Authority—(CLA, the DCGI), and for corresponding permissions being issued by CLA. Key updates/ changes in NDCTR from the past regulations are given below:

- Several new definitions have been included like that for academic CTs, orphan drugs, biomedical and health research, BA/BE studies, etc. Similarly, some definitions have been modified like that for "new drug" and IND.
- CT application fee has been increased multi-fold. There have also been new categories for which fee-based review has been included. Provision has been kept for complete fee waiver, provided the CT is to be conducted by a person of an institution or organization, funded or owned, by the central government or a state government. For Micro Small Medium Enterprises (MSME), there is a 50% fee waiver in applications being filed towards conduct of CT, approval of new drug, and pre- as well as post-submission meetings. The fee hike is expected to supplement the associated cost of reviewing submissions and inspecting facilities in a more efficient and effective manner.
- Clearly defined timelines for review and CT and BA/BE study submissions have been laid down, as it was missing in erstwhile Schedule Y.
- While defining academic CTs (done on permitted drug formulation, initiated solely for academic research purpose for a new indication or new route or new dosage form), the new rules have made it clear that results/observations derived/ data generated are not intended for seeking approval of CLA or regulatory authority of any country for marketing or commercial/promotional purpose. They can be initiated after approval by the respective ECs and are required to be conducted in accordance with the ethical principles specified in ICMR guidelines for biomedical research on human participants. However, when in doubt on the nature of the study or in the event of a possible overlap between the academic CT and CT, ECs should formally refer the protocol to CLA for their opinion, who in turn should issue the necessary clarification within 30 working days from date of receipt of application.
- ECs reviewing biomedical and health research proposals are required to be registered with the authority designated by the Department of Health Research (DHR) and not with CDSCO within the MoHFW.
- To avoid any bias, it's been mandated that EC should consist of at least 50% of its members, who are not affiliated with the institution or organization in which the committee is constituted. Further, each EC member needs to undergo such training and development programs, as may be specified by CLA from time to time to keep up to date with the regulatory and ethical requirements. Also, any change in membership or constitution of registered EC is to be notified to CLA in

writing. ECs reviewing "biomedical and health research" proposals are required to register with the authority designated by the DHR, within the MoHFW.

- The approving ECs and the CT site or BA/BE center, as the case may be, are required to be located within the same city or within a radius of 50 km to have better study oversight. Independent ECs can now review and approve regulated CTs as long as they comply with the NDCTR. Sponsors are mandated to notify CLA of the EC approval within 15 days of its issuance.
- NDCTR has clearly differentiated the requirement for Phase IV CT; post-marketing surveillance study; and Post-Marketing Surveillance (PMS).
  - Phase IV CT would include additional drug–drug interactions, dose-response or safety studies and trials designed to support use of new drug under the approved indications, under an approved protocol by CDSCO. In such trials, the ethical aspects for protection of rights, safety, and well-being of the trial subjects shall be followed as per the regulatory provisions including that for compensation in case of CT-related injury or death and GCP guidelines. In such trials, the study drug may be provided to the trial subject free of cost, unless otherwise justified to the satisfaction of CDSCO.
  - Post-marketing surveillance study or observational or non-interventional studies are to be done as part of active surveillance under approved indications of new drug. Though the study protocol has to be approved by CDSCO, the regulatory provisions and guidelines applicable for CT of a new drug are not applicable. Also the study drugs are the part of treatment of patient in the wisdom of the prescriber included in the protocol.
  - Post-Marketing Surveillance (PMS) are mandatory for 4 years post approval of new drugs in the country, in line with what was earlier specified under Schedule Y.
- Provision has been kept for waiver or relaxation in requirements relating to animal toxicology, reproduction studies, teratogenic studies, perinatal studies, mutagenicity, and carcinogenicity in case of new drugs approved and marketed for more than 2 years in other countries, subject to CLA being satisfied that there is adequate published evidence regarding the safety of the drug. This will help to avoid repetition of many of the non-clinical studies.
- Provision has been included for waiver of local CT requirement for an imported new drug for marketing under specific conditions. The conditions may be relaxed by CLA where the drug is indicated for life-threatening or serious diseases or diseases of special relevance to Indian health scenario, or for a condition which is an unmet need in India, or for the rare diseases for which drugs are not available, or available at a high cost, or if it is an orphan drug.
- Expedited approval process and a complete fee waiver for CT filing of orphan drugs (a drug intended to treat a condition which affects not more than 500,000 persons in India).
- Provision of pre-submission and post-submission meetings of the sponsor/applicant with the CLA or an authorized officer by CLA for seeking formal written guidance and clarification about the regulatory requirements and procedures.

- Manufacturing of drugs for BA/BE studies (and not just CTs) carried under CLA permission is mandated to be done in accordance with the principles of GMP.
- Labelling requirements have been spelt out for any new drug or IND manufactured for the purpose of a CT or BA/BE study. No alteration of the label or container of any new drug manufactured shall be allowed without CLA permission.
- Quarterly enrollment status of trial subjects and six monthly status reporting to be done for each CT to CLA.
- Data and records are to be maintained for 5 years after completion of the CT/BA/BE study or at least 2 years after the expiration date of the batch of the new drug product, whichever is later.
- All samples of test and reference drug products used in a BA/BE study are to be retained for 5 years after the conduct of the study or 1 year after the expiry of the drug, whichever is later.
- Provision kept for post-trial access of investigational drug to be provided by the sponsor free of cost, if the drug has been found to be beneficial to a trial subject during CT. However, the sponsor shall have no liability for post-trial use of the drug.

Sponsors/applicants need to follow the NDCTR for submission of CT application with new drug. If manufacturing of the new drug is involved, then the firm needs to take manufacturing test license as well. Similarly, if import of the reference or test drug is involved for the purpose of CT, import test license is required to be taken. For biological drugs, additional approvals need to be taken from Institutional Biosafety Committee (IBSC) and Review Committee on Genetic Manipulation (RCGM) under DBT for conducting preclinical studies, prior to seeking CDSCO approval for CT. Upon successful completion of clinical studies, an application for market authorization can be submitted for review and approval. The NDCTR, 2019 are intended to make drugs more rapidly accessible to the Indian population and prevent repetition of studies based on data generated from other countries. With fast and committed turn-around review period, the new rules ensure that the overall clinical development timelines get reduced. Clinical research in newer areas of science like stem cell derived products and gene therapeutic products will now be more effectively regulated. The new rules also encourage sponsors to consider qualifying for one or more expedited programs and accelerated approval process for orphan drugs, serious or life-threatening disorders with unmet medical needs. An amendment has been proposed in May 2020 to NDCTR for permitting the usage of unapproved new drugs for compassionate use in treatment of patients, be it by import into India or by local manufacturing of the same. However, no part of the imported/manufactured quantity shall be sold in the market or supplied to any other person, agency or institution and should be used specifically for which the permission is granted by CLA. While the new rules give an opportunity for early access to innovative medicines, it ensures ethical principles are followed and patients' rights are protected.

## 12.8 Quality and Compliance in Clinical Trials

With increasing number of newer drugs under development, globalization of CTs has been a preferred option sought by sponsors to cut short on timelines. However, containing the spiraling cost of CTs continues to be a challenge. Some of factors which contribute to this are the ever-increasing time and financial pressures on clinician-cum-investigators, stringent regulatory requirements, the increased complexity of CT protocols, and on-site clinical data-quality monitoring to check for compliance. There are also challenges in the core capabilities of CT sites, such as in the entire EC review process, study start-up activities, patient recruitment process, and quality of research data. Regulators have increasingly been observing violations involving manufacturing, quality control and data integrity during CT and BA/BE studies during regulatory inspections. The EMA started conducting GCP inspections from 2006, while U.S. FDA has been conducting GCP inspections under the bioresearch monitoring program since 1977. The FDA clinical reviewers essentially look for (a) compliance with the protocol (admission criteria, blinding, randomization, treatment, assessment, and analysis); (b) checks for data consistency (comparison among centers in multicenter trials, comparison of data across formats—tables, summaries, and listings); (c) checks on clinical judgment calls (such as cause of death, cause of an adverse event, etc.). The violations reported in warning letters issued by the U.S. FDA are also posted on their website. Broad themes under which some of the frequently cited violations could be grouped, have been cited (in alphabetical order) below:

- Deviation from investigational plan as per the study protocol
- Failure of investigators to personally supervise the study
- Failure to address conflict of interest among institutional review board (IRB/EC) members
- Failure to follow SOPs and maintain documentation
- Failure to protect subject safety, and report AEs
- Falsification of data
- Inability to retain records or produce records for inspection
- Inadequate or lack of systems for monitoring CTs
- Inappropriate EC membership, quorum issues, misuse of expedited review
- Informed consent discrepancies
- Irregularities in dispensing and accountability of investigational product

Warning letters from FDA can serve as a learning tool for sponsors, investigators, and ECs to take effective measures to help improve systems to ensure regulatory compliance and standards while safeguarding trial subjects. Based on the repetitive findings, sponsors should draft appropriate training modules for orienting the stakeholders—be it ECs, investigators, or other study team members, in their respective areas for improvement.

Indian regulators have started GCP inspections from the last few years and have been issuing show-cause notices to sponsors/CROs and CT investigators for gross violations. Though CDSCO is encouraging self-regulation, giving more latitude to stakeholders, it is expected to come down heavily on enforcement of rules if repetitive non-compliances are observed. If any sponsor including their employees, subsidiaries and branches, their agents, contractors and sub-contractors, investigators or personnel conducting CTs at sites fail to comply with any of the regulatory requirements, the CLA may take one or more of the following actions:

1. Issue warning letter giving details of deficiency found during the inspection.
2. Recommend that the CT may be rejected or discontinued.
3. Suspend or cancel the CT permission.
4. Debar the investigator(s), sponsor including their employees, subsidiaries and branches, their agents, contractors and sub-contractors to conduct any CT in the future.

In case, the CLA issues any order of suspension or revocation or cancellation of any permission or license or registration granted under the NDCTR, such order shall be made available in the public domain immediately by uploading it on the website of CDSCO. This is in line with the way U.S. FDA has been posting warning letters and import alerts on its website, serving as a deterrent to defaulters.

Organizational culture also plays a vital role in shaping people's insights and actions in terms of decision-making related to compliance be it standards, laid down guidelines or regulations. Some of the key elements include: (1) having quality management and internal; auditing system in place, (2) adopting information technology (IT) based systems and processes to ensure compliance; (3) implementing risk-based *monitoring* (RBM) and risk-based quality management approaches in CTs on the lines of ICH E6 (R2) guidelines. Implementation of RBM program requires strong IT based systems with centralized data analytics support. While the cost of transitioning from paper to an electronic mode (be it with workflow management, case report forms, CT batch manufacturing records, analytical testing of samples, investigational product supplies) may be high, in the long term it will result in reduction in error rates, overall improvement in data quality, increased productivity, and ultimately cost savings. Regulators too expect to see more of these systems with proper validations and audit trails, as it reflects firm's intention for greater transparency and drive for quality.

Adequately trained clinical research team is critical for success of any CT and generating quality data which could withstand the test of time. Training needs should be customized as per the stakeholder requirements, like for (a) site staff; (b) investigators; (c) ethics committee, etc. While sponsors have their own GCP courses, training programs conducted by accredited third parties can reduce sponsor-organized trainings, which could be limited to study-specific requirements. Apart from orienting on GCP and applicable regulatory requirements, emphasis should also be laid to develop soft skills like communication, negotiation, project management, inter-personnel skills, etc. The impact of the training should be evaluated using metrics such as reduction in number of audit observations, queries, data entry errors, etc. Continuous updating is essential key in successful operation of any undertaking and GCP is no exception.

## 12.9 Going Forward

The clear intent of the GoI through MoHFW to perform, reform, and transform the CT sector is very evident from the tangible benefits which have started showing up. The NDCTR, 2019 are expected to promote robust data generation in clinical research in a more predictable and transparent manner while reinforcing public trust in CTs. Multinational companies having their drugs already approved and marketed in any of European Union member state, or in other developed markets like USA, Australia, Canada, and Japan, stand to gain in terms of additional clinical waivers. The measures which have been instituted have definitely facilitated and speeded up the regulatory process apart from giving more responsibility and accountability to the stakeholders be it sponsors, ECs, or CT investigators. However, there are still some areas for improvement which include:

- The Indian GCP guidelines of 2001 could go in for an upgrade by adopting some of the points from ICH E6 R2 Amendment. These could include (a) implementing quality management system (QMS) for CT quality systems; (b) adopting risk-based monitoring (RBM) approach and implementing centralized monitoring; (c) additional emphasis on data integrity; (d) having computerized validated systems for handling data, etc. In fact, CDSCO should be taking cue from both the European Medicines Agency (EMA) and the U.S. FDA who have implemented it from June 2017 and March 2018, respectively.
- Apart from mandatory registration of EC with CDSCO for being eligible to approve CTs, accreditation by either international or national (third party) bodies will help to raise the level of quality and standard of clinical research and boost confidence among all stakeholders. Presently, EC accreditation is not mandated by regulations and is purely voluntary. Deliberations are also on as to whether or not to allow central ECs for oversight of multicenter CTs. The advantages include (1) wider panel of expert members; (2) accreditation; (3) standardized requirements for application and review which could oversee CTs.
- Despite the urgent need, development of drugs for children significantly lags behind that for adults. At present, CTs in children are done by and large after the drug is approved for adults, causing delay in providing access to the new drug to children. Parallel clinical development across various age groups can address this concern. Off-label use of drugs in pediatric patients is also widely practiced. Development of well-designed CTs with validated end points or surrogate markers in children should be part of the clinical development plan, as is the case in EU and the USA. So also, centers of excellence for pediatric CTs need to be developed with support from government, academia, pharmaceutical industry, and professional associations/organizations.

- Women differ from men in genetic, physiological, social, economic, and environmental context in the diseases they suffer, and to some extent their access to treatment and response to it. Clinical research and regulations can have great impact on health promotion, disease prevention and management in women. However, percentage of recruitment of women in CTs is quite low. One way of bridging the gap is for regulations to mandate analysis of data based on gender differences.
- CTs typically include patients in the age group 18–65 years. It is also well known that pharmacokinetics and pharmacodynamics (and hence efficacy and safety) substantially change after the age of 75 years. With elderly patients usually excluded from trials, there is usually insufficient data available on safety and efficacy of the prescribed drugs in geriatric population. Former Schedule Y and the current NDCTR 2019 specify inclusion of geriatric patients in Phase III CTs (and in Phase II trials, at sponsor's discretion), if the disease affects them. However, it does not insist CT results to be disaggregated by age. Conscious efforts should be made to design drug trials catering to the elderly and incentivize such studies by probably giving fast track review or even substantial reduction in application fees.
- It is often noted that the Package Insert (PI) does not contain information about the clinical and non-clinical studies done by the sponsor as part of development of the new drug, rather only what has been published by the innovator, is cited. Regulations should mandate inclusion of summary of analysis of results from preclinical studies and CTs done for new drugs and INDs in India (including data for gender and age differences, if any) as part of PI and Summary of Product Characteristics (SPC).
- With the expansion of new drug definition under NDCTR to include stem cell derived products and gene therapeutic products, the CTs undertaken with these products shall get regulated by the CLA. Close on the heels of releasing the national guidelines for doing research on stem cell derived products, ICMR, in consultation with DBT and CDSCO has now released the national guidelines for gene therapy product development and CTs. However, this should be followed by developing disease-specific guidelines for different conditions.

In times to come, CTs are likely to become more complex (e.g., multi-arm multi-stage, adaptive design type of CTs for oncology) both in design and execution. Thus maintaining high ethical standards and stringent quality assurance will become exceedingly important to ensure that potential regulatory inspection observations/violations are minimized and the rights, safety, and well-being of subjects/patients in research are protected. Sponsors need to identify, develop CT sites in line with ICH GCP standards for conducting studies. CROs also need to gear up both in terms of competency and capacity to cope up with research activities. Steps taken by the MoHFW towards harmonization of standards and convergence of regulatory practices for CTs are truly credible and equally laudable. The Indian CTs market size was estimated at USD 1.6 billion in 2017 and is anticipated to expand at a CAGR of 8.7% over the next few years. It is hoped that the new rules would bring

India to the forefront of global clinical research and propel the country as one of the preferred destinations for scientific and ethical CTs. It must be understood that it is easier said than done in our country, but the approach is on!

## Bibliography


NABH (2015) Accreditation standards for clinical trial in India ethics committee, Investigator and clinical trial site. National Accreditation Board for Hospitals and Healthcare Providers, New Delhi

Bajpai V (2013) Rise of clinical trials industry in India: an analysis. ISRN Public Health 2013:167059

Banach M, Limaye N, Young S, Ancukiewicz T, Wilson S, Proeve J (2019) A roadmap for implementing risk-based monitoring and quality management. Appl Clin Trials

Barnes M, Flaherty J, Caron M, Naqvee A, Bierer B (2018) The evolving regulatory landscape for clinical trials in India. Food Drug Law J 73:601–623

Bhatt A (2017) International Council for Harmonization E6 (R2) addendum: challenges of implementation. Perspect Clin Res 8(4):162–166

Bhave A, Menon S (2019) Subject expert committees: past, present, and future. Perspect Clin Res 10(1):1–3

Bhosale N, Nigar S, Das S, Divate U, Divate P (2014) Protection of human research participants: accreditation of program's in the Indian context. Indian J Med Ethics 11:55–59

Bowalekar SK, Talele R (2012) Warning letter study. Appl Clin Trials 21(10)

Clinical Trials Registry of India (n.d.). http://ctri.nic.in/

Davis S, Poonam S, Murtuza B, Vrunda P, Shilpi S (2017) Ethics committees and the changed clinical research environment in India in 2016: a perspective. Perspect Clin Res 8:17–21

Vohora D (2018) Chapter 14 - Ethical considerations in clinical research. In: Pharmaceutical medicine and translational clinical research. Elsevier BV, Philadelphia, PA, pp 265–283. ISBN 978-0-12-802103-3

CDSCO (2013a) G.S.R. 53(E). Rule 122 DAB – 'Compensation in case of injury or death during clinical trial'. Published by the Central government under the MoHFW

CDSCO (2013b) G.S.R. 63(E). Rule 122 DAC – 'Permission to conduct clinical trial'. Published by the Central government under the MoHFW

CDSCO (2013c) G.S.R. 72(E). Rule 122 DD – 'Registration of ethics committees'. Published by the Central government under the MoHFW

George B (2020) GCP and clinical research standards in India. In: Earl W Hulihan (ed) Good clinical practice: a question & answer reference guide, vol Section 20, Barnett International Publishers, A division of Cambridge Healthcare Institute, Needham, MA, pp 653–670 ISBN: 978-0-9963462-8-3

George B (2019) A new dawn in the history of Clinical Research for India. IDMA Bull L(17):08–14

Gogtay NJ, Doshi BM, Kannan S, Thatte U (2011) A study of warning letters issued to clinical investigators and institutional review boards by the United States FDA. Indian J Med Ethics VIII (4):211–214

Gogtay NJ, Ravi R, Thatte UM (2017) Regulatory requirements for clinical trials in India: what academicians need to know? Ind J Anesth 61(3):192–199

CDSCO (2019) GSR. 227(E). New drugs and clinical trials rules. Published by the Central government under the MoHFW

ICMR, CDSCO, DBT (2019) National guidelines for gene therapy product development and clinical trials

ICMR, CDSCO (2017) Handbook for applicants and reviewers of clinical trials of new drugs in India

ICMR (2008) Guidelines for good clinical laboratory practices

ICMR (2017) National ethical guidelines for biomedical and health research involving human participants. ISBN: 978-81-910091-94

Johnston SC, Lewis-Hall F, Bajpai A, Blumenfeld TH, Goldschmidt P, Koski G, Lee G, Schneider R, Tenaerts P, Womack A (2017) It's time to harmonize clinical trial site standards. Expert voices in health & health care. National Academy of Medicine, Washington, DC

Kondal A, Krishna GVM, Bansal D (2015) Clinical trial regulations in India: progress and challenges arising from recent amendments to Schedule Y of the Drugs and Cosmetics (D&C) Act 1940 (D&C Rules 1945). Pharmaceut Med 30:1

Kshirsagar NA, Drici M, Ferner RE, Jackson SH, Fernandopulle R, Anker J, Bhowmik S, Mehta D, Khetrapal M, Gupta M, Supe A, Pravin K (2017a) Safer prescriptions for the elderly: challenges and solutions from India. The Lancet Global Health Blog

Kshirsagar NA, Pai R, Pandit B, Patel M, Medhi B, Gupta YK, Roy M, Ferner R, Bhowmik S, Mehta D, Somani VG (2016) Regulatory and ethical issues in clinical research in women in developing countries. Br Med J 355:i6316

Kshirsagar NA, Swaminathan S, Jog P, Dalwai S, Mathur R, Shekhar C, Meibohm B, Gupta YK, Shafiq N, Sunkara G, Somani VG, Kulkarni P, Tseng B, Mehta D (2017b) Regulatory and ethical issues in pediatric clinical research: recommendations from a panel discussion. J Clin Pharmacol 57(8):943–946

Kulkarni R, Saraiya U (2015) Accreditation of ethics committees in India: experience of an ethics committee. Indian J Med Ethics 12:241

Lahiry S, Sinha R, Choudhury S, Mukherjee A, Chatterjee S (2018) Paradigm shift in clinical trial regulations in India. Indian J Rheumatol 13:51–55

Mandal SC (2017) Bioequivalence requirements in various global jurisdictions. Chapter 5. In: Kanfer I (ed) AAPS advances in pharmaceutical sciences, vol 28. Springer Nature, Cham, pp 101–125

Naik K (2017) Clinical trials in India: history, current regulations and future considerations. Master of Science project. School of Health Sciences, Eastern Michigan University

Sharma H, Parekh S (2012) Clinical trial regulations in India. Pharmaceut Regul Affairs 1(4):1–3

Sivanandan S, Jain K, Plakkal N, Bahl M, Sahoo T, Mukherjee S, Gupta YK, Agarwal R (2019) Issues, challenges, and the way forward in conducting clinical trials among neonates: investigators perspective. J Perinatol 39(Suppl 1):20–30

## *Websites Referred*

www.cdsco.nic.in
www.clinicalleader.com
www.cliniexperts.com
www.iscr.org
www.issuesinmedicalethics.org

**Bobby George** M. Pharm, PhD, CCRA, PGDHHM is serving as the Group Head, Vice President Regulatory Affairs at Reliance Life Sciences, Navi Mumbai. He is responsible for regulatory services across the company's business verticals (plasma proteins, biosimilars, pharmaceuticals, etc.) for both domestic and export markets. He holds a Ph.D in Pharmacology and has close to 24 years of industry experience. He is also an avid speaker and writer. Dr. George has to his credit 40 publications in peer reviewed journals and has written four book chapters. He has also authored and published an interesting book on Pharma healthcare titled The Act that Wasn't®.

**Shrinivas Krishnarao Kulkarni,** M. Pharm, PhD, FAMS, FNASc, FPAS, FIPA former Pro-Vice Chancellor (DUI) Panjab University, Chandigarh is presently Emeritus professor of Pharmacology. He was formerly the Dean and Chairperson of the Faculty of Pharmaceutical Sciences. A well-known pharmacologist of long-standing. Professor Kulkarni's research has been extensively cited (Google Citation index > 13,000; H factor-61, i10-index > 172; books-13). His recent book (co-authored) entitled "Pharmacologists of India" covers pharmacology research in India during the last seven decades. He is a Fellow of National Academy of Medical Sciences, National Academy of Sciences, and the Indian Pharmaceutical Association. A Fulbright Visiting Scholar of US-Education Foundation, consultant to WHO and member of Indian Pharmacopeia Committee, Dr. Kulkarni was the General President of the 53rd Indian Pharmaceutical Congress, and Secretary General of Indian Pharmacological Society.

**Nilima A. Kshirsagar**, MD, DNB, PhD, FNAMS, FNAS, FACCP (USA), FRCP (UK), former Dean, Director Medical Education Health, Vice Chancellor, State University, is presently National Chair Clinical Pharmacology, Indian Council of Medical Research (ICMR), Govt. of India.

Under guidance from Dr. Balram Bhargava, DG ICMR, Secretary DHR, she has conceptualized and spearheaded development of National Centre for Clinical Pharmacology with network of centers for Rational Use of Medicines and network for Product Development. She is recipient of numerous international (e.g., Nathaniel Kwit Award of American College of Clinical Pharmacology) and national (B. C. Roy, Vasvik, Mayor of Mumbai) awards for her contributions and industrial research. She is member of WHO Advisory committee on Safety of Medicinal products, chairperson of core training panel of Pharmacovigilance Program of India, subcommittee for FDC, Scientific Advisory Group of BMS ICMR, and member of DTAB Govt. of India. She has over 200 publications, written chapters, edited books, e.g., on drug development and clinical research. Her areas of expertise include clinical trial, drug development, tropical diseases, and medical education.

# New Drug Discovery and Development: Indian Pharmaceutical Industry

# 13

Nagarajan Kuppuswamy, Srinivas Nanduri, and Venkateswarlu Akella

## 13.1 Introduction

The beginning of the pharmaceutical industry in India based on allopathic medicine at the turn of the twentieth century can be traced to Acharya P.C. Ray's establishment of Bengal Chemicals and Pharmaceuticals Ltd. on April 12, 1901. Earlier to it, people were resorting to indigenous systems of medicine including Ayurveda, Siddha, Unani, etc., which even today enjoy high popularity in the country. Starting with one factory in Kolkata in 1905, three more factories were built. Due to unsatisfactory performance, the company was nationalized by the government of India in March 1981. Sadly, the company's growth never matched the national fervor of Ray (Joseph 2011).

The Alembic Chemical Works Co limited incorporated in 1907 in Vadodara initially dealt with cough syrups, vitamins, tonics, and sulfa drugs. They became pioneers in establishing fermentation facilities—penicillin (1961), streptomycin (1968), and erythromycin (1971) using Eli Lilly expertise and continue to specialize in antibiotics ("Database of Company Financial Annual Reports," n.d.; Joseph 2011). Unichem laboratories founded by Amrut Modi in 1944 represented an effort to collaborate with innovative pharma MNC, UCB of Belgium. Unichem bought bulk drugs from them for developing and sale of formulations in India. Later they also licensed UCB's unique drug delivery system for Levetiracetam (KEPRA).

N. Kuppuswamy
Hikal Ltd., Bangalore, India
e-mail: k_nagarajan@hikal.com

S. Nanduri
Department of Pharmaceutical Technology (Process Chemistry), National Institute of Pharmaceutical Education & Research, Hyderabad, India
e-mail: nanduri.niperhyd@gov.in

V. Akella (✉)
Dr Reddy's Institute of Life Sciences, Hyderabad, India



M. Dikshit (ed.), *Drug Discovery and Drug Development*,
https://doi.org/10.1007/978-981-15-8002-4_13

Sarabhai chemicals was started in 1943 in Baroda for manufacturing chemicals and pharmaceuticals. In collaboration with ERSquib of USA, Symbiotic of Sarabhai was the largest producer of antibiotics in India ("Profile," n.d.).

Ranbaxy Laboratories was a private Indian pharmaceuticals company started in 1961. It went public in 1971. After an extremely successful stint, it was bought over by Daichi and later by Sun Pharma. Its disappearance is a tragic chapter in the history of Indian Pharma (Carroll 2008). The first major innovative pharma company in India was Ciba Pharmaceuticals, incorporated on December 13, 1947. Ciba (now Novartis) imported their bulk drugs into India for formulation and sale like other foreign MNCs. Hoechst (now Sanofi) started their operations in India in 1956, SKF in 1961, Merck in 1967 and there were others engaged in similar activities. The first three companies were notable for engaging in new drug development in India for several years before these activities were wound up. Astra Zeneca, a late entrant MNC also had a vibrant NCE research center which was closed in 2014 (Suresh et al. 2015).

The Indian government with its emphasis on self-reliance in the early decades after independence in critical areas sought to encourage the growth of Indian Pharma in multiple ways. Hindustan Antibiotics Limited was started in the public sector in Pune in March 1954 to give India self-reliance and expertise in the fermentation of antibiotics. During its glorious years, it manufactured several antibiotics like penicillin, streptomycin, gentamycin and semi synthetics like ampicillin and also their formulations. Sadly, it lost the race to China in development of better yielding strains. The same fate befell other Indian players like SPIC and TORRENT.

Indian Drugs and Pharmaceuticals Limited was a public sector company set up in April 1961 to manufacture and sell essential synthetic drugs using Russian technology. It diversified into formulations later. Unfortunately, it did not keep up with the fast-paced introduction of new drugs and vastly improved processes for synthetic drugs. Nevertheless, it served the important cause of training a number of chemists and chemical engineers who later went on to establish a hoard of pharma companies in and around Hyderabad.

On the institutional side, Government of India established a number of laboratories to create self-reliance in various areas. The Central Drug Research Institute was established in 1951 in Lucknow under the aegis of Council of Scientific and Industrial Research with a multi-disciplinary approach to discover drugs from nature as well as through synthesis. The institute was also equipped to device processes for known drugs side stepping innovator routes (Gupta 2005; McGrath 2007). In the latter area other CSIR labs, RRL Hyderabad (IICT) in 1944, RRL, Jammu (IIIM) in 1941, NCL Pune (1952) were also entrusted with chemical process development. CSIR had envisaged close interaction of these laboratories with industry. In addition, introduction of national policies like MRTP in 1969 (foreign companies to have less than 40% holding, for manufacture of drugs in India and give away 10% to non-associated formulators), the Indian patent act of 1970 (no product patent, only process patent with severely limited protection) equally fuelled the rapid growth of Indian Pharma ("The Patents Act, 1970," n.d.). These encouraged rapid blooming of a number of Indian Pharma Companies who successfully side-stepped infringing processes of MNCs, produced a large number of formulations and sold them at competitive prices, aided by less expensive Indian labor. These were sold as branded generics locally and also in non-regulated markets. With these initiatives, India came to be known as a pharmacy of the world.

In this favorable atmosphere, the Hatch-Waxman Act of 1984 of United States gave a God-sent opportunity for the leading Indian generic companies to get into the lucrative US generic market through the ANDA route and occasionally exploiting para 4 challenges (Mossinghoff 1999). As their Coffers grew by leaps and bounds, a number of these Indian pharma companies started venturing into new drug development, while the fortunes of MNC pharma's like Novartis, Sanofi, Merck, etc. either stagnated or grew in low single digits. Another matter of great significance for new drug development in India was the Indian Government's changes in the patent laws as a part of compliance with TRIPS. The membership of TRIPS obliged countries to align their patent laws with acceptable requirements by year 2005, beginning to be made by 1995. Accordingly, the patent act of 1970 offering only process but not product protection was to be scrapped and replaced by the act of 2005 giving full protection of products for 20 years. There were other clauses, like no protection for formulations or new forms unless these had improved efficacy over the original ones. Compulsory licensing was provided in conditions of national emergency. These changes were welcome to the Indian pharma industry in their efforts in NCE research but were only partly acceptable to the MNCs (Kiran and Mishra 2011).

### 13.1.1 Indian Pharma Companies

Table 13.1 (Differding 2017) contains the list of companies, year of their incorporation, year of start of NCE research, and also the closure or downsizing of NCE research in some cases.

The succeeding section of this chapter gives a critical account of the performance of these pharmaceutical companies and pubic funded research organizations/ institutions in their Drug Discovery & Development efforts. They are categorized in different therapeutic areas.

## 13.2 New Drug Discovery and Development: Indian Contributions

### 13.2.1 Brief History of Indian Contributions and Present Status

The advent of modern drug discovery in India dates back to early 1900s with the efforts to develop drugs for treating visceral leishmaniasis (also known as Kala-azar) at the Campbell Medical College, Calcutta. This led to the development of **Urea Stibamine** by Professor Upendranath Brahmachari in 1922, for which he was nominated for the Nobel Prize in 1929. The toxicity of inorganic pentavalent antimony salts was minimized by synthesizing their organometallic aniline

**Table 13.1** Indian Pharmaceutical Industry: Status of NCE research

| S. No. | Name of the company | Year of establishment | Year of start of NCE research | Year of closure of NCE research |
|---|---|---|---|---|
| Pharmaceutical companies | | | | |
| 1 | Dr. Reddy's Laboratories | 1984 | 1994 | 2009 |
| 2 | Ranbaxy Laboratories | 1961 | 1995 | 2008 |
| 3 | Torrent Pharmaceuticals Ltd. | 1972 | 1997 | |
| 4 | Wockhardt Ltd. | 1967 | 1997 | |
| 5 | Piramal Life Sciences | 1988 | 1998 | 2014 |
| 6 | Dabur Research Foundation—now Fresenius-Kabi Oncology | 1979 | 1998 | 2010 |
| 7 | Sun Pharma | 1993 | 1999 | |
| 8 | Alembic Ltd | 1907 | 1999 | |
| 9 | Zydus Cadila (Cadila Healthcare) | 1995 | 2000 | |
| 10 | Cadila Pharmaceuticals | 1995 | 2000 | |
| 11 | FDC Ltd | 1940 | 2000 | |
| 12 | JB Chemicals & Pharmaceuticals | 1976 | 2000 | 2006 |
| 13 | Cipla | 1935 | 2000 | |
| 14 | Glenmark Pharmaceuticals | 1977 | 2001 | |
| 15 | Lupin Ltd. | 1972 | 2001 | 2017 |
| 16 | Reliance Life Sciences | 2001 | 2001 | 2010 |
| 17 | Orchid Pharma | 1992 | 2002 | |
| 18 | Suven Life Sciences | 1989 | 2003 | |
| 19 | Natco Pharma | 1981 | 2004 | |
| 20 | Panacea Biotech | 1984 | 2005 | |
| 21 | Matrix Laboratories—now Mylan | 2000 | 2005 | 2006 |
| 22 | Hetero Drugs | 1993 | 2006 | |
| 23 | Jubilant Life Sciences | 1978 | 2007 | |
| 24 | Elder Pharmaceuticals | 1989 | 2008 | |
| 25 | IPCA Laboratories | 1949 | 2009 | |
| 26 | Mankind Pharma | 1995 | 2011 | |
| 27 | Alkem Laboratories | 1973 | 2013 | 2015 |
| 28 | Emcure | 1981 | 2014 | |
| Contract research companies with proprietary projects | | | | |
| 29 | Advinus | 2005 | 2005 | |
| 30 | Anthem Biosciences | 2009 | 2009 | |
| 31 | Aurigene | 2002 | 2002 | |
| 32 | GVK Bio | 2001 | 2001 | |
| Biotech companies | | | | |
| 33 | Kareus Therapeutics | 2007 | 2007 | |
| 34 | Connexios Life Sciences | 2003 | 2008 | |

(continued)

**Table 13.1** (continued)

| S. No. | Name of the company | Year of establishment | Year of start of NCE research | Year of closure of NCE research |
|---|---|---|---|---|
| 35 | Rhizen Pharmaceutics/Incozen Therapeutics | 2008 | 2008 | |
| 36 | Sphaera Pharma | 2008 | 2008 | |
| 37 | Curadev | 2010 | 2010 | |
| 38 | Shantani Proteome Analytics | 2010 | 2010 | |
| 39 | Vyome Biosciences | 2010 | 2010 | |
| 40 | Vitas Pharma | 2011 | 2011 | |
| 41 | Invictus Oncology | 2011 | 2011 | |
| 42 | Krish Biotech | 2015 | 2015 | |

derivatives. Later, Brahmachari, Kikuth, Schmidt, and others had come up with less toxic pentavalent antimonials like antimony gluconate (Solustibosan) and sodium stibogluconate in 1937 and 1945, respectively. Currently, sodium antimony gluconate (SAG; manufactured by Albert-David, Kolkata, India) is one of the most widely used organic compounds of antimony (Brahmachari 1922; Goodwin 1995; Kikuth and Schmidt 1937; Singh and Sivakumar 2004).

Subsequent to the above development, the drug discovery efforts in India were virtually unknown till the establishment of Central Drug Research Institute (CDRI) in Lucknow to lead the country's efforts in drug research and development (1951) followed later by other public institutions such as the Regional Research Laboratories in Hyderabad (1956, now Indian Institute of Chemical Technology, IICT) and in Jammu (1957, now Indian Institute of Integrative Medicine, IIIM). CDRI's main focus was to identify lead molecules for tropical diseases and population control measures initially from medicinal plants and later expanded to include synthetic small molecules. These efforts were also strengthened by notable discoveries made by multinational companies in India, viz., CIBA Research Centre (1960s) and Hoechst Research Centre (1970s) in Mumbai (then Bombay), Smith Kline & French (in 1970s and 1980s) and AstraZeneca in Bangalore (Dikshit and Dikshit 2016).

With the rise of Indian Generic Industry through the introduction of Indian Patent Act in 1970, the drug discovery efforts also gained momentum with many of the generic players such as Dr. Reddy's Laboratories Ltd. (DRF), Ranbaxy, Sun Pharma (SPARC), Wockhardt, Glenmark, Lupin, Zydus Cadila, Torrent, Dabur, Cipla, Natco, Biocon in collaboration with BMS (BBRC), Panacea, Suven Pharma, and Hetero Drugs initiating drug discovery programs. This period also saw the increase of collaborative drug discovery efforts by many CROs, viz., Anthem, Aurigene, Jubilant, GVK, and Advinus with various multinational pharmaceutical industries. A number of preclinical and clinical candidates were developed through these efforts largely in the areas of anti-microbial, anti-tubercular, antimalarial, drugs for metabolic disorders, anti-inflammatory, anti-cancer, drugs acting on central nervous system and reproductive systems.

An attempt has been made to collate the outcome of these efforts in different therapeutic categories. The drug candidates covered include registered, approved drugs and also under clinical development. Several anthelmintics, drugs for amebiasis, protozoal diseases, antimalarial, anti-tubercular and anti-bacterial agents including glycopeptide antibiotics were developed by Ciba Geigy, SmithKline Beckmann, CDRI, Ranbaxy & MMV, AstraZeneca & MMV, IIIM, Lupin, Sphaera & ICGEB, Napo Pharma, Glenmark, Wockhard, Aurigene, Vyome Biosciences, Vitas Pharma, Orchid, and Allecra Therapeutics (Table 13.2).

Significant efforts have been made in pursuing discovery of drugs for metabolic disorders. From the isolation of Peruvoside in Andhra University as cardiotonic agent in 1970s, Forskolin by CDRI & Hoechst Research Centre, Gugulipid by CDRI and Malti-Chem Research Centre, the Indian Industries and Institutes have paid major attention in this area. Discovery of various PPAR inhibitors and AMPK modulators for the treatment of diabetes and other cardiovascular agents by Dr. Reddy's Laboratories was considered as a major breakthrough by an Indian Company, though not successful in reaching the market. Triggered with the above success, Ranbaxy Laboratories, Torrent Pharmaceuticals, Piramal, Cadila, Glenmark, Lupin, Orchid, Panacea Biotec, Matrix Laboratories, Alkem, Advinus, Kareus, Connexious, Shantani, and Rhizen Pharmaceuticals have made important contributions in this area. Even though many of the above discoveries are still in the preclinical/clinical stages, Zydus was successful in introducing Lipaglyn (a dual PPAR $\alpha/\gamma$ agonist) in the market for the treatment of diabetes (Table 13.3).

With the development of Tinazoline by CIBA as a vasoconstrictor and Tromaril (enfenamic acid) by RRL, Hyderabad (licensed to Unichem Laboratories) discovery and development of several anti-inflammatory agents were pursued. Efforts by Ranbaxy, Aurigene, Piramal, Sun Pharmaceuticals, J.B. Chemicals & Pharmaceuticals, Glenmark, Lupin, Orchid Chemicals, Matrix Laboratories, Incozen therapeutics, and Reliance Life Sciences on a number of anti-inflammatory targets are noteworthy (Table 13.4).

The pursuit for anti-cancer agents in India began with the synthesis of Alvocidib (Flavopiridol), a synthetic analog of natural product Rohitukine which was initially extracted from Amoora rohituka [syn. "Aphanamixis polystachya"] and later from Dysoxylum binectariferum. Clinical development of new camptothecin analogs DRF 1042 and 1644 was pursued by Dr. Reddy's Research Foundation/Dr. Reddy's Laboratories. A number of other agents were pursued by Ranbaxy, Piramal, Dabur, Sun Pharma, Zydus Cadila, Lupin, Orchid, Natco, Jubilant, Anthem, Aurigene, GVK Bio, Rhizen Pharmaceuticals, and Invictus Oncology (Table 13.5). Discovery of several CNS acting drugs was pursued by CDRI, Ciba Geigy, Suven Life Sciences, and Kareus Therapeutics (Table 13.6). In search of fertility regulating agents many compounds related to triphenylethylene structure were synthesized at CDRI, of which 3,4-diphenylchromenes and chromans proved to be of particular interest, which led to the discovery of Ormeloxifene (Centchroman). Centchroman is world's first non-steroidal oral contraceptive pill. It was licensed to Hindustan Latex, Trivandrum, in 1989 and is in the market with the trade name SAHELI and Chaya (Table 13.7).

**Table 13.2** Anti-infective drugs

| S. No. | Name and structure of the drug | Indication | Developed by | Licensed to | Status |
|---|---|---|---|---|---|
| 1. | Amoscanate 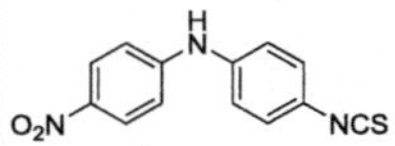 (Doshi et al. 1977; Shapiro et al. 1986; Singh et al. 1981; Singh 2014) | Broad spectrum anthelmintic. | Ciba-Geigy | | Registered as a drug for use in hookworm infections, but has not been marketed. |
| 2. | CWI-785 (Verma et al. 1986, 1987; Cavalleri and Parenti 2000) | Glycopeptide antibiotic | SmithKline Beckmann Corporation | | |
| 3. | Elubaquine 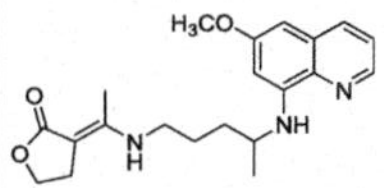 (CDRI, R. C. A. 2003; Dutta et al. 1989; Puri et al. 1989) | Antimalarial | Central Drug Research Institute (CDRI) | Nicholas Piramal | Marketed as Aablaquin |
| 4. | Arteether (the ethyl ether of dihydroartemisinin) 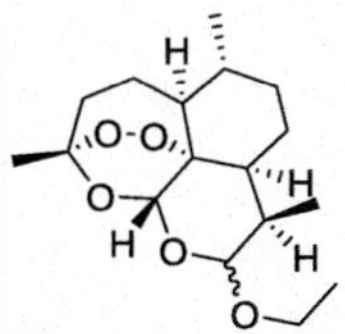 (Sethi et al. 1988; Tripathi et al. 1990; Tripathi et al. 1997) | For treating cerebral malaria | CDRI and Central Institute of Medicinal and Aromatic Plants (CIMAP) | Themis Medicare Limited | |

(continued)

**Table 13.2** (continued)

| S. No. | Name and structure of the drug | Indication | Developed by | Licensed to | Status |
|---|---|---|---|---|---|
| | **Note:** Mixture of alpha and beta enantiomers much more easily made than single anomer—exported (WHO, MMV recognized) | | | | |
| 5. | Arterolane (OZ277, RBx 11160) 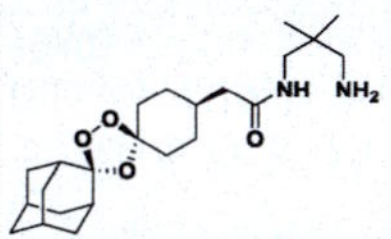 (Dong et al. 2009; “Ranbaxy Laboratories Limited (RLL) Collaborates with Medicines for Malaria Venture (MMV),” 2006) | For treating *Plasmodium falciparum* malaria. | Licensed from Medicines for Malaria Venture (MMV) by Ranbaxy | | |
| 6. | Synriam, a fixed dose combination of arterolane maleate and piperaquine phosphate (“Ranbaxy Launches Synriam—India’s First New Drug,” 2012) | Antimalarial | Licensed from Medicines for Malaria Venture (MMV) by Ranbaxy | | |
| 7. | CDRI-97/78 (“IPCA New Drug Discovery/Development,” 2012) | Antimalarial ozonide | CDRI | IPCA laboratories | Phase 1 |

| | | | | | |
|---|---|---|---|---|---|
| 8. | CDRI-99/411 (“IPCA New Drug Discovery/Development,” 2012) | Antimalarial ozonide | CDRI | IPCA laboratories | Presumed to have been discontinued at the preclinical stage |
| 9. | MMV253 (or AZ13721412) (Hameed et al. 2015; “MMV-Supported Projects: MMV253” 2018) | A fast killing and long-acting antimalarial. Acts by inhibiting triaminopyrimidine V-type H+ ATPase | Astra Zeneca India and MMV. | Following the termination of R&D in Astra Zeneca’s Bangalore in 2014, MMV and Zydus Cadila further developed the drug. | |
| 10. | Satranidazole 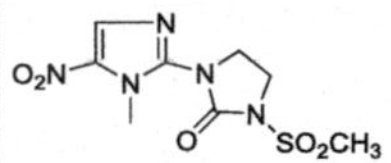 (Nagarajan 2006; Ray et al. 1983; Ray et al. 1984) | For treating luminal and hepatic amoebiasis | Ciba-Geigy Research Centre | Alkem Labs introduced after expiry of Ciba-Geigy patents, they have a good selling combination of satranidazole with ofloxacin | |
| 11. | CGI 17341 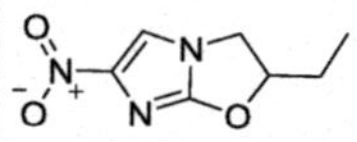 (Ashtekar et al. 1993; Nagarajan et al. 1989) | For treating infections due to several anaerobic bacteria and protozoa and also exhibits anti-mycobacterial activity | Ciba-Geigy Research Centre | | |
| 12. | OPC 67683[a] (also known as Delamanid) 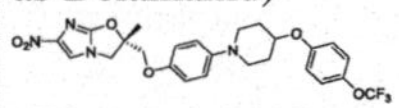 (Gler et al. 2012; Matsumoto et al. 2006; “WHO Model List of Essential Medicines,” 2015) | Potent activity against drug-resistant and drug-susceptible TB. Acts by interfering with the synthesis of mycolic acid, thus disrupting the cell wall. | Otsuka Pharmaceuticals | | Approved for medical use in 2014 in Europe, Japan, and South Korea. It is on the World Health Organization’s List of Essential Medicines. |

(continued)

**Table 13.2** (continued)

| S. No. | Name and structure of the drug | Indication | Developed by | Licensed to | Status |
|---|---|---|---|---|---|
| 13. | PA-824[a] (Pretomanid) 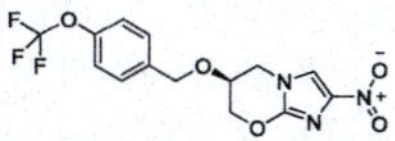 (Manjunatha et al. 2009; Stover et al. 2000) | Anti-tubercular. | Pathogenesis Corporation | TB Alliance | Pretomanid, developed by the non-profit TB Alliance, has received US approval in combination regimen with bedaquiline and linezolid for people with XDR-TB or treatment-intolerant/non-responsive MDR-TB (https://www.tballiance.org/news/fda-approves-new-treatment-highly-drug-resistant-forms-tuberculosis) |
| 14. | "Risorine"-Bioenhancer. Contains reduced dose (200 mg) of rifampicin, isoniazid (300 mg), and piperine (10 mg) ("CSIR NEWS progress, promise and prospects," 2010) | Anti-tubercular. | Indian Institute of Integrative Medicine (IIIM), Jammu, and Cadila Pharmaceuticals Ltd | | Launched in November 2009 |

| | | | | | |
|---|---|---|---|---|---|
| 15. | TBA-7371<br>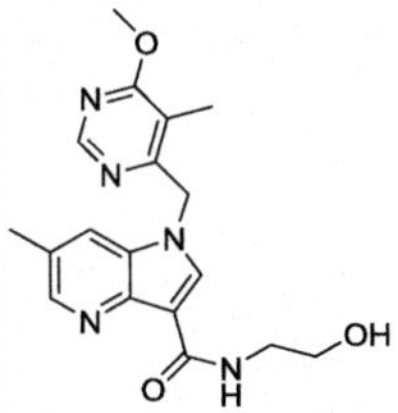<br>(Chatterji et al. 2014; Cooper 2016; Fonteilles-Drabek et al. 2017) | Potent inhibitor of DprE1 (a vital enzyme for mycobacterial cell wall synthesis) | Astra Zeneca and "TB Alliance" | As a result of shutdown of AstraZeneca's Bangalore R&D center in 2014, TB Alliance continued the preclinical development of TBA-7371 with Lilly | |
| 16. | LL 3858 (Sudoterb)<br>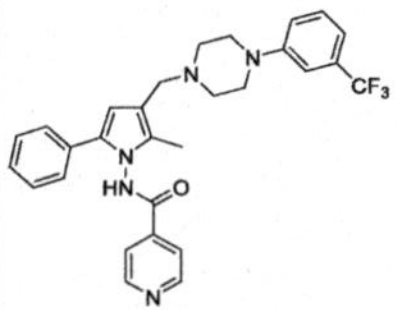<br>(Rivers and Mancera 2008; Arora et al. 2004; "Sudoterb," 2015c) | Anti-tubercular agent with an undefined mechanism of action | Lupin | | Completed Phase 2 studies for tuberculosis in India in 2013<br>No further progress on clinical development reported |
| 17. | SPR113<br>(Dugar et al. 2015; "Sphaera Pharma, THSTI & Wellcome Trust to Jointly Develop Drug for Resistant TB," 2015; "SPR-113," 2016) | Niacin receptor 1 (NIACR1) inhibitor for treating drug sensitive and drug-resistant tuberculosis | Sphaera and ICGEB (International Centre for Genetic Engineering and Biotechnology) | | Under advance preclinical evaluation |

(continued)

**Table 13.2** (continued)

| S. No. | Name and structure of the drug | Indication | Developed by | Licensed to | Status |
|---|---|---|---|---|---|
| 18. | Crofelemer—commercial name Mytesi 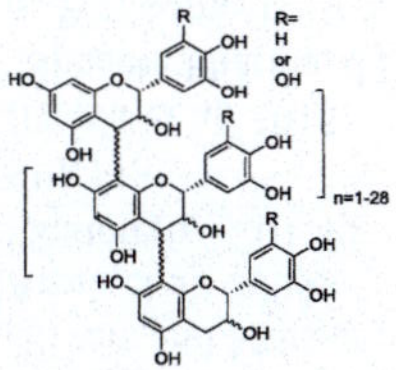 (Gabriel et al. 1999; Sax 2013; Tradtrantip et al. 2010) | To relieve symptoms of non-infectious diarrhea in HIV-infected patients taking antiretroviral therapy (ART), which acts by decreasing chloride secretion in the intestine by inhibiting the cystic fibrosis transmembrane conductance regulator (CFTR), as well as calcium-activated chloride channel | Napo Pharmaceuticals | Glenmark Pharmaceuticals | Granted FDA approval on December 31, 2012<br>It was initially developed by Napo Pharmaceuticals, which licensed it to Glenmark Pharmaceuticals to market in 140 emerging markets and to Salix Pharmaceuticals in the USA, EU, and some other markets. A Phase III clinical trial for diarrhea in HIV patients was completed in 2012, and the drug was approved by the US Food and Drug Administration (FDA) on 31 December 2012<br>https://en.wikipedia.org/wiki/Crofelemer |
| 19. | RBx 7644 (Ranbezolid) 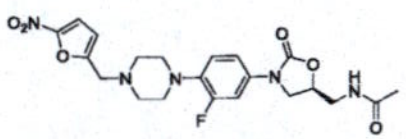 (Das et al. 2005; Naruganahalli et al. 2006; "Ranbezolid," 2017d) | Competitive inhibition of monoamine-oxidase A (MAO-A). | Ranbaxy | Sun pharmaceuticals | Preclinical development stage |

| | | | | | |
|---|---|---|---|---|---|
| 20. | WCK 771—the arginine salt of the S enantiomer of the levonadifloxacin 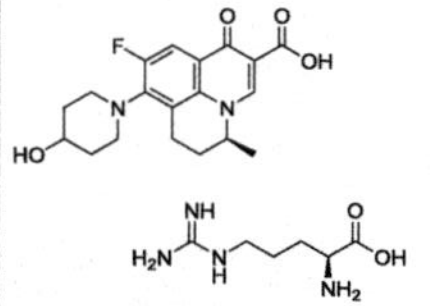 (Al-Lahham et al. 2005; de Souza et al. 2005; "Levonadifloxacin—Wockhardt," 2017b; Patel et al. 2004) | For treating skin and soft tissue infections and MRSA infections by targeting bacterial DNA-gyrase along with topoisomerase-IV. | Wockhardt | | Currently under Phase-3 clinical trials. Was granted QIDP (Qualified Infectious Disease Product) status from USFDA in 2014. |
| 21. | WCK 2349 or Alalevonadifloxacin—oral amino acid ester prodrug of levonadifloxacin. 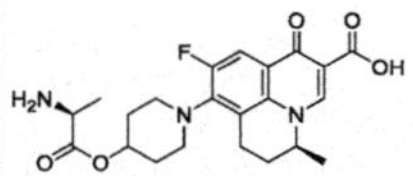 ("Alalevonadifloxacin—Wockhardt," 2018b; de Souza Mendes and de Souza Antunes 2013) | For treating skin and soft tissue infections and MRSA infections. | Wockhardt | | Phase 3 clinical trials for skin and soft tissue infections in India and is in Phase 2 for MRSA infections. Was granted QIDP (Qualified Infectious Disease Product) status from USFDA in 2014 |
| 22. | WCK 1152 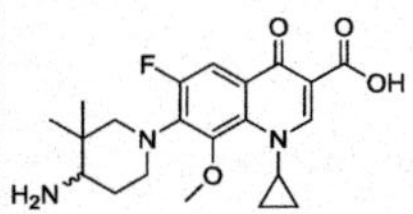 (Al-Lahham et al. 2005; "WCK 1152," 2016g) | Respiratory tract infections | Wockhardt | | Abandoned |

(continued)

**Table 13.2** (continued)

| S. No. | Name and structure of the drug | Indication | Developed by | Licensed to | Status |
|---|---|---|---|---|---|
| 23. | WCK 4873 or Nafithromycin 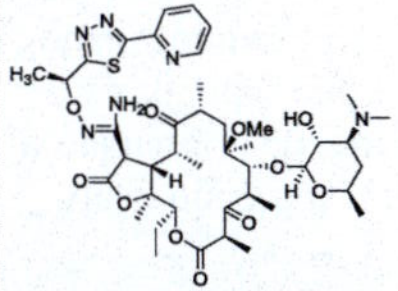 (“Annual Reports 1997–2016,” Wockhardt Ltd. 1997–2016; “Nafithromycin,” 2015a; “Nafithromycin—Wockhardt,” 2018e; “Wockhardt Receives Qualified Infectious Disease Product (QIDP) Designation for its New Drug WCK 4873 from U.S. FDA,” 2015) | MDR—pneumococci | Wockhardt | | Phase 2 studies for community-acquired pneumonia in the USA. |
| 24. | WCK 5107 or Zidebactam 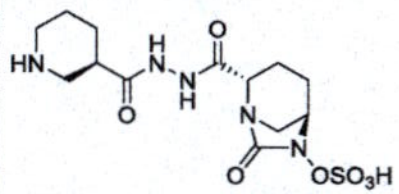 (Huband et al. 2016) | Dual penicillin-binding protein 2 (PBP2) and beta-lactamase inhibitory action. | Wockhardt | | Currently in clinical development stages |

| | | | | | |
|---|---|---|---|---|---|
| 25. | WCK-5222—combination of zidebactam and cefepime. ("Cefepime/Zidebactam—Wockhardt," 2019b) | | Wockhardt | | Preclinical development for gram-negative infections. |
| 26. | AZD5847 (Posizolid) 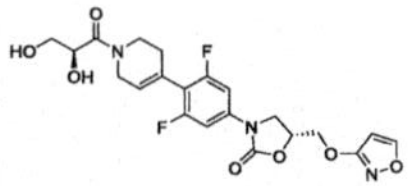 ("Annual Reports 1999–2016," AstraZeneca 1999–2016; Laughon and Balganesh 2010; Balasubramanian et al. 2014; "Posizolid," 2018f) | Oxazolidinone antibiotic | Astrazeneca India and National Institute of Allergy and Infectious Diseases | | Phase 1 for tuberculosis and gram positive infections in 2009 but was discontinued. |
| 27. | AEA16 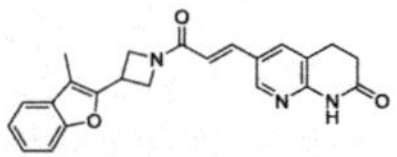 (Takhi et al. 2014) | Inhibitor of FabI enzyme (enoyl-acyl carrier protein reductase) | Aurigene | | |
| 28. | VB-1953 ("VB 1953," 2019e; "Vyome Biosciences Administers First In-human Dose of VB 1953 in U.S. Phase 1 Clinical Study in Patients with Facial Acne Vulgaris," 2016) | Dual acting compound with a combination of a nitro-imidazole antibiotic and a fluoroquinolone joined with a linker moiety | Vyome biosciences | | Currently undergoing Phase-2 clinical trials in the USA for acne as of February 2019. |

(continued)

**Table 13.2** (continued)

| S. No. | Name and structure of the drug | Indication | Developed by | Licensed to | Status |
|---|---|---|---|---|---|
| 29. | VT-02-00068 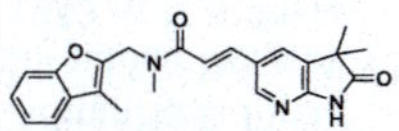 (Rangarajan et al. 2014) | Prevents fatty acid biosynthesis by inhibiting FabI enzyme | Vitas Pharma | | Currently undergoing safety and toxicity testing indicated for treating MRSA infections. |
| 30. | OCID 5090 (Lamonica et al. 2010; "AAI 101," 2019a; "New Anti-Infectives Company Allecra Therapeutics Created," 2013; Palanisamy et al. 2010; Sundria 2013) | Beta-lactamase inhibitor | Orchid | Allecra Therapeutics | Phase 1 clinical trials for treating gram negative infections in the USA as of January 2019 |
| 31. | AAI202, a combination of cefepime and AAI101/OCID5090 ("AAI 101/cefepime—Allecra Therapeutics," 2018a; Papp-Wallace and Bonomo 2016; Palanisamy et al. 2014) | Urinary tract infections | Allecra Therapeutics | | Phase 3 studies for urinary tract infections in Hungary in 2018. |

[a]Entries 12 and 13 are inspired by 11, not discovered in India

**Table 13.3** Drugs acting on metabolic disorders

| S. No. | Name and structure of the drug | Indication | Developed by | Licensed to | Clinical status |
|---|---|---|---|---|---|
| 1. | Peruvoside—marketed as "Encordin" in West Germany. 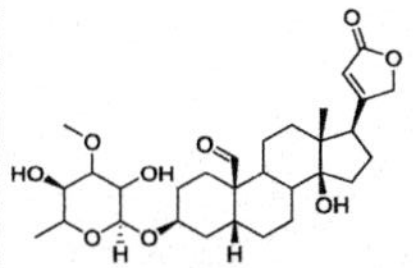 (Arora and Rangaswami 1972; Bhatia et al. 1970; Nesy and Mathew 2014) | Cardiotonic agent useful in congestive heart failure | Andhra University | | Peruvoside was a former drug in Germany |
| 2. | Colforsin 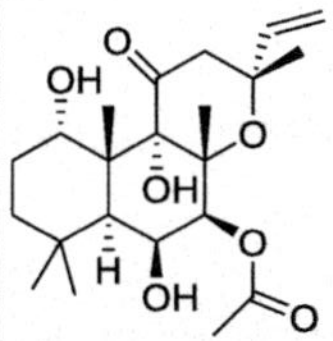 (Nagarajan 2014; Saksena et al. 1985) | Activates adenylate cyclase, hence has positive inotropic effect. It also acts as bronchodilator and reduces intraocular pressure in glaucoma. | CDRI (Coleonol) Hoechst Research Center (Forskolin). So, designated as Colforsin. | Insoluble molecule has been solubilized as a cyclodextrin complex by Sami Laboratories, Bangalore. | Approved in India for the treatment of glaucoma. |
| 3. | Gugulipid—marketed as GUGLIP 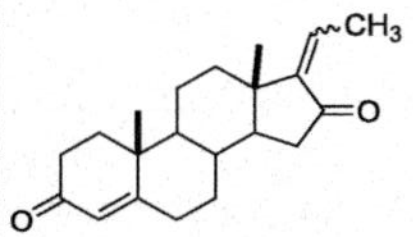 | Hypolipidemic agent | CDRI and Malti-Chem Research Centre (Baroda) | Cipla | |

(continued)

**Table 13.3** (continued)

| S. No. | Name and structure of the drug | Indication | Developed by | Licensed to | Clinical status |
|---|---|---|---|---|---|
| 4. | DRF-2593 (Balaglitazone) 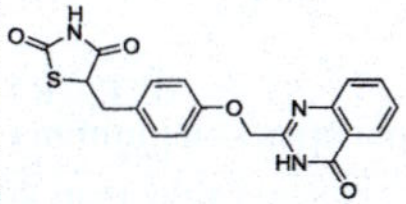 (Agrawal et al. 2012; Maji and Samanta 2015) | Partial PPARγ (Peroxisome proliferator-activated receptor gamma) agonist activity, indicated for the treatment of type-2 diabetes mellitus. | Dr. Reddy's Laboratories Ltd. (DRL) | Returned to DRL in 2004. It was further developed by DRL in collaboration with Rheosciences in 2005. | Terminated in Phase 3 trials due to the withdrawal of Rosiglitazone (with similar mechanism of action) in 2010, from European and Indian market due to increased cardiovascular side effects. |
| 5. | DRF 2725 (Ragaglitazar) 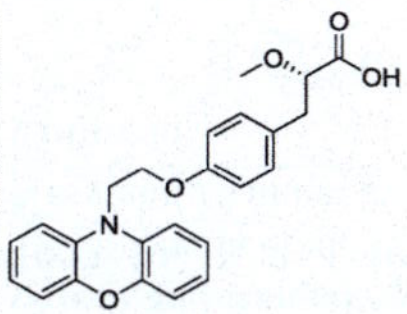 (Chakrabarti et al. 2003; Maji and Samanta 2015) | Partial PPARγ (Peroxisome proliferator-activated receptor gamma) agonist activity is indicated for the treatment of type-2 diabetes mellitus. | Dr. Reddy's Laboratories Ltd. (DRL) | Novo Nordisk | Abandoned in 2002 due to reports of adverse events in toxicology studies |
| 6. | DRF 4158 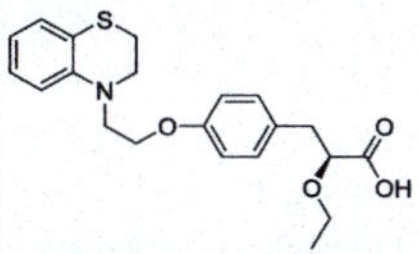 (Lohray et al. 1999) | Dual PPARα/γ agonist | DRL | Licensed to Novartis but was subsequently returned to DRL | Abandoned in preclinical stages in 2003 |

| | | | | | |
|---|---|---|---|---|---|
| 7. | RUS 3108 ("Dr. Reddy's Announces the Formation of Perlecan Pharma," 2005; Yeleswarapu et al. 2005; "New Anti-Diabetic Agent from Reddy's Labs Being Tested," 2006) | Perlecan inducer | DRL and Perlecan Pharma | | Abandoned in Phase-1 clinical trials |
| 8. | DRF 10945 ("Dr. Reddy's Announces the Formation of Perlecan Pharma," 2005; Yeleswarapu et al. 2005; "New Anti-Diabetic Agent from Reddy's Labs Being Tested," 2006) | PPARα agonist | DRL and Perlecan Pharma | | Abandoned in Phase-1 clinical trials |
| 9. | DRL 11605 ("Dr. Reddy's Announces the Formation of Perlecan Pharma," 2005; Yeleswarapu et al. 2005; "New Anti-Diabetic Agent from Reddy's Labs Being Tested," 2006) | PPARα/δ/γ agonist | DRL and Perlecan Pharma | | Abandoned in Phase-1 clinical trials |
| 10. | DRL 16536 ("Dr. Reddy's Announces the Formation of Perlecan Pharma," 2005; Yeleswarapu et al. 2005; "New Anti-Diabetic Agent from Reddy's Labs Being Tested," 2006) | AMPK modulator | DRL and Perlecan Pharma | | Abandoned in Phase-1 clinical trials |

(continued)

**Table 13.3** (continued)

| S. No. | Name and structure of the drug | Indication | Developed by | Licensed to | Clinical status |
|---|---|---|---|---|---|
| 11. | DRL 21994 ("Annual Reports 1997–2016") | Undefined mechanism of action | DRL | | Discontinued |
| 12. | DRL 21995 ("Annual Reports 1997–2016") | Undefined mechanism of action | DRL | | Discontinued |
| 13. | DRL 17822 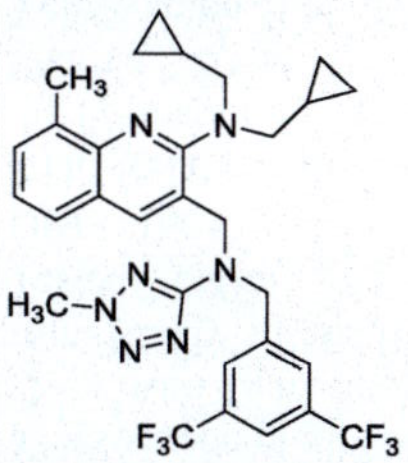  (Boruah and Alikunju 2013) | CETP (cholesteryl ester transfer protein) inhibitor for treatment of atherosclerosis therapy lipoprotein disorders. | DRL | | Abandoned in 2013 in Phase 2. |
| 14. | RBx 10558 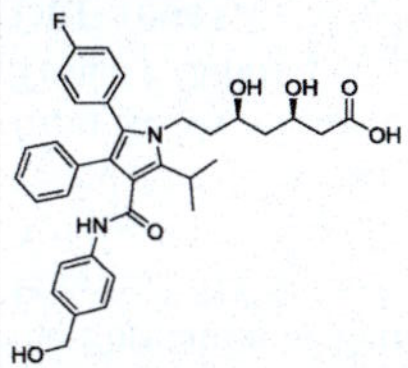  (Davenport et al. 2011; "RBx 10558," 2014b; Kumar et al. 2004) | HMG-CoA reductase inhibitor | Ranbaxy Laboratories | PPD (Pharmaceutical Product Development). Subsequently, its spin-off Furiex Pharmaceuticals undertook the clinical studies. | Abandoned in 2011 due to lack of efficacy in Phase 2b clinical studies. |

| | | | | | |
|---|---|---|---|---|---|
| 15. | TRC150094 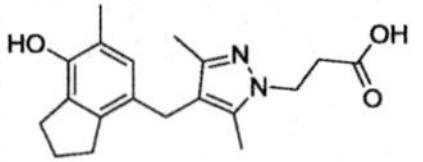 (Cioffi et al. 2010; "TRC 150094," 2018h) | T2 mimetic (diiodothyronine mimetic) indicated for the treatment of cardiometabolic risks and diabetes mellitus. | Torrent Pharmaceuticals | | Undergoing Phase 3 clinical trials as of January 2018 |
| 16. | TRC-4186 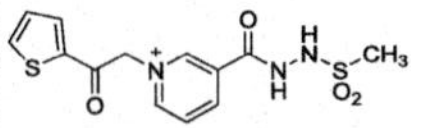 (Joshi et al. 2009; "TRC 4186," 2018g) | Advanced glycosylation end product inhibitor for treating diabetic complications and heart failure associated with hyperlipidemia or diabetes. | Torrent | Gave option to Novartis but was returned to Torrent. | Phase 2 clinical trials as of February 2018 |
| 17. | TRC-282 (Gupta et al. 2013; Sankaranarayanan 2001) | NO donor for treating cardiovascular disorders. | Torrent | | Failed at preclinical stage |
| 18. | TRC-8156 (Gupta et al. 2013; Sankaranarayanan 2001) | DPP-IV (dipeptidyl peptidase-IV) inhibitor. | Torrent | | Failed at preclinical stage |
| 19. | TRC 120038 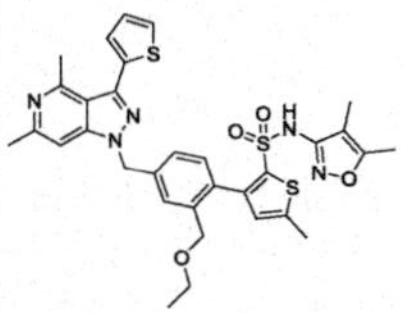 (Differding 2014; Mohanan et al. 2011) | Dual angiotensin II receptor type 1 (AT1 receptor) and Endothelin type A (ETA) receptor antagonists for treating cardio metabolic risks and diabetic nephropathy. | Torrent | | No progress has been reported. |

(continued)

Table 13.3 (continued)

| S. No. | Name and structure of the drug | Indication | Developed by | Licensed to | Clinical status |
|---|---|---|---|---|---|
| 20. | P 1201 ("Annual Reports, 1997–2016," Piramal 1997–2016, "P 1201," 2016c) | Undisclosed mechanism of action. | Lilly | Out-licensed to Piramal | Entered Phase 1 clinical trials in 2008 and no further development was reported since. |
| 21. | P2202 ("Annual Reports, 1997–2016," Piramal 1997–2016) | 11β-hydroxysteroid dehydrogenase type 1 (betaHSD1) inhibitor indicated for the treatment of diabetes. | Lilly | Out-licensed to Piramal | Entered Phase 2 clinical trials in India and Canada, but further studies were terminated in 2013. |
| 22. | P 1736 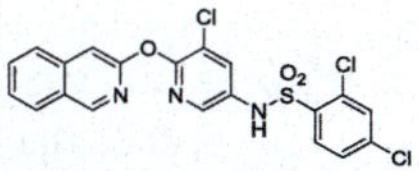 (Anthony et al. 2013, "P 1736," 2015a) | Non-thiazolidinedione insulin sensitizer for the treatment of diabetes. | Piramal | | Phase-1 studies in 2008 |
| 23. | P7435 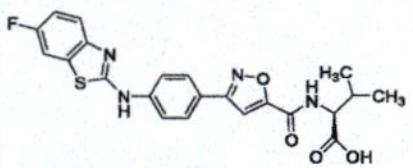 ("Annual Reports, 1997–2016," Piramal 1997–2016) | DGAT1 (diacylglycerol-O-acyltransferase 1) inhibitor for the treatment of diabetes and lipid metabolism disorders | Piramal | | |
| 24. | P11187 ("Annual Reports, 1997–2016," Piramal 1997–2016) | GPR40 (G-protein-coupled receptor 40) agonist for type-2 diabetes | Piramal | | Completed Phase 1 studies in 2014 in the USA. No further development has been reported. |

| | | | | | |
|---|---|---|---|---|---|
| 25. | Polycap ("About Polycap," n.d.-a; "Polycap," n.d.-b) | Five-in-one fixed dose combination of existing drugs to treat heart attacks. | Cadila | | |
| 26. | GRC 8200 (Melogliptin) 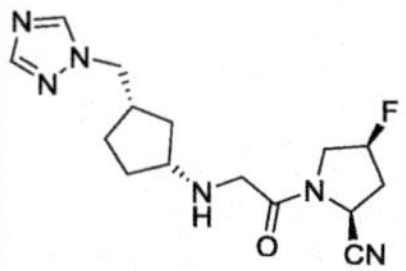 (Thomas et al. 2006; "JP Morgan Healthcare Conference," 2013; Kushwaha et al. 2014) | DPPIV inhibitor for type 2 diabetes | Glenmark | Out-licensed to Merck KGaA for Europe, Japan and N. America markets, but reacquired the rights in 2008. | Entered Phase-3 clinical trials in 2015 and no further information was published since. |
| 27. | LL 6531 ("Annual Reports, 2001–2016") | PPAR modulator | Lupin | | Abandoned at preclinical stage. |
| 28. | LNP1892 ("LNP 1892," 2018d; "A Study to Test the Safety/ Tolerability of Increasing Doses of LNP1892 Versus Placebo in Healthy Male/ Female Subjects," 2014) | Calcium sensing receptor (CaSR) modulator for the treatment of hyperparathyroidism. | Lupin | | Phase 2 clinical trials in India in August 2017. |
| 29. | BLX-1002 (Lee and Kim 2010; "Orchid's anti-diabetic molecule gets US nod," 2004) | Non-PPAR AMPK activator for the treatment of diabetes. | Orchid and Bexel Biotechnology | | Completed the limited phase-2 proof of concept studies in 2004 in Europe but no further reports are published. |

(continued)

**Table 13.3** (continued)

| S. No. | Name and structure of the drug | Indication | Developed by | Licensed to | Clinical status |
|---|---|---|---|---|---|
| 30. | DPOC-4088 (or DP-4088) ("Diakron Licenses Anticoagulant Drug Candidate from Merck & Co., Inc.; Orchid Pharma Partners with Diakron," 2008) | Anticoagulant drug which is a direct thrombin inhibitor. | Orchid in partnership with Diakron Pharmaceuticals developed Merck, Sharp & Dohme's DPOC-4088 | | |
| 31. | PBL 1427 (Jain et al. 2009; "Study of Single Ascending Doses of PBL 1427 in Healthy Volunteers," 2012b) | DPPIV inhibitor | Panacea Biotec | | Entered Phase 1 clinical trials in India in 2012. |
| 32. | MX-6001 (Gopalan et al. 2007; Gopalan 2012b) | DPPIV inhibitor | Matrix Laboratories | | |
| 33. | Alkem-43 (Bandodkar et al. 2014; Nagaraj et al. 2016a, b) | Cathepsin K inhibitor for the treatment of osteoporosis. | Alkem laboratories | | |
| 34. | GKM-001 ("Advinus' GK-Activator Achieves Early POC for Diabetes," 2011; Filipski et al. 2012; Mookhtiar 2015) | Glucokinase activator | Advinus | | Successfully completed a 14-day Phase 2 proof-of-concept study |
| 35. | GKM-002 (*Business Forum* 2015; "Discovery Pipeline," Advinus Therapeutics n.d.) | Glucokinase activator | Advinus | | Preclinical level |

| | | | | | |
|---|---|---|---|---|---|
| 36. | KU-5039 (Khanna and Pillarisetti 2014; "KU 5039," 2016b) | Adenylate kinase stimulant, for the treatment of diabetes. | Kareus Therapeutics | | Phase-1 clinical trials in the USA in 2016. |
| 37. | CNX-012-570 (Rao et al. 2014a, b) | AMPK activator for the treatment of type 2 diabetes | Connexios | Boehringer Ingelheim | Currently in preclinical development stage as of 2017. |
| 38. | CNX-011-67 (Rao et al. 2012) | GPR 40 agonist for treating Type-2 Diabetes | Connexios | | Entered preclinical development in 2014–2015, but no further reports were published since. |
| 39. | CNX-013-B2 (Ranga et al. 2015) | Activator of Retinoid X Receptor α,β,γ for treating dyslipidemia | Connexios | | |
| 40. | CNX-010-49 (Rao et al. 2013) | Beta hydroxysteroid dehydrogenase inhibitor for diabetes | Connexios | | |
| 41. | NDS100179 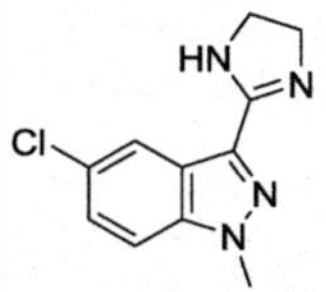 (Reddy et al. 2015; Saxena 2015) | Indazole compound for treating type 2 diabetes. | Shantani Proteome Analytics together with National Chemical Laboratory-CSIR | | |
| 42. | KBR2001 (KBRPL2001) (Anupindi et al. 2016) | GPR120 agonist, for treating diabetes and metabolic disorders. | Piramal | Krish Biotech | |

(continued)

**Table 13.3** (continued)

| S. No. | Name and structure of the drug | Indication | Developed by | Licensed to | Clinical status |
|---|---|---|---|---|---|
| 43. | RP-3128 (Muthuppalaniappan et al. 2011; Sutovska et al. 2016) | CRAC channel inhibitor, for the treatment of asthma. | Rhizen Pharmaceuticals | | Currently in clinical development |
| 44. | ZYH1 (INN Saroglitazar Marketed as LIPAGLYN) 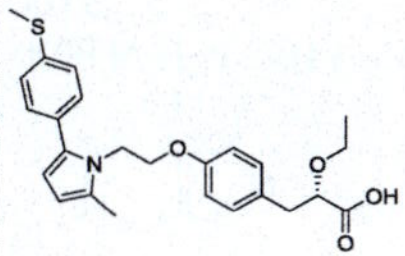 (Manoria et al. 2013; Ramakrishnan 2015) | Dual alpha/gamma agonist, indicated for the treatment of diabetic dyslipidemia and hypertriglyceridemia with Type 2 diabetes mellitus not controlled by statin therapy. It is currently under development for other indications, including non-alcoholic steatohepatitis (NASH). | Zydus Cadila | | |
| 45. | ZY H7 ("Annual Reports, 2000–2016. Corporate presentations," 2000–2016; "ZY H7" 2016h) | PPARα agonist for treating dyslipidemia, | Zydus Cadila | | Entered Phase 2 studies in 2014 in India |
| 46. | ZYDPLA1 (Desai et al. 2014; "ZYDPLA1" 2017e) | DPPIV inhibitor | Zydus Cadila | | Discontinued in 2017 in Phase-1 clinical trials in the USA. |

| | | | | | |
|---|---|---|---|---|---|
| 47 | Trequinsin<br>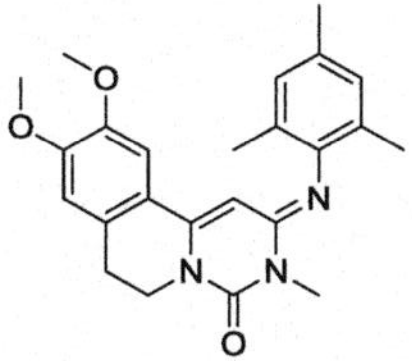<br>(Lal et al. 1984) | Antihypertensive vasodilator | Hoechst | | Reported to be in Phase 1 clinical trials. No further development reported. |
| 48 | Imidazoles, pyrazoles and isothiazoles as potent antihypertensive agents<br>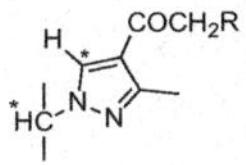 101 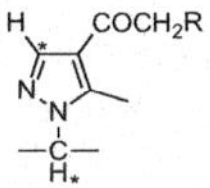 102<br>(Nagarajan et al. 1986)<br>(Nagarajan 2006) | Potent antihypertensive agents | Ciba-Geigy | | All compounds in these series failed in phase 2 trials |

**Table 13.4** Anti-inflammatory drugs

| S. No. | Name and structure of the drug | Indication | Developed by | Licensed to | Clinical status |
|---|---|---|---|---|---|
| 1. | Tinazoline 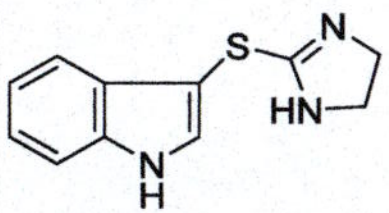  (Nagarajan et al. 1981) | Vasoconstrictor, useful as a nasal decongestant | Ciba | | |
| 2. | Tromaril (enfenamic acid) 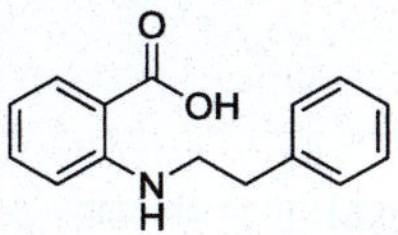  (Mehra and Agrawal 1984) | To relieve pain, reduce swellings and improve joint mobility and is used in the treatment of inflammations caused by sprains, bruises, or surgical incisions | RRL | Unichem Laboratories commercialized Tromaril while the patent rests with the CSIR | |
| 3. | RBx-4638 (Clafrinast) ("RBx 4638," 2006b; Sattigeri et al. 2006; Palle et al. 2006) | VLA-4 (Very Late Antigen-4) antagonist to treat asthma and COPD | Ranbaxy | | Most probably discontinued |
| 4. | RBx 7796 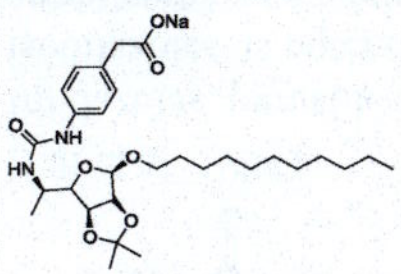  ("RBx 7796," 2015b; Sattigeri et al. 2006; Palle et al. 2006) | 5-LO (5-Lipoxygenase) inhibitor against asthma and allergic rhinitis. | Ranbaxy | | Reached clinical development. |
| 5. | RBx 10017609 ("MMP-9," n.d.; "RBx 10017609," 2016e; Palle et al. 2006) | Dual matrix metalloproteinase-9 and metalloproteinase-12 (MMP-9/MMP-12) inhibitor, for the treatment of COPD | Ranbaxy and GSK | | Abandoned in Phase 2 clinical trials. |

| | | | | | |
|---|---|---|---|---|---|
| 6. | AUR-101 ("Aurigene Announces First in Human Dosing with RoRγt inverse agonist AUR-101," 2018b) | Potent inverse agonist of RoRγt (retinoic acid-related orphan receptor gamma) inverse agonists for the treatment of inflammatory disorders. It is designed for the treatment of IL17-driven immunological conditions, including psoriasis. | Aurigene | | Currently undergoing Phase-1 first-in-human dosing studies as of 2018 |
| 7. | P 3914 ("P 3914," 2016d; Satyam 2011; "Study of P3914 to Evaluate the Safety, Tolerability, Food Effect & Pharmacokinetics in Healthy Male Subjects and Efficacy & Safety of P3914 in Patients With Acute Dental Pain," 2011b) | Treatment of dental pain | Piramal | | Entered clinical development stage in 2011 but no further development has been reported since. |
| 8. | P7170 (Panulisib) 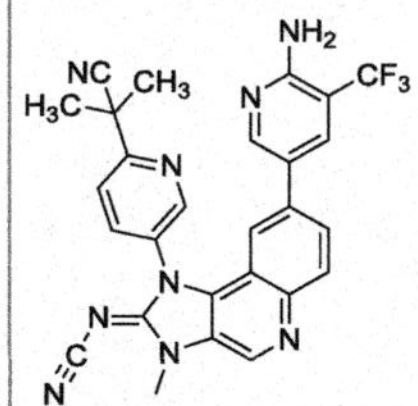  ("Clinical Study of Oral PI3K/mTOR Inhibitor in Patients With Advanced Refractory Solid Tumors," 2013; Kumar et al. 2013) | PI3K (Phosphatidylinositol 3 kinase) inhibition/mTOR (mammalian target of rapamycin) inhibition, indicated for treatment of inflammation and solid tumors. | Piramal and MSD | | Phase-1 currently underway in India |

(continued)

**Table 13.4** (continued)

| S. No. | Name and structure of the drug | Indication | Developed by | Licensed to | Clinical status |
|---|---|---|---|---|---|
| 9. | SUN 1334H 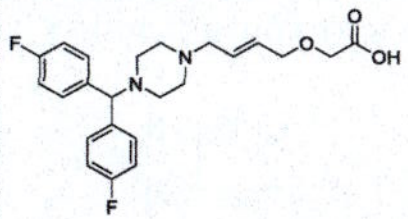 (Midha et al. 2003; Mandhane et al. 2008; "SUN 1334H," 2016f) | Histamine H1 receptor antagonist, for the treatment of allergic conjunctivitis, perennial allergic rhinitis, seasonal allergic rhinitis, and urticaria. | Sun Pharmaceuticals | | Phase 2 clinical trials but has been abandoned in 2013. |
| 10. | SUN 461 (Badorrek et al. 2015; Patel et al. 2014; Thennati et al. 2014) | Glucocorticoid receptor agonists | SPARC (Sun Pharma Advanced Research Company) | | Abandoned in Phase 1 |
| 11. | SUN 597 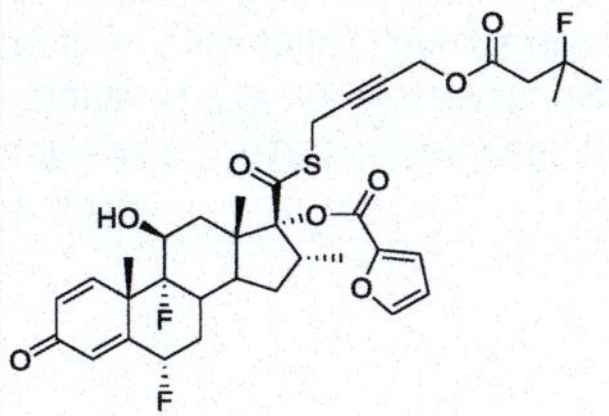 (Badorrek et al. 2015; Patel et al. 2014; Thennati et al. 2014) | Glucocorticoid receptor agonists | SPARC (Sun Pharma Advanced Research Company) | | Phase 1 studies for atopic dermatitis and psoriasis in the USA, but is most probably abandoned for seasonal allergic rhinitis |

| | | | | | |
|---|---|---|---|---|---|
| 12. | SUN L731<br>(Rathod et al. 2013) | Oral LTD4 (Leukotriene D4) antagonist | SPARC (Sun Pharma Advanced Research Company) | | Entered preclinical development, but has been abandoned in 2016 |
| 13. | JB-7/G<br>("JB Chemicals Files for Patent on 20 NCEs in NSAIDs Slot," 2002) | COX-2 inhibitor | JBCPL<br>(J.B. Chemicals & Pharmaceuticals Ltd.) | | Further development was discontinued as a result of withdrawal of rofecoxib due to increased cardiovascular risks |
| 14. | GRC 3886 (Oglemilast)<br>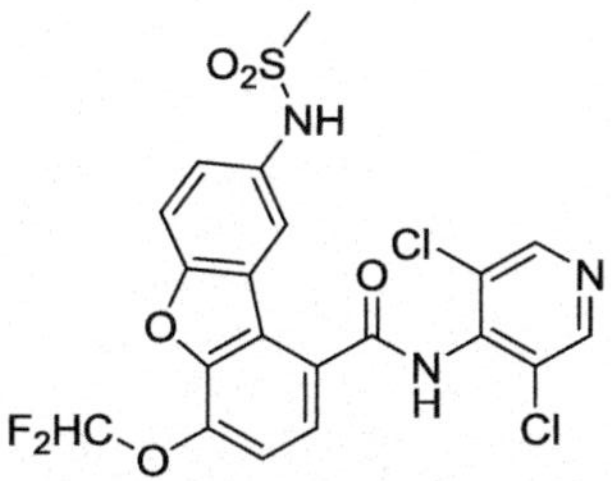<br>(Dyke 2007; Gharat et al. 2015; Pagès et al. 2009) | Type 4 cyclic nucleotide phosphodiesterase type 4 (PDE4) inhibitor for the treatment of asthma and COPD | Glenmark | Out-licensed to Forest Laboratories in 2004 for the North American market and to Teijin in 2005 for Japan | Abandoned in Phase 2 in 2009 |
| 15. | GRC 4039 (Revamilast)<br>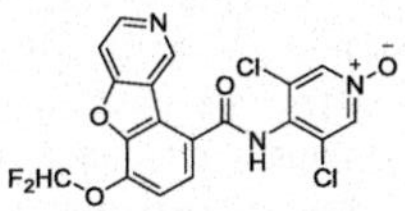<br>(Dyke 2007; Gharat et al. 2015; Pagès et al. 2009; "Revamilast," 2010a) | Type 4 cyclic nucleotide phosphodiesterase type 4 (PDE4) inhibitor for the treatment of asthma and COPD | Glenmark | | Abandoned in Phase 2 in 2012 |

(continued)

**Table 13.4** (continued)

| S. No. | Name and structure of the drug | Indication | Developed by | Licensed to | Clinical status |
|---|---|---|---|---|---|
| 16. | GRC 15300 (Unnikrishnan and Ananthanarayanan 2014; "Efficacy and Safety of SAR292833 Administration for 4 weeks in Patients With Chronic Peripheral Neuropathic Pain (Alchemilla)," 2011a) | First-in-class TRPV3 (transient receptor potential cation channel type V3) antagonist for the treatment of pain. | Glenmark | Out-licensed to Sanofi in 2010 | Further development of GRC 15300 was discontinued due to unconvincing Phase-2 proof-of-concept results |
| 17. | GRC 27864 ("Glenmark Pharmaceuticals provides update on clinical development of GRC 27864," 2018; Gharat et al. 2012a, b; Gharat et al. 2015) | mPGES1 (microsomal prostaglandin E synthase-1) inhibitor for the treatment of moderate osteoarthritic pain | Glenmark | | Undergoing Phase-2b studies as of 2018 |
| 18. | GRC 10693 (Tedalinab) 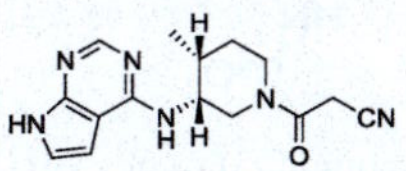 (Narayanan et al. 2007; "Tedalinab," 2010b, 2015b) | Cannabinoid CB2 receptor agonist for the treatment of inflammatory and neuropathic pain. | Glenmark | | Most probably abandoned |
| 19. | GRC 17536 ("GRC 17536," 2017a; Anupindi et al. 2010; Skerratt 2017) | Selective TRPA1 (transient receptor potential ankyrin-1) antagonist, for the treatment of cough and neuropathic pain | Glenmark | | Completed Phase 2a studies in 2017 |

| | | | | | |
|---|---|---|---|---|---|
| 20. | LNP1955<br>("LNP 1955," 2017c; "A Phase II, Dose Ranging, Exploratory Clinical Study to Assess the Efficacy, Pharmacodynamics, Pharmacokinetics, and Safety of LNP1955 in Patients with Moderate to Severe Rheumatoid Arthritis," 2017) | Immunomodulator for the treatment of autoimmune diseases like rheumatoid arthritis and plaque psoriasis | Lupin | | Completed a Phase-2 trail for plaque psoriasis in Hungary as of October 2017 |
| 21. | OCID 2987<br>(Narayanan et al. 2010; "OCID 2987" 2014a; "Orchid Pharma completes Phase I trial of PDE4 inhibitor molecule, OCID 2987," 2012) | PDE4 inhibitor for the treatment of inflammatory disorders such as COPD | Orchid Chemicals and Pharmaceuticals Ltd | | Completed Phase-1 clinical trials in 2012 but no further development is reported. |
| 22. | MX-4007<br>(Gopalan et al. 2008; Gopalan 2012a) | Phosphodiesterase (PDE) inhibitor | Matrix Laboratories | | |
| 23. | RP-6503<br>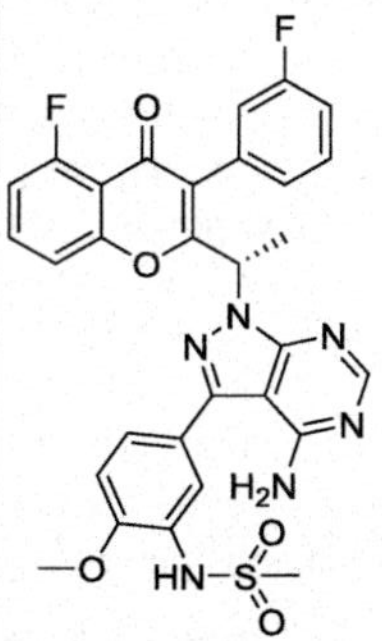<br>(Bhavar and Vakkalanka 2015) | Dual PI3Kγ/δ (Phosphatidylinositol 3 kinase) inhibitor | Incozen Therapeutics/ Rhizen Pharmaceutics | Licensed to Novartis in 2015 | |

(continued)

**Table 13.4** (continued)

| S. No. | Name and structure of the drug | Indication | Developed by | Licensed to | Clinical status |
|---|---|---|---|---|---|
| 24. | KBRPL1001<br>(Sharma et al. 2016) | RoRγt (retinoic acid-related orphan receptor gamma t) antagonist used for the treatment of autoimmune disorders. | Piramal | Out-licensed to Krish Biotech. | |
| 25. | RSCL-0409 (a gluco-disaccharide)<br>(Addepalli et al. 2010; Kalluri et al. 2010) | Inhibits Toll-Like Receptor (TLR) signalling pathways | Reliance | | Further research on these compounds is probably been abandoned when the focus deviated on to siRNA-mediated approaches to treat cancer in 2010. |
| 26. | RSCL-0520 (a phenanthrene obtained from Eulophia ochreata)<br>(Addepalli et al. 2010; Datla et al. 2010) | Inhibits Toll-Like Receptor (TLR) signalling pathways | Reliance | | |

**Table 13.5** Anticancer drugs

| S. No. | Name and structure of the drug | Indication | Developed by | Licensed to | Clinical status |
|---|---|---|---|---|---|
| 1. | Alvocidib (Flavoperidol) 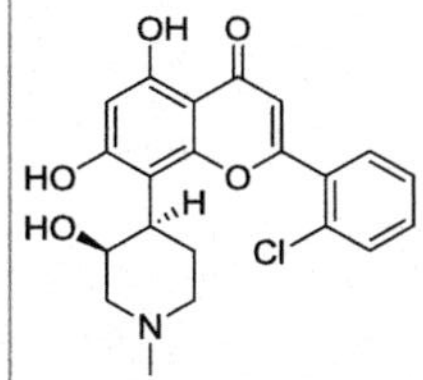 (Lakdawala et al. 1988) | Inhibits the cyclin-dependent kinase and is used for the treatment of chronic lymphocytic leukemia | Hoechst Research Centre | | Tolero Pharmaceuticals Inc. announced that the FDA has granted orphan drug designation for Alvocidib, its cyclin-dependent kinase small molecule inhibitor, for the treatment of patients with acute myeloid leukemia (https://www.toleropharma.com/ourresearch/alvocidib) |
| 2. | DRF 1042 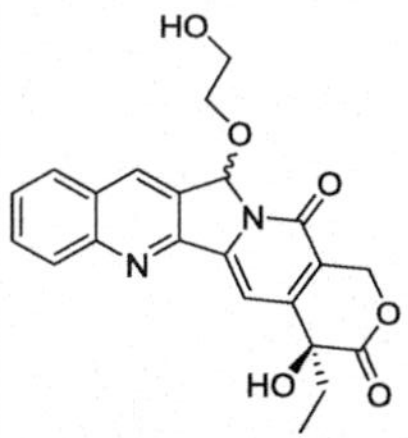 ("Dr. Reddy's Laboratories Limited," n.d.; Rajagopal et al. 2004; Nekkanti et al. 2011) | Camptothecin analog to treat solid tumors | DRL and ClinTec | | Most probably abandoned in its Phase-1 clinical trials |

(continued)

**Table 13.5** (continued)

| S. No. | Name and structure of the drug | Indication | Developed by | Licensed to | Clinical status |
|---|---|---|---|---|---|
| 3. | DRF-1644 ("Dr Reddy's cancer molecule completes Phase-I trials," 2002; "DRF 1644" 2008) | Camptothecin analog to treat solid tumors | DRL and ClinTec | | Discontinued in its Phase-1 studies in 2008 |
| 4. | RBx 2258 (Pamirosin) 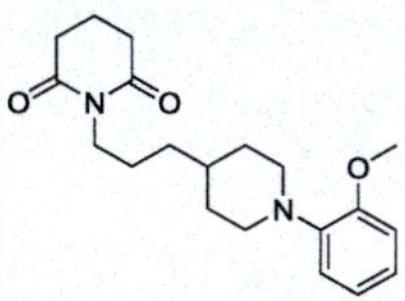  ("RBx 2258," 2006a; "Trial of prostate drug stopped—Schwarz move may impact Ranbaxy deal," 2004) | Adrenergic alpha-1 receptor antagonist, indicated for the treatment of benign prostatic hyperplasia | Ranbaxy | Schwarz Pharma | Discontinued due to unconvincing Phase 2 results |
| 5. | P-276 (Riviciclib) 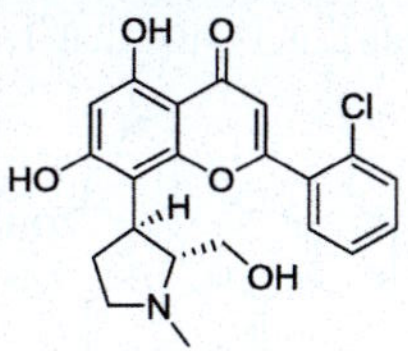  (Butler et al. 2014) | cdk4 inhibitor for the treatment of cancer | Piramal | | Terminated in 2013 after a setback in Phase 2 clinical trials |

| | | | | | |
|---|---|---|---|---|---|
| 6. | P1446 (Voruciclib) 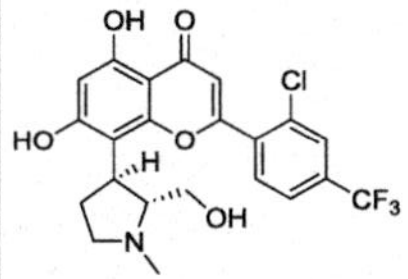  (Butler et al. 2014) | cdk4 inhibitor | Piramal | | Entered clinical development in 2008 |
| 7. | PL225B (Roychowdhury et al. 2013; Abdel-Magid 2012; Balachandran et al. 2012; Chennamsetty et al. 2012) | Insulin-like growth factor 1 receptor (IGF1R) inhibitor | Piramal and MSD | | Reached Phase 1 but was probably abandoned in 2014 |
| 8. | P1961A ("Annual Reports, 1997–2016," Piramal 1997–2016) | Dual CDK4/HIF1alpha inhibitor | Piramal | | |
| 9. | P2745 (Sivakumar et al. 2011) | Potent inhibitor of the transforming growth factor beta (TGF-β) pathway for the treatment of hematological malignancies. | Piramal | | Most probably discontinued in Phase 1 in 2013 |
| 10. | DRF 7295 (Burman et al. 2003) | Anti-cancer vaccine composed of peptides derived from bombesin, Vasoactive Intestinal Peptide (VIP), substance P and somatostatin, which are over-expressed in various cancers. | Dabur | | Though it cracked Phase 2 trials initiated in 2004, since 2008 it had shown no progress and possibly been abandoned. |

(continued)

**Table 13.5** (continued)

| S. No. | Name and structure of the drug | Indication | Developed by | Licensed to | Clinical status |
|---|---|---|---|---|---|
| 11. | SUN K706 ("Annual Reports 2007–2016"; "Investor Update on R&D Pipeline," 2016; Sengupta et al. 2009; Sengupta et al. 2012) | Bcr-abl tyrosine kinase inhibitor, devoid of arterial thrombosis as side effect. | Sun | | |
| 12. | SUN K954 ("Annual Reports 2007–2016"; "Investor Update on R&D Pipeline," 2016; Sengupta et al. 2009; Sengupta et al. 2012) | More selectivity towards the T351M mutant | Sun | | |
| 13. | ZYTP1 ("Minutes of IND Committee Meeting Held on 16.12.2015," 2015) | Poly (ADP-ribose) polymerase (PARP) inhibitor | Zydus Research Centre | | Phase 1 trials in India |
| 14. | Mycidac-C (Balganesh et al. 2014; "Cadi 05," n.d.) | Lung cancer vaccine and also treats leprosy and is being tested for TB. | National Institute of Immunology and Cadila | | Undergoing Phase III clinical trials against Tuberculosis. Marketed for Leprosy, Non-small cell lung cancer |
| 15. | LNP3794 ("LNP3794 in patients with Advanced Solid Tumours Having Mutations," 2014) | MEK Inhibitor | Lupin | | Completed a Phase 1 study in the UK on terminally ill patients |

| | | | | | |
|---|---|---|---|---|---|
| 16. | OCID 4681 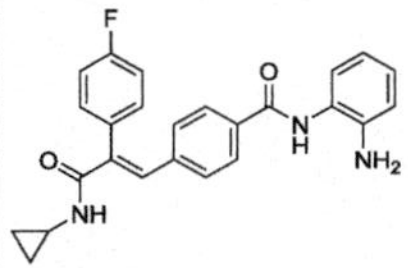 ("Clinical Trial Details," 2012a; Rajagopal et al. 2009) | Histone deacetylase (HDAC) inhibitor against cancer | Orchid Chemicals & Pharmaceuticals | | Approved for Phase 1 studies in 2011 for solid tumors in India, but most likely to be abandoned as there are no reports on progress. |
| 17. | OCID 5005 ("Annual Reports 2002–2016") | STAT-3/IL-6 inhibitor for cancer | Orchid Chemicals & Pharmaceuticals | | Abandoned at preclinical stage. |
| 18. | MALT1 inhibitor ("Lupin developing new products to treat cancer, other diseases," 2019a) | MALT1 (Mucosa-Associated Lymphoid Tissue Lymphoma Translocation Protein1) inhibitor for the treatment of hematological cancers. | Lupin partnered with AbbVie in December 2018 | | |
| 19. | NRC-AN-019 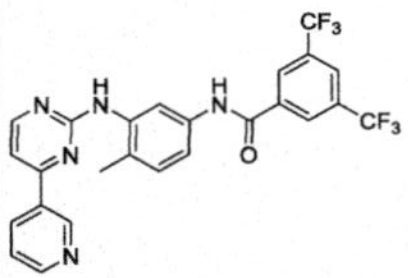 (Amala et al. 2013) | Synthetic analog of imatinib, categorized as orphan drug for treatment against chronic myelogenous leukemia (CML), pancreatic cancer and glioma in 2011 by the USFDA | Natco | | Phase 2 trials in India |
| 22. | NRC 2694 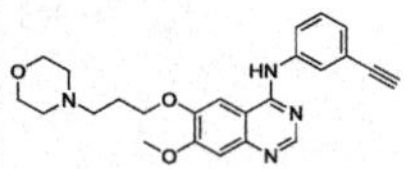 (Prasad et al. 2011) | Erlotinib's analog, EGFR kinase inhibitor indicated for advanced stage solid tumors | Natco | | Phase 1 trials in India |

(continued)

**Table 13.5** (continued)

| S. No. | Name and structure of the drug | Indication | Developed by | Licensed to | Clinical status |
|---|---|---|---|---|---|
| 23. | NRC/AN/015 ("Annual Reports 2003–2016") | Bcr-Abl tyrosine kinase inhibitor | Natco | | Abandoned at the preclinical stage |
| 24. | CGI-1842 (JI-101) (Boyd et al. 2014) | Tyrosine kinase inhibitor targeting selectively (VEGFR2), platelet-derived growth factor receptor beta (PDGFRβ) and ephrin type-B receptor 4 (EphB4) | Jubilant and CGI Pharmaceuticals | | Entered Phase 1/Phase 2 stage trial in the USA in 2012 for solid tumors, but has been discontinued due to lack of efficacy |
| 25. | JBI-097 (Sivanandhan et al. 2015) | Dual lysine-specific histone demethylase (LSD1)-histone deacetylase inhibitor | Jubilant | | |
| 26. | JIEM-0186 (Sivanandhan et al. 2014) | EGFR kinase inhibitor | Jubilant | | |
| 27. | JBET-050 ("Targeting Epigenetic Readers and Chromatic Remodelers," 2015) | Bromodomain and extra-terminal motif (BET) inhibitor | Jubilant | | |
| 28. | CK-103 ("CK-103 BET Inhibitor" n. d.) | BET BRD4 inhibitor | Jubilant | Checkpoint Therapeutics | |
| 30. | PAT-1102 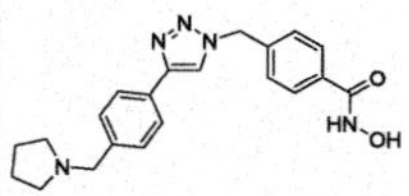  (Hiriyan et al. 2015; Natesan et al. 2011) | HDAC inhibitor | Anthem Biosciences | | Preclinical development |

| | | | | | |
|---|---|---|---|---|---|
| 31. | PAT-1118 (Hiriyan et al. 2015; Natesan et al. 2011) | | Anthem Biosciences | | Preclinical development |
| 32. | Debio 1142 ("2015 Investor Day," 2015; "Debiopharm and Aurigene sign agreement for the development and commercialisation of Debio 1142, a novel inhibitor of an undisclosed oncology pathway," 2011) | Inhibitor of a non-disclosed oncology pathway | Aurigene and Debiopharm | | |
| 34. | AUNP-12/W016A (Differding 2014a, b; "Immuno-oncology: licensing agreement between aurigene and pierre fabre pharmaceuticals," 2014) | PD-1 pathway inhibitor | Aurigene | Pierre Fabre Médicaments but returned in 2015 | |
| 35. | CA-170/AUPM-170 ("Aurigene and its partner curis announce the dosing of CA-170 in phase II India trial," 2018a; "Curis Licenses CA-170, IRAK4 Programs from Aurigene," Genetic Engineering and Biotechnology News 2015a, b; "Overview and Path for Growth: Aurigene Strategic Collaboration," 2015) | Dual PD-L1 (Programmed cell death-1 ligand-1)/VISTA (V-domain Ig suppressor of T-cell activation) inhibitor for the treatment of cancer | Aurigene | Curis | Phase-2 clinical trials in India in 2018 |

(continued)

**Table 13.5** (continued)

| S. No. | Name and structure of the drug | Indication | Developed by | Licensed to | Clinical status |
|---|---|---|---|---|---|
| 36. | CA-327/AUPM-327 ("Curis and Aurigene Announce Collaboration, License and Option Agreement to Discover, Develop and Commercialize Small Molecule Antagonists for Immuno-Oncology and Precision Oncology Targets," 2015; "Overview and Path for Growth: Aurigene Strategic Collaboration," 2015) | Dual PD-L1/TIM-3 (T-cell immunoglobulin and mucin-domain containing molecule-3) inhibitory action | Aurigene | Curis | Preclinical candidate |
| 37. | CA-4948/AU-4948 ("Curis and Aurigene Announce Collaboration, License and Option Agreement to Discover, Develop and Commercialize Small Molecule Antagonists for Immuno-Oncology and Precision Oncology Targets," 2015) | IRAK-4 (Interleukin-1 Receptor-Associated Kinase 4) inhibitor | Aurigene | Curis | |
| 38. | ODM-207 (Aurigene 2014; Björkman et al. 2016; Samajdar et al. 2015) | pan BET inhibitor (BRD4/BRD2/BRD3) | Aurigene | Orion | Currently undergoing IND enabling studies |
| 39. | AU-4869 (Chikanna et al. 2016; Satyam et al. 2016a, b) | Nicotinamide phosphoribosyltransferase (NAMPT) inhibitor | Aurigene | | |

| | | | | | |
|---|---|---|---|---|---|
| 40. | Small molecule inhibitors of an undisclosed cancer metabolism target (Aurigene 2017) | Small molecule inhibitors of an undisclosed cancer metabolism target | Aurigene and Agios Pharmaceuticals | | |
| 41. | GBO-006-1 (Arnab Roy Chowdhury 2014) | Novel first-in-class PLK2 inhibitor for the treatment of breast cancer | GVK Bio and Onconova | | Most probably discontinued |
| 42. | GVK-TrkI (Nagaswamy and Thirunagaru 2016; Tirunagaru et al. 2016) | Selective Tropomyosin receptor kinase A (TrkA) inhibitor for the treatment of cancer | GVK Bio | | |
| 43. | GVK01406 (Chowdhury et al. 2016) | PI3Kβ inhibition | GVK Bio | | |
| 44. | RP-5264 (TGR-1202 or Umbralisib)<br>RP-5264<br>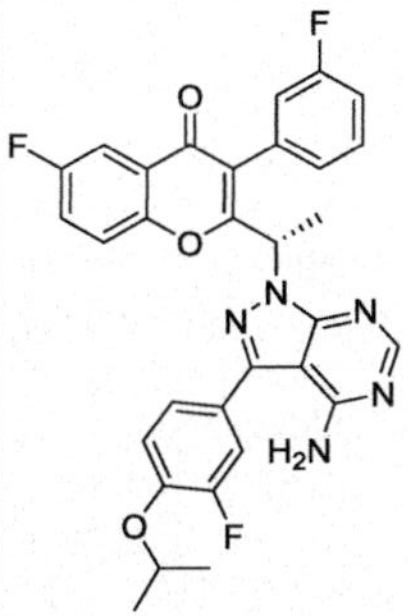<br>("Press Releases," 2019; "Umbralisib—TG Therapeutics," 2019d) | Selective PI3Kδ inhibitor for treating hematological lymphomas<br>Got the status of "breakthrough therapy" for treating Marginal zone B-cell lymphoma in the USA and was categorized as an orphan drug. | Rhizen | TG Therapeutics | Currently in Phase 2 studies |

(continued)

**Table 13.5** (continued)

| S. No. | Name and structure of the drug | Indication | Developed by | Licensed to | Clinical status |
|---|---|---|---|---|---|
| 45. | TGR-1202 in combination with TGR-1101 (Burris et al. 2015; Muthuppalaniappan et al. 2012; "Umbralisib—TG Therapeutics," 2019d) | Chronic lymphocytic leukemia | Rhizen | TG Therapeutics | Undergoing Phase 3 studies. |
| 46. | RP-6530 (Tenalisib) 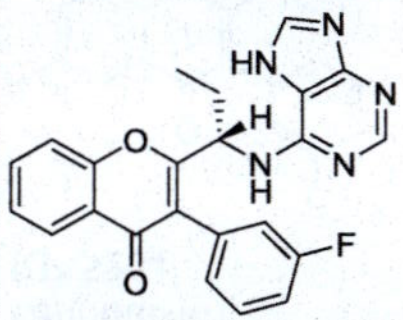  (Muthuppalaniappan et al. 2012; Sneha et al. 2016; "Tenalisib—Rhizen Pharmaceuticals," 2019c) | Dual PI3Kγ/δ inhibitor for the treatment of Non-Hodgkin's lymphoma and hematological malignancies. Received orphan drug status for cutaneous and peripheral T-cell lymphoma | Rhizen | | Phase-2 studies for treatment of Non-Hodgkin's lymphoma and in Phase-1 for hematological malignancies |
| 47. | RP-1400 (Vakkalanka et al. 2013) | c-Met kinase inhibitor | Rhizen | | Preclinical candidate. Abandoned in 2015 |
| 48. | RV1001 ("Onco Tx, Vet Tx," n.d.) | PI3Kδ inhibitor for veterinarian use in canine Non-Hodgkin lymphoma | Rhizen | | Phase-2 studies. |
| 49. | SPR965 (Reena Arora et al. 2014; Larsen et al. 2015; Dugar et al. 2012) | PI3K/mTOR inhibitors | Sphaera | | Preclinical evaluation stage |

| | | | | | |
|---|---|---|---|---|---|
| 50. | IO-125<br>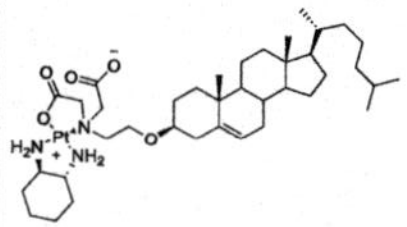<br>(Paraskar et al. 2010; Roy et al. 2014; Sengupta et al. 2015a, b) | Platinum chemotherapeutic used for treating breast cancer | Invictus Oncology | | Preclinical development |
| 51. | Disarib<br>(Nair 2017) | Kills cancer cells overproducing a protein called BCL2. | Indian Institute of Science in collaboration with other institutes | | |

**Table 13.6** Drugs acting on central nervous system

| S. No. | Name and structure of the drug | Indication | Developed by | Licensed to | Clinical status |
|---|---|---|---|---|---|
| 1. | Methaqualone (Quaalude) 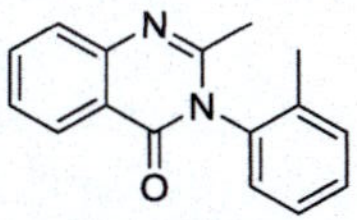 (Angelos and Meyers 1985; Kacker and Zaheer 1951) | Sedative-hypnotic | Initially, developed by Lucknow University. Later continued at the Regional Research Laboratory. | | Banned |
| 2. | Mipimazole 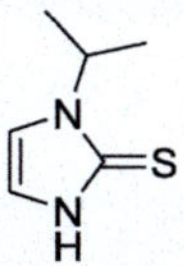 (Karkun and Anand 1962) | Anti-thyroid activity | CDRI | Unichem Laboratories, Mumbai. | |
| 3. | Nonaperone 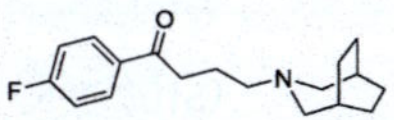 (Nagarajan and Arya 1982) | Antipsychotic drug useful in treating schizophrenia at doses which do not elicit Extrapyramidal side effects | Ciba-Geigy Research Centre | | Registered in India by Ciba-Geigy, but not sold |
| 4. | Buriperone (Centbutindole) 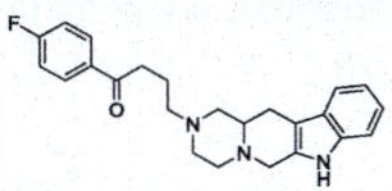 (Saxena et al. 1973) | Dopamine antagonist and 5HT2 blocker. | CDRI | Chemosyn Pvt Ltd & Merind Ltd | |

| | | | | | |
|---|---|---|---|---|---|
| 5. | Bucricaine (Centbucridine)—marketed as Centoblok 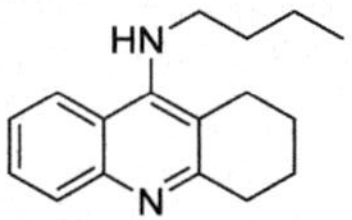 (Patnaik and Vohra 1966) | Local anesthetic | CDRI | Themis Chemicals, Mumbai. | |
| 6. | Nitroxazepine Hydrochloride (SINTAMIL) 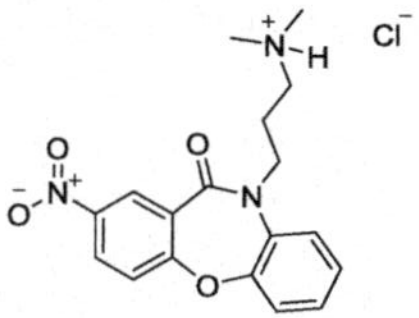 (Nagarajan and Arya 1982) | Antidepressant for the treatment of all grades and types of depression and also indicated for nocturnal enuresis | Ciba-Geigy Research Centre | | Ciba-Geigy (Novartis) sold in India as Sintamil |
| 7. | Chandonium iodide 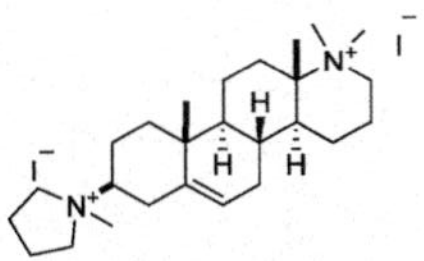 (Singh and Paul 1974) | Non-depolarizing neuromuscular blocking agent | Punjab University and CDRI | Ranbaxy Laboratories and Cipla | |
| 8. | Centpropazine 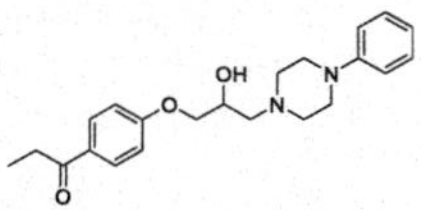 (Rastogi et al. 1972) | Serotonin uptake inhibitor | CDRI | Merind Ltd | |

(continued)

**Table 13.6** (continued)

| S. No. | Name and structure of the drug | Indication | Developed by | Licensed to | Clinical status |
|---|---|---|---|---|---|
| 9. | LND 101001 ("LND 101001," 2018c; Sinha et al. 2014) | Nootropic for treating cognitive deficits such as in Alzheimer's, alpha-7 nAChR (Alpha7 nicotinic acetylcholine receptor modulators) modulator | Lupin | | Currently in Phase-2 development |
| 10. | SUVN-502 ("Discovery Pipeline," 2018; Nirogi et al. 2016) | Selective 5-HT6 serotonin receptor antagonist for the treatment of mild cognitive impairment associated with CNS diseases such as AD, Parkinson's disease (PD), or schizophrenia | Suven Life Sciences | | Phase 2a trials were conducted in 2015. |
| 11. | SUVN-G3031 (Bhyrapuneni et al. 2010; "Discovery Pipeline," 2018) | Histamine H3 receptor antagonist for treating cognitive disorders | Suven Life Sciences | | Completed Phase 1 studies in 2015 in the USA |
| 12. | SUVN-D4010 ("Discovery Pipeline," 2018; Jayarajan et al. 2016; "SUVN-D4010 for Cognition in Alzheimer's disease commenced Phase 1 Clinical Trial in USA under US-IND 126099," 2015) | Partial 5HT4 agonist for treating cognitive disorders | Suven Life Sciences | | Entered Phase 1 trials in the USA in 2015 |
| 13. | SUVN-911 ("Discovery Pipeline," 2018; "SUVN-911," 2018) | Alpha-4 beta-2 nAChR antagonist for treating depression | Suven Life Sciences | | Entered Phase 1 clinical trials in 2018 in the USA |

| | | | | | |
|---|---|---|---|---|---|
| 14. | KU-046 ("KU-046," 2016; "KU 046," 2016a; Sharma 2014) | Amyloid β-protein modulator | Kareus Therapeutics | | Phase 1 studies for the treatment of Alzheimer's disease and in preclinical development for multiple sclerosis and fragile X syndrome. |

**Table 13.7** Drugs acting on reproductive system

| S. No. | Name and structure of the drug | Indication | Developed by | Licensed to | Clinical status |
|---|---|---|---|---|---|
| 1. | Ormeloxifene (Centchroman)—world's first non-steroidal oral contraceptive pill.<br>Trade names: SAHELI (oral contraceptive) NOVADEX (DUB) CENTRON (breast cancer, mastalgia and mastitis, breast pain, nodular breast, and osteoporosis).<br>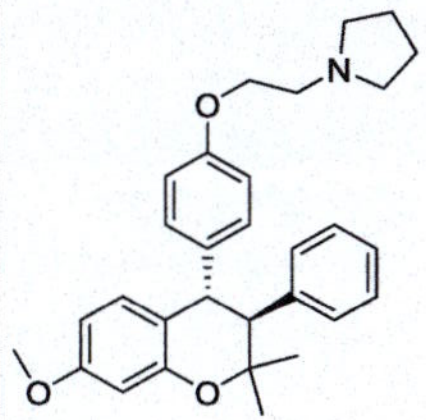<br>(Nityanand and Anand 1996; Ray et al. 1976) | Post-coital/weekly oral contraceptive and also indicated for dysfunctional uterine bleeding (DUB) | CDRI | Hindustan Latex (SAHELI and NOVADEX) and Torrent (CENTRON) | Marketed |

Despite the impressive list of potential drug candidates developed by Indian pharmaceutical companies, many of these candidates are at either preclinical/clinical stages of development or abandoned or out-licensed but none has reached the market except Lipaglyn (a dual PPAR α/γ agonist, Saroglitazar) discovered and developed by Zydus Cadila. From the early discoveries made by CIBA-Geigy Research Centre, Central Drug Research Institute (CDRI), Central Institute of Medicinal and Aromatic Plants (CIMAP), RRL Hyderabad (now Indian Institute of Chemical Technology), and Indian Institute of Integrative Medicine (IIIM), Jammu, antimalarial agents Elubaquine and Arteether, Satranidazole for treating luminal and hepatic amebiasis, Tinazoline (a vasoconstrictor), Tromaril (enfenamic acid) a potent anti-inflammatory agent, Mipimazole (an anti-thyroid drug), Nonaperone (antipsychotic drug), Buriperone (Centbutindole) a dopamine antagonist, Bucricaine (Centbucridine) a local anesthetic, Nitroxazepine Hydrochloride (SINTAMIL) an antidepressant, Chandonium iodide and Centpropazine (CNS acting drugs) were some of the significant contributions. Natural product based Peruvoside—marketed as "Encordin" in West Germany useful in congestive heart failure, Gugulipid—marketed as GUGLIP as hypolipidemic agent have also been useful in patients with cardiovascular diseases. Ormeloxifene (Centchroman)—world's first non-steroidal oral contraceptive pill SAHELI discovered by CDRI, "Risorine"—a bioenhancer, developed by Indian Institute of Integrative Medicine (IIIM), Jammu and Cadila Pharmaceuticals Ltd launched in November 2009 in the treatment of tuberculosis were other notable inventions from India (Dikshit and Dikshit 2016).

### 13.2.2 Drug Discovery Research in Indian Industry: Current Trends (from Recent Press Announcements)

In spite of the significant progress made by the Pharmaceutical Industry/Institutes in new drug discovery research (NDDR), several of the organizations have slowed down their NCE R&D efforts in the recent past and many of them have closed down their operations. However, consistent and continued efforts by few of the organizations in this direction holds a glimmer of hope for NCE research in India. Some of the out-licensing deals announced by these Pharmaceutical Companies, significant progress in clinical development and also the new symbiotic approaches being adopted holds promise.

1. MALT-1 (Mucosa-Associated Lymphoid Tissue Lymphoma Translocation Protein 1) is a protein involved in T-cell and B-cell lymphocyte activation. Lupin's MALT-1 is a first-in-class drug discovery program delivered exclusively by Lupin right from concept generation through the various stages of drug discovery and development. Lupin partnered with AbbVie in December 2018 for the development and commercialization of MALT1 inhibitor for the treatment of hematological cancers with limited treatment options. The agreement was hailed as one of the major partnerships by an Indian Company ("Lupin and AbbVie Announce Partnership to Develop and Commercialize Novel Oncology Drug to

treat Hematological Cancers," 2018). Subsequently, Lupin partnered with Germany's Boehringer Ingelheim for the development of novel cancer drugs by combining the targeted oncology molecules of the two companies to develop first-of-its-kind treatment for certain types of lung and gastrointestinal cancers. The approach has a strategic goal to focus on patients with cancers defined by KRAS mutations. KRAS is a protein that regulates cell growth and its mutation can result in uncontrolled cell growth leading to cancer. KRAS mutations occur in one in seven of all human metastatic cancers (Phadnis 2019).

2. Aurigene Discovery Technologies Limited developed a number of proprietary in-house drug discovery projects using both small molecule and peptide or peptidomimetic approaches. It was successful in licensing many of them. From the first licensing agreement with Debiopharm on Debio 1142 (an inhibitor of a non-disclosed oncology pathway) in 2011 ("2015 Investor Day," 2015; "Debiopharm and aurigene sign agreement for the development and commercialisation of Debio 1142, a novel inhibitor of an undisclosed oncology pathway," 2011), licensing of AUNP-12/W016A {a programmed cell death 1 (PD1) signalling pathway inhibitor 2014} in 2014 to Pierre Fabre Médicaments (returned in 2015) (Differding 2014a, b; "Immuno-oncology: licensing agreement between aurigene and Pierre Fabre Pharmaceuticals," 2014), small molecule peptidomimetics for the treatment of cancer (CA-170/AUPM-170 and CA-327/AUPM-327) and an orally active IRAK-4 (Interleukin-1 Receptor-Associated Kinase 4) inhibitor CA-4948/AU-4948 for non-Hodgkin's Lymphoma to Curis (Aurigene 2015; "Aurigene and its partner Curis announce the dosing of CA-170 in phase II India trial," 2018a; Curis n.d.; "Curis Licenses CA-170, IRAK4 Programs from Aurigene," 2015a, b), licensing of pan BET inhibitors (BRD4/BRD2/BRD3), such as ODM-207 to Orion (Aurigene 2014; Björkman et al. 2016; Samajdar et al. 2015). Aurigene continued its success journey and announced partnership with Agios Pharmaceuticals for the development of small molecule inhibitors of an undisclosed cancer metabolism target (Aurigene 2017) in 2017. In a significant development, EXELIXIS and AURIGENE have entered into exclusive collaboration, option and license agreement to discover and develop novel therapies for cancer ("Exelixis and Aurigene enter into exclusive collaboration, option and license agreement to discover and develop novel therapies for cancer," 2019).

3. The drug discovery team of Suven Life Sciences exclusively focused on treatment of CNS disorders and developed various compounds for treating dementia. Suven filed its first Investigational New Drug (IND) application in 2007 for lead compound SUVN-502, a selective 5-HT6 serotonin receptor antagonist. After successful completion of Phase 1 trials in 2009, the Phase 2a trials were conducted in 2015 ("Discovery Pipeline," 2018; Nirogi et al. 2016). SUVN-G3031, a histamine H3 receptor antagonist for treating cognitive disorders completed Phase 1 studies in 2016 in the USA (Bhyrapuneni et al. 2016; "Discovery Pipeline," 2018). SUVN-D4010, a partial 5HT4 agonist for the same indication entered Phase 1 trials in the USA in 2015 ("Discovery Pipeline," 2018; Jayarajan et al. 2016; "SUVN-D4010 for Cognition in Alzheimer's disease

commenced Phase 1 Clinical Trial in USA under US-IND 126099," 2015). Suven's alpha-4 beta-2 nAChR antagonist SUVN-911, for treating depression is undergoing Phase 1 clinical trials in the USA as of 2018 ("Discovery Pipeline," 2018; "SUVN-911," 2018).

4. Researchers at Dr. Reddy's Laboratories announced that its candidate drug (PPC-06) has emerged as a therapeutic option for treating psoriasis. The treatment routines followed are creams, oral medicines, and light therapy. In the Phase 2b studies, the drug PPC-06 provided positive clinical data in patients with moderate to severe plaque psoriasis. It met the prescribed treatment standards after 24 weeks of oral treatment (Somasekhar 2019).
5. In their continued efforts on the development of Saroglitazar Magnesium in the treatment of non-alcoholic steatohepatitis, Zydus Cadila has recently concluded enrolment in EVIDENCES 4 Phase 2 clinical trial ("Zydus Announces Positive Results From EVIDENCES IV Phase 2 Clinical Trial of Saroglitazar Magnesium in NAFLD and NASH," 2019).
6. Interestingly, Glenmark Pharma was also actively involved in in-licensing of development compounds. Napo Pharmaceutical's anti-diarrheal product Crofelemer in 2005 and Kissei Pharmaceutical's Remogliflozin etabonate in 2019 were in-licensed by Glenmark. Remogliflozin is a SGLT2 (sodium glucose co-transporter-2) inhibitor useful for the treatment of Type 2 diabetes. Glenmark completed Phase III clinical trials for Remogliflozin in India after obtaining rights through a licensing collaboration with BHV Pharma. Remogliflozin was discovered and developed by Kissei Pharmaceutical Co, Japanese firm and subsequently developed by GlaxoSmithKline Plc and BHV Pharma, a wholly owned subsidiary of Avolynt Inc. Glenmark was the first in the world to launch Remogliflozin and India was the first country to get access to this drug ("Press Release," 2019).
7. In a significant development for drugs against antimicrobial resistance (AMR), Wockhardt has reported the progress of its NCEs in various phases of clinical development. Recognition of WCK 4282 by Chinese regulator National Medical Products Administration (NMPA) for the unmet needs in China, Phase 3 CABP (community-acquired bacterial pneumonia) study approval for WCK 4873 in India by Drug Controller General of India (DCGI), completion of phase 3 acute bacterial skin and skin structure infection (ABSSSI) study in India for WCK 771/WCK 2349, acceptance of WCK 6777s IND application for a unique once-a-day MDR (multi-drug resistant) gram-negative product by the United States Food and Drug Administration (USFDA) are some of the examples cited ("New antibiotic drugs under development to have positive impact in future: Wockhardt's Khorakiwala," 2019).
8. In a trendsetting move, the Pharma major Sun Pharmaceutical Industries signed a licensing pact with the Hyderabad based CSIR-IICT for "patents related to certain compounds having potential therapeutic activity across multiple therapeutic indications in the pharma major's speciality focus areas." Announced in a press release on Aug 15, 2019, Sun Pharma gets an exclusive global license for development, regulatory filings, manufacturing and commercialization of the potential products arising from these CSIR-IICT patents and any other future

patents covered in the agreement. Sun Pharma claims that its collaboration with CSIR-IICT under this pact signifies its efforts to "bring innovations from Indian research institutes to the market to address the unmet medical needs of patients globally." This initiative appears to be the most ambitious instance of collaboration between government funded institutions and pharma industry in the country ("Sun Pharma signs licensing pact with CSIR-IICT," 2019b).

### 13.2.3 Drug Discovery Research in Indian Industry: More Recent Strategies

In an effort to improve the chances of success in NCE research, some companies are embracing new approaches by synergizing with the strengths and skill set available outside the country.

1. Glenmark Pharma is now spinning off its R&D business into a separate company that would be based in the USA. A former executive from Gilead & Novartis has been hired to head the new company. All innovative molecules in the pipeline, including preclinical assets and technology, the R&D centers in Switzerland, Paramus, and Navi Mumbai, and the biologics manufacturing facility in Switzerland, along with 400 employees associated with innovative drug discovery will form part of this new company (Somvanshi 2019).
2. **Sun Pharma:** In 2019, Sun Pharma Advanced Research Company (SPARC) collaborated with HitGen, China's biotech firm to identify novel small molecule leads for targets of interest. Under this collaboration, SPARC will utilize HitGen's advanced technology platform, based on DNA-encoded library design, synthesis and screening, to discover novel leads ("Sun pharma ties up with China's HitGen for molecular discovery," 2019).

## 13.3 Reasons for Slowing Down NDDR Across Indian Pharma Industry

With humble and unremarkable beginnings in the 1970s and 1980s of the last century, the Indian pharmaceutical industry is currently the most highly organized and technologically advanced sector in all aspects of the business of "generics" for domestic as well as global markets. The Indian Patent Act 1970 has proven to be a landmark event that enabled this industry to develop into a powerhouse of generic pharmaceuticals comprising drug substances (APIs) and drug products (formulated dosage forms). Other legislations such as MRTP, FERA, and IDR Act have also contributed to the rapid rise of Indian pharma industry in generics business.

### 13.3.1 Profits from Generics Business Inadequate to Support NDDR

Despite stiff competition from international generic companies (Mylan, Teva, Sandoz, Hospira, etc.), the generics business by Indian companies continues to be profitable and attractive. However, India's contribution of exports to the global generic market (USD 315 billion in 2017) is very low (5–6%, approx. USD 18.9 billion) ("Commerce Ministry Data," 2019). While the Indian industry is striving to improve its export performance, it is facing increasingly difficult and complex regulatory clearances for pharma exports. In the domestic market, the industry is experiencing diminishing returns due to escalating costs of development and manufacture, unusually high competition coupled with price control regime. Thus, the profits from generics business could come under pressure and prove inadequate to fuel New Drug Discovery Research (NDDR) which is a high risk (success rate 1 in 10,000 new compounds), high cost (bench idea to drug product in market: USD >2.5 billion), and a long drawn endeavor (10–12 years from bench idea to market). Such high costs are attributed to complex and expensive clinical trials, costs of promotion and distribution, patent maintenance costs, unsuccessful R&D, and high failure rates (only 1 successful NCE out of 10–12 NCEs in development pipeline). The estimates of drug discovery and development costs by Tufts Centre for the Study of Drug Development (TCSDD) have been regarded as the benchmark estimates (USD 802 million in 2003, equal to USD 1.0 billion in 2013; and the cost of drug development in 2013 was estimated to be USD >2.5 billion) (Mullin 2014).

### 13.3.2 Knowledge Gap

New Drug Discovery Research (NDDR) is a highly knowledge intensive, IP driven, interdisciplinary effort involving researchers drawn from diverse disciplines of modern biology, chemistry, and pharmacology. Drug development calls for expertise in toxicology, clinical pharmacology, clinical research, medical science and regulatory affairs as well as process chemistry, technology development, and scale-up expertise. Indian Pharma Industry must be complimented for embarking on NDDR as early as 1994. Over a dozen large companies have pursued NDDR during the period 1994–2015, largely using "analog" research coupled with fast-follower approaches. Although these research efforts were productive culminating in the discovery of a number of patented new chemical entities (NCEs), very few have made it to the low-key Indian market and not a single NCE has entered global market.

For NCE research involving rational drug design and discovery, our industry has to embrace modern approaches based on genomics and proteomics, target selection and validation, structural biology and systems biology, pharmacology and pharmacokinetics, and cutting edge technologies drawn from artificial intelligence and big data management. We see knowledge gaps and lack of advanced skill set in most of these areas. Given the possibilities of using CRO expertise for outsourcing some of these components of research, it may be possible to progress the indigenously discovered NCEs through preclinical stage or perhaps as far as pre-IND stage. Further development of NCEs will involve clinical trials (Phases I, II, and III) followed by New Drug Application (NDA) for market approval, and post-marketing follow-up studies. Our industry at this time lacks in-house competencies for organizing and overseeing all such activities which require intense coordination with clinical research organizations and extensive networking for regulatory interface. The regulatory framework in the country at this point of time is not very conducive for clinical development of NCEs within the country. Outsourcing this activity to CROs overseas could be prohibitively expensive while our experience in out-licensing of NCE assets for milestone payments and royalties hitherto has not been rewarding. Thus, our inability to take forward NCE assets past preclinical stage is a dampener for even highly successful generic companies to divert resources to NDDR at this time. However, the recent Govt notification of "New Drugs and Clinical Trials Rules, 2019" aimed at promoting clinical research in the country, when implemented, should change the situation for good ("PTI Newsletter," 2019).

### 13.3.3 Resource Pressures Due to China Competition

India appears to have an edge over China in terms of quality, regulatory compliance and production of both APIs and finished formulations under cGMP conditions in USFDA approved manufacturing plants. Consequently, Indian companies have been exporting the generic pharmaceuticals to highly regulated markets of the USA, Europe, and Japan despite stiff competition from international generic companies, including those set up by research-driven MNCs (*Indian Pharmaceutical Alliance (IPA) report* 2019).

China is proficient in low-cost manufacture and supply of fine chemicals and chemical intermediates in bulk quantities. India is heavily dependent on China for the intermediates required for manufacture of bulk APIs. It is a matter of concern that China may leverage on this chemicals advantage in course of time to dominate the generic exports once its quality systems and regulatory compliance efforts are upgraded. In this event, the profits from generic exports might come under pressure and consequently the Indian generic companies will not have adequate resources to invest in NDDR. Thus, there is an urgent need for Indian generic industry to deploy resources to develop indigenous know-how and production of fine chemicals and chemical intermediates required for API manufacture. In order to address this problem and as a part of Make in India mindset, the country's Council of Scientific and Industrial Research (CSIR) has launched a Mission Mode Project on "Innovative Processes and Technologies for Indian Pharmaceutical and Agrochemical Sector Industries (INPROTICS-Pharma and Agro)." The project aims to develop cost-effective, profitable, indigenous processes for key drugs and agrochemicals. In case of pharmaceuticals, new or non-infringing processes that are free to operate shall be developed ("Year End Review: CSIR," 2018).

### 13.3.4 Other Important Factors for Slowing Down NDDR

Indian pharmaceutical companies in general appear to be slow to the idea of discovery research and innovation. The research-based companies forming part of the Indian Pharmaceutical Alliance (IPA), with their promoters having passion for innovation, are an exception. Just over a dozen such companies have started NDDR with ambitious projects and investment plans during the period 1994–2000. By 2015, these companies have discovered a number of interesting NCEs, some of which have entered development pipeline. However, a majority of these companies have either downsized or rationalized or closed down these projects for a variety of reasons which are as follows (see Table 13.1 in the previous section of this chapter):

1. Slow progress in research due to lack of critical mass, lack of trained manpower, and limited availability of specialist research scientists for pursuit of modern research
2. Inadequate laboratory infrastructure and lack of uninterrupted supply of basic requirements: quality water and stable electricity; and gaps in sophisticated scientific instruments and equipment profile
3. Complex approval system for import of animal models for in vivo screening in discovery research and preclinical stages
4. Lack of research ecosystem and poor industry–academia interactions
5. Diminishing government supported incentives (e.g., lack of tax breaks, weighted deduction of 200% on R&D spend said to be tapering off by 2021, etc.)
6. Uphill task of going forward post the discovery research/preclinical development due to (a) complexity of regulatory framework for clinical trials in the country, (b) lack of in-house expertise in the area of clinical trials, (c) exorbitant costs of clinical development overseas, and (d) inadequate exposure to out-licensing opportunities

## 13.4 Enabling Measures to Boost NDDR in India

Pharmaceutical industry is a knowledge driven, R&D intensive and highly regulated industry sector. As mentioned above in this chapter, the Indian pharma industry has registered phenomenal growth post the Indian Patent Act 1970. Focused on generics comprising both drug substances (APIs) and drug products (formulated products), the Indian pharmaceutical industry has grown to the status of leadership in global markets, ranking 3rd by volume of production and 8th by value of sales (*Indian Pharmaceutical Alliance (IPA) report* 2019). The top ten companies have registered sales turnover in the range of INR 5200 to 28,000 crores in the year ending March 2018 (*Pharmaceuticals Report* 2018). Thus, the Indian pharma companies today are not only much larger than when the same companies started NDDR during 1994–2004, but also are proficient in operating cGMP manufacturing facilities, cost-effective product development and process innovation as well as marketing worldwide, including the most highly regulated markets of the USA and Europe.

Consequently, the pharma industry in the country with its vibrant ecosystem in generics space is well poised today to embark on NDDR towards hardcore innovation and IP generation (Vishwakarma 2014). The following are some of the measures that would accelerate the launch of ambitious and sustainable NDDR programs afresh:

1. Improve industry–academia interactions to provide research ecosystem for advancement of interdisciplinary and translational research. A case in point here is the Industry-Institutional Collaborative R&D Projects supported by DST under its "Drugs and Pharmaceuticals Research Program" (DRPP) wherein the project funding is shared by DST and industry partner in the ratio of 70% and 30%, respectively, while the industry partner is entitled to get the rights to any intellectual property generated in the project research. More such schemes will help enhance industry–academia interactions and facilitate industry-institutional collaborative research on a much larger scale.
2. Promote networking between national institutes and pharma companies for collaborative research opportunities. The most recent licensing pact signed between CSIR-IICT and Sun Pharma stands out as an outstanding example of such a collaboration (see Sect. 13.2.2h above). We need more such collaboration pacts and any measures to support networking between industry and academic/research institutions are most welcome.
3. Incentivize innovation and IP generation in academic institutions by empowering faculty researchers to seek patents for their inventions emanating from government funded research projects and assigning rights to the host academic institution (e.g., Bayh–Dole Act of 1980, USA). The IP so generated in academic institutions can be leveraged for industry collaboration and commercialization.
4. Improve risk capital environment and set up incubation parks to encourage entrepreneurs and facilitate start-ups to help promote "invent in India" culture in tune with the "Make in India" mission. We see this trend already in place with the recent establishment of Atal Incubation Centre at CCMB, Hyderabad, and DBT supported BioNest incubation center at University of Hyderabad, Hyderabad. More such centers at other equally reputed institutions in the country should help create a critical mass of entrepreneurial drive and innovation in the country.
5. Government may launch a finishing school type arrangement for emerging MSc's and PhDs to continue research work for at least 1 year in custom-designed collaborative research projects between industry and academic institutions.
6. Reverse the "**brain drain**" by instituting a special scheme to retain emerging PhDs in the country (cf. "CSIR Pool Officer Scheme" operated in the past) to pursue industry relevant research and thereby develop industrial post doc pool within the country; also a similar very special scheme to attract experienced researchers of Indian origin from abroad ("**brains in reserve**") to return to India and participate in cutting edge approaches to NDDR.
7. The recently announced regulatory framework be implemented earliest for promoting clinical development of new molecular entities (NMEs) within the

country and thereby bring inventions to market at affordable pricing ("Govt. notifies new rules for drugs and clinical trials," 2019)

In conclusion, the authors wish to observe that the estimates of ever increasing costs of NDDR and drug development are for western models (mostly the USA) and are very complex. Thus the Tufts Centre's estimate of USD 802 million in the year 2003 is said to be equal to approx. USD 1.0 billion in 2013. In the subsequent estimate in 2013, the Tufts Centre has come up with the mind boggling figure of USD 2.558 billion comprising USD 1.4 billion out-of-pocket costs and USD 1.2 billion "Time Costs" towards expected returns that investors forego while a drug is in development. The Tuft Centre's analysis was based on the information drawn from ten pharma companies on 106 randomly selected drugs in clinical development during the period 1995–2007 and the estimated figures included the costs due to unsuccessful projects (Dickson and Gagnon 2004; DiMasi et al. 1991; Mullin 2014; "Govt. notifies new rules for drugs and clinical trials," 2019). However, it has been observed that the cost of drug discovery research alone (i.e., the "R" component of R&D) is estimated to be 1/3rd of the cost of R&D, while the cost of drug development (i.e., the "D" component) is 2/3rd of the total cost of R&D.

It is unfortunate that we do not have the cost estimates of drug discovery and development in India due to lack of adequate experience in this field. Given the experience of Indian industry in the cost-effective development and production of a large number of generics (both API and formulated drug products) which are sold globally at affordable prices, let us assume that we will realize similar performance outcomes from drug discovery research and drug development carried out within the country.

The R&D expenditure in Indian pharma industry has been steadily rising post the TRIPS Agreement in 1995 and more so since 2005 with the introduction of Product Patent Regime in the country. Thus, the R&D spend in the industry has been reported to have increased from 5.5% in FY 2011 to 9% in FY 2017 (Das 2017; "IBEF India Brand Equity Foundation," n.d.). But it is not clear as to how much of this R&D spend goes into new discovery and development, perhaps very little!

The situation needs to be vastly improved, keeping in view the trends in the global companies in the West (ca. 15–19% of sales revenue in drug discovery and development) (Banerji and Suri 2017; Bedi et al. 2013; "Pharmaceuticals Report," 2018).

The growth of Indian pharmaceutical industry in the generics space is a success story built upon the strengths of process innovation, technology development and cGMP manufacture of high quality APIs as well as formulated products for sale globally at affordable prices. The industry is continuing to grow with aggressive investments for expansion and is taking initiatives for entry into "complex generics" space. Since the implementation of Indian Patent Act 1970, it took nearly 4.5 decades of hard work and diligent navigation through complex regulatory processes for the Indian pharma industry to attain the present status of global leadership in generics. In analogy, we expect the Indian Pharma to perform equally well in NDDR and Drug Development in the country in course of time. Given the nature of more complex, expensive and long drawn processes in drug discovery research and drug development, it may take a longer time for our industry to demonstrate success and sustainable progress. After all, it is just about 15 years since the introduction of the Product Patent Regime in 2005 in the country. In anticipation of this product patent era in the country, over a dozen Indian pharma companies have started NCE research projects during the period 1994–2005 but these were not sustainable due to a variety of reasons as outlined in Sect. 13.3 of this chapter. Despite limited success, a few of these companies are continuing the NCE R&D programs in a low profile activity. It may be observed that "**best is yet to come**" from the Indian pharma industry and hopefully the top 10–15 Indian pharma companies will show the way forward for NDDR and demonstrate successful, sustainable and IP-driven research innovation in the country.

## References


Abdel-Magid AF (2012) Insulin-like growth factor-1 receptor inhibitors: a potential cancer treatment: patent highlight. ACS Med Chem Lett 4:20

ACROBiosystems (n.d.) MMP-9. https://www.acrobiosystems.com/L-378-MMP-9.html

Addepalli M, Ray K, Kumar B, Ramnath R, Chile S, Rao H (2010) RNAi-mediated knockdown of AURKB and EGFR shows enhanced therapeutic efficacy in prostate tumor regression. Gene Ther 17(3):352

Adisinsight (2006a) RBx 2258. https://adisinsight.springer.com/drugs/800011920

Adisinsight (2006b) RBx 4638. https://adisinsight.springer.com/drugs/800013680

Adisinsight (2008) DRF 1644. https://adisinsight.springer.com/drugs/800013100

Adisinsight (2014a) OCID 2987. https://adisinsight.springer.com/drugs/800041825

Adisinsight (2014b) RBx 10558. https://adisinsight.springer.com/drugs/800025331

Adisinsight (2015a) P 1736. https://adisinsight.springer.com/drugs/800028473

Adisinsight (2015b) RBx 7796. https://adisinsight.springer.com/drugs/800018165

Adisinsight (2015c) Sudoterb. https://adisinsight.springer.com/drugs/800031647

Adisinsight (2016a) KU 046. https://adisinsight.springer.com/drugs/800037276

Adisinsight (2016b) KU 5039. https://adisinsight.springer.com/drugs/800039913

Adisinsight (2016c) P 1201. https://adisinsight.springer.com/drugs/800028475

Adisinsight (2016d) P 3914. https://adisinsight.springer.com/drugs/800034351

Adisinsight (2016e) RBx 10017609. https://adisinsight.springer.com/drugs/800029478

Adisinsight (2016f) SUN 1334H. https://adisinsight.springer.com/drugs/800026836

Adisinsight (2016g) WCK 1152. https://adisinsight.springer.com/drugs/800019222

Adisinsight (2016h) ZY H7. https://adisinsight.springer.com/drugs/800031967

Adisinsight (2017a) GRC 17536. https://adisinsight.springer.com/drugs/800032789

Adisinsight (2017b) Levonadifloxacin - Wockhardt. https://adisinsight.springer.com/drugs/800016734

Adisinsight (2017c) LNP 1955. https://adisinsight.springer.com/drugs/800047506

Adisinsight (2017d) Ranbezolid. https://adisinsight.springer.com/drugs/800017411

Adisinsight (2017e) ZYDPLA 1. https://adisinsight.springer.com/drugs/800039138

Adisinsight (2018a) AAI 101/cefepime - Allecra therapeutics. https://adisinsight.springer.com/drugs/800041174

Adisinsight (2018b) Alalevonadifloxacin - Wockhardt. https://adisinsight.springer.com/drugs/800038027

Adisinsight (2018c) LND 101001. https://adisinsight.springer.com/drugs/800041369

Adisinsight (2018d) LNP 1892. https://adisinsight.springer.com/drugs/800040907

Adisinsight (2018e) Nafithromycin - Wockhardt. https://adisinsight.springer.com/drugs/800042488

Adisinsight (2018f) Posizolid. https://adisinsight.springer.com/drugs/800013296

Adisinsight (2018g) TRC 4186. http://adisinsight.springer.com/drugs/800017743

Adisinsight (2018h) TRC 150094. https://adisinsight.springer.com/drugs/800033334

Adisinsight (2019a) AAI 101. https://adisinsight.springer.com/drugs/800042043

Adisinsight (2019b) Cefepime/Zidebactam - Wockhardt. https://adisinsight.springer.com/drugs/800044946

Adisinsight (2019c) Tenalisib - Rhizen Pharmaceuticals. https://adisinsight.springer.com/drugs/800039373

Adisinsight (2019d) Umbralisib - TG Therapeutics. https://adisinsight.springer.com/drugs/800037325

Adisinsight (2019e) VB 1953. https://adisinsight.springer.com/drugs/800044288

Adisinsight (n.d.) Cadi 05. https://adisinsight.springer.com/drugs/800026303

Advinus Therapeutics (n.d.) Discovery pipeline. http://www.advinus.com/pipeline.asp

Agrawal R, Jain P, Dikshit N, S. (2012) Balaglitazone: a second generation peroxisome proliferator-activated receptor (PPAR) gamma (γ) agonist. Mini-Rev Med Chem 12(2):87–97

Al-Lahham A, De Souza NJ, Patel M, René Reinert R (2005) Activity of the new quinolones WCK 771, WCK 1152 and WCK 1153 against clinical isolates of Streptococcus pneumoniae and Streptococcus pyogenes. J Antimicrob Chemother 56(6):1130–1133

Allecrra (2013) New anti-infectives company Allecra Therapeutics created. http://www.allecra.com/news/new-anti-infectives-company-allecra-therapeuticscreated-with-15-million-series-a-financing

Amala K, Bhujanga Rao A, Gorantla B, Gondi CS, Rao JS (2013) Design, synthesis and preclinical evaluation of NRC-AN-019. Int J Oncol 42(1):168–178

Angelos S, Meyers J (1985) The isolation and identification of precursors and reaction products in the clandestine manufacture of methaqualone and mecloqualone. J Forensic Sci 30 (4):1022–1047

Anthony J, Kelkar A, Wilankar C, Ranjith V, Bhumra SK, Mutt S et al (2013) Discovery of p1736, a novel antidiabetic compound that improves peripheral insulin sensitivity in mice models. PLoS One 8(10):e77946

Anupindi R, Mukhopadhyay I, Thomas A, Kumar S, Chaudhari S, Kulkarni A et al (2010) GRC 17536, a novel, selective TRPA1 antagonist for potential treatment of respiratory disorders. Eur Respir J 36:E5645

Anupindi R, Sharma R, Husain R (2016) KBRPL2001, a novel GPR120 agonist, improves insulin sensitivity, glucose tolerance and decreases hepatic steatosis in rodent models of type 2 diabetes. In: ADA 2016 - American Diabetes Association 76th Scientific Sessions, New Orleans, LA, USA

Arnab Roy Chowdhury AJD, Maddi SR, Deb C, Mitra S, Reddy MVR, Jatiani S, Maniar M, Cosenza SC, Padgaonkar A, Reddy EP (2014) Pharmacokinetics of PLK2 inhibitor GBO-006-1, developed as a novel first-in-class molecule to treat triple negative breast cancer. Cancer Res 74:4537

Arora SK, Sinha N, Jain S, Upadhayaya RS, Jana G, Ajay S, Sinha RK (2004) WO 2004/026828 A1. W. I. P. Organization, Geneva

Arora R, Rangaswami S (1972) Peruvoside and other cardiotonic glycoside [s] of thevetia neriifolia juss: chemical, pharmacological, and clinical studies, vol 1. Thomson Press (India) Limited, New Delhi

Arora R, Dutta BK, Goel R, Hollinger FP, Kulia B, Mahajan D et al (2014) SPR965: an oral PI3K/mTOR C1/C2 inhibitor for the treatment of solid tumors. Cancer Res 74:4515

Ashtekar DR, Costa-Perira R, Nagrajan K, Vishvanathan N, Bhatt A, Rittel W (1993) In vitro and in vivo activities of the nitroimidazole CGI 17341 against Mycobacterium tuberculosis. Antimicrob Agents Chemother 37(2):183–186

AstraZeneca (1999–2016) Annual reports, 1999–2016. https://www.astrazeneca.com/investor-relations/annual-reports.html

Aurigene (2011) Debiopharm and aurigene sign agreement for the development and commercialisation of DEBIO 1142, a novel inhibitor of an undisclosed oncology pathway. http://www.aurigene.com/debiopharm-and-aurigene-sign-agreement-for-the-development-and-commercialisation-of-debio-1142-a-novel-inhibitor-of-an-undisclosed-oncology-pathway/

Aurigene (2014) Aurigene and Orion Corporation Announce research, collaboration and option agreement for rights to Aurigene's epigenetics program for Pan BET and selective BET bromodomain inhibitors. http://www.aurigene.com/aurigene-and-orion-corporation-announce-research-collaboration-and-option-agreement-for-rights-to-aurigenes-epigenetics-program-for-pan-bet-and-selective-bet-bromodomain-inhibitors/

Aurigene (2015) Curis and aurigene announce collaboration, license and option agreement to discover, develop and commercialize small molecule antagonists for immuno-oncology and precision oncology targets. http://www.aurigene.com/curis-and-aurigene-announce-collaboration-license-and-option-agreement-to-discover-develop-and-commercialize-small-molecule-antagonists-for-immuno-oncology-and-precision-oncology-targets/

Aurigene (2017) Agios and Aurigene enter into exclusive license agreement for novel small molecules for cancer metabolism target. http://www.aurigene.com/agios-and-aurigene-enter-into-exclusive-license-agreement-for-novel-small-molecules-for-cancer-metabolism-target/

Aurigene (2018a) Aurigene and its partner Curis announce the dosing of CA-170 in Phase II India trial. http://www.aurigene.com/aurigene-and-its-partner-curis-announce-the-dosing-of-ca-170-in-phase-ii-india-trial/

Aurigene (2018b) Aurigene announces first in human dosing with RoRγt inverse agonist AUR-101. http://www.aurigene.com/aurigene-announces-first-in-human-dosing-with-ror%CE%B3t-inverse-agonist-aur-101/

Aurigene (2019) Exelixis and Aurigene enter into exclusive collaboration, option and license agreement to discover and develop novel therapies for cancer

Badorrek P, Hohlfeld JM, Krug N, Joshi A, Raut A (2015) Efficacy and safety of a novel nasal steroid, S0597, in patients with seasonal allergic rhinitis. Ann Allergy Asthma Immunol 115 (4):325–329.e321

Balachandran S, Dinsmore CJ, Roychowdhury A, Sharma R, Vishwakarma RA (2012) WO2012/145471 A1. W. I. P. Organization, Geneva

Balasubramanian V, Solapure S, Iyer H, Ghosh A, Sharma S, Kaur P et al (2014) Bactericidal activity and mechanism of action of AZD5847, a novel oxazolidinone for treatment of tuberculosis. Antimicrob Agents Chemother 58(1):495–502

Balganesh T, Kundu TK, Chakraborty TK, Roy S (2014) Drug discovery research in India: current state and future prospects. ACS Publications, Washington, DC

Bandodkar B, Mukherejee K, Nagaraj H, Banerjee T, Jayaraman R, Singh JK et al (2014) Discovery of novel cathepsin k inhibitors as leads for osteoporosis therapy. http://www.eposters.net/pdfs/discovery-of-novel-cathepsin-k-inhibitors-as-leadsfor-osteoporosis-therapy.pdf

Banerji U (2014) LNP3794 in patients with Advanced Solid Tumours Having Mutations. http://www.hra.nhs.uk/news/researchsummaries/lnp3794-in-patients-with-advanced-solid-tumours-having-mutations/

Banerji A, Suri FK (2017) Patents, R&D expenditure, regulatory filings and exports in Indian pharmaceutical industry. J Intellect Prop Rights 22:136

Bedi N, Bedi P, Sooch BS (2013) Patenting and R&D in Indian pharmaceutical industry: post-TRIPS scenario. J Intellect Prop Rights 18(2):105

Bhatia M, Manchanda S, Roy SB (1970) Haemodynamic studies with peruvoside in human congestive heart failure. Br Med J 3(5725):740–743

Bhavar PK, Vakkalanka SKVS (2015) WO2015/198289 A1. W. I. P. Organization, Geneva

Bhyrapuneni G, Muddana NR, Kandikere V, Benade V, Ponnamaneni RK, Mudigonda K, Nirogi R (2010) The novel and selective α4β2 agonist, SUVN-911 enhances acetylcholine levels in frontoparietal cortex. Alzheimers Dement 6(4):S563

Bhyrapuneni G, Mudigonda K, Penta KK, Palacharla VRC, Muddana N, Jayarajan P et al (2016) SUVN-G3031, a potent and selective h3 receptor inverse agonist: safety, tolerability and pharmacokinetics in humans. Alzheimers Dement 12(7):618–619

BioSpace (2019) Zydus announces positive results from evidences IV phase 2 clinical trial of saroglitazar magnesium in NAFLD and NASH. https://www.biospace.com/article/releases/zydus-announces-positive-results-from-evidences-iv-phase-2-clinical-trial-of-saroglitazar-magnesium-in-nafld-and-nash/

Björkman M, Mattila E, Riikonen R, Abbineni C, Jaleel M, Marappan S et al (2016) ODM-207, a novel BET-bromodomain inhibitor as a therapeutic approach for the treatment of prostate and breast cancer. Paper presented at the Cancer Research

Boruah A, Alikunju S (2013) WO2013/024358 A2. W. I. P. Organization, Geneva

Boyd AW, Bartlett PF, Lackmann M (2014) Therapeutic targeting of EPH receptors and their ligands. Nat Rev Drug Discov 13(1):39

Brahmachari U (1922) A new form of cutaneous leishmaniasis—dermal leishmanoid. Indian Med Gaz 57(4):125

Burman AC, Mukherjee R, Prasad S, Jaggi M, Singh AT (2003) WO2003/002203 A1. W. I. P. Organization, Geneva

Burris HA, Patel MR, Fenske TS, O'Connor OA, Deng C, Brander DM et al (2015) Clinical activity and safety profile of TGR-1202, a novel once daily PI3Kδ inhibitor, in patients with CLL and B-cell lymphoma. J Clin Oncol 33:7069

Business Forum (2015) Paper presented at the BIO International Convention, Philadelphia

Business Standard (2018) Glenmark Pharmaceuticals provides update on clinical development of GRC 27864. https://www.business-standard.com/article/news-cm/glenmark-pharmaceuticals-provides-update-on-clinical-development-of-grc-27864-118011500958_1.html

Business Wire (2005) Dr. Reddy's Announces the Formation of Perlecan Pharma. http://www.businesswire.com/news/home/20050927006290/en/Dr.-Reddys-Announces-Formation-Perlecan-Pharma-Perlecan#.VFofh00tC70

Business Wire (2016) Vyome biosciences administers first in-human dose of VB 1953 in U.S. phase 1 clinical study in patients with facial acne vulgaris. http://www.businesswire.com/news/home/20161031005368/en/Vyome-Biosciences-Administers-In-human-Dose-VB-1953

Butler MS, Robertson AA, Cooper MA (2014) Natural product and natural product derived drugs in clinical trials. Nat Prod Rep 31(11):1612–1661

Cadila Healthcare Limited (2000–2016) Annual reports, 2000–2016. Corporate presentations. http://zyduscadila.com/

Carroll J (2008) Daiichi Sankyo acquires ranbaxy in $4.6B deal. http://www.fiercebiotech.com/biotech/daiichi-sankyo-acquiresranbaxy-4-6b-deal

Cavalleri B, Parenti F (2000) Glycopeptides (dalbaheptides). In: Kirk-Othmer encyclopedia of chemical technology

CDRI, R. C. A (2003) New antimalarial drug development in india: radical curative agents CDRI 80/53 (Elubaquine) and WR 238605 (Tafenoquine). Proc Indian Natl Sci Acad B69(6):871–882

CDSCO (2015) Minutes of IND Committee Meeting held on 16 December 2015. http://www.cdsco.nic.in/writereaddata/IND%20Committee%20Meeting%20Minutes%20held%20on%2016_12_2015.pdf

Chakrabarti R, Vikramadithyan RK, Misra P, Hiriyan J, Raichur S, Damarla RK et al (2003) Ragaglitazar: a novel PPARα & PPARγ agonist with potent lipid-lowering and insulin-sensitizing efficacy in animal models. Br J Pharmacol 140(3):527–537

Chatterji M, Shandil R, Manjunatha M, Solapure S, Ramachandran V, Kumar N et al (2014) 1, 4-Azaindole, a potential drug candidate for treatment of tuberculosis. Antimicrob Agents Chemother 58(9):5325–5331

Checkpoint Therapeutics (n.d.) CK-103 BET inhibitor. http://checkpointtx.com/pipeline/ck-103-bet-inhibitor/

Chennamsetty SMB, Joshi K, Chinchwade Y, Hulawale Y, Paramasivan S, Sivakumar M, Hariharan S (2012) WO2012/143879 A1. W. I. P. Organization, Geneva

Chikanna D, Lakshminarasimhan A, Khairnar V, Panigrahi S, Ramanathan A, Rao N et al (2016) Abstract A72: novel NAMPT inhibitors for the treatment of pancreatic cancer. Mol Cancer Res 14:A72

Chowdhury AR, Athisayamani JD, Mitra S, Maddi S, Adluri RS, Bethi SR, Joshi H (2016) Discovery of selective PI3Kβ inhibitors for PTEN deficient tumors and other indications. Cancer Res 14:373

Cioffi F, Zambad SP, Chhipa L, Senese R, Busiello RA, Tuli D et al (2010) TRC150094, a novel functional analog of iodothyronines, reduces adiposity by increasing energy expenditure and fatty acid oxidation in rats receiving a high-fat diet. FASEB J 24(9):3451–3461

ClinicalTrials.gov (2011a) Efficacy and safety of SAR292833 administration for 4 weeks in patients with chronic peripheral neuropathic pain (Alchemilla). https://clinicaltrials.gov/ct2/show/NCT01463397

ClinicalTrials.gov (2011b) Study of P3914 to evaluate the safety, tolerability, food effect & pharmacokinetics in healthy male subjects and efficacy & safety of P3914 in patients with acute dental pain. https://clinicaltrials.gov/ct2/show/NCT01339156

ClinicalTrials.gov (2012a) Clinical trial details. http://www.ctri.nic.in/Clinicaltrials/pmaindet2.php?trialid=2976

ClinicalTrials.gov (2012b) Study of single ascending doses of PBL 1427 in healthy volunteers. https://clinicaltrials.gov/ct2/show/NCT01554293

ClinicalTrials.gov (2013) Clinical study of oral PI3K/mTOR inhibitor in patients with advanced refractory solid tumors. https://clinicaltrials.gov/ct2/show/NCT01762410

ClinicalTrials.gov (2014) A study to test the safety/tolerability of increasing doses of LNP1892 versus placebo in healthy male/female subjects. https://clinicaltrials.gov/ct2/show/NCT02174237

Commerce Ministry Data (2019)

Cooper CB (2016) TB alliance drug discovery and development - harnessing global resources to address a global disease. In Streamlining Drug Discovery and Development Symposium. http://www.optibrium.com/downloads/TBAlliance-Harnessing-Global-Resources.pdf

CSIR (2010) CSIR NEWS progress, promise and prospects. 60(5–6)

CSIR (2018) Year end review [Press release]

Curis (n.d.) Pipeline. http://www.curis.com/pipeline/

Curis, Inc. (2015) Overview and path for growth: Aurigene strategic collaboration. http://files.shareholder.com/downloads/ABEAChemMedChem5QPVEJ/500566877x0x804254/4F1C08B4-25DC-4B17-B512-B006FFA2EB63/Curis%20Aurigene%20Collaboration_21Jan2015_Final.pdf

Daiichi Sankyo (2012) Ranbaxy launches Synriam – India's First New Drug. https://www.daiichisankyo.com/media_investors/media_relations/press_releases/detail/005898.html

Das S (2017) Indian drug firms' R&D spending rises as growth slows in US, domestic mkts. https://www.business-standard.com/article/companies/indian-drug-firms-r-d-spending-rises-as-growth-slows-in-us-domestic-mkts-117071001081_1.html

Das B, Rudra S, Yadav A, Ray A, Rao AR, Srinivas A et al (2005) Synthesis and SAR of novel oxazolidinones: discovery of ranbezolid. Bioorg Med Chem Lett 15(19):4261–4267

Datla P, Kalluri MD, Basha K, Bellary A, Kshirsagar R, Kanekar Y et al (2010) 9, 10-Dihydro-2, 5-dimethoxyphenanthrene-1, 7-diol, from Eulophia ochreata, inhibits inflammatory signalling mediated by Toll-like receptors. Br J Pharmacol 160(5):1158–1170

Davenport JM, Covington P, McIntyre G (2011) WO2011/153247 A1. W. I. P. Organization, Geneva

Desai RC, Bahekar R, Jadav P, Goswami A, Patel P (2014) WO2014/061031 A1. W. I. P. Organization, Geneva

Dickson M, Gagnon JP (2004) Key factors in the rising cost of new drug discovery and development. Nat Rev Drug Discov 3(5):417

Differding E (2014a) AUNP-12 - a novel peptide therapeutic targeting PD-1 immune checkpoint pathway for cancer immunotherapy – structure activity relationships & peptide/peptidomimetic analogs. http://www.differding.com/data/AUNP_12_A_novel_peptide_therapeutic_targeting_PD_1_immune_checkpoint_pathway_for_cancer_immunotherapy.pdf

Differding E (2014b) Drug discovery alliances in India—indications, targets, and new chemical entities. ChemMedChem 9(1):43–60

Differding E (2017) The drug discovery and development industry in India—two decades of proprietary small-molecule R&D. ChemMedChem 12(11):786–818

Dikshit M, Dikshit D (2016) Drug discovery research in India. Curr Sci 111(2):252

DiMasi JA, Hansen RW, Grabowski HG, Lasagna L (1991) Cost of innovation in the pharmaceutical industry. J Health Econ 10(2):107–142

Discovery on Target (2015) Targeting epigenetic readers and chromatic remodelers

Dong Y, Wittlin S, Sriraghavan K, Chollet J, Charman SA, Charman WN et al (2009) The structure–activity relationship of the antimalarial ozonide Arterolane (OZ277). J Med Chem 53(1):481–491

Doshi JC, Vaidya AB, Sen HG, Mankodi NA, Nair C, Grewal R (1977) Clinical trials of a new anthelmintic, 4-isothiocyanato-4′-nitrodiphenylamine (C. 9333-Go/CGP 4540), for the cure of hookworm infection. Am J Trop Med Hyg 26(4):636–639

Dr. Reddy's Laboratories Ltd (1997–2016) Annual reports, 1997–2016. http://www.drreddys.com/investors/reports-and-filings/annual-reports/

Dr. Reddy's Laboratories Ltd (2015) 2015 Investor Day. https://www.drreddys.com/media/40948/investor-day-2015.pdf

Dugar S, Mahajan D, Deokar RC, Hollinger FP, Kapoor KK (2012) WO2012/101654 A2. W. I. P. Organization, Geneva

Dugar S, Mahajan D, Rai SK, Rao K, Singh V (2015) WO2015/181837 A2. W. I. P. Organization, Geneva

Dutta G, Bajpai R, Vishwakarma R (1989) Antimalarial efficacy of arteether against multiple drug resistant strain of Plasmodium yoelii nigeriensis. Pharmacol Res 21(4):415–419

Dyke HJ (2007) Novel 5, 6-dihydropyrazolo-[3, 4-E][1, 4] diazepin-4 (1H)-one derivatives for the treatment of asthma and chronic obstructive pulmonary disease: AstraZeneca: WO2007040435. Expert Opin Ther Patents 17(9):1183–1189

Economic Times (2019a) Lupin developing new products to treat cancer, other diseases. https://economictimes.indiatimes.com/industry/healthcare/biotech/pharmaceuticals/lupin-developing-new-products-to-treat-cancer-other-diseases/articleshow/67510017.cms

Economic Times (2019b) Sun Pharma signs licensing pact with CSIR-IICT. https://health.economictimes.indiatimes.com/news/pharma/sun-pharma-signs-licensing-pact-with-csir-iict/70674585

ETHealthworld (2019) New antibiotic drugs under development to have positive impact in future: Wockhardt's Khorakiwala. https://health.economictimes.indiatimes.com/news/pharma/new-antibiotic-drugs-under-development-to-have-positive-impact-in-future-wockhardts-khorakiwala/70259665

Eurofins Advinus (2011) Advinus' GK-activator achieves early POC for diabetes. http://www.advinus.com/wp-content/uploads/2016/04/November-29-2011.pdf

Fierce Biotech (2008) Diakron licenses anticoagulant drug candidate from Merck & Co., Inc.; Orchid Pharma partners with Diakron. http://www.fiercebiotech.com/biotech/diakron-licenses-anticoagulant-drug-candidate-frommerck-co-inc-mrk-orchid-pharma-partners

Filipski KJ, Stevens BD, Pfefferkorn JA (2012) Glucokinase activators in development. New Strat Type 2:88–100

Financial Express (2002) JB Chemicals Files for Patent on 20 NCEs in NSAIDs Slot. http://www.financialexpress.com/archive/jb-chemicals-files-forpatent-on-20-nces-in-nsaids-slot/37338/

Financial Express (2019) Sun pharma ties up with China's HitGen for molecular discovery. https://www.financialexpress.com/industry/sun-pharma-ties-up-with-chinas-hitgen-for-molecular-discovery/1514492/

Fonteilles-Drabek S, Reddy D, Wells TN (2017) Managing intellectual property to develop medicines for the world's poorest. Nat Rev Drug Discov 1fc6(4):223

Gabriel S, Davenport S, Steagall R, Vimal V, Carlson T, Rozhon E (1999) A novel plant-derived inhibitor of cAMP-mediated fluid and chloride secretion. Am J Physiol Gastrointest Liver Physiol 276(1):G58–G63
Genetic Engineering and Biotechnology News (2015a) Curis licenses CA-170, IRAK4 programs from aurigene. https://www.genengnews.com/topics/drug-discovery/curis-licenses-ca-170-irak4-programs-from-aurigene/81251868/
Genetic Engineering and Biotechnology News (2015b) Curis licenses CA-170, IRAK4 programs from aurigene. http://www.genengnews.com/gen-news-highlights/curislicenses-ca-170-irak4-programs-from-aurigene/81251868/
Gharat LA, Gajera JM, Narayana L, Khairatkar-Joshi N, Kattige VG (2012a) WO 2012/110860 A1. W. I. P. Organization, Geneva
Gharat LA, Kattige VG, Khairatkar-Joshi N, Muthukaman N, Narayana L (2012b) WO2012/055995 A1. W. I. P. Organization, Geneva
Gharat LA, Muthukaman N, Tambe MS, Pisal D, Khairatkar-Joshi N, Kattige VG (2015) WO2015/059618 A1. W. I. P. Organization, Geneva
Glenmark Pharmaceuticals (2019) Press Release. https://www.glenmarkpharma.com/sites/default/files/attachment%201.pdf
Gler MT, Skripconoka V, Sanchez-Garavito E, Xiao H, Cabrera-Rivero JL, Vargas-Vasquez DE et al (2012) Delamanid for multidrug-resistant pulmonary tuberculosis. N Engl J Med 366 (23):2151–2160
Goodwin L (1995) Pentostam® (sodium stibogluconate); a 50-year personal reminiscence. Trans R Soc Trop Med Hyg 89(3):339–341
Gopalan B (2012a) Solapur, India: ISCBC-2012, p 57
Gopalan B (2012b) ISCBC-2012. http://www.iscbindia.com/index.php/downloads/proceedings?download=36%3A17th-iscbcinterna-tional-conference-iscbc-2012
Gopalan B, Ravi D, Rasheed M, Sreedhara SKH, Ishtiyaque A (2007) WO2007/113634 A1. W. I. P. Organization, Geneva
Gopalan B, Ravi D, Shrikant H, Sreedhara SKH (2008) WO2008/032171 A1. W. I. P. Organization, Geneva
Gupta CM (2005) Drug discovery and development: India. In: Gupta CM (ed) Examples of the development of pharmaceutical products from medicinal plants, vol 10. UNDP Special Unit for South-South Cooperation, New York, NY, pp 29–44
Gupta RC, Chhipa L, Mandhare AB, Nadkarni SS, Joshi D, Zambad S, Pathak P, Chauthaiwale V, Dutt C (2013) Design and synthesis of novel thiazolidine and pyrrolidine derivatives as DPP-IV inhibitors. http://www.eposters.net/pdfs/design-andsynthesis-of-novel-thiazolidine-and-pyrrolidine-derivatives-as-dpp-iv-inhibitors-.pdf
Hameed S, Solapure S, Patil V, Henrich PP, Magistrado PA, Bharath S et al (2015) Triaminopyrimidine is a fast-killing and long-acting antimalarial clinical candidate. Nat Commun 6:6715
Hiriyan J, Shivarudraiah P, Gavara G, Annamalai P, Natesan S, Sambasivam G, Sukumaran SK (2015) Discovery of PAT-1102, a novel, potent and orally active histone deacetylase inhibitor with antitumor activity in cancer mouse models. Anticancer Res 35(1):229–237
Huband MD, Castanheira M, Jones RN, Rhomberg PR, Watters AA, Sader HS (2016) WCK 5222 (Cefepime-Zidebactam) in vitro time-kill studies against pseudomonas aeruginosa and acinetobacter baumannii isolates with characterized β-lactamases. https://www.jmilabs.com/data/posters/Microbe16-WCK-5222-Friday-477.pdf
IBEF India Brand Equity Foundation (n.d.). https://www.ibef.org/
India Brand Equity Foundation (2018) Pharmaceuticals report. https://www.ibef.org
IPA (2019) Indian Pharmaceutical Alliance (IPA) report
IPCA (2012) New drug discovery/development. https://www.ipca.com/active-pharmaceutical-ingredients-research-development.html
IPO (n.d.) The Patents Act 1970 http://www.ipindia.nic.in/writereaddata/Portal/IPOAct/1_31_1_patent-act-1970-11march2015.pdf

Jain R, Trehan S, Das J, Singh N, Nanda GK, Magadi SK, Sharma SK (2009) WO2009/093269 A1. W. I. P. Organization, Geneva

Jayarajan P, Bhyrapuneni G, Mudigonda K, Penta KK, Muddana N, Palacharla VRC et al (2016) SUVN-D4010, a potent and selective 5-HT4 receptor partial agonist: safety, tolerability and pharmacokinetics in humans. Alzheimers Dement 12(7):823–P824

Joseph R (2011) The R&D scenario in Indian pharmaceutical industry. Research and Information System for Developing Countries, New Delhi

Joshi D, Gupta R, Dubey A, Shiwalkar A, Pathak P, Gupta RC et al (2009) TRC4186, a novel AGE-breaker, improves diabetic cardiomyopathy and nephropathy in Ob-ZSF1 model of type 2 diabetes. J Cardiovasc Pharmacol 54(1):72–81

JP Morgan Healthcare (2013) JP Morgan Healthcare Conference. https://www.glenmarkpharma.com/sites/default/files/Glenmark_-_JPM_-Jan_2013.pdf

Kacker I, Zaheer S (1951) Synthesis of substituted 4-quinazolones. J Indian Chem Soc 28:344

Kalluri MD, Datla P, Bellary A, Basha K, Sharma A, Sharma A et al (2010) Novel synthetic gluco-disaccharide RSCL-0409–a lipopolysaccharide-induced Toll-like receptor-mediated signalling antagonist. FEBS J 277(7):1639–1652

Karkun J, Anand N (1962) Antithyroid activity of some 1-alkyl imidazolidine 2-thiones. Cell Mol Life Sci 18(4):181–183

Khanna I, Pillarisetti S (2014) US8.623.897 B2. U. S. P. a. T Office, Alexandria, VA

Kikuth W, Schmidt H (1937) Contribution to the progress of antimony therapy of kala-azar. Chin Med J 52(3)

Kiran R, Mishra S (2011) Research and development, exports and patenting in the Indian pharmaceutical industry: a post TRIPS analysis. Eurasian J Bus Econ 4(7):53–67

Kumar Y, Aryan RC, Sattigeri J, Salman M, Shankar G, Bhushan KH, Pandya BR, Sharma R (2004) WO2004/005250 A1. W. I. P. Organization, Geneva

Kumar S, Sharma R, Zahler R, Sahu B, Agarwal VR, Naik N (2013) US2013/0116248 A1. U. S. P. a. T Office, Alexandria, VA

Kushwaha R, Haq W, Katti S (2014) Sixteen-years of clinically relevant dipeptidyl peptidase-IV (DPP-IV) inhibitors for treatment of type-2 diabetes: a perspective. Curr Med Chem 21 (35):4013–4045

Laboratoires Pierre Fabre (2014) Immuno-oncology: licensing agreement between Aurigene and Pierre Fabre Pharmaceuticals. https://www.pierre-fabre.com/en/news/immuno-oncology-licensing-agreement-between-aurigene-and-pierre-fabre-pharmaceuticals

Lakdawala A, Shirole M, Mandrekar S, Dohadwalla A (1988) Immunopharmacological potential of rohitukine-a novel compound isolated from the plant dysoxylum-binectariferum. Asia Pac J Pharmacol 3(2):91–98

Lal B, Dohadwalla AN, Dadkar NK, D'Sa A, De Souza NJ (1984) Trequinsin, a Potent New Antihypertensive Vasodilator in the Series of 2-(Arylimino)-3-alkyl-9, 10-dimethoxy-3, 4, 6, 7-tetrahydro-2H-pyrimido [6, 1-a] isoquinolin-4-ones. J Med Chem 27(11):1470–1480

Lamonica A, Forzatti M, Biondi S (2010) WO2015/067787 A1. W. I. P. Organization, Geneva

Larsen JT, Ramakrishnan V, Haug J, Kimlinger T, Sen S, Mahajan D et al (2015) Preclinical activity of dual PI3K/mTOR inhibitor SPR965 in multiple myeloma. Cancer Res 15:2653

Laughon BE, Balganesh TS (2010) Status of the TB drug pipeline. http://www.tballiance.org/events/downloads/of4/presentations/Laughon-Status-of-the-TBDrug-Pipeline.pdf

Lee WH, Kim SG (2010) AMPK-dependent metabolic regulation by PPAR agonists. PPAR Res 2010:549101

Lohray VB, Lohray BB, Paraselli RB, Ramanujam R, Chakrabarti R (1999) WO1999/008501 A2. W. I. P. Organization, Geneva

Lupin (2001–2016) Annual reports, 2001–2016. http://www.lupin.com/archives-annual-report.php

Lupin (2018) Lupin and AbbVie announce partnership to develop and commercialize novel oncology drug to treat hematological cancers [Press release]. https://www.lupin.com/portfolio/lupin-and-abbvie-announce-partnership-to-develop-and-commercialize-novel-oncology-drug-to-treat-hematological-cancers/

Maji D, Samanta S (2015) A review on the role of peroxisome prolifertor-activated receptor-γ agonists and hybrids in type 2 diabetes and cardiomyopathy. Asian J Pharm Clin Res 8:26–31

Mandhane SN, Ayer UB, Midha AS, Rao CT, Rajamannar T (2008) Preclinical efficacy and safety pharmacology of SUN-1334H, a potent orally active antihistamine agent. Drugs R & D 9 (2):93–112

Manjunatha U, Boshoff HI, Barry CE (2009) The mechanism of action of PA-824: novel insights from transcriptional profiling. Communicat Integr Biol 2(3):215–218

Manoria P, Chopra H, Parashar S, Dutta A, Pinto B, Mullasari A, Prajapati S (2013) The nuances of atherogenic dyslipidemia in diabetes: focus on triglycerides and current management strategies. Indian Heart J 65(6):683–690

Matsumoto M, Hashizume H, Tomishige T, Kawasaki M, Tsubouchi H, Sasaki H et al (2006) OPC-67683, a nitro-dihydro-imidazooxazole derivative with promising action against tuberculosis in vitro and in mice. PLoS Med 3(11):e466

McGrath P (2007) Central Drug Research Institute, Lucknow, India - Excellence in science: profiles of research institutions in developing countries. TWAS, Trieste

Mehra K, Agrawal N (1984) Role of tromaril--a new anti-inflammatory drug in eye diseases. Ann Ophthalmol 16(2):132–133

Midha AS, Chokshi HA, Chitturi TR, Thennati R (2003) WO2003/079970 A2. W. I. P. Organization, Geneva

MMV (2018) MMV-supported projects: MMV253. https://www.mmv.org/research-development/mmv-supported-projects

Mohanan A, Gupta R, Dubey A, Jagtap V, Mandhare A, Gupta RC et al (2011) TRC120038, a novel dual AT 1/ET A receptor blocker for control of hypertension, diabetic nephropathy, and cardiomyopathy in ob-ZSF1 rats. Int J Hypertens 2011:751513

Mookhtiar KA (2015) Liver selective glucokinase activation for treating type 2 diabetes: translation from mouse to man. In: Ramanbhai Foundation 7th International Symposium on Current Trends in Pharmaceutical Sciences, Ahmedabad, India

Mossinghoff GJ (1999) Overview of the Hatch-Waxman Act and its impact on the drug development process. Food Drug Law J 54:187

Mullin R (2014) Estimates of drug discovery and development costs by Tufts Center for the Study of Drug Development (TCSDD) for the year 2013, Chemical & Engineering News

Muthuppalaniappan M, Viswanadha S, Varanasi KV, Merikapudi GS, Vakkalanka SKVS (2011) WO2011/042798 A1. W. I. P. Organization, Geneva

Muthuppalaniappan M, Viswanadha S, Babu G, Vakkalanka SKVS (2012) US2012/0059001 A1. U. S. P. a. T Office, Alexandria, VA

Nagaraj HKM, Bandodkar BS, Ravilla L, Yellapu S, Rudresha AS, Singh JK, Vaidyanathan G (2016a) WO2016/027285 A2. W. I. P. Organization, Geneva

Nagaraj HKM, Bandodkar BS, Yellapu S, Sugandham SR, Goutham M, Mukherjee K et al (2016b) WO2016/027284 A2. W. I. P. Organization, Geneva

Nagarajan K (2006) Creative research in the chemical industry–four decades in retrospect. J Chem Sci 118(4):291–309

Nagarajan K (2014) History of natural products chemistry in India. Indian J Hist Sci 49:377–398

Nagarajan K, Arya V (1982) Two decades of medicinal chemistry research in India. J Sci Ind Res 41:232–240

Nagarajan K, Arya V, Kaul C, David J, Grewal R (1981) Structure activity relations among cyclic & acyclic S-(3-indolyl) isothioureas-development of a potent vasoconstrictor, tinazoline, 3-(2-imidazolin-2-yl-thio) indole. Indian J Exp Biol 19(12):1150–1153

Nagarajan K, Arya V, Shenoy S (1986) J Chem Res B25:697

Nagarajan K, Shankar RG, Rajappa S, Shenoy SJ, Costa-Pereira R (1989) Nitroimidazoles XXI 2, 3-dihydro-6-nitroimidazo [2, 1-b] oxazoles with antitubercular activity. Elsevier Masson, Paris

Nagaswamy K, Thirunagaru VG (2016) WO2016/116900 A1. W. I. P. Organization, Geneva

Nair SS (2017) A cancer therapeutic molecule from India: Indian researchers have discovered possibly the best BCL2 inhibitor yet. https://researchmatters.in/mr/node/213

Narayanan S, Amrutkar D, Gupta S, Gullapalli S (2007) GRC 10693 - a selective, orally active CB2 agonist for treatment of neuropathic pain. In: 37th Annu. Meet. Soc. Neurosci., San Diego, USA, Abstr. 287.18

Narayanan S, Vishwakarma S, Saxena S, Honnegowda S, Ganesan K, Ahamed F et al (2010) OCID 2987: a novel, chemically distinct orally active PDE4 inhibitor. J Allergy Clin Immunol 125(2): AB49

Naruganahalli KS, Shirumalla RK, Bansal V, Gupta JB, Das B, Ray A (2006) Ranbezolid, a novel oxazolidinone antibacterial: in vivo characterisation of monoamine oxidase inhibitory potential in conscious rats. Eur J Pharmacol 545(2-3):167–172

Natco Pharma (2003–2016) Annual reports, 2003–2016. http://natcopharma.co.in/investors/annual-reports/

Natesan S, Govinda RG, Annamalai P, Sambasivam G (2011) WO2011/021209 A1. W. I. P. Organization, Geneva

Nekkanti VK, Karatgi PJ, Paithankar M, Pillai RS, Venkateswarlu A, Shanvas A, Kumar RA, Ramesh M, Raju S, Subrahmanyam D, Sriram R (2011) US2011/0177161 A1. U. S. P. a. T Office, Alexandria, VA

Nesy E, Mathew L (2014) Detection and quantification of cardiotonic drug peruvoside using HPTLC from thevetia neriifolia, juss seed extracts. Int J Pharm Sci Invent 3:11–16

Nirogi R, Mudigonda K, Penta KK, Bhyrapuneni G, Benade V, Muddana N et al (2016) SUVN-502, potent and pure 5-HT6 receptor antagonist: proof-of-concept study design in moderate Alzheimer's disease patients. Alzheimers Dement 12(7):P434

Nityanand S, Anand N (1996) Centchroman: a nonsteroidal antifertility agent. FOGSI Focus:8–10

Oneindia (2006) New anti-diabetic agent from Reddy's labs being tested. http://www.oneindia.com/2006/12/26/new-anti-diabetic-agent-from-reddys-labs-beibngtested-1167147168.html

Orchid Pharma (2002–2016) Annual reports, 2002–2016. http://www.orchidpharma.com/

Pagès L, Gavaldà A, Lehner MD (2009) PDE4 inhibitors: a review of current developments (2005–2009). Expert Opin Ther Patents 19(11):1501–1519

Palanisamy SU, Gnanaprakasam A, Ganapathy P, Gohain M, Hariharan V, Rajagopal S et al (2010) US7687488 B2. U. S. P. a. T Office, Alexandria, VA

Palanisamy SU, Paul-Satyaseela M, Narayanan S, Balasubramania G, Appu A, Manickam S, Periasamy H (2014) US2014/0057888 A1. U. S. P. a. T Office, Alexandria, VA

Palle VP, Sattigeri VJ, Khera MK, Voleti SR, Ray A, Dastidar SG (2006) WO2006/090235 A1. W. I. P. Organization, Geneva

Papp-Wallace KM, Bonomo RA (2016) New β-lactamase inhibitors in the clinic. Infect Dis Clin 30 (2):441–464

Paraskar AS, Soni S, Chin KT, Chaudhuri P, Muto KW, Berkowitz J et al (2010) Harnessing structure-activity relationship to engineer a cisplatin nanoparticle for enhanced antitumor efficacy. Proc Natl Acad Sci 107(28):12435–12440

Patel MV, De Souza NJ, Gupte SV, Jafri MA, Bhagwat SS, Chugh Y et al (2004) Antistaphylococcal activity of WCK 771, a tricyclic fluoroquinolone, in animal infection models. Antimicrob Agents Chemother 48(12):4754–4761

Patel JR, Patel GC, Sheth GS, Mandhane SN, Rao CT, Thennati R (2014) WO2014/192027 A1. W. I. P. Organization, Geneva

Patnaik G, Vohra M (1966) Compounds acting on the central nervous system. IV. 4-Substituted 2, 3-polymethylenequinolines. J Med Chem 9(4):483–488

Phadnis A (2019) Lupin, Boehringer Ingelheim ink $700-mn deal for anti-cancer drug. https://www.business-standard.com/article/companies/lupin-boehringer-ingelheim-ink-700-mn-deal-for-anti-cancer-drug-119090401132_1.html

Pharmabiz.com (2012) Orchid Pharma completes phase I trial of PDE4 inhibitor molecule, OCID 2987. http://pharmabiz.com/NewsDetails.aspx?aid=66836&sid=1

Pharmacodia (2016) KU-046. https://www.pharmacodia.com/yaodu/html/v1/chemicals/6e1e62335a75d469de0550c35fbc45ba.html

Piramal (1997–2016) Annual reports, 1997–2016. http://www.piramal.com/investors/financial-reports#parentVerticalTab2

Polycap (n.d.-a) About Polycap. http://www.polycap.org/about-polycap.html

Polycap (n.d.-b). https://en.wikipedia.org/wiki/Polycap

Prasad RJ, Sathya BRA, Rao BN, Chowdary NV (2011) US2011/0039844 A1. U. S. P. a. T Office, Alexandria, VA

PTI (2019) Govt. notifies new rules for drugs and clinical trials. PTI Newsletter

Puri S, Srivastava R, Pandey V, Sethi N, Dutta G (1989) Methemoglobin toxicity and hematological studies on malaria anti-relapse compound CDRI 80/53 in dogs. Am J Trop Med Hyg 41 (6):638–642

Rajagopal S, Kumar RA, Subramanya D, Sharma V, Sastry T, Houghton PJ et al (2004) Preclinical evaluation of the anticancer activity of DRF-1042, a novel camptothecin analog targeting Topoisomerase-I. Cancer Res 64:710

Rajagopal S, Kachhadia V, Ponpandian T, Keeri AR, Anandhan K, Rajagopal S, Praveen R, Daivasigamani P (2009) WO2009/053808 A2. W. I. P. Organization, Geneva

Ramakrishnan S (2015) From 'Make in India' to 'Made in India': the saroglitazar story. Indian Heart J 67(1):8–10

Ranga MG, Rao JMR, Gudla CS (2015) WO2015/107549 A1. W. I. P. Organization, Geneva

Rangarajan R, Kumar R, Prabhakar BV, Chandrasekhar P, Mallikarjuna P, Banerjee A (2014) US2014/0249170 A1. U. S. P. a. T Office, Alexandria, VA

Rao JMR, Arumugam N, Ansari MM, Gudla C, Pachiyappan S, Ramalingam M et al (2012) WO2012/011125 A1. W. I. P. Organization, Geneva

Rao JMR, Venkatesham U, George J, Fernand G, Doppalapudi SR, Madhavan GR et al (2013) WO2013/128465 A1. W. I. P. Organization, Geneva

Rao JMR, Ranga MG, Pachiyappan S (2014a) WO 2014/202528 A1. W. I. P. Organization, Geneva

Rao JMR, Ranga MG, Pachiyappan S (2014b) WO 2014/202580 A1. W. I. P. Organization, Geneva

Rastogi SN, Anand N, Prasad C (1972) Agents acting on the central nervous system. 14. 1-(p-alkanoylphenoxy)-3-(N4-arylpiperazinyl) propan-2-ols. New class of antidepressants. J Med Chem 15(3):286–291

Rathod R, Bhatt T, Joshi K, Dole B, Murty KVSN, Thennati R (2013) WO2013/051024 A2. W. I. P. Organization, Geneva

Ray S, Grover PK, Kamboj VP, Setty B, Kar AB, Anand N (1976) Antifertility agents. 12. Structure-activity relation of 3, 4-diphenylchromenes and-chromans. J Med Chem 19 (2):276–279

Ray D, Shrivastava V, Tendulkar J, Dutta A, Bhopale K, Chatterjee D, Nagarajan K (1983) Comparative studies on the amoebicidal activity of known 5-nitroimidazole derivatives and CG 10213-Go in golden hamsters, Mesocricetus auratus, infected in the liver or caecum or both with trophozoites of Entamoeba histolytica. Ann Trop Med Parasitol 77(3):287–291

Ray D, Tendulkar J, Shrivastava V, Datta A, Nagarajan K (1984) A metronidazole-resistant strain of Trichomonas vaginalis and its sensitivity to Go 10213. J Antimicrob Chemother 14 (4):423–426

Reddy DS, Saxena C, Komirishetty K (2015) WO2015/015519 A1. W. I. P. Organization, Geneva

Rediff.com (2004) Orchid's anti-diabetic molecule gets US nod. https://www.rediff.com/money/report/orchid/20040927.htm

ReportJunction (n.d.) Database of company financial annual reports. http://www.reportjunction.com/

Rhizen Pharmaceuticals S.A. (n.d.) Onco Tx, Vet Tx. http://www.rhizen.com/research.html

Rivers EC, Mancera RL (2008) New anti-tuberculosis drugs in clinical trials with novel mechanisms of action. Drug Discov Today 13(23-24):1090–1098

RLL & MMV (2006) Ranbaxy Laboratories Limited (RLL) collaborates with Medicines for Malaria Venture (MMV). Asia Pac Biotech News 10(1):34–35. http://www.asiabiotech.com/publication/apbn/10/english/preserved-docs/1001/0034_0035.pdf

Roy M, Hossain SS, Sarkar A, Sengupta A, Gupta N, Hussain S et al (2014) IO125, a novel Pt-based supramolecular therapeutic exhibits increased anti-cancer efficacy compared with oxaliplatin. Cancer Res 74:4483

Roychowdhury A, Sharma R, Mascarenhas M, B-Rao C, Patil SM, Chavan S et al (2013) WO2013/148227 A1. W. I. P. Organization, Geneva

Saksena AK, Green MJ, Shue H-J, Wong JK, McPhail AT, Gross PM (1985) Identity of coleonol with forskolin: structure revision of a base-catalysed rearrangement product. Tetrahedron Lett 26(5):551–554

Samajdar S, Abbineni C, Sasmal S, Hosahalli S (2015) WO2015/104653 A1. W. I. P. Organization, Geneva

Sankaranarayanan A (2001) US6.232.331 B1. U. S. P. a. T Office, Alexandria, VA

Sarabhai Chemicals (n.d.) Profile. http://sarabhai-chemicals.co.in/Profile.html

Sattigeri VJ, Palle VP, Soni A, Naik KP, Ray A, Dastidar SG (2006) WO2006/090234 A1. W. I. P. Organization, Geneva

Satyam A (2011) US2011/0263526 A1. U. S. P. a. T Office, Alexandria, VA

Satyam LK, Chikkanna D, Khairnar VV, Reddy S, Durgaprasad V, Radhakrishna K et al (2016a) Identification of novel covalent inhibitors of K-Ras G12C that are efficacious in a xenograft model of NSCLC. Cancer Res 76:339

Satyam LK, Poddutoori R, Mukherjee S, Marappan S, Gopinath S, Ramachandra R et al (2016b) Potent and selective inhibition of CDK7 by novel covalent inhibitors. Cancer Res 76:3070

Paul E. Sax (2013) Crofelemer is approved for noninfectious diarrhea in HIV-infected patients. https://www.jwatch.org/ac201301070000001/2013/01/07/crofelemer-approved-noninfectious-diarrhea-hiv

Saxena C (2015) A 'First-In-Class' pre-IND compound for management of type-2-diabetes. BIO International Convention, Philadelphia, PA

Saxena AK, Jain PC, Anand N, Dua P (1973) Agents acting on the central nervous system. 15. 2-Substituted 1, 2, 3, 4, 6, 7, 12, 12a-octahydropyrazino [2′, 1′: 6, 1] pyrido [3, 4-b] indoles. New class of central nervous system depressants. J Med Chem 16(5):560–564

Sengupta P, Puri CS, Chokshi H, Chitturi TR, Thennati R (2009) WO2009/109991 A2. W. I. P. Organization, Geneva

Sengupta P, Chokshi HA, Puri CS, Chimanwala SY, Mehta VA, Desai DM et al (2012) WO2012/098416 A1. W. I. P. Organization, Geneva

Sengupta S, Roy M, Hossain S, Sengupta A, Mylavarapu S (2015a) WO2015/153345 A1. W. I. P. Organization, Geneva

Sengupta A, Mylavarapu S, Kumari S, Hossain S, Gupta N, Sarkar A et al (2015b) IO-125: a novel supramolecular platinum chemotherapy for triple negative breast cancer. J Clin Oncol 33:153

Sethi N, Srivastava S, Murthy P, Singh R (1988) Systemic toxicity study of a new schizontocidal antimalarial drug-arteether in rats and monkeys. Indian J Parasitol 12(2):223–235

Shapiro TA, Were JB, Talalay P, Bueding E, Rocco L, Danso K et al (1986) Clinical evaluation of amoscanate in healthy male volunteers. Am J Trop Med Hyg 35(5):945–953

Sharma EK (2014) Chasing a dream. https://www.businesstoday.in/magazine/focus/kareus-therapeutics-anji-reddy-dr-reddys-laboratories/story/211533.html

Sharma R, Anupindi R, Husain R, Sahu B, Almeida A, Chakor N (2016) Discovery of novel ursolic acid derivatives as ROR-gamma inhibitors. In: 251st ACS National Meeting, San Diego, CA, USA

Singh H (2014) Medicinal chemistry research in India. Indian J Hist Sci 49(413):e423

Singh H, Paul D (1974) Steroids and related studies. Part XXV. Chandonium iodide (17a-methyl-3β-pyrrolidino-17a-aza-D-homoandrost-5-ene dimethiodide) and other quaternary ammonium steroid analogues. Journal of the Chemical Society. Perkin Trans 1:1475–1479

Singh S, Sivakumar R (2004) Challenges and new discoveries in the treatment of leishmaniasis. J Infect Chemother 10(6):307–315

Singh D, Bala-Subramaniam R, Bhatia V, Kumar V, Chandrasekar S (1981) Study of the efficacy of compound Go. 9333 (Ciba-Geigy) in hookworm infestation. Chemotherapy 27(3):220–223

Sinha N, Karche NP, Tilekar AR, Palle VP, Kamboj RK (2014) WO2014/195848 A1. W. I. P. Organization, Geneva

Sivakumar M, Joshi KS, Aware VS, Sarde AG, Bagul SM, Manohar SM (2011) WO2011/104680 A1. W. I. P. Organization, Geneva

Sivanandhan D, Parikh PK, Sheshachalam A, Chandregowda V, Bhakthavatchalam R, Gupta M et al (2014) Abstract LB-232: potent, novel small molecule inhibitors targeting EGFR L858R/T790M mutation for non-small cell lung cancer. Cancer Res 74:LB-232

Sivanandhan D, Rajagopalan S, Nair S, Dewang P, Kumar DP, Mulakala C et al (2015) Novel dual inhibitors of LSD1-HDAC for treatment of cancer. Cancer Res 78:1382

Skerratt S (2017) Recent progress in the discovery and development of TRPA1 modulators. In: Progress in medicinal chemistry, vol 56. Elsevier, Amsterdam, pp 81–115

Sneha S, Nivetha R, Viswanadha S, Vakkalanka S, Ganesan T (2016) RP6530, a dual PI3K δ/γ inhibitor attenuates cancer stem cell proliferation in serous adenocarcinoma of ovary. Cancer Res 76:3331

Somasekhar M (2019) Dr Reddy's Laboratories announces drug to treat plaque psoriasis. https://www.thehindubusinessline.com/companies/dr-reddys-laboratories-announces-drug-to-treat-plaque-psoriasis/article27762934.ece

Somvanshi KK (2019) Can Glenmark Pharma taste success where its peers have not? https://economictimes.indiatimes.com/markets/stocks/news/can-glenmark-pharma-taste-success-where-its-peers-have-not/articleshow/68554865.cms?from=mdr

de Souza Mendes C, de Souza Antunes A (2013) Pipeline of known chemical classes of antibiotics. Antibiotics 2(4):500–534

de Souza NJ, Gupte SV, Deshpande PK, Desai VN, Bhawsar SB, Yeole RD et al (2005) A chiral benzoquinolizine-2-carboxylic acid arginine salt active against vancomycin-resistant Staphylococcus a ureus. J Med Chem 48(16):5232–5242

SPARC (2007–2016) Annual reports, 2007–2016

SPARC (2016) Investor update on R&D pipeline. http://www.sunpharma.in/SPARC_Investor%20Presention_4th%20August_Final.pdf

Sphaera Pharma (2015) THSTI & Wellcome Trust to jointly develop drug for resistant TB. http://www.biospectrumindia.com/biospecindia/news/221936/sphaera-pharma-thstiwellcome-trust-jointly-develop-drug-resistant-tb

Stover CK, Warrener P, VanDevanter DR, Sherman DR, Arain TM, Langhorne MH et al (2000) A small-molecule nitroimidazopyran drug candidate for the treatment of tuberculosis. Nature 405 (6789):962

Sundria S (2013) Orchid chem out-licenses molecule to German firm. http://www.dnaindia.com/money/report-orchid-chem-out-licensesmolecule-to-german-firm-1824296

Suresh N, Chandan S, Venkatesh A (2015) How india lost an iconic biotech lab. http://www.biospectrumindia.com/biospecindia/views/220303/how-india-lost-iconicbiotech-lab

Sutovska M, Kocmalova M, Franova S, Vakkalanka S, Viswanadha S (2016) Pharmacodynamic evaluation of RP3128, a novel and potent CRAC channel inhibitor in guinea pig models of allergic asthma. Eur J Pharmacol 772:62–70

Suven, Inc. (2018) Discovery pipeline. http://www.suvenincusa.com/

Suven Life Sciences (2018) SUVN-911. http://www.suven.com/pdf/SUVN-911.pdf

Takhi M, Sreenivas K, Reddy CK, Munikumar M, Praveena K, Sudheer P et al (2014) Discovery of azetidine based ene-amides as potent bacterial enoyl ACP reductase (FabI) inhibitors. Eur J Med Chem 84:382–394

TG Therapeutics (2019) Press Release. http://ir.tgtherapeutics.com/press-releases

The EU Clinical Trials Register (2017) A phase II, dose ranging, exploratory clinical study to assess the efficacy, pharmacodynamics, pharmacokinetics, and safety of LNP1955 in patients with moderate to severe rheumatoid arthritis. https://www.clinicaltrialsregister.eu/ctr-search/trial/2016-001532-35/HU

The Hindu Business Line (2002) Dr Reddy's cancer molecule completes Phase-I trials. https://www.thehindubusinessline.com/2002/05/14/stories/2002051401130400.htm

The Hindu Business Line (2004) Trial of prostate drug stopped — Schwarz move may impact Ranbaxy deal. https://www.thehindubusinessline.com/2004/11/07/stories/2004110701360200.htm

The Hindu Business Line (2015) SUVN-D4010 for cognition in Alzheimer's disease commenced phase 1 clinical trial in USA under US-IND 126099. https://www.thehindubusinessline.com/companies/announcements/others/article7607101.ece

Thennati R, Khanna A, Khanna M, Sonaiya T, Mehta T, Mehta K et al (2014) Safety, tolerability, pharmacokinetics, and pharmacodynamics of compound SFDAC by intranasal administration of multiple escalating dose in healthy male subjects. Clin Pharmacol Drug Dev 3(6):428–438

Thomas A, Gopalan B, Lingam PRVS, Shah DM (2006) WO2006/040625 A1. W. I. P. Organization, Geneva

Tirunagaru VG, Nagaswamy K, Mitra S, Maddi S, Adluri RS, Joshi H, Gupta JB (2016) Discovery and characterization of a selective TrkA inhibitor for the treatment of TrkA-driven cancers. Cancer Res 76:2643

Tradtrantip L, Namkung W, Verkman AS (2010) Crofelemer, an antisecretory antidiarrheal proanthocyanidin oligomer extracted from Croton lechleri, targets two distinct intestinal chloride channels. Mol Pharmacol 77(1):69–78

Tripathi R, Dutta G, Vishwakarma R (1990) Gametocytocidal Activity of the Antimalarial α/β Arteether against Plasmodium cynomolgi B. Am J Trop Med Hyg 43(6):571–575

Tripathi R, Vishwakarma R, Dutta G (1997) Plasmodium fragile: efficacy of arteether (α/β) against cerebral malaria model. Exp Parasitol 87(3):290–292

Unnikrishnan CH, Ananthanarayanan R (2014) Sanofi junks Glenmark's painkiller molecule. https://www.livemint.com/Companies/RdthRWQs9wY8NC8h2jVMAP/Sanofi-junks-Glenmarks-painkiller-molecule.html

Vakkalanka SKVS, Nagarathnam D, Viswanadha S, Muthuppalaniappan M, Babu G, Bhavar PK (2013) WO2013/144737 A2. W. I. P. Organization, Geneva

Verma AK, Goel AK, Arjuna Rao V, Venkateswarlu A, Sitrin RD (1986) US4742045A. U. S. P. a. T Office, Alexandria, VA

Verma AK, Goel AK, Arjuna Rao V, Venkateswarlu A, Sitrin RD, Christensen SB, Iv. (1987) EP0255299A2. E. P Office, Munich

Vishwakarma R (2014) Modern drug discovery and the Indian opportunity. Curr Sci 107(3):335–336

WHO (2015) Model list of essential medicines, 19th edn. WHO, Geneva. https://www.who.int/medicines/publications/essentialmedicines/EML_2015_FINAL_amended_NOV2015.pdf?ua=1

WHO (n.d.) Dr. Reddy's Laboratories Limited. https://www.who.int/intellectualproperty/events/en/Reddy.pdf

WHO Drug Information (2010a) Revamilast, vol 24. WHO, Geneva, p 278

WHO Drug Information (2010b) Tedalinab, vol 24. WHO, Geneva, p 169

WHO Drug Information (2015a) Nafithromycin, vol 29. WHO, Geneva, p 545

WHO Drug Information (2015b) Tedalinab, vol 29. WHO, Geneva, p 570

Wockhardt Ltd. (1997–2016) Annual reports, 1997–2016. http://www.wockhardt.com/investor-connect/annual-reports.aspx

Wockhardt Receives Qualified Infectious Disease Product (QIDP) (2015) Designation for its new drug WCK 4873 from U.S. FDA [Press release]. http://www.wockhardt.com/pdfs/Press%20Release.pdf

Working Group for New TB Drugs (2016) SPR-113. http://www.newtbdrugs.org/pipeline/compound/spr-113

Yeleswarapu KR, Manojit P, Manohar SV, Akella V, Pillarisetti S, Padakanti S, Kalleda SR (2005) WO2005/040163 A1. W. I. P. Organization, Geneva

**Nagarajan Kuppuswamy** had experience in new drug development for more than two decades as Head of Medicinal Chemistry at CIBA-Geigy Research Centre, Bombay from 1963 to 1984. The Centre had got registration from Drug Controller, India for four new drugs, antidepressant nitroxazepine, antipsychotic, nonaperone, anthelmintic, amoscanate and antiprotozoal, satranidazole. Out of these, the first drug is sold in India by Novartis as Sintamil and the second, by Alkem Laboratories, Bombay as Satrogyl. Dr. Nagarajan has been associated with Searle, Alkem, BPRL and Astra-Zeneca Research Centre earlier and is now with Hikal Ltd., Bombay. He has been given the Bhatnagar Prize and Life Time Achievement Award for his researches.

**Srinivas Nanduri** is experienced in drug discovery and development as Senior Director at the R&D center of Dr. Reddy's Laboratories, Hyderabad (1995–2008). Led and executed drug discovery programs in the areas of oncology and metabolic disorders using natural products and synthetic small molecules. Dr. Nanduri has been associated with Biocon BMS Research Centre, Bangalore, Aurigene Discovery Technologies Ltd., Albany Molecular Research Hyderabad in senior management positions and also as a faculty at Sri Sathya Sai Institute of Higher Learning, Prasanthinilayam, A.P. Currently, he is a Professor in the Department of Pharmaceutical Technology (Process Chemistry) at NIPER Hyderabad.

**Venkateswarlu Akella** is a Harvard-trained chemist with >45 years of Pharmaceutical Industry experience in Drug Discovery Research and Research Management. As the Founding-President of Dr. Reddy's Research Foundation (1991–2001), he successfully initiated and directed drug discovery research program culminating in significant NCE discoveries in India. He also made key contributions to the establishment and management of Aurigene Discovery Technologies and Dr. Reddy's Institute of Life Sciences. Earlier, he served Smith Kline & French India (1974–1987) and Atul Products (1987–1991) in top R&D positions. His discovery and development of TBDMS protecting group in his postdoctoral research at Harvard (Mentor: E J Corey) has turned out to be a milestone contribution in organic synthesis.

The manufacturer's authorised representative in the EU is Springer Nature Customer Service Centre GmbH, Europaplatz 3, 69115 Heidelberg, Germany. If you have any concerns regarding our products, please contact ProductSafety@springernature.com

Printed and bound by CPI Group (UK) Ltd, Croydon, CR0 4YY

03/10/2026

02227493-0001